AF361895

Eurosensors 2023 Selected Papers

Eurosensors 2023 Selected Papers

Editors

Bruno Ando
Pietro Siciliano
Luca Francioso

Basel • Beijing • Wuhan • Barcelona • Belgrade • Novi Sad • Cluj • Manchester

Editors
Bruno Ando
Department of Electric
Electronic and Information
Engineering (DIEEI)
University of Catania
Catania
Italy

Pietro Siciliano
CNR-IMM
Institute for Microelectronics
and Microsystems
Lecce
Italy

Luca Francioso
CNR-IMM
Institute for Microelectronics
and Microsystems
Lecce
Italy

Editorial Office
MDPI AG
Grosspeteranlage 5
4052 Basel, Switzerland

This is a reprint of articles from the Special Issue published online in the open access journal *Sensors* (ISSN 1424-8220) (available at: https://www.mdpi.com/journal/sensors/special_issues/2G0C9FJ5E6).

For citation purposes, cite each article independently as indicated on the article page online and as indicated below:

Lastname, A.A.; Lastname, B.B. Article Title. *Journal Name* **Year**, *Volume Number*, Page Range.

ISBN 978-3-7258-2379-6 (Hbk)
ISBN 978-3-7258-2380-2 (PDF)
doi.org/10.3390/books978-3-7258-2380-2

Contents

Article

Thermal Behavior of Biaxial Piezoelectric MEMS-Scanners

Laurent Mollard *, Christel Dieppedale, Antoine Hamelin, Gwenael Le Rhun, Jean Hue, Laurent Frey and Gael Castellan

University Grenoble Alpes, CEA, Leti, F-38000 Grenoble, France; christel.dieppedale@cea.fr (C.D.); antoine.hamelin@cea.fr (A.H.); gwenael.le-rhun@cea.fr (G.L.R.); jean.hue@cea.fr (J.H.); laurent.frey@cea.fr (L.F.); gael.castellan@cea.fr (G.C.)
* Correspondence: laurent.mollard@cea.fr

Abstract: This paper presents the thermal behavior of non-resonant (quasi-static) piezoelectric biaxial MEMS scanners with Bragg reflectors. These scanners were developed for LIDAR (LIght Detection And Ranging) applications using a pulsed 1550 nm laser with an average power of 2 W. At this power, a standard metal (gold) reflector can overheat and be damaged. The Bragg reflector developed here has up to 24 times lower absorption than gold, which limits heating of the mirror. However, the use of such a reflector involves a technological process completely different from that used for gold and induces, for example, different final stresses on the mirror. In view of the high requirements for optical power, the behavior of this reflector in the event of an increase in temperature needs to be studied and compared with the results of previous studies using gold reflectors. This paper shows that the Bragg reflector remains functional as the temperature rises and undergoes no detrimental deformation even when heated to 200 °C. In addition, the 2D-projection model revealed a 5% variation in optical angle at temperatures up to 150 °C and stability of 2D scanning during one hour of continuous use at 150 °C. The results of this study demonstrate that a biaxial piezoelectric MEMS scanner equipped with Bragg reflector technology can reach a maximum temperature of 150 °C, which is of the same order of magnitude as can be reached by scanners with gold reflectors.

Keywords: thermal behavior; 2D MEMS mirror; piezoelectric; Bragg reflector; high optical power management

Citation: Mollard, L.; Dieppedale, C.; Hamelin, A.; Rhun, G.L.; Hue, J.; Frey, L.; Castellan, G. Thermal Behavior of Biaxial Piezoelectric MEMS-Scanners. *Sensors* **2023**, *23*, 9538. https://doi.org/10.3390/s23239538

Academic Editors: Bruno Ando, Luca Francioso and Pietro Siciliano

Received: 16 October 2023
Revised: 15 November 2023
Accepted: 24 November 2023
Published: 30 November 2023

1. Introduction

For most applications, including pico-projection and automotive and biomedical applications, MEMS scanners are of great interest due to their small size, low cost and low power consumption compared to standard mechanical beam-scanning systems. Numerous publications present potential applications for MEMS scanners, as well as the related requirements [1,2]. There has been particular focus recently on the development of LIDAR systems for autonomous driving systems. For long-range applications (distances greater than 100 m), the MEMS-LIDAR approach, based on a MEMS-scanner and using an incident laser wavelength of 1550 nm, currently appears to be the best option [3,4]. The main reason for using a 1550 nm laser is that the maximum allowable exposure for human eyes is higher at this wavelength than at 905 nm.

The 2D MEMS-scanner was developed with PZT (lead zirconium titanate) piezoelectric actuators on 8-inch silicon wafers using VLSI (very large-scale Integration) technology. The scanner's design is compact compared with electromagnetic scanners, which require a bulky magnet to be integrated into the packaging. In addition, the PZT actuator is low-voltage (<25 V) compared to other types, such as electrostatic actuators (>150 V). Using this actuator, we obtained a total optical angle close to 8°, with a maximum driving voltage of 20 V [5]. As will be discussed later, the optical angle can be improved by pushing the design-drawing rules. In addition to these advantages, PZT actuators offer fast response times and low power consumption.

Long-range LIDAR detection requires high incident power. Our specification is an average incident optical power, at 1550 nm, of 2 W on the 2×2 mm^2 mirror, which corresponds to a power density of 5000 W/m^2. With a pulse laser, this specification corresponds to a few kW of peak power.

Thus, the absorption (A) of the reflector is key to handling this significant incident optical power while also limiting heating of the mirror. Indeed, the power absorbed (P_a) inside the mirror is proportional to the reflector's absorption (A): $P_a = A \times P_i$, where P_i is the incident optical power. When the mirror is at thermal equilibrium, this absorbed optical power is equal to the power dissipated (P_d). Incident power can be dissipated in three ways: by surface radiation, by conduction through the arms or the air surrounding the scanner, or by convection. Free convection is often neglected because of the mirror's dimensions, and dissipation is mainly attributed to conduction [6,7]. For our specific design, the main route of power dissipation was conduction through the arms. Air conduction is much lower because of the distances (of the order of 700 μm) between the mirror and the substrate. The thermal power dissipated by conduction (Q_{cond}) is proportional to the mirror's temperature: $Q_{cond} = \frac{(T_m - T_s)}{R}$, where T_m and T_s are the temperature of the mirror and silicon substrate (considered at ambient temperature), respectively. R is the overall thermal resistance of the PZT arm (R_{arm}) and of the hinge (R_{hinge}), which are in series, as shown in Figure 1. This approximation leads to the following equations: $P_a \cong Q_{cond}$ and $A \times R \times P_i \propto T_m - T_s$.

Figure 1. MEMS mirror, top view (**left**) and zoom on hinge and PZT arm (**right**).

R can be expressed as follows: $R = R_{arm} + R_{hinge} = \frac{e_{arm}}{\lambda * S_{arm}} + \frac{e_{hinge}}{\lambda * S_{hinge}}$, with $R_{arm} = \frac{e_{arm}}{\lambda * S_{arm}}$ and $R_{hinge} = \frac{e_{hinge}}{\lambda * S_{hinge}}$. In this equation, $e_{arm/hinge}$ are the length of the arm and hinge, respectively, and $S_{arm/hinge}$ is the cross-section area (product of width and thickness) of the arm and hinge, respectively. The value of λ can be defined, initially, as the thermal conductivity of silicon in view of the very low thickness of the other layers. In the specific case of our mirror, with an optical angle of $8°$, R_{arm} is $15\times$ lower than R_{hinge}, and consequently $R \approx R_{hinge}$.

Finally, the temperature of the mirror T_m can be expressed as follows:

$$T_m - T_s \propto A \times R_{hinge} \times P_i$$

One way of increasing the optical angle in the near future will be to limit the stiffness of the hinge and therefore further reduce its width, which will result in a greater R_{hinge}. This increase in the value of R_{hinge} will limit the power dissipated by conduction through the arms and therefore lead to an increase in mirror temperature. One way of limiting this temperature rise is to reduce the reflector's absorption in order to reconcile greater angular deflection with limited mirror heating.

As a potential solution, a low-absorption Bragg reflector compatible with piezoelectric actuation was previously developed [5]. The optical absorption by this Bragg reflector was up to 24-fold lower than absorption by a standard gold reflector. This Bragg reflector consists of repeated (n times) dielectric bilayers of amorphous silicon (a-Si, 110 nm) and oxide (SiO$_2$, 305 nm).

However, the impact of increased temperature on scanner performance needs to be assessed in the context of high optical power or high-temperature environments. The impact of increasing scanner temperature on the performance and characteristics (flatness, absorption, laser-induced damage threshold, etc.) has been studied elsewhere [6–9]. However, only technological processes using metal reflectors have been reported. Because the technological process with the Bragg reflector is different, the final stress of the multilayers will also be different, as will the way they change with increasing temperature. Their performance as the temperature rises must be investigated.

The main aims of the study are to ensure that the mirror remains functional even when the temperature rises, to study the deformations a rise in temperature could induce and to determine the maximum operating temperature for this Bragg scanner. The results will enable us not only to extend the use of such a mirror to a high-temperature environment, but also to consider applications other than the LIDAR that might require higher optical power.

2. Results and Discussion

The Bragg reflector was manufactured by physical vapor deposition (PVD) to avoid degradation of the PZT performance. It consists of several (*n*) bilayers of amorphous silicon (110 nm) and oxide (305 nm), depending on the target reflectivity and absorption. A two-bilayer (*n* = 2) Bragg reflector was chosen for the LIDAR application.

In the PZT process, process used was the most efficient in terms of material thickness, annealing and electrodes [10]. The 535 nm PZT piezoelectric film was deposited by sol-gel chemical solution deposition (CSD) on 8-inch silicon wafers. This film consists of 10 layers of a commercial PZT (52/48) solution from Mitsubishi Materials Corporation. Each layer of PZT was spun, dried at 130 °C and calcinated at 360 °C. Crystallization and densification annealing were performed at 700 °C under oxygen for 1 min using a Jipelec RTA (Rapid Thermal Annealing) furnace from Annealsys (Fr.). A first round of annealing was done on the first layer after it was coated to promote the desired (100) orientation. Subsequently, annealing was performed after coating of every three layers. This standard process does not include a hot-poling step after the deposition of PZT.

Figure 2 shows a schematic cross-section view of our scanner. The fabrication process flow and PZT characterizations have been previously reported [5].

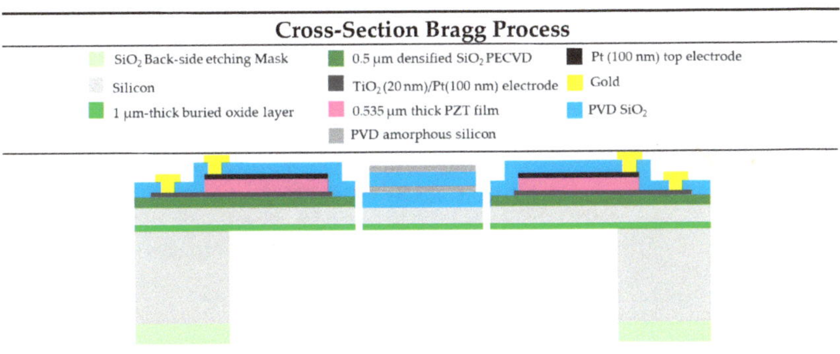

Figure 2. Cross-section of scanner with Bragg reflector.

To create the Bragg bilayers, an Endura PVD magnetron sputtering chamber from Applied Material (U.S.A.) was used, having been loaded with a silicon target for the 110 nm amorphous silicon layers. The 305 nm silicon dioxide layers were deposited in the same chamber by reactive sputtering of DC-generated plasma.

The evolution of the mirror's flatness and of the 2D-scan patterns with increasing temperature were experimentally characterized. These characterizations were carried out with the scanner thermalized on a hot plate. This experimental protocol has two principal advantages: it allows precise control of the scanner temperature, and it is compatible with a wide range of temperatures (from ambient to 200 °C).

2.1. Mirror Planarity with Temperature

Due to residual stresses in the multiple layers making up the mirror, the mirror was not perfectly flat. A concave or convex deformation of the mirror is often observed after fabrication, and this deformation affects beam reflection. The deformation can be minimized through technological means, for example, by changing the thickness and stress of the layers on the top and bottom of the mirror. The deposition process was first optimized to minimize internal multilayer stress. The average internal stress of the Bragg reflector bilayers ($n = 2$) was measured at -155 MPa on a complete wafer after deposition. This low compressive stress makes it possible to minimize mirror deformation induced by stress.

At the same time, the scanner's Z-deformation was measured after the Bragg process, using an Altisurf 520 tool from Altimet (Fr.). Figure 3 shows an overview of the scanner's deformation along the Z axis at ambient temperature.

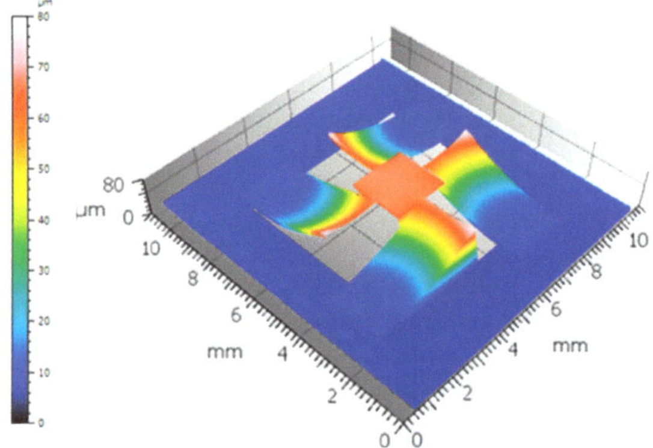

Figure 3. Scanner deformation of a Bragg reflector ($n = 2$) along the Z-axis near ambient temperature (30 °C).

The mirror clearly displayed a convex deformation at 30 °C. The static radius of curvature (SRC), measured along the diagonal of the mirror over 2 mm and centered on the mirror, was estimated to be 230 mm. The 4 mm-long actuators had a positive deflection of 57 μm, which was induced by compressive stress affecting the whole multilayer stack.

The evolution of Z deflection by the actuators and the mirror, as a function of temperature, was characterized over the temperature range from 30 °C to 200 °C. To carry out this evaluation, the scanner was thermalized on the heating plate integrated into the Altisurf 520 tool. As shown in Figure 4, increasing scanner temperature resulted in a 10% decrease in the Z-deflection, which reached a value close to 50 μm at 150 °C.

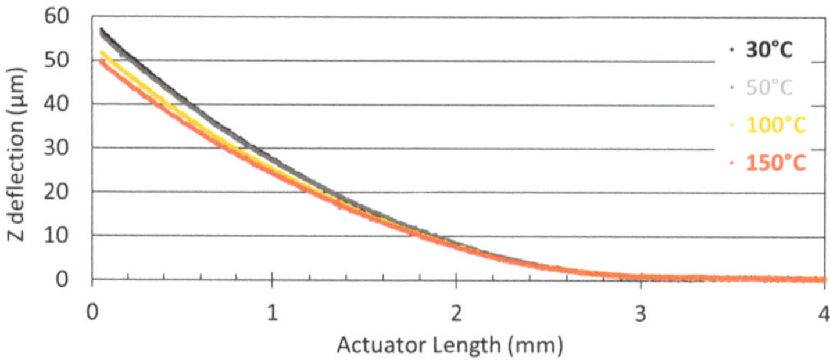

Figure 4. Z deflection of the PZT arms decreases with increasing temperature.

This expected evolution is consistent with published analytical models of multilayers [11,12]. The thermal stress (σ_{Ti}) between a film (i) and a substrate (s) is defined by the equation

$$\sigma_{Ti} = E_i'(\alpha_i - \alpha_s)\left(T_{dep} - T_{sub}\right),$$

where $\alpha_{i,S}$ are the thermal expansion coefficients of the film (i) and substrate (s), $E_{i,S}' = \dfrac{E_{i,S}}{(1 - \vartheta_{i,S})}$ corresponds to their biaxial Young modulus, $\vartheta_{i,S}$ is the Poisson ratio and T_{dep} and T_{sub} are, respectively, the temperatures of deposition and operation.

Taking the values of $E_{i,s}'$ and $\alpha_{i,s}$ reported in [8] and our T_{dep} values, as shown in Table 1, the formula clearly shows that increasing the operating temperature, T_{sub}, from 30 °C to 200 °C induces a decrease in overall stress, which in turn reduces beam deflection.

Table 1. Values of material characteristics from [8] and our actuator technology process.

Material	$E_{i,s}'$ [GPa] [8]	$\alpha_{i,s}(10^{-6})$ [8]	$t_{i,s}$ (µm)	T_{dep} [°C]
Pt	271	9	0.1	450
PZT	119	6	0.55	700
Pt	271	9	0.1	450
SiO$_2$	100	0.6	0.5	250
Si	236	4.3 (30 °C) 5.5 (150 °C)	20	N/A
SiO$_2$	100	0.6	1	1050 [8]

The radius of curvature of a multilayer beam can be estimated using the following formula [11]:

$$\frac{1}{r} = \sum_{i=1}^{n} \frac{6E_i't_i(\alpha_i - \alpha_S)\Delta T}{E_s't_S^2}$$

where $t_{i,S}$ the thickness of the layer (i) or substrate (s). This last equation is a first-order approximation that omits terms with orders higher than t_i. This omission is consistent with the t_i layer thicknesses being at least 20-fold thinner than the substrate thickness. However, these models do not account for any bending moment on the beams. This assumption does not apply in our case because the beams are connected to the mirror. Nevertheless, the model allowed us to determine that the largest $E_i't_i$ contributions were related to the bottom 1 µm-SiO$_2$ film, which was present under the substrate. To a lesser extent, the PZT layer contributed as well. Some layers, like TiO$_2$, may not be taken into account because of their negligible thickness. As a result, the overall radius of curvature of the actuators, and the degree of deflection, is mainly to the result of the compressive 1 µm-bottom SiO$_2$ film present under the mirror. It should be noted that the initial deformation of the actuators had no direct impact on the optical angle attained; it affected only the Z-position of the mirror.

Figure 5 shows changes to the mirror's flatness as a function of temperature. The multilayer stack on the mirror is different from that present on the actuator. The mirror tends to flatten as the temperature increases. The increase in Z-signal noise with increasing temperature is linked to the measurement tool and does not reflect a change in the reflector surface, as the reflector was deposited at a higher temperature (250 °C).

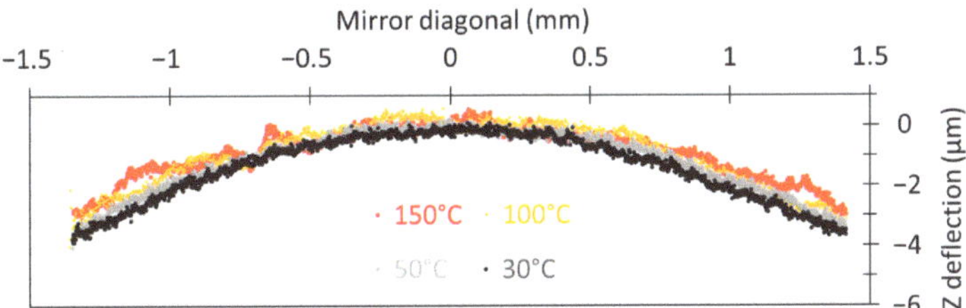

Figure 5. Bragg ($n = 2$) mirror planarity at different temperatures. The mirror was scanned along its diagonal.

Despite this noise, from experimental measurements, the SRC was estimated to be close to 340 mm at 150 °C. In future developments, this mirror curvature will need to be reduced to achieve the Rayleigh criteria: a Z-deformation of less than $\lambda/4$. This result could be achieved by, for example, increasing the thickness of the SiO_2 layer present under the mirror or by adding a tensile layer on top of the silicon substrate.

2.2. 2D Projected Scan with Temperature

A key parameter to characterize is how the 2D scan of the reflected beam evolves as a function of scanner temperature. This measurement is used to determine and extrapolate the potential variation in the total optical angle depending on the mirror's temperature in the context of a LIDAR application. The X and Y coordinates of the optical angle determine the size of the scene that the LIDAR can image.

The scanner was thermalized at temperatures between 30 °C and 150 °C, and the optical angles were measured using a class-2 visible laser reflected by the mirror and projected onto a screen. A camera was used to track the position of the laser on the screen by continuously measuring its barycenter. If the optical path (l) (around 130 mm) between the screen and the scanner and the position (x, y) of the barycenter of the laser on the screen are known, the X and Y (θ_x, θ_y) coordinates of the optical angle of the spot can be deduced using $\tan\left(\theta_{x\ or\ y}\right) = \frac{x\ or\ y}{l}$. The absolute uncertainty of the optical angle, which is linked to how precisely the optical path is measured, was estimated to be 0.2° on our test bench. The scanner temperature was limited to 150 °C because electrical breakdowns are often observed when our technological process is used at temperatures exceeding 200 °C.

A 2D scan was performed with a 200 Hz sinusoidal driving signal (X-axis) for the fast axis and a 4 Hz ramp for the slow axis (Y-axis). The PZT supply voltage was swept from 0 V to a maximum of 20 V with $V_{AC} = V_{DC} = 10$ V. The total optical angle θ_x and θ_y was measured at y = 0 and x = 0, respectively. In addition, for each of the four temperatures tested, the total duration of the measurement was at least 30 min, including thermalization and the time required for 2D scanning. Figure 6 shows the optical angles measured for one of our scanners at 30 °C and 150 °C. A slight decrease can be observed in both axes of the scanning pattern. Additionally, it is worth noting a minor distortion in the 2D scan at 150 °C. This distortion will require further examination.

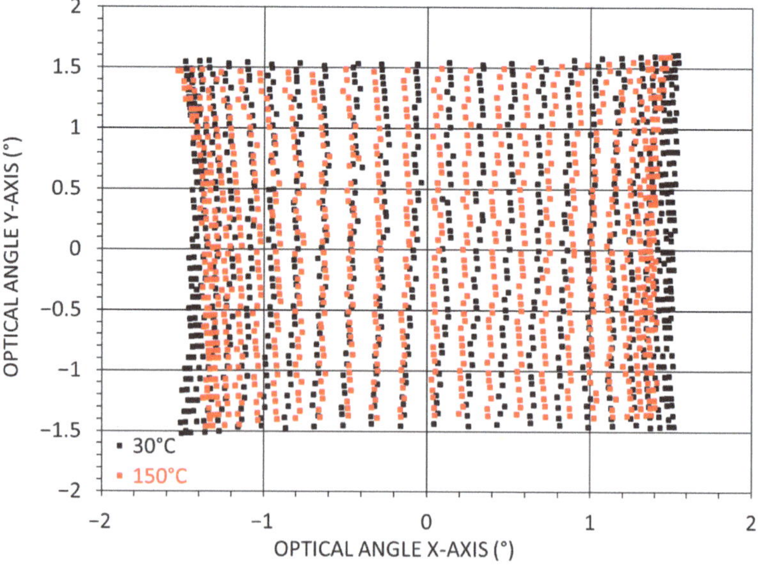

Figure 6. 2D-scanning representation at 30 °C (black) and 150 °C (red).

Four identical Bragg mirrors were used for this test, as shown in Figure 7. These mirrors have exactly the same design, the same Bragg reflector and the same technological process flow. The evolution of the effective optical angle, $\theta_e = \sqrt{\theta_x \theta_y}$, as a function of changes in scanner temperature from 30 °C to 150 °C is reported. The changes observed correlated well with previously published results [8,13–15].

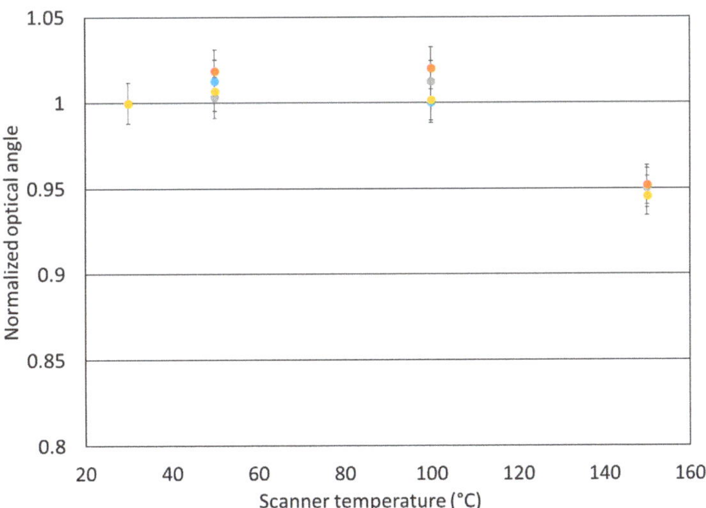

Figure 7. Optical angle θ_e (normalized relative to angle measured at 30 °C) as a function of scanner temperature done with four identical mirrors.

Mirror deflection thus results from the movement of the PZT actuators and, consequently, of the effective transverse piezoelectric coefficient $e_{31,f}$, based on the inverse piezoelectric effect. Indeed, the piezoelectric stress (σ_P) can be expressed as $\sigma_P = -e_{31,f}E_z$ [13], where E_z is the transverse electric field. In addition, $e_{31,f}$ is proportional to P_S, the saturation polarization, and ε_{33}, the relative permittivity measured out of plane, as $e_{31,f} \propto -2\varepsilon_{33}P_3$. This coefficient is highly dependent on the operating temperature of the mirror. Theoretical calculations [14] predict that $e_{31,f}$ will increase with temperature. In contrast, other authors [8,15] have reported experimental results showing that $e_{31,f}$ increases to a maximum (near 40 °C) and then gradually decreases at higher temperatures (up to 200 °C).

After deposition, ferroelectrics like PZT may not be polarized, in which case they will have poor piezoelectric properties. Poling treatment, which involves applying a voltage well above the coercive voltage, orients the piezoelectric domains in the same direction. However, after poling, the piezoelectric coefficient slowly decays due to rearrangement of the domains towards their equilibrium state. If poling is performed at a high temperature (hot poling), the mobility of the ferroelectric domains increases significantly, allowing a better alignment in the direction of the poling field [10]. Nevertheless, it should be noted that this increase in $e_{31,f}$ after hot poling has been observed only at ambient temperature. Another study [15] with an initial poling step of 25 V at ambient temperature and intermediate poling steps at higher temperatures found that $e_{31,f}$ decreases with increasing temperature. The two curves converged at high temperatures (near 200 °C).

This case is different, as the temperature of the scanner increases progressively with continuous polarization of the PZT. A similar case has been reported elsewhere [8]. Dahl-Hansen et al. followed the deflection tilt of a unipolar micro-mirror driven by a 20 V signal at a frequency of 1.5 kHz while varying the mirror temperature stepwise from ambient temperature to 175 °C. Our study shows a similar trend, with an increase in the optical angle between 50 °C and 100 °C and a decrease at 150 °C. The value of the maximum is more difficult to determine given the uncertainty of the measurements. In addition, one of our samples underwent an electrical breakdown at 100 °C, reflecting degradation at high temperature. This decrease in optical angle with increasing temperature has been attributed mainly to the decrease in P_s as the scanner temperature approaches the Curie temperature (near 350 °C in our case) [8,16,17]. Alteration of film stress is also reported to have a potential impact [8] but is considered to contribute only slightly to the evolution of $e_{31,f}$ [17].

The decrease in optical angle observed in Figure 7 was close to 5%, which is consistent with values reported elsewhere [8]. However, this evolution of the optical angle, which is linked to the actuation power, even if it is limited to a change of few percent, must be taken into account in the context of future use within a LIDAR system. This value has direct implications for the global area scanned from the scene. Limiting the temperature increase by means of a low-absorption reflector is a first response to this issue, but that solution could also be coupled to others, like those based on the influence of PZT poling or hot-poling conditions, as has been reported [9]. These and other studies, such as investigations focusing on the evolution of P_s, should be explored further in the future.

The last parameter checked was the stability of the 2D scan at an operating temperature of 150 °C. To conduct this investigation, an image of four spots, as shown in Figure 8, was continuously projected onto the screen for one hour. An image of the screen was taken every minute by a 720-by-540-pixel camera (X and Y) and the barycenter of each of the spots was calculated. The shift in the X and Y positions of the barycenter, is reported in pixels.

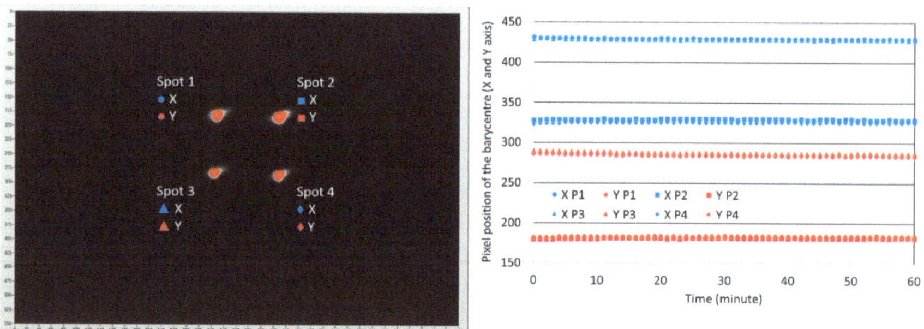

Figure 8. Projected image of the four spots on the screen (**left**) and shift in the barycenter (X and Y) of the four spots (**right**).

The standard deviation S_{n-1} of the barycenter positions of the four spots, calculated in pixels for both X and Y axis, was estimated using the following formula: $S_{n-1} = \sqrt{\frac{\sum (x - \bar{x})^2}{n-1}}$, where x is the value for one sample, $\bar{x}$ is the mean value of all the samples and n is the number of samples. In our case, $n = 60$. The maximum value of S_{n-1} is close to 1 pixel. This value is the smallest deviation that can be measured using our test bench. Thus, in our case, the relative deviation of the optical angle was close to 0.03°.

In conclusion, no significant deviation of the optical angle at a scanner temperature of 150 °C was observed for continuous one-hour measurements, demonstrating the stable performance of the PZT actuator.

Further studies are needed to verify that PZT performance is not affected by a return to ambient temperature after operation at different temperatures. The potential impact of a hot-poling step remains to be investigated as well.

3. Conclusions

In this paper, we report on the thermal behavior of a biaxial piezoelectric MEMS-scanner with integrated Bragg-reflector technology. We first demonstrate that the flatness of the mirror and actuators compensates for increases in internal stresses in the multilayer stack with increasing temperature. Better adaptation of the thickness of the layers making up the mirror could further improve initial planarity to provide better optical quality of the reflected beam. We then looked at the evolution of the 2D projected pattern with increasing scanner temperature. An optical angle variation close to 5% was observed at 150 °C, which is consistent with previously reported values. Such results could be taken into account in the case of application at elevated temperature to adapt the FoV of the scanner. Finally, no significant deviation of the optical angle was observed for continuous measurements lasting one hour at a scanner temperature of 150 °C. The results presented here demonstrate that a biaxial piezoelectric MEMS scanner equipped with Bragg reflector technology can reach a maximum operating temperature near 150 °C, which is of the same order of magnitude as scanners equipped with a gold reflector. Moreover, thanks to its low-absorption Bragg reflector, this technology will make it possible in the future to increase angular deflection by reducing the width of the mirror hinge while limiting mirror heating. This technology thus paves the way for applications requiring high optical power or high-temperature environments. Finally, now that the maximum operating temperature is known, future simulation models will be able to estimate the maximum permissible incident power using the design of the scanner and the absorption of the Bragg reflector.

Author Contributions: L.M.: Conceptualization, Methodology, Writing—original draft, Project administration, Visualization. C.D.: Investigation. A.H.: Investigation. L.F.: Validation, Investigation. G.L.R.: Investigation. J.H.: Investigation. G.C.: Investigation. All authors have read and agreed to the published version of the manuscript.

Funding: This research was funded by ECSEL Joint Undertaking (JU) grant number No. 826600 (project VIZTA).

Institutional Review Board Statement: Not applicable.

Informed Consent Statement: Not applicable.

Data Availability Statement: Data are contained within the article.

Acknowledgments: Part of this work, performed on the Platform for NanoCharacterisation (PFNC) at CEA, was supported by the "Recherche Technologique de Base" Program funded by the French Ministry for Research and Higher Education. The authors would like to thank Catherine Brunet-Manquat (University Grenoble Alpes, CEA, Leti) for characterizing the micro-mirrors.

Conflicts of Interest: The authors declare no conflict of interest.

References

1. Holmstrom, S.T.S.; Baran, U.; Urey, H. MEMS Laser Scanners: A Review. *J. Microelectromech. Syst.* **2014**, *23*, 259–275. [CrossRef]
2. Song, Y.; Panas, R.M.; Hopkins, J.B. A Review of Micromirror Arrays. *Precis. Eng.* **2018**, *51*, 729–761. [CrossRef]
3. Wang, D.; Watkins, C.; Xie, H. MEMS Mirrors for LiDAR: A Review. *Micromachines* **2020**, *11*, 456. [CrossRef] [PubMed]
4. Kasturi, A.; Milanovic, V.; Lovell, D.; Hu, F.; Ho, D.; Su, Y.; Ristic, L. Comparison of MEMS Mirror LIDAR Architectures. In Proceedings of the MOEMS and Miniaturized Systems XIX Conference, San Francisco, CA, USA, 1–6 February 2020. [CrossRef]
5. Mollard, L.; Riu, J.; Royo, S.; Dieppedale, C.; Hamelin, A.; Koumela, A.; Verdot, T.; Frey, L.; Le Rhun, G.; Castellan, G.; et al. Biaxial Piezoelectric MEMS Mirrors with Low Absorption Coating for 1550 Nm Long-Range LIDAR. *Micromachines* **2023**, *14*, 1019. [CrossRef] [PubMed]
6. Zhang, J.; Lee, Y.-C.; Tuantranont, A.; Bright, V.M. Thermal Analysis of Micromirrors for High-Energy Applications. *IEEE Trans. Adv. Packag.* **2003**, *26*, 310–317. [CrossRef]
7. Burns, D.M.; Bright, V.M. Optical Power Induced Damage to Microelectromechanical Mirrors. *Sens. Actuators A Phys.* **1998**, *70*, 6–14. [CrossRef]
8. Dahl-Hansen, R.P.; Tybell, T.; Tyholdt, F. Performance and Reliability of PZT-Based Piezoelectric Micromirrors Operated in Realistic Environments. In Proceedings of the 2018 IEEE ISAF-FMA-AMF-AMEC-PFM Joint Conference (IFAAP), Hiroshima, Japan, 27 May–1 June 2018; pp. 1–4.
9. Bevilacqua, M.F.; Casuscelli, V.; Costantini, S.; Ferrari, P.; Pedaci, I. High Temperature Operating Lifetime Test on Piezo-MEMS Devices. In Proceedings of the 2021 IEEE 6th International Forum on Research and Technology for Society and Industry (RTSI), Naples, Italy, 6–9 September 2021; pp. 452–456.
10. Wague, B.; Dieppedale, C.; Le Rhun, G. Effects of Hot Poling Treatment on Properties and Behavior of PZT Based Piezoelectric Actuators. In Proceedings of the 2019 IEEE International Symposium on Applications of Ferroelectrics (ISAF), Lausanne, Switzerland, 14–19 July 2019; pp. 1–4.

11. Hsueh, C.-H. Modeling of Elastic Deformation of Multilayers Due to Residual Stresses and External Bending. *J. Appl. Phys.* **2002**, *91*, 9652. [CrossRef]

12. Defaÿ, E. (Ed.) *Integration of Ferroelectric and Piezoelectric Thin Films: Concepts and Applications for Microsystems*; ISTE; Wiley: Hoboken, NJ, USA, 2011; ISBN 978-1-84821-239-8.

13. Damjanovic, D. Ferroelectric, Dielectric and Piezoelectric Properties of Ferroelectric Thin Films and Ceramics. *Rep. Prog. Phys.* **1998**, *61*, 1267. [CrossRef]

14. Haun, M.J.; Furman, E.; Jang, S.J.; Cross, L.E. Thermodynamic Theory of the Lead Zirconate-Titanate Solid Solution System, Part V: Theoretical Calculations. *Ferroelectrics* **1989**, *99*, 63–86. [CrossRef]

15. Rossel, C.; Sousa, M.; Abel, S.; Caimi, D.; Suhm, A.; Abergel, J.; Le Rhun, G.; Defay, E. Temperature Dependence of the Transverse Piezoelectric Coefficient of Thin Films and Aging Effects. *J. Appl. Phys.* **2014**, *115*, 034105. [CrossRef]

16. Benedetto, J.M.; Moore, R.A.; McLean, F.B. Effects of Operating Conditions on the Fast-Decay Component of the Retained Polarization in Lead Zirconate Titanate Thin Films. *J. Appl. Phys.* **1994**, *75*, 460–466. [CrossRef]

17. Wolf, R.A.; Trolier-McKinstry, S. Temperature Dependence of the Piezoelectric Response in Lead Zirconate Titanate Films. *J. Appl. Phys.* **2004**, *95*, 1397–1406. [CrossRef]

 *sensors*

Article

A Flexible Printed Circuit Board Based Microelectromechanical Field Mill with a Vertical Movement Shutter Driven by an Electrostatic Actuator

Tao Chen and Cyrus Shafai *

Department of Electrical & Computer Engineering, University of Manitoba, Winnipeg, MB R3T 5V6, Canada; tao.chen@umanitoba.ca
* Correspondence: cyrus.shafai@umanitoba.ca; Tel.: +1-204-474-6302

Abstract: Micromachined electric field mills have received much interest for the measurement of DC fields; however, conventional designs with lateral moving shutters could have shutter lifting in the presence of strong fields, which affects their performance. This paper presents a MEMS electric field mill utilizing a vertical movement shutter to address this issue. The sensor is designed and fabricated based on a flexible PCB substrate and is released using a laser-cutting process. The movement of the shutter is driven by an electrostatic actuator. When the driving signal is a sine wave, the shutter moves in the same direction during both the positive and negative half-periods. This facilitates the application of a lock-in amplifier to synchronize with the signal at twice the frequency of the driving signal. In experimental testing, when the vertical shutter is driven at a resonance of 840 Hz, the highest sensitivity of the sensor is achieved and is measured to be 5.1 V/kVm^{-1}. The sensor also demonstrates a good linearity of 1.1% for measuring DC electric fields in the range of 1.25 kV/m to 25 kV/m.

Keywords: electric field sensor; electric field mill; flexible PCB; MEMS; micromachining; electrostatic actuator; laser micromachining

Citation: Chen, T.; Shafai, C. A Flexible Printed Circuit Board Based Microelectromechanical Field Mill with a Vertical Movement Shutter Driven by an Electrostatic Actuator. *Sensors* **2024**, *24*, 439. https://doi.org/10.3390/s24020439

Academic Editors: Bruno Ando, Luca Francioso and Pietro Siciliano

Received: 30 November 2023
Revised: 31 December 2023
Accepted: 6 January 2024
Published: 11 January 2024

1. Introduction

A DC (direct current) electric field sensor (EFS) is a device designed to measure and quantify the strength of static or slowly changing electric fields in a given environment. They are useful for applications that involve static or slowly changing electric fields. In power utilities, they can be used to evaluate the electromagnetic environment in an electric power transmission system [1] and the design of insulators [2], and to ensure the safety of power workers [3]. In industrial manufacturing processes, they can be used to prevent electrostatic discharges (ESDs) to protect electronic equipment [4]. In atmospheric science, they are used to study the mechanisms of thunderstorms [5] and predict lightning [6]. They are also useful in the study of climate and geophysics [7].

Common techniques to measure DC electric fields include induction probes, optical sensors, and electric field mills (EFMs). An induction probe has a sensing electrode placed in an electric field, and when it is in equilibrium status, the voltage on the electrode is proportional to the field, which can be measured by using a high-input impedance electrical meter. Some benefits of induction probe type EFSs include a low cost and small size, but they require re-zeroing in shielding conditions, which is inconvenient in long-term applications. Optical sensors are usually based on Pockel's effect, where an electric field can change a light waveguide crystal's birefringent properties, and thus, the polarity of the light transversing the light waveguide will be changed [8]. Optical sensors cause minimal distortion of the electric field to be measured, but the optical equipment is often expensive and not stable in a variable temperature environment [9]. An EFM usually has a rotating grounded shutter over a set of sensing electrodes. The rotation of the shutter exposes and covers the sensing electrodes periodically, which generates an AC current with an amplitude proportional to the strength of the electric field. Traditional EFMs can stably work outside in various weather conditions, but they are bulky and require high power consumption and frequent maintenance. Besides these three common measurement techniques, some other types of electric field sensors reported in the literature include a graphene-based device [10], a ferroelectric material-based sensor [11], and a steered-electron sensor [12].

Recently, microelectromechanical field mills (MEFM) have been reported by multiple research groups [13–18]. They offer the benefits of small size, low cost, light weight, low power consumption, and minimal maintenance requirements. Similar to a traditional EFM, MEFMs employ grounded shutters vibrating above or adjacent to sensing electrodes. However, when these sensors are exposed to a strong electric field, grounded shutters will lift toward the field, which can affect their shielding ability, significantly reducing the sensor sensitivity. This is because the induced charge on MEFM electrodes is very sensitive to the gap between the shielding shutter and the underlying sensor electrodes. A vertical vibrating shutter type EFM is a promising design to overcome this issue, as the lifting of the shutter can be compensated for by adjusting the vibration range. The concept was first simulated by C. Gong et al. in 2004 [19], where they demonstrated that a vertical movement shutter has a similar shielding effect. However, until now, no MEFM has been reported using a vertical vibration shutter, except S. Ghionea et al., who reported a device that only works for measuring AC fields in 2013 [20].

In this work, we designed a vertical movement shutter type of MEFM by using a flexible PCB (FPCB) substrate. FPCB has the benefit of leveraging a commercialized manufacturing process that is faster and lower cost, making it an ideal choice for academic research in prototyping many MEMS devices without the need for a cleanroom. Recently, an increasing number of devices based on FPCB have been implemented [21–24]. In comparison to traditional silicon-based devices, their FPCB-based counterparts exhibit comparable performance levels while concurrently showcasing a marked improvement in durability. In this design, the shutter movement is driven by an electrostatic actuator. A simulation was performed to investigate the micro-spring spring constant, resonant frequency, and interference from the driving signal. After the simulation, the sensor was fabricated and tested, and it demonstrated a sensitivity of 5.1 V/kVm^{-1} when operating at resonance.

2. Sensor Design

2.1. Working Principle

Figure 1 illustrates the working principle of this sensor. It consists of a grounded shutter and grounded sensing electrodes. The shutter is supported by micro-springs, which are not shown in Figure 1. In the center of the shutter, there is a grounded area to form an electrostatic actuator together with an electrode under it. When a voltage is applied to this electrode, the shutter will move. Obviously, if the electrode is closer to the shutter, a stronger electrostatic force can be generated. If the voltage provided is a sine wave $V = V_0\sin(\omega t)$, where $\omega = 2\pi f$, at $t = 0$ and $t = \pi$, the voltage on the actuator is $V = 0$, and the shutter has no movement (Figure 1a). At $t = \frac{\pi}{2}$, the voltage on the actuator is V_0, and at $t = \frac{3\pi}{2}$, the voltage on the actuator is $-V_0$. In both situations, the actuator will drive the shutter to move downward for a distance of d (Figure 1b). If the shutter moves up and down periodically within an electric field, the sensing electrodes detect variations in the field caused by the changes in the fringing effect. As a result, varying charges are induced, leading to the generation of an alternating current. Based on Gauss's law, the total induced charge on the sensing electrodes is equal to the total electric flux. At $t = 0$ and $t = \pi$, the electric field on the sensing electrodes is E_1, and the induced charge can be calculated as:

$$Q_1 = \oint_S E_1 \varepsilon_0 ds \tag{1}$$

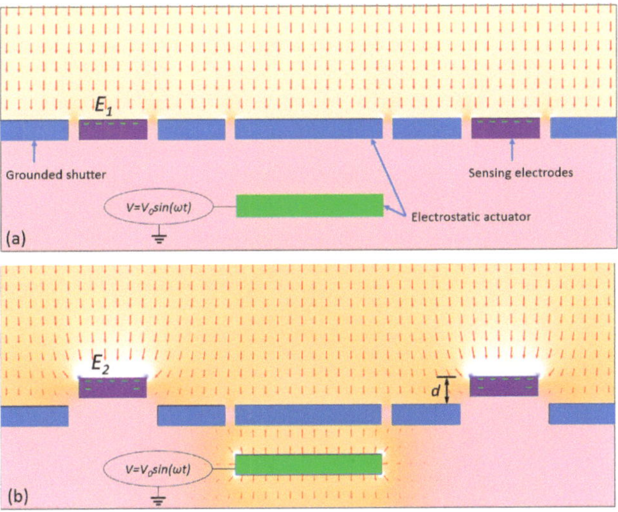

Figure 1. Sensor working principle. When an AC voltage is applied to the electrostatic actuator: (**a**) At $t = 0$ and $t = \pi$, the shutter is at the same height as the sensing electrodes. (**b**) At $t = \frac{\pi}{2}$ and $t = \frac{3\pi}{2}$, the shutter is pulled down and the sensing electrodes are exposed to a stronger field and more surface area is exposed.

At $t = \frac{\pi}{2}$ and $t = \frac{3\pi}{2}$, the shutter movement is d, the electric field on the sensing electrodes is E_2, and the induced charge can be calculated as:

$$Q_2 = \oint_S E_2 \varepsilon_0 ds \tag{2}$$

where $\varepsilon_0 \approx 8.85 \times 10^{-12}$ F/m is the vacuum dielectric constant and S is the surface area of the sensing electrodes.

Considering the shutter has the same direction of movement for both the positive and negative half period of the sine wave on the actuator, the induced charge equation can be written as:

$$\Delta Q = |(Q_1 - Q_2)\sin(\omega t)| \tag{3}$$

This is equivalent to $(Q_1 - Q_2)\sin(\omega t)$ modulated with an amplitude 1 square wave at the same frequency. Substituting the Fourier series of the ideal square wave into (3), then:

$$\Delta Q = \frac{4}{\pi}(Q_1 - Q_2)\sin(\omega t) \sum_{n=1}^{\infty} \frac{1}{2n}\sin((2n-1)\omega t) \tag{4}$$

Expanding Equation (4) we obtain:

$$\Delta Q = \frac{4}{\pi}(Q_1 - Q_2)\sin(\omega t)\left[\sin(\omega t) + \frac{1}{3}\sin(3\omega t) + \frac{1}{5}\sin(5\omega t) + \cdots\right]$$

$$\Delta Q = \frac{4}{\pi}(Q_1 - Q_2)\left[\sin(\omega t)\sin(\omega t) + \frac{1}{3}\sin(\omega t)\sin(3\omega t) + \frac{1}{5}\sin(\omega t)\sin(5\omega t) + \cdots\right]$$

$$\Delta Q = \frac{4}{\pi}(Q_1 - Q_2)\left\{\frac{1}{2}(1 - \cos(2\omega t)) - \frac{1}{6}[\cos(4\omega t) - \cos(2\omega t)] - \frac{1}{10}[\cos(6\omega t) - \cos(4\omega t)] + \cdots\right\} \tag{5}$$

Ignoring all high frequency components in Equation (5), and substituting (1) and (2) into (5), if only considering components at frequency 2ω, Equation (5) can be simplified as:

$$\Delta Q_{2\omega} = -\frac{4\cos(2\omega t)}{3\pi} \oint_S (E_1 - E_2)\varepsilon_0 ds \tag{6}$$

And so, the generated current at frequency 2ω is:

$$I_{2\omega} = \frac{d\Delta Q_{2\omega}}{dt} \tag{7}$$

Substituting (6) into (7), we get the equation for the current at frequency 2ω:

$$I_{2\omega} = \frac{8\omega\sin(2\omega t)}{3\pi} \oint_S (E_1 - E_2)\varepsilon_0 ds \tag{8}$$

This current signal is detectable after being amplified using a high impedance amplifier and then extracted from the noise by using a lock-in amplifier.

2.2. Working Principle

COMSOL Multiphysics software 6.0 was used for the simulations. The COMSOL material library was employed for defining the material properties of copper and polyimide. The 3D model created in COMSOL is shown in Figure 2a, and its dimension parameters are listed in Table 1. Figure 2b illustrates the schematic diagram of the model, demonstrating the arrangement of the sensor components. As we can see, the sensing electrodes and shutter are on one polyimide substrate, while the driving electrode, guard line, and bottom shielding electrode are on another polyimide substrate. In practice, they can be separated by using a spacer, but in the simulation, a spacer is not used in the model. The grounded shutter is supported by four S-shaped micro-springs and is placed 100 µm over the driving electrode. The sensing electrodes are at the same height as the shutter. The first simulation is to find the spring constant and shutter movement distance driven by the actuator. Figure 3a shows the shutter's downward movement when a voltage of 200 V is applied on the actuator, which results in a peak deflection of 5.59 µm. Cross-section views of the shutter downward movement are shown in Figure 3b–d. Figure 4 plots the simulated deflection for the drive voltages of 50 V, 100 V, 150 V, and 200 V, resulting in electrostatic forces on the shutter of 4.82×10^{-6} N, 1.97×10^{-5} N, 4.6×10^{-5} N, and 8.69×10^{-5} N, respectively. The overall spring constant is calculated to be 15.5 N/m.

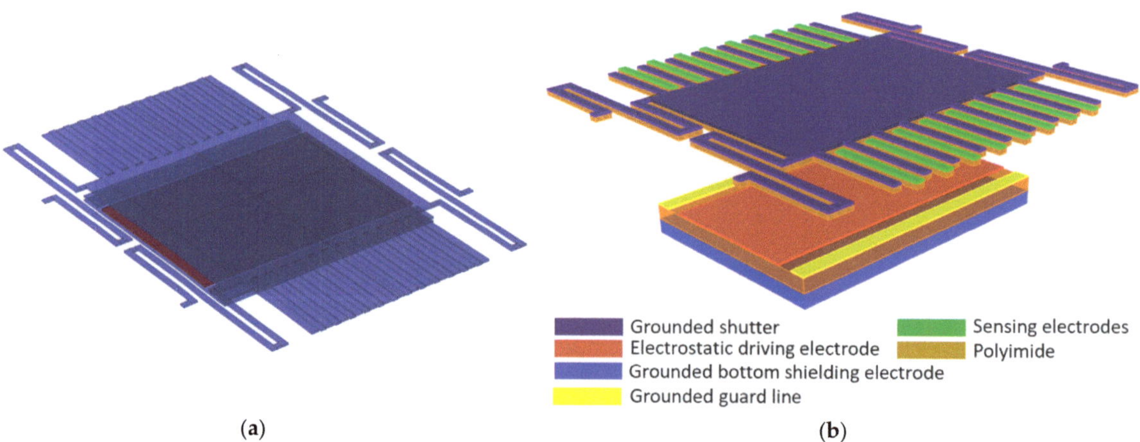

Figure 2. Sensor 3D model. (**a**) COMSOL 3D model. (**b**) 3D model schematic diagram.

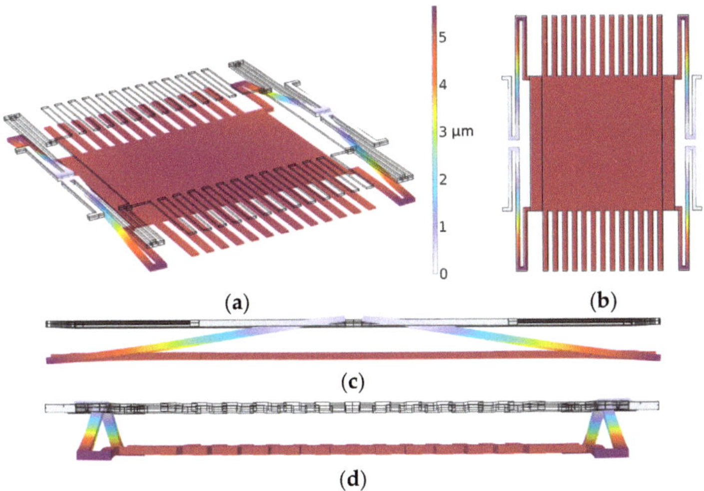

Figure 3. Shutter movement simulation. (**a**) Shutter movement under force of 86.9 μN. (**b**) x-y plane view. (**c**) y-z plane view. (**d**) x-z plane view.

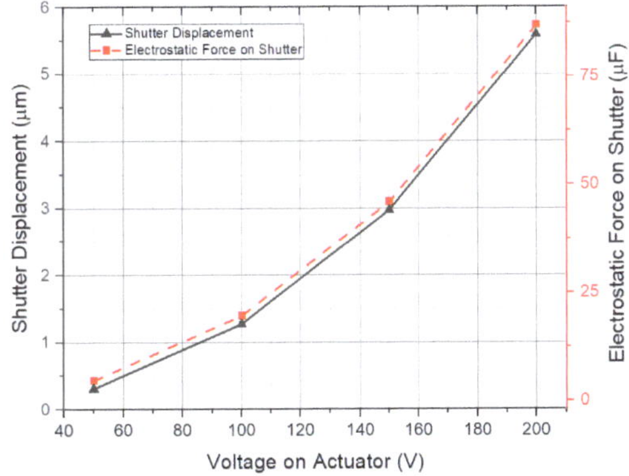

Figure 4. Simulation results for the shutter displacement and electrostatic force on the shutter when driven by the actuator.

Table 1. COMSOL 3D model properties.

Property	Value (μm)
Spring length	2000
Spring width	60
Copper thickness	18
Polyimide thickness	25
Sensing electrodes length	1000
Sensing electrodes width	60
Shutter finger length $\times$ width	1000 $\times$ 60
Gap between sensing electrodes and shutter finger	10
Electrostatic driving electrode width	2120
Electrostatic driving length	1920
Guard line length $\times$ width	1920 $\times$ 120
Distance from electrostatic driving electrode to shutter	100

The resonant frequency of a micro-springs supported shutter is also simulated. It has multiple orders of vibration modes, and Table 2 lists the results of the first three modes. As we can see, the first order is up and down at 493 Hz, the second order is tilt along the x-axis at 568 Hz, and the third order is tilt along the y-axis at 1045 Hz. Both the secondary and third orders have half of the structure move up and half move down, where the induced charges on each side will cancel each other; therefore, no signal will be generated. The first-order vibration mode will have a maximized induced charge.

Table 2. Resonant frequency simulation results.

Mode	Picture	Frequency (Hz)
1	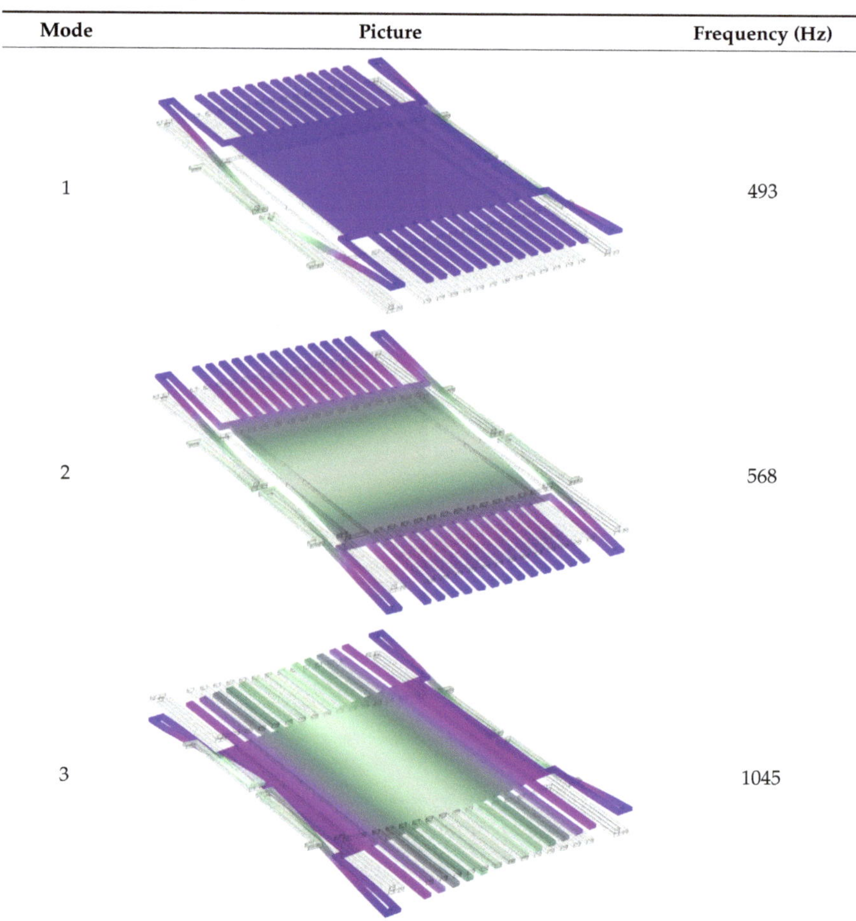	493
2		568
3		1045

The high voltage on the driving electrode is a concern in that it may produce extra induced charges on the sensing electrodes. To minimize this interference, in the design, two grounded guard lines are placed on each side of the driving electrode to separate it from the sensing electrode, and another grounded electrode is placed under the driving electrode and guard lines to prevent an electric field from emanating from below (on the other side of the PCB). Figure 5 depicts the distribution of the electric field between the driving electrode and the surrounding grounded structures. We can see that the sensing electrodes are minimally affected by the driving voltage. For example, when 200 V at 246.5 Hz is applied to the driving electrode (the shutter has a resonant frequency of 493 Hz), the induced charge on the sensing electrodes is 2.5 fC. The generated current is:

$$I_\omega = \frac{d\Delta Q}{dt} = \omega(Q_1 - Q_2)\cos(\omega t) \qquad (9)$$

Using Equation (9), the amplitude of the current is calculated to be 3.9 pA. In an electric field of 10 kV/m, the movement of the shutter (driven by 200 V) results in an induced charge of 0.17 pC, while at rest (0 V drive), the induced charge is 0.153 pC. Using Equation (8), the amplitude of the generated current is calculated to be 44.7 pA. Compared to the interference from the driving electrode, the sensing signal is 11.5 times stronger than the driving signal interference, and the frequency is two times higher.

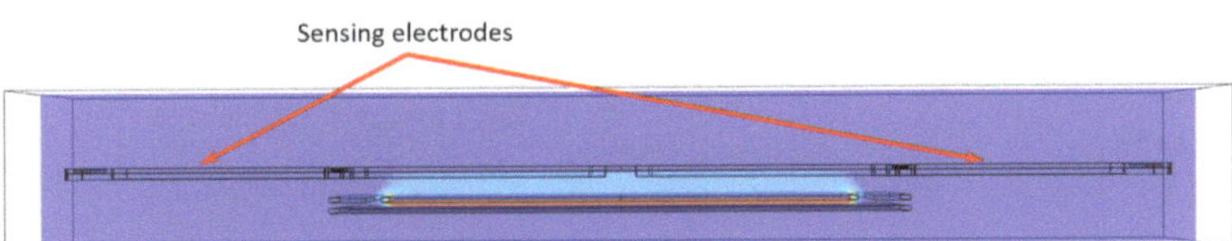

Figure 5. Driving electric field distribution.

3. Sensor Fabrication

After the simulation, the sensor patterns were designed using a free and open-source PCB design tool named KiCAD 7.0 The generated Gerber and drill files are demonstrated in Figure 6a. These files were sent to a PCB manufacturer called PCBway (https://www.pcbway.com/, accessed on 30 November 2023) to fabricate the structure. The flexible PCB samples are shown in Figure 6b,c, where Figure 6b is the shutter and sensing electrodes, and Figure 6c is the actuator driving electrode. These samples have two conductive layers, where Figure 6b,c shows the front view, and the black features on these two pictures are the features on the back side. According to the manufacturer's datasheet, the thickness of the copper is 18 μm and that of the polyimide is 25 μm, and minimum line width is 60 μm.

The shutter sample needs to be released to separate the sensing electrodes. This was completed by using a laser etching process. The equipment used is the A Series Laser Micromachining system, which is an ultraviolet (355 nm) diode-pumped solid-state picosecond laser dicer from Oxford Lasers Ltd. (Didcot, Oxfordshire, UK). The cutting paths were created by using AlphaCAM 2023 R2 software. The laser beam peak pulse energy was set to 0.12 mJ and the pulse duration was 6 ps. The laser pulse frequency setting was 400 Hz, cutting speed 1 mm/s, and the diameter of the laser beam was 10 μm. Using 30% of this laser power, 15 cutting passes were required to cut through the polyimide-only areas and 30 passes were required for areas having both copper and polyimide.

The released micro-springs, shutter, and sensing electrodes are shown in Figure 6e, where two corners are cut to expose the electrodes on the layer below it. After soldering wires to the sample, then the sensor is assembled by aligning the released sample on top of the actuator driving electrode and using taps to fix all of the parts. As the polymer substrate is partially transparent, in the alignment process, the cross-shaped alignment marks on both layers will make the process easier.

Figure 7a shows the picture of the released micro-springs, shutter, and sensing electrodes. The picture was taken using a Leica DM6 M microscope (Leica Microsystems, Wetzlar, Germany). As we can see, the micro-springs, shutter, and sensing electrodes have slight deformations, but the shutter and the sensing electrodes are still aligned very well. After assembly, the separation distance from the shutter to the driving electrode under it is measured to be 105 μm by using a microscope. Figure 7b shows the initial location of the shutter fingers and sensing electrodes in the zoomed-in area of Figure 7a. As an illustration of the shutter movement test, the application of 200 V to the driving electrode results in noticeable shutter finger movement, shown in Figure 7c. We can see that the sensing electrodes remain focused and unchanged, while the shutter fingers move downward, altering the focal status. By refocusing the shutter finger and observing the displacement of the fine object adjustment knob of the microscope, the movement of the shutter fingers was measured as 10 μm. As detailed in Section 2.2, the electrostatic force exerted on the shutter at 200 V is 8.69×10^{-5} N. Therefore, the spring constant of the shutter's supporting springs is calculated to be 8.7 N/m.

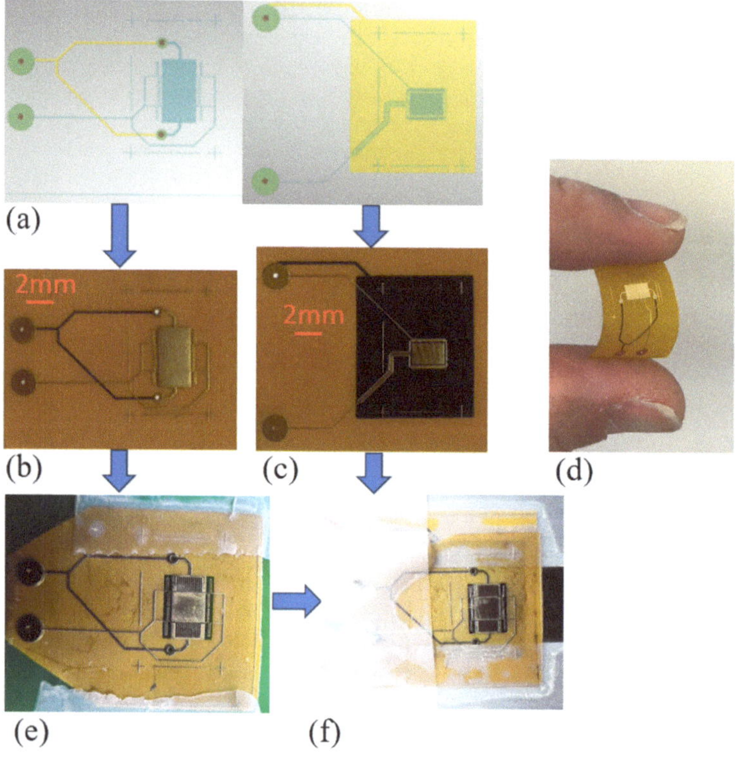

Figure 6. Sensor fabrication process. (**a**) The view of the design PCB and generate Gerber files. (**b**) Manufactured shutter and sensing electrodes. (**c**) Manufactured actuator driving electrodes. (**d**) Bending the sample manually to show the flexibility. (**e**) After laser cutting, the micro-springs are released and the shutter and sensing electrodes are separated. (**f**) Place (**e**) on top of (**c**) to form the sensor.

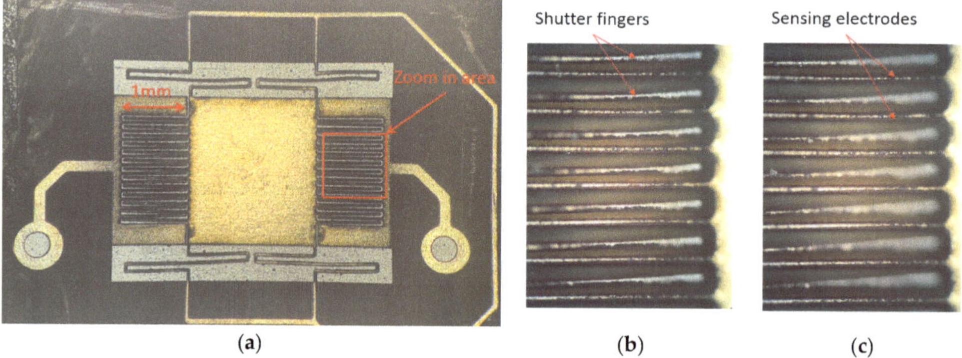

Figure 7. Released shutter and sensing electrodes under the microscope. (**a**) Whole area. (**b**) Zoom picture of the electrodes at the initial rest position. (**c**) Zoom picture after applying 200 V on the driving electrode.

4. Sensor Testing

4.1. Test Setup

The fabricated sensor was tested in the laboratory. Figure 8 shows a functional diagram of the test setup. The sensor was placed on a grounded metal plate, and the distance from the sensor surface to the metal plate was measured to be 0.5 mm. Another metal plate was placed above the sensor, and the two metal plates were separated by two 4.5 mm thick spacers on each side, then the distance from the top metal plate to the surface of the sensor was 4 mm. A DC power supply was connected to the top metal plate to generate a test DC electric field. The actuator driving sine wave was generated by an Agilent 33120A signal generator (Agilent Technologies, Santa Clara, CA, USA), followed by three 1:7 transformers in parallel in the primary side and in series in secondary side. When an amplitude 10 V sine wave was provided by the signal generator, the output driving signal amplitude was measured to be 193 V. The output of the sensing electrodes was sent to a transimpedance amplifier (TIA), with a gain of 10^7. After the TIA, an interfering signal was detected, which had the same frequency as the actuator driving signal and overloaded the lock-in amplifier. This is because the deformation of the shutter after release and the flatness of the electrodes was not perfect, which caused the electric field from the driving electrode to reach the sensing electrodes more easily.

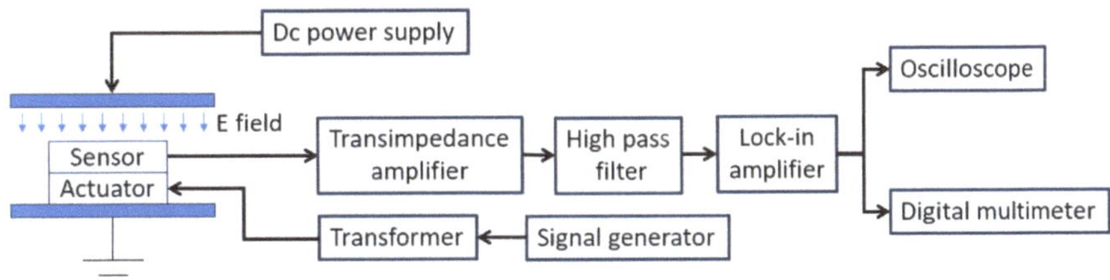

Figure 8. Test function diagram.

To minimize this interference while not attenuating the sensing signal, a sixth-order high-pass Butterworth filter with a cutoff frequency of 500 Hz was employed. The gain for infinite frequency was set to 1, and the quality factor Q was selected to be 0.707. The frequency response of this filter was simulated by using Cadence Orcad 22.1 Pspice (Cadence, San Jose, CA, USA), and Figure 9 shows the result. Since the driving signal frequency was only half the sensing frequency, it was attenuated more. After the high-pass filter, the signal was fed to an SR510 lock-in amplifier. During the test, the lock-in amplifier integration time was set to be 1 s, and the lock-in frequency was set to be 2*f*. The output of the lock-in amplifier was sent to an oscilloscope and a digital multimeter. By adjusting the lock-in amplifier sensitivity, different ranges of electric field can be measured.

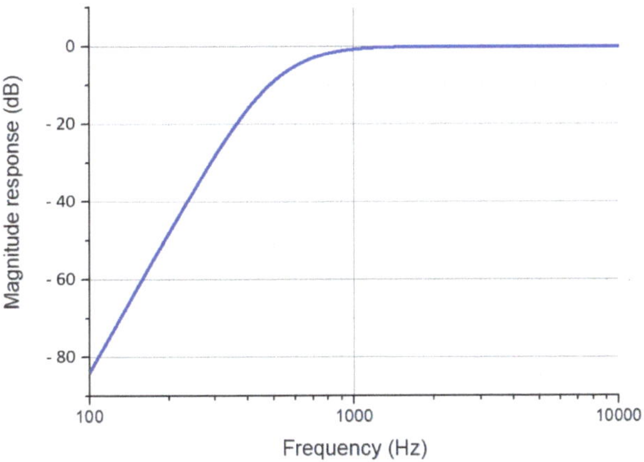

Figure 9. High pass filter frequency response simulation.

4.2. Frequency Response Test

After setting up the test apparatus, the first test performed was to find the actual resonant frequency, where the sensor has its maximum response. In this test, we selected the lock-in amplifier sensitivity to be 2 mV. Then we set the signal generator amplitude to be 8 V. After transformer amplification, the amplitude of the signal driving the actuator was 153 V. The frequency was swept from 160 Hz to 580 Hz with an increment of 10 Hz. The DC power supply output voltage was set to be 100 V, which can create a test electric field of 25 kV/m at each frequency, enabling and disabling the output of the DC power supply. The responses of the sensor shown on the digital multimeter were recorded and are plotted in Figure 10. The analog output of the SR510 lock-in amplifier is given by the equation:

$$V_{out} = 10A_e(A_v V_i \cos \varnothing + V_{OS})$$ (10)

where A_e is the expanded setting on the lock-in amplifier panel; A_v equals the reciprocal of the sensitivity setting, which is 2 mV in this test; V_i is the magnitude of the input signal to the lock-in amplifier; Φ is the phase difference between the signal and reference; and V_{OS} is the offset. However, the SR510 only gives the in-phase component X of V_{out}. The amplitude of V_{out} can be derived by both the in-phase and quadrature component Y:

$$Z = \sqrt{X^2 + Y^2}$$ (11)

In this test, for each frequency, we measured the X component by setting the phase to be $0°$, and measured the Y component by setting the phase to be $90°$, then calculated the Z component using Equation (11). The test results are shown in Figure 10. The Z component maximum occurs at 420 Hz, which indicates that the shutter resonant frequency is 840 Hz.

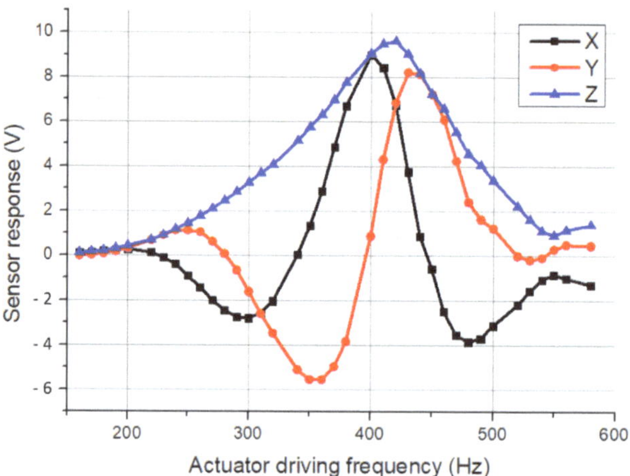

Figure 10. Sensor frequency response test results.

4.3. Sensor Sensitivity Test

In this section, we explore the sensor sensitivity. For a linear sensor, the transfer function can be described by:

$$E = A + Bs$$ (12)

where s is the input signal, E is the output signal, A is the output signal E at zero input signal $s = 0$, and B is the slope of the line. B is also called sensitivity [25]. Therefore, the sensor sensitivity is:

$$B = \frac{E - A}{s}$$ (13)

For this test, we set the signal generator amplitude to be 10 V, and after transformer amplification, the amplitude of the signal driving the actuator was 193 V. As the in-phase X component peaked at 800 Hz (the driving signal at 400 Hz), we chose the lock-in amplifier to display the X component, the phase was set to be 0°, and we set the signal generator output frequency to be 400 Hz. The DC power supply output voltage was set to be 10 V, which provided a 2.5 kV/m test electric field when turned on. To explore the highest sensitivity of the sensor, the SR510 lock-in amplifier sensitivity was pushed to the lowest before overload to be 100 µV. Figure 11 shows the test results. As we can see, when the 2.5 kV/m field turns on and off, the sensor output is averaged to be 12.8 V. The sensor sensitivity can be calculated to be 5.1 V/kVm^{-1}. The noise level of the sensor is 0.3 V, and the resolution of the sensor is calculated to be 62.5 V/m. Table 3 compares the EFMs presented in this paper and some reported in recent years. This sensor demonstrates significantly higher sensitivity and comparable resolution.

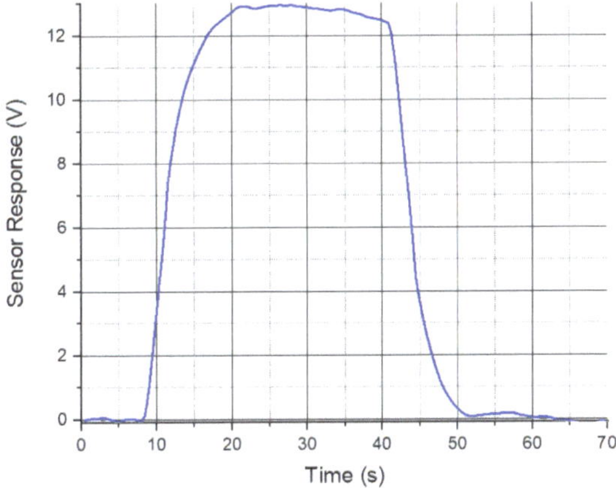

Figure 11. Sensor sensitivity test results.

Table 3. Comparison between references and this work.

Reference	Year	Sensitivity	Resolution
MEMS EFM [13]	2001	40 µV (kV/m)$^{-1}$	N/A
Thermal actuator EFM [14]	2009	0.1 mV (kV/m)$^{-1}$	42 V/m
Thermal actuator EFM [15]	2006	0.4 mV (kV/m)$^{-1}$	240 V/m
Torsional Resonance EFM [16]	2018	5 mV (kV/m)$^{-1}$	N/A
Comb drive EFM [18]	2017	10 mV (kV/m)$^{-1}$	N/A
EFM in this paper	2023	5.1 V (kV/m)$^{-1}$	62.5 V/m

4.4. Sensor Response Linearity Test

In order to explore the sensor response linearity, a test was performed to measure the sensor response from 1.25 kV/m to 25 kV/m. In this test, the driving signal amplitude was 193 V, and the frequency 400 Hz. The lock-in amplifier sensitivity was selected to be 1 mV. When the DC power supply swept the voltage from 5 V to 100 V, electric fields from 1.25 kV/m to 25 kV/m were generated. Figure 12 plots the sensor response (each reading has an error of ±0.03 V). The linearity of the sensor is calculated to be 1.1%, with an average sensitivity of 0.49 V/kVm^{-1}.

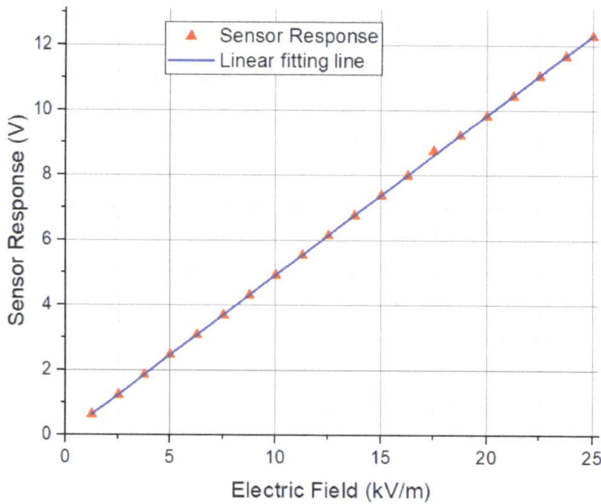

Figure 12. Sensor response to DC electric fields from 1.25 kV/m to 25 kV/m.

5. Conclusions

This paper introduces an MEFM employing a vertical movement shutter based on a flexible PCB substrate. The shutter is driven by an electrostatic actuator. The sensor structure was fabricated by a commercial PCB manufacturing process followed by laser cutting for release. The fabricated sensor demonstrated a linear response in electric fields ranging from 1.25 kV/m to 25 kV/m. The highest sensitivity was measured to be 5.1 V/kVm^{-1}. This sensor has the potential to compensate for the shutter lift-up issue of common MEFMs by providing an initial bias voltage on the actuator. This allows this type of sensor to be applied in a wide range of DC electric field measurements from sub-kV/m to over MV/m.

Author Contributions: Conceptualization, C.S.; methodology, T.C.; design and fabrication, T.C.; validation, T.C.; formal analysis, T.C.; writing—original draft preparation, T.C.; writing—review and editing, C.S. and T.C.; visualization, T.C.; supervision, C.S.; project administration, C.S.; funding acquisition, C.S. All authors have read and agreed to the published version of the manuscript.

Funding: This research was funded by the Natural Sciences and Engineering Research Council (NSERC) of Canada (grant no. RGPIN-05019-2022), Mitacs (Canada) (grant no. IT17156), and Manitoba Hydro International (Winnipeg, Canada).

Institutional Review Board Statement: Not applicable.

Informed Consent Statement: Not applicable.

Data Availability Statement: Data sharing does not apply to this article as no new data were created or analyzed in this study.

Acknowledgments: The authors would like to thank Dwayne Chrusch, Zoran Trajkoski, Sinisa Janjic, and James Dietrich for their technical support in this work.

Conflicts of Interest: The authors declare no conflicts of interest.

References

1. Cui, Y.; Song, X.; Wang, C.; Zhao, L.; Wu, G. Ground-level DC electric field sensor for overhead HVDC/HVAC transmission lines in hybrid corridors. *IET Gener. Transm. Distrib.* **2020**, *14*, 4173–4178. [CrossRef]
2. Yang, X.; Jia, Y.; Gao, L.; Ji, S.; Li, Z. A method for measuring surface electric field intensity of insulators based on electroluminescent effect. *Energy Rep.* **2020**, *6*, 1537–1543. [CrossRef]
3. Mu, X.; Sheng, Q.; Li, J.; Yu, S.; Zhang, J. Fault detection system of single-phase grounding based on electric field sensor. In Proceedings of the IEEE Industrial Electronics and Applications Conference (IEACon), Kuala Lumpur, Malaysia, 3–4 October 2022.

4. Tan, X.; Sun, H.; Suo, C.; Wang, K.; Zhang, W. Research of electrostatic field measurement sensors. *Ferroelectrics* **2019**, *549*, 172–183. [CrossRef]

5. Antunes de Sá, A.; Marshall, R.; Sousa, A.; Viets, A.; Deierling, W. An Array of Low-Cost, High-Speed, Autonomous Electric Field Mills for Thunderstorm Research. *Earth Space Sci.* **2020**, *7*, e2020EA001309. [CrossRef] [PubMed]

6. Korovin, E.A.; Gotyur, I.A.; Kuleshov, Y.V.; Shchukin, G.G. Lightning discharges registration by the electric field mill. *IOP Conf. Ser. Mater. Sci. Eng.* **2019**, *698*, 44047. [CrossRef]

7. Nicoll, K.A.; Harrison, R.G.; Barta, V.; Bor, J.; Brugge, R.; Chillingarian, A.; Chum, J.; Georgoulias, A.K.; Guha, A.; Kourtidis, K.; et al. A global atmospheric electricity monitoring network for climate and geophysical research. *J. Atmos. Sol.-Terr. Phys.* **2019**, *184*, 18–29. [CrossRef]

8. Wang, H.; Zeng, R.; Zhuang, C.; Lyu, G.; Yu, J.; Niu, B.; Li, C. Measuring AC/DC hybrid electric field using an integrated optical electric field sensor. *Electr. Power Syst. Res.* **2020**, *179*, 106087. [CrossRef]

9. Han, Z.; Xue, F.; Hu, J.; He, J. Micro Electric Field Sensors: Principles and Applications. *IEEE Ind. Electron. Mag.* **2021**, *15*, 35–42. [CrossRef]

10. Kareekunnan, A.; Agari, T.; Hammam, A.M.M.; Kudo, T.; Maruyama, T.; Mizuta, H.; Muruganathan, M. Revisiting the Mechanism of Electric Field Sensing in Graphene Devices. *ACS Omega* **2021**, *6*, 34086–34091. [CrossRef] [PubMed]

11. Andò, B.; Baglio, S.; Bulsara, A.R.; Marletta, V. A Ferroelectric-Capacitor-Based Approach to Quasistatic Electric Field Sensing. *IEEE Trans. Instrum. Meas.* **2010**, *59*, 641–652. [CrossRef]

12. Willianms, K.R.; De Bruyker, D.P.H.; Limb, S.J.; Amendt, E.M. Vacuum Steered-Electron Electric field Sensor. *J. Microelectromechanical Syst.* **2014**, *23*, 157–167. [CrossRef]

13. Horenstein, M.N.; Stone, P.R. A micro-aperture electrostatic field mill based on MEMS technology. *J. Electrost.* **2001**, *51–52*, 515–521. [CrossRef]

14. Wijeweera, G.; Bahreyni, B.; Shafai, C.; Rajapakse, A.; Swatek, D.R. Micromachined Electric Field Sensor to Measure ac and dc Fields in Power Systems. *IEEE Trans. Power Deliv.* **2009**, *24*, 988–995. [CrossRef]

15. Chen, X.; Peng, C.; Tao, H.; Ye, C.; Bai, Q.; Chen, S.; Xia, S. Thermally driven micro-electrostatic fieldmeter. *Sens. Actuators A* **2006**, *132*, 677–682. [CrossRef]

16. Chu, Z.; Peng, C.; Ren, R.; Ling, B.; Zhang, Z.; Lei, H.; Xia, S.A. A High Sensitivity Electric Field Microsensor Based on Torsional Resonance. *Sensors* **2018**, *18*, 286. [CrossRef] [PubMed]

17. Yang, P.F.; Peng, C.; Zhang, H.Y.; Liu, S.G.; Fang, D.M.; Xia, S.H. A high sensitivity SOI electric-field sensor with novel comb-shaped microelectrodes. In Proceedings of the 16th International Solid-State Sensors, Actuators and Microsystems Conference, Beijing, China, 5–9 June 2011.

18. Ma, Q.; Huang, K.; Yu, Z.; Wang, Z. A MEMS-Based Electric Field Sensor for Measurement of High-Voltage DC Synthetic Fields in Air. *IEEE Sens.* **2017**, *17*, 7866–7876. [CrossRef]

19. Gong, C.; Xia, S.H.; Deng, K.; Bai, Q.; Chen, S.F. Design and simulation of miniature vibrating electric field sensors. *Proc. IEEE Sens.* **2004**, *3*, 1589–1592.

20. Ghionea, S.; Smith, G.; Pulskamp, J.; Bedair, S.; Meyer, C.; Hull, D. MEMS Electric-Field Sensor with Lead Zirconate Titanate (PZT)-Actuated Electrodes. In Proceedings of the IEEE Sensors Conference, Baltimore, MD, USA, 3–6 November 2013.

21. Dutta, G.; Regoutz, A.; Moschou, D. Commercially fabricated printed circuit board sensing electrodes for biomarker electrochemical detection: The importance of electrode surface characteristics in sensor performance. *Proc. Eurosens.* **2018**, *2*, 741.

22. Xiao, S.Y.; Che, L.F.; Li, X.X.; Wang, Y.L. A novel fabrication process of MEMS devices on polyimide flexible substrates. *Microelectron. Eng.* **2008**, *85*, 452–457. [CrossRef]

23. Petropoulosa, A.; Pagonis, D.N.; Kaltsas, G. Flexible PCB-MEMS Flow Sensor. *Procedia Eng.* **2012**, *47*, 236–239. [CrossRef]

24. Chen, T.; Hill, B.; Isik, S.; Shafai, C.; Shafai, L. DC electric field sensor based on polyimide substrate. In Proceedings of the 2021 IEEE Sensors, Sydney, Australia, 31 October–3 November 2021; pp. 1–4.

25. Fraden, J. *Handbook of Modern Sensors: Physics, Designs, and Applications*, 5th ed.; Springer International Publishing AG: Cham, Switzerland, 2016; pp. 15–21.

 *sensors*

Article

Waste Material Classification: A Short-Wave Infrared Discrete-Light-Source Approach Based on Light-Emitting Diodes

Anju Manakkakudy, Andrea De Iacovo *, Emanuele Maiorana, Federica Mitri and Lorenzo Colace

Department of Industrial, Electronic and Mechanical Engineering, Roma Tre University, 00146 Rome, Italy; anju.manakkakudykumaran@uniroma3.it (A.M.); emanuele.maiorana@uniroma3.it (E.M.); federica.mitri@uniroma3.it (F.M.); lorenzo.colace@uniroma3.it (L.C.)
* Correspondence: andrea.deiacovo@uniroma3.it

Abstract: Waste material classification is a challenging yet important task in waste management. The realization of low-cost waste classification systems and methods is critical to meet the ever-increasing demand for efficient waste management and recycling. In this paper, we demonstrate a simple, compact and low-cost classification system based on optical reflectance measurements in the short-wave infrared for the segregation of waste materials such as plastics, paper, glass, and aluminium. The system comprises a small set of LEDs and one single broadband photodetector. All devices are controlled through low-cost and low-power electronics, and data are gathered and managed via a computer interface. The proposed system reaches accuracy levels as high as 94.3% when considering seven distinct materials and 97.0% when excluding the most difficult to classify, thus representing a valuable proof-of-concept for future system developments.

Keywords: discrete spectroscopy; feature selection; SWIR; material classification; optical sensor

Citation: Manakkakudy, A.; De Iacovo, A.; Maiorana, E.; Mitri, F.; Colace, L. Waste Material Classification: A Short-Wave Infrared Discrete-Light-Source Approach Based on Light-Emitting Diodes. *Sensors* **2024**, *24*, 809. https://doi.org/10.3390/s24030809

Academic Editors: Bruno Ando, Luca Francioso and Pietro Siciliano

Received: 19 December 2023
Revised: 17 January 2024
Accepted: 24 January 2024
Published: 26 January 2024

1. Introduction

Material classification plays a crucial role in various industries such as manufacturing, healthcare, environmental monitoring, and security. The ability to accurately identify and classify different materials is essential for ensuring product quality, safety, and compliance with industry standards. In recent years, the importance of material classification has been further amplified by the growing emphasis on sustainability and environmental conservation [1].

In this era of rapid technological advancements, sensors have emerged as indispensable tools for material classification. Unlike traditional methods that often rely on manual inspection or chemical analysis, sensors offer real-time, non-destructive, and precise material identification capabilities. For this aim, various technologies such as spectroscopy, imaging, and X-ray analysis can be employed [2]. Among such approaches, those relying on optical sensors stand out, due to their ability to use infrared light to gather information about material characteristics that are often invisible to the human eye. Optical sensors offer remarkable precision and speed in distinguishing materials, without the need for extensive sample preparations or prolonged laboratory tests [3]. Moreover, these sensors operate in non-contact mode, facilitating the analysis of moving materials.

The integration of optical sensors with cutting-edge technologies such as hyperspectral imaging and Raman spectroscopy has further elevated their capabilities. Machine learning algorithms, fuelled by data collected from optical sensors, enable intelligent pattern recognition, resulting in highly accurate material classification outcomes [4]. Industries spanning pharmaceuticals, food processing, aerospace, and environmental monitoring have significantly benefited from the advanced capabilities of optical sensors [5–7].

Optical sensors typically exploit spectral information to identify materials based on their reflectance and/or transmittance fingerprint. Different spectral ranges can be employed, depending on the materials of interest and on the morphological characteristics of the samples. Short-wave infrared (SWIR) spectroscopy stands out as a beacon of precision. Operating within the wavelengths from 1000 to 2500 nm, SWIR spectroscopy offers an unparalleled depth of analysis. Its ability to penetrate materials, coupled with its sensitivity to various molecular vibrations, makes it indispensable in material classification [8]. SWIR spectroscopy and SWIR-based hyperspectral imaging have been extensively employed in the last decade for material classification purposes, when used together with classification algorithms [9–11].

Spectral analysis and hyperspectral imaging can reach very high performance in material classification, but, on the other hand, they rely on cumbersome and expensive instrumentation, thus hampering their adoption for high-volume applications. Recently, a simplified approach to spectral measurements has been proposed, employing imaging systems and photodetectors with a large spectral response and quasi-monochromatic light sources (LEDs), selected in order to investigate only those wavelengths where target materials express specific reflectance signatures [12–14]. The extension of such a discrete-light-source spectroscopy approach to SWIR wavelengths can further enhance the achievable performance, allowing for better material recognition.

In this study, we explore SWIR-based material classification using discrete LED sources for spectral analysis. Our preliminary study utilized 10 carefully chosen LEDs, achieving 98% accuracy in classifying waste materials [15]. The current research aims to further improve the proposed system by revisiting the initial set of LEDs and optimizing the achievable accuracy while reducing the number of employed LEDs. The proposed system measures the reflectance of solid samples at specific wavelengths, acquiring discrete reflectivity spectra while minimizing data redundancy. The goal is to create a cost-effective, user-friendly instrument for material sorting, applicable in recycling and waste management. Despite potential performance trade-offs, our approach ensures quick and efficient implementation, making it ideal for distributed waste segregation, enabling the accurate differentiation of waste materials.

Figure 1 presents a schematic overview of the principle of operation of the system. In this configuration, LEDs function as light sources, sequentially controlled by electronic drivers, while the photodetector measures the reflected intensities. These signals undergo conditioning through a trans-impedance amplifier, before being acquired through a commercial data acquisition interface. The MATLAB suite is employed for the classification of the acquired data, corresponding to different materials.

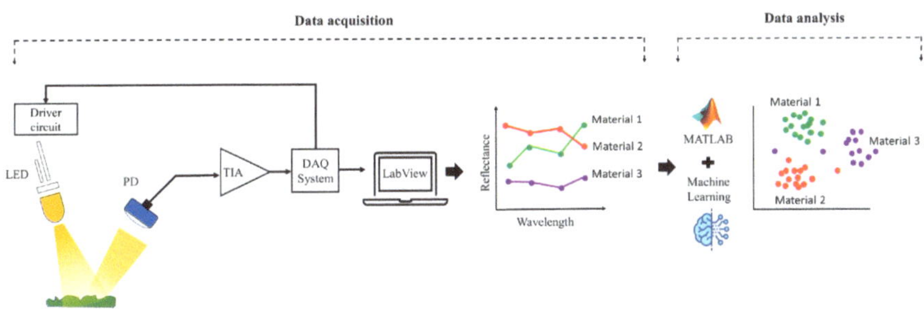

Figure 1. Schematic representation of the system working principle.

2. Materials and Methods

2.1. Materials

In our study, we analysed seven materials, including polyethylene terephthalate (PET), polypropylene (PP, transparent and white), a composite polymer made of polylactic acid (PLA) and polybutylene succinate (i.e., AB400L), glass, paper, and aluminium. It is important to note that these samples represent only a subset of the materials found in industrial and domestic waste, and they were utilized in the forms of flakes and pellets, which might not fully represent real waste materials in terms of physical properties. Hence, our system serves as a proof of concept, demonstrating the potential of discrete spectroscopy for waste segregation. We specifically selected transparent or white materials as reference samples, except for light-grey PET, since SWIR spectroscopy still poses challenges in analysing dark-coloured samples due to their high absorption levels [16]. In addition, cutting the materials into macro-sized flakes provided several advantages, such as an increased sample surface area, material homogenization, and reduced effects of extraneous reflections like specular reflections.

2.2. SWIR Spectral Acquisition

The spectra for each material were collected using two distinct Ocean Insight compact spectrometers (Orlando, FL, USA) operating in the VIS (450 nm to 1030 nm) and SWIR (954 nm to 1710 nm) ranges, with a step size of 1 nm. Subsequently, the individual VIS and NIR spectra were rescaled, merged, and presented as a unified spectrum spanning the range from 450 nm to 1710 nm. A total of 50 spectra were recorded for each material. Figure 2 illustrates the combined averaged reflectance spectra obtained from the two compact spectrometers. The right side of the figure shows photographs of each material.

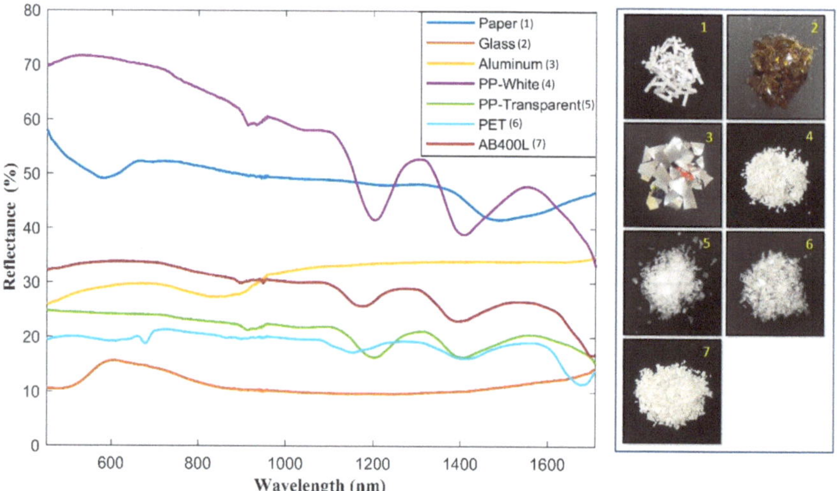

Figure 2. Spectra of all materials using compact spectrometers. The numbers in the photographs align with the labels provided in the plot legends.

The reflectance spectra reveal that glass distinguishes itself from other materials by virtue of its exceptional transparency, characterized by a solitary peak in the 600 nm wavelength range. The white PP material exhibits elevated reflectance values in comparison to other materials, featuring two prominent peaks within the wavelength range of 1200 nm to 1600 nm. Transparent PP shares peaks with white PP, but its reflectance values are significantly lower, and the peaks appear broader. The paper displays distinctive characteristics with high reflectance values, featuring two notable troughs at 550 nm and 1550 nm in the spectrum. Much like transparent PP, AB400L unveils distinct features solely in the SWIR range, displaying two peaks, with one exhibiting a broader profile. Additionally, it shows a nearly constant spectrum in the NIR and VIS ranges. In contrast, aluminium exhibits a consistent spectrum in the SWIR range with high reflectance values, whereas in the NIR range, it features a notable wide reduction between 800 nm to 950 nm. PET exhibits a low reflectance level, akin to transparent PP, with closely aligned spectra. This similarity can be attributed to the selected materials being both white and more transparent than the others.

It is evident from Figure 2 that each material exhibits unique characteristics in the SWIR range as opposed to the VIS range. Therefore, despite the available data in the spectral range between 450 nm to 1710 nm, in our data analysis, we exclusively considered the spectral values within the SWIR range from 810 nm to 1710 nm.

2.3. LED Selection Method

The LED selection was executed by simulating the system response for all the different LEDs employing the spectral data reported in Figure 2 and the spectral emissions of the LEDs as obtained from the datasheets.

Figure 3 illustrates the process of LED selection, encompassing data preparation and optimization methods. Initially, fourteen commercially available LEDs ranging from 890 nm to 1700 nm (as shown in Figure 4) were selected, spanning the SWIR spectrum. Subsequently, a hybrid approach was employed to narrow down the selection to the top four LEDs. This hybrid method [17] was instrumental in ensuring that the eventual experimental setup would be equipped with the most fitting and efficient LEDs to achieve precise and reliable results.

Figure 3. LED optimization process.

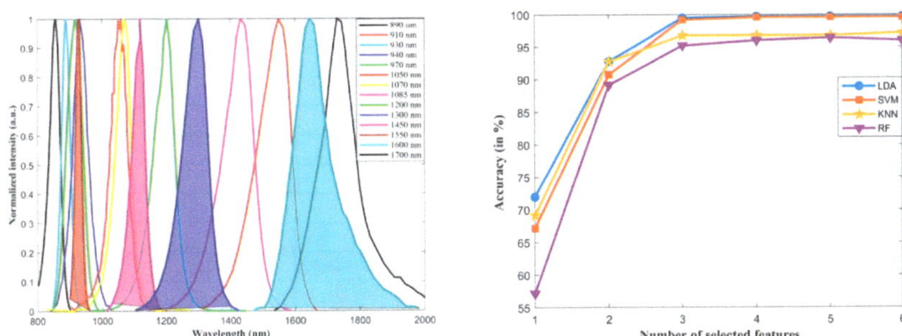

Figure 4. Spectral distribution of 14 LEDs with peak emission of wavelengths reported in the legend (**left**), and accuracy achieved with different classifiers for an increasing number of LEDs (**right**).

The data preparation process initiated with the collection of reflectance spectra for each material using compact spectrometers, as explained in Section 2.2. Subsequently, the 14 LED spectral irradiance data were gathered from datasheets and normalized based on the intensity at the peak wavelength. The normalized irradiance spectra were then interpolated to be aligned with the wavelength range and step of the materials' reflectance datasets. The theoretical amount of light reflected by each sample, when illuminated with a single LED, was calculated as the overlap integral of the LED irradiance spectrum and the material reflectance spectrum measured using the compact spectrometers. The coefficients derived from this process indicated the ratio of reflected light from the sample to the total incident light from the LED. This produced, for each measurement, a reduced set with 14 features that captured essential information from the reflectivity spectrum of each material. For each of the seven considered materials, the 50 collected reflectance spectra were employed for feature (i.e., LED) selection purposes.

In our hybrid approach, we adhered to a standard methodology for effective feature selection and classification model construction. We started by splitting the dataset into distinct training (80%) and testing subsets (20%). The training set became the focal point for employing the Correlation-based Feature Selection (CFS) algorithm [18], which carefully evaluates features to identify their relevance. The algorithm scores features based on their relevance, assigning priorities. Subsequently, the top 10 features with the highest scores were preselected. This subset then served as input for a Sequential Forward Selection (SFS) wrapper method [19]. Resorting to a 5-fold cross-validation, this wrapper-based optimization ensured the selection of a focused and informative feature subset comprising only 4 LEDs. Subsequently, the chosen features were utilized to construct a classification model. For a proper performance assessment, the model underwent evaluation using the untouched testing data, offering insights into its accuracy and generalization capabilities. To perform a comprehensive analysis, our methodology was applied by considering widely used classification algorithms such as Support Vector Machine (SVM) [20], K-Nearest Neighbours (KNN), Linear Discriminant Analysis (LDA), and Random Forest (RF) [21]. This standardized approach ensured a thorough exploration of feature relevance and model efficacy across diverse classifiers, contributing to the reliability and versatility of our mixed method.

After thorough evaluation, each method consistently achieved high accuracy, exceeding 95% with the set of the four selected LEDs. Notably, LDA achieved an accuracy of 99.70%, closely followed by SVM at 99.60%. We strategically selected LEDs at wavelengths of 1085 nm, 1600 nm, 1300 nm, and 910 nm due to their significant and consistent accuracy across the LDA, SVM, KNN, and RF classifiers. This choice was based on the selection of each of these LEDs within the set of the best four, for at least three classifiers out of the four considered ones, as shown in Table 1, highlighting the robustness of these LEDs in the optimization approach. Figure 4 visually presents the spectra of all 14 selected LEDs on the left side, with the filled area representing the spectra of the LEDs chosen after optimization. On the right side, the graph illustrates the LEDs alongside their corresponding accuracies at each step of the selection process for the four classification methods. Notably, as additional features were introduced, the accuracy levelled off and remained stable after selecting 6 LEDs in each method, confirming the chosen 4 LEDs were sufficient to guarantee the desired classification performance.

Table 1. Ranked list of LEDs (expressed in terms of emission peak wavelength, in nm) selected for each classifier.

Methods	LED 1	LED 2	LED 3	LED 4
SVM	1085 nm	1600 nm	910 nm	1300 nm
KNN	1085 nm	910 nm	1600 nm	1300 nm
LDA	1085 nm	1600 nm	1450 nm	930 nm
RF	1085 nm	910 nm	1450 nm	1300 nm

2.4. Simulation Results

To provide further details on the performed analysis, and to assess the robustness of our proposed spectral acquisition and classification analysis using the selected four features (LEDs), we report in Figure 5 the confusion matrix obtained considering the LDA classifier, which demonstrated superior performance compared to the SVM, KNN, and RF algorithms. Notably, the overall accuracy rate reached 98.6%, with only a minor misclassification observed for aluminium, glass, and paper.

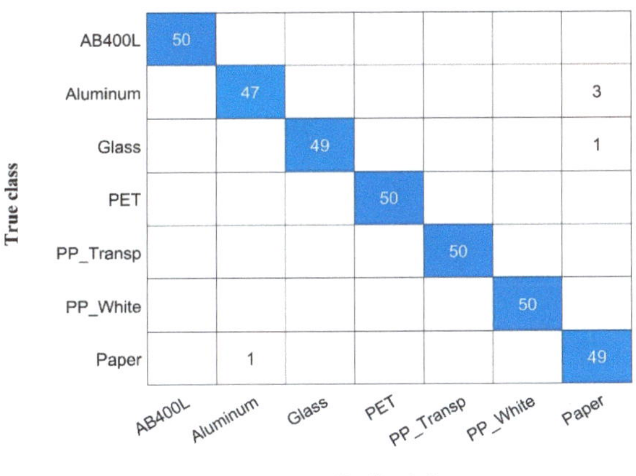

Figure 5. Confusion matrix obtained during the feature selection process to simulated LED data.

2.5. Methodology

Following the analysis performed to select the best four LEDs, an experimental setup was built, as depicted in the schematic block diagram shown in Figure 6. This system was composed of three main components: a sensor head, an analogue front-end, and a multifunction interface connected with a computer. The sensor head was housed in a 3D-printed case with dimensions of 5 cm in diameter and 1.5 cm in height. Within this casing, there was a germanium photodetector surrounded by four strategically positioned LEDs. To prevent direct illumination, the LEDs were evenly spaced. These LED light sources are commercially available and come equipped with epoxy and glass lenses, measuring 4 mm in diameter, and have a maximum current of 25 mA. All the LEDs were biased with a custom circuit in order to provide a mean optical power of 2 mW $\pm$ 10%. The germanium photodiode operated within the spectral range of 700–1800 nm, with a peak responsivity of 0.85 A/W at 1550 nm.

The analogue front-end of the system consisted of a 4-channel current amplifier responsible for driving the LEDs and a variable-gain transimpedance amplifier designed for the photodiode. This analogue front-end was connected to a computer through a multifunction input/output interface, leveraging the National Instruments DAQ USB6009 (Austin, TX, USA). The USB6009 facilitated the sequential activation of the LED current drivers and performed analogue-to-digital conversion of the output from the photodiode amplifier. This conversion was achieved at a precision of 14 bits with a sample rate of 1 kS/s. To streamline the measurement process, a user-friendly graphic interface, developed using LabVIEW (version 2023.Q1), was employed. This interface allowed for the management of measurements, the real-time monitoring of each light source's status, the configuration of measurement parameters (such as sampling rate, number of readings to be averaged, duration of LED illumination, and measurement timing), and the display and storage of sensor readings.

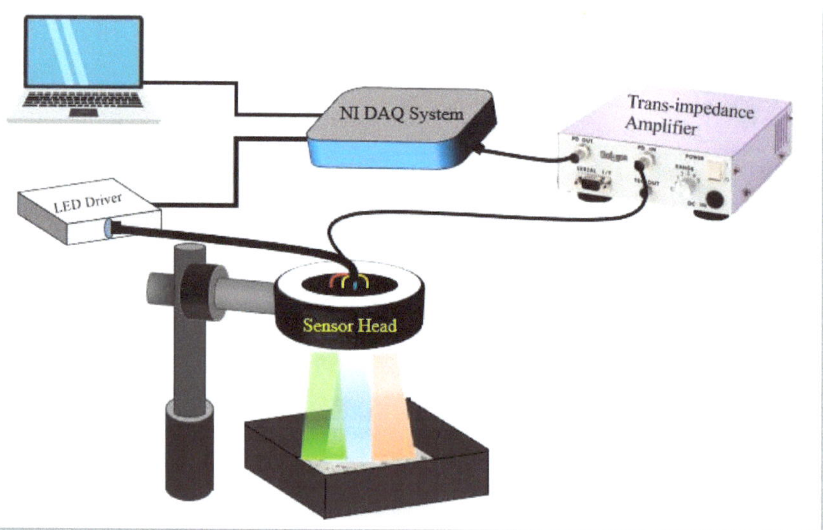

Figure 6. Schematic representation of the system set-up.

Each material sample was placed and stored in a separate black box. The samples underwent exposure to each LED light source for a duration of 10 ms. During this time, reflectance values were captured and used to calculate the ratio of reflected light to incident light. The total incident light amount was pre-calibrated and stored in a look-up table. The system output was presented graphically in LabVIEW. Maintaining a uniform illumination of the sample was ensured by placing the sensor at a fixed distance of 7 cm. A 5 V power supply was utilized to power the entire system which, along with the sample under measurement, was enclosed in a dark box. To enhance measurement accuracy, each sample was measured 10 times and averaged. Before each reading, the sample box was shaken to minimize artifacts associated with preferential reflection angles from the grains.

Upon exposure to each LED, the sample material underwent partial reflection of the incident optical radiation. The diffused reflection was captured by the photodiode, generating a photocurrent that was proportional to the optical power. This photocurrent was then amplified and converted to a voltage using the trans-impedance amplifier before being acquired by the DAQ. This acquisition process was repeated for all the different LEDs, resulting in a set of 4 values. To account for variations in the emission intensity of the 4 LEDs, all measurements were normalized with respect to a calibration dataset. This dataset was obtained by substituting the sample with a metal mirror during the calibration process. The sample's reflectivity was calculated using Equation (1), where $V0(\lambda)$ represents the voltage value obtained during the calibration with a flat mirror (proportional to the incident light intensity), and $Vs(\lambda)$ is the voltage acquired while illuminating the sample under analysis (proportional to the reflected light intensity).

$$R(\lambda) = Vs(\lambda)/V0(\lambda) \tag{1}$$

The measurements were conducted on the seven considered materials over a period of three days. Specifically, 50 spectra were collected from each material daily, amounting to a total of 350 spectra per day. Across the entire study, 1050 spectra were gathered, creating a comprehensive dataset that encompassed a broad range of variations and measurements for subsequent analysis. This extensive dataset enabled the consideration of the potential impact of environmental conditions on the proposed setup. In the subsequent section, various classifiers were employed on the collected data. The objective was to identify the most effective approach and to explore the relationship between achievable performance and the intra-class variability observed in the conducted measurements.

3. Results

The collected data were classified using various methods through a MATLAB (version 2022a) script. Table 2 below presents the classification accuracy achieved on each of the three days, employing an 80%–20% division for training and test data, and the four considered classification methods. Better results are typically obtained when using an SVM classifier, which provides the best average performance when considering data from all the three different acquisition sessions.

Table 2. Classification accuracies achieved using four optimized LEDs across four different classification methods over the considered three-day dataset. Values in bold indicate the best results obtained for each acquisition session.

Data Set	SVM	LDA	KNN	RF
1st day	95.7%	**96.0%**	94.3%	92.9%
2nd day	**94.6%**	93.4%	92.6%	88.6%
3rd day	**93.4%**	92.9%	89.7%	90.0%

Overall, classification performances were similar for the different days, but some slight differences could be observed, with data from day 3 providing the worst classification results irrespective of the classification algorithm. This behaviour was expected, since we purposely did not control environmental conditions (e.g., temperature, room illumination, relative humidity) during measurements. It is therefore evident that the specific acquisition conditions may have had a non-negligible impact on the reliability of the measurements and on the resulting classification performance.

For a better assessment of the actual capabilities of the proposed system and to enhance its generalization capabilities, additional data analysis was conducted by splitting the collected samples according to two distinct modalities, that is, day-wise split and random split. The day-wise split means that data from one specific day were exclusively assigned for training, and data from another day were reserved for testing. This approach allowed for a targeted assessment of the model's performance on unseen data from a different day, providing insights into its ability to generalize and handle variations specific to distinct timeframes. Evaluating the model under such day-wise conditions helps gauge its adaptability to changing circumstances and ensures a comprehensive understanding of its performance across different temporal contexts. The random split modality implies that the data obtained during the 3 days are mixed and then randomly divided into training (80%) and test (20%) samples in a 5-fold cross validation. Random split helps ensure a representative distribution of data across training and test sets, reducing bias and enhancing the model's generalization to unseen data.

Figure 7a–d show the confusion matrices associated with four out of the seven considered scenarios, that is, 1st vs. 2nd day (with LDA classifier), 1st vs. 3rd day (with SVM), 3rd vs. 2nd day (SVM), and shuffled data (SVM). The accuracies achieved in all considered tests are detailed in Table 3, where the results related to the conditions considered in Figure 7 are marked in bold. When checking the reported confusion matrices, certain noteworthy misclassifications are evident. Figure 7a reveals a significant error rate, i.e., 62%, in classifying aluminium with other materials. AB400L follows with a 40% misclassification, and PP (transparent) exhibits a 22% error rate. In Figure 7b, a persistent 62% misclassification is observed between aluminium and other materials, accompanied by a 16% misclassification between PP (transparent) and AB400L, emphasizing the complexity of accurate classification for these material pairs. Figure 7c,d continue to show a consistent trend of high misclassification for aluminium materials with other substances. Notably, in Figure 7c, there is a 38.0% misclassification with other materials, and in the shuffled dataset, a similar pattern is observed with a 23% misclassification rate. These findings highlight the difficulty in accurately categorizing aluminium using classification models applied to training/test data collected in different scenarios, which significantly impacts the overall accuracy. On the other hand, consistently high accuracy, close to 100%, is achieved for both glass and PP (white) across all the considered situations. These results suggest that these two materials have highly characterizing spectral signatures that allow effective differentiation within a classification framework.

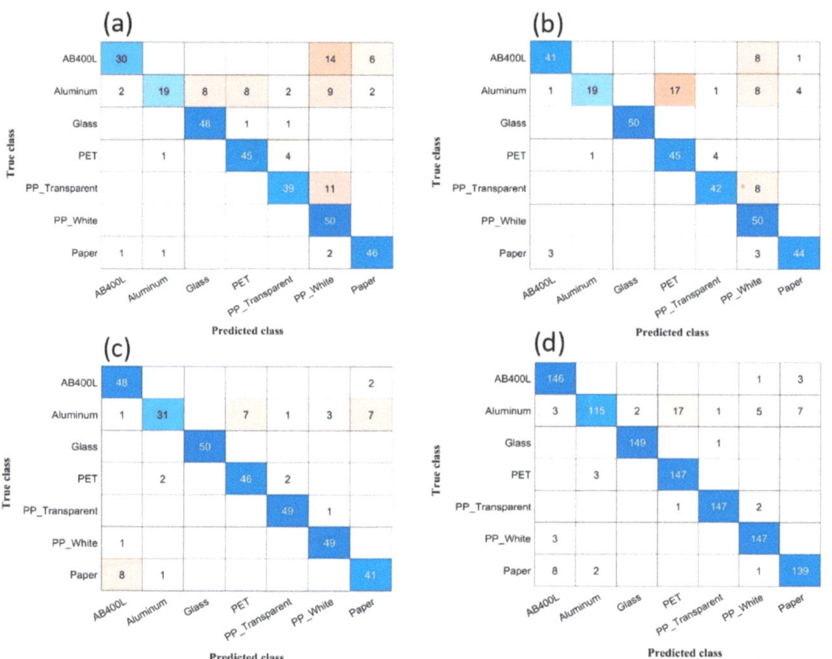

Figure 7. (**a**) Confusion matrix of 1st (training) vs. 2nd (test) acquisition sessions for LDA classifier. (**b**) Confusion matrix of 1st (training) vs. 3rd (test) acquisition sessions for SVM classifier. (**c**) Confusion matrix of 3rd (training) vs. 2nd (test) acquisition sessions for SVM classifier. (**d**) Confusion matrix of shuffled data acquisition sessions for SVM classifier.

The misclassification of aluminium in comparison to other materials can be attributed to the distinct surface characteristics of the materials, where one side exhibits a shiny appearance while the other is coloured. Notably, these aluminium samples were sourced from used soft drink cans obtained from the environment. To address this challenge, we opted to exclude aluminium from our dataset and focused on classifying the remaining materials. The results revealed a promising improvement in our data analysis. Actually, Table 3 outlines a comparison of the performance obtained when employing SVM and LDA classifiers, and both with and without aluminium, for all the considered data divisions. The findings from Table 3 clearly indicate that the dataset excluding aluminium exhibits substantial improvements across all scenarios. Particularly noteworthy is the fact that, out of the seven considered scenarios, five of them consistently provided overall classification accuracy above 90% without aluminium. Figure 8a–d show the confusion matrices obtained when excluding aluminium for the four scenarios considered in Figure 7. Actually, the considered scenarios, with the associated classifiers, are those showing the largest differences between considering or excluding aluminium.

Table 3. Comparative analysis of classification accuracy using SVM and LDA methods in different data sets. Values in bold indicate the scenarios considered in Figures 7 and 8.

Scenario	All Materials (SVM)	Without Aluminium (SVM)	All Materials (LDA)	Without Aluminium (LDA)
1st (Training) vs. 2nd (Testing)	79.4%	87.3%	**79.0%**	**92.0%**
1st (Training) vs. 3rd (Testing)	**83.0%**	**91.3%**	67.7%	88.7%
2nd (Training) vs. 1st (Testing)	76.3%	82.7%	66.6%	83.0%
2nd (Training) vs. 3rd (Testing)	80.0%	91.7%	81.4%	88.0%
3rd (Training) vs. 1st (Testing)	86.0%	95.3%	73.0%	87.7%
3rd (Training) vs. 2nd (Testing)	**82.6%**	**90.0%**	74.0%	89.7%
Shuffled data	**94.3%**	**97.0%**	93.0%	96.0%

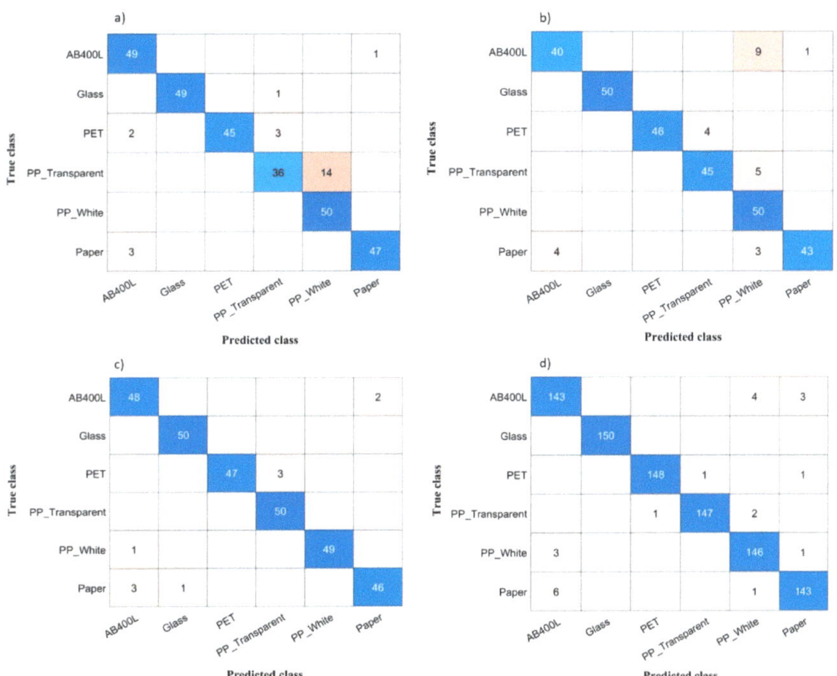

Figure 8. (**a**) Confusion matrix of 1st (training) vs. 2nd (test) acquisition sessions, without aluminium, for LDA. (**b**) Confusion matrix of 1st (training) vs. 3rd (test) acquisition sessions, without aluminium, for SVM. (**c**) Confusion matrix of 3rd (training) vs. 2nd (test) acquisition sessions, without aluminium, for SVM. (**d**) Confusion matrix of shuffled data acquisition sessions, without aluminium, for SVM.

The observed improvement underscores the positive impact of excluding aluminium on the model's ability to accurately classify materials in various scenarios. Likewise, data shuffling facilitates improved the training of the classifier, enhancing its generalization capability and resulting in an overall higher accuracy of 94.3% when considering all materials, with a further improvement to 97.0% when excluding aluminium, further highlighting the positive impact of data shuffling on the model's performance. Finally, it can be seen that the performance of the SVM classifier is consistently higher than that attained by LDA.

Despite removing aluminium from the dataset, the confusion matrices in Figure 8 highlight significant misclassifications, particularly among AB400L, PP (transparent), and paper. The classification models encounter challenges in distinguishing these three materials, particularly in the case of PP (transparent), which consistently experiences minor misclassifications with PP (white) due to their shared material properties. Glass, on the other hand, proves to be the easiest material to classify, achieving almost 100% accuracy across all cases, thanks to its distinctively low reflectance values. Furthermore, PP (white) stands out as another material with high accuracy within this system setup, attributed to its distinctive high reflectance characteristics, as evident in Figure 1, which depicts the continuous spectra of the materials. While our current system setup with 4 LEDs has demonstrated good performance in material classification, our previous analysis using a larger number of LEDs revealed superior classification for "critical" materials. This underscores the significance of achieving finer spectral discrimination for enhanced results in material classification. The utilization of a more extensive set of LEDs seems to boost the model's capacity to differentiate and classify materials, highlighting the importance of spectral precision in achieving accurate results, particularly for critical materials.

4. Conclusions

This paper introduces a cost-effective approach for waste material classification through discrete optical analysis in the short-wave infrared (SWIR) range. The system integrates four LEDs and a single photodetector, controlled by basic electronic drivers and a transimpedance amplifier, respectively. The conditioned photodetector signal is acquired via a commercial data acquisition interface. With respect to our previous work using 10 LEDs and a 98% accuracy rate, our new feature selection method focuses on four LEDs, attaining 97% accuracy with support vector machines. The system is applied to classify diverse materials, including glass, paper, and three types of plastic. Challenges arise in classifying aluminium due to its reflective properties and the presence of plastic coatings on the samples.

The proposed system should be interpreted as a proof-of-concept and has been tested with a small number of samples and in a reduced complexity framework. More specifically, no coloured plastics were employed, and all the samples shared similar surface characteristics in terms of roughness. In order to assess the real extent of the proposed classification approach, further systematic analyses including samples with different colours and surface characteristics should be executed. However, it should be considered that, apart from black pigments, plastic colourings do not typically show any peculiar absorption or reflection in the SWIR range. In addition, the scattering characteristics of a certain material depend both on its surface roughness and on its complex refractive index; thus, different finishes can act as confounding parameters in the classification system, but this issue could be addressed by increasing the amount of samples and data employed for the training.

It may also be noted that, due to the employment of a single photodetector, our system may fail in the classification of samples made of mixed materials (e.g., glass bottles with plastic or metallic caps) or when dealing with very small samples with respect to the field of view of the detector. A straightforward approach to improve the performance of the system consists of the implementation of an imaging sensor as a replacement for the single photodetector, thus allowing for the acquisition of images with embedded spectral data. However, this performance enhancement would come at the cost of a more expensive measurement set-up. Despite such constraints, our research demonstrates that the proposed system is promising and highlights the potentialities of discrete spectrometric analysis for material recognition.

In summary, our research contributes to the development of an affordable and reliable system for material classification. While refining the accuracy of aluminium classification remains a challenge, our system demonstrates significant potential for accurately categorizing plastic, paper, and glass waste materials.

Author Contributions: Conceptualization, L.C. and A.D.I.; methodology, L.C., E.M. and A.D.I.; software, A.M. and F.M.; validation, A.M., L.C., E.M. and A.D.I.; formal analysis, A.M.; investigation, A.M. and F.M.; resources, L.C.; data curation, A.M. and F.M.; writing—original draft preparation, A.M.; writing—review and editing, L.C., E.M., F.M. and A.D.I.; visualization, A.M.; supervision, L.C.; project administration, L.C.; funding acquisition, L.C. All authors have read and agreed to the published version of the manuscript.

Funding: This research was funded by Ministero dell'Università e della Ricerca under grant F83C21000170001.

Institutional Review Board Statement: Not applicable.

Informed Consent Statement: Not applicable.

Data Availability Statement: The data presented in this study are available on request from the corresponding author. The data are not publicly available due to their large dimensions.

Acknowledgments: We thank Massimiliano Barletta and Clizia Aversa for providing the plastic samples.

Conflicts of Interest: The authors declare no conflicts of interest.

References

1. Kassim, S.M. The Importance of Recycling in Solid Waste Management. *Macromol. Symp.* **2012**, *320*, 43–50. [CrossRef]
2. Gundupalli, S.P.; Hait, S.; Thakur, A. A Review on Automated Sorting of Source-Separated Municipal Solid Waste for Recycling. *Waste Manag.* **2017**, *60*, 56–74. [CrossRef] [PubMed]
3. Vrancken, C.; Longhurst, P.J.; Wagland, S.T. Critical Review of Real-Time Methods for Solid Waste Characterisation: Informing Material Recovery and Fuel Production. *Waste Manag.* **2017**, *61*, 40–57. [CrossRef] [PubMed]
4. Kroell, N.; Chen, X.; Greiff, K.; Feil, A. Optical Sensors and Machine Learning Algorithms in Sensor-Based Material Flow Characterization for Mechanical Recycling Processes: A Systematic Literature Review. *Waste Manag.* **2022**, *149*, 259–290. [CrossRef] [PubMed]
5. Zhao, J.; Tian, G.; Qiu, Y.; Qu, H. Rapid Quantification of Active Pharmaceutical Ingredient for Sugar-Free Yangwei Granules in Commercial Production Using FT-NIR Spectroscopy Based on Machine Learning Techniques. *Spectrochim. Acta Part A Mol. Biomol. Spectrosc.* **2021**, *245*, 118878. [CrossRef] [PubMed]
6. Ozturk, S.; Bowler, A.; Rady, A.; Watson, N.J. Near-Infrared Spectroscopy and Machine Learning for Classification of Food Powders during a Continuous Process. *J. Food Eng.* **2023**, *341*, 111339. [CrossRef]
7. Kussul, N.; Lavreniuk, M.; Skakun, S.; Shelestov, A. Deep Learning Classification of Land Cover and Crop Types Using Remote Sensing Data. *IEEE Geosci. Remote Sens. Lett.* **2017**, *14*, 778–782. [CrossRef]
8. Hansen, M.P.; Malchow, D.S. *Overview of SWIR Detectors, Cameras, and Applications*; Vavilov, V.P., Burleigh, D.D., Eds.; SPIE: Orlando, FL, USA, 2008; p. 69390I.
9. Mauruschat, D.; Plinke, B.; Aderhold, J.; Gunschera, J.; Meinlschmidt, P.; Salthammer, T. Application of Near-Infrared Spectroscopy for the Fast Detection and Sorting of Wood–Plastic Composites and Waste Wood Treated with Wood Preservatives. *Wood Sci. Technol.* **2016**, *50*, 313–331. [CrossRef]
10. Bonifazi, G.; Capobianco, G.; Serranti, S. A Hierarchical Classification Approach for Recognition of Low-Density (LDPE) and High-Density Polyethylene (HDPE) in Mixed Plastic Waste Based on Short-Wave Infrared (SWIR) Hyperspectral Imaging. *Spectrochim. Acta Part A Mol. Biomol. Spectrosc.* **2018**, *198*, 115–122. [CrossRef] [PubMed]
11. Serranti, S.; Fiore, L.; Bonifazi, G.; Takeshima, A.; Takeuchi, H.; Kashiwada, S. Microplastics Characterization by Hyperspectral Imaging in the SWIR Range. In *SPIE Future Sensing Technologies*; Valenta, C.R., Kimata, M., Eds.; SPIE: Tokyo, Japan, 2019; p. 38.
12. Dong, Y.; Liu, X.; Mei, L.; Feng, C.; Yan, C.; He, S. LED-Induced Fluorescence System for Tea Classification and Quality Assessment. *J. Food Eng.* **2014**, *137*, 95–100. [CrossRef]
13. Lopez-Ruiz, N.; Granados-Ortega, F.; Carvajal, M.A.; Martinez-Olmos, A. Portable Multispectral Imaging System Based on Raspberry Pi. *Sens. Rev.* **2017**, *37*, 322–329. [CrossRef]
14. Sato, M.; Yoshida, S.; Olwal, A.; Shi, B.; Hiyama, A.; Tanikawa, T.; Hirose, M.; Raskar, R. SpecTrans: Versatile Material Classification for Interaction with Textureless, Specular and Transparent Surfaces. In Proceedings of the 33rd Annual ACM Conference on Human Factors in Computing Systems, Seoul, Republic of Korea, 18 April 2015; ACM: New York, NY, USA, 2015; pp. 2191–2200.
15. Manakkakudy, A.; De Iacovo, A.; Maiorana, E.; Mitri, F.; Colace, L. Material Classification Based on a SWIR Discrete Spectroscopy Approach. *Appl. Opt.* **2023**, *62*, 9228–9237. [CrossRef] [PubMed]
16. Rozenstein, O.; Puckrin, E.; Adamowski, J. Development of a New Approach Based on Midwave Infrared Spectroscopy for Post-Consumer Black Plastic Waste Sorting in the Recycling Industry. *Waste Manag.* **2017**, *68*, 38–44. [CrossRef] [PubMed]
17. Hsu, H.-H.; Hsieh, C.-W.; Lu, M.-D. Hybrid Feature Selection by Combining Filters and Wrappers. *Expert Syst. Appl.* **2011**, *38*, 8144–8150. [CrossRef]

18. Gopika, N.; ME, A.M.K. Correlation Based Feature Selection Algorithm for Machine Learning. In Proceedings of the 2018 3rd International Conference on Communication and Electronics Systems (ICCES), Coimbatore, India, 15–16 October 2018; pp. 692–695.
19. Kohavi, R.; John, G.H. Wrappers for Feature Subset Selection. *Artif. Intell.* **1997**, *97*, 273–324. [CrossRef]
20. Soman, K.P.; Loganathan, R.; Ajay, V. *Machine Learning with SVM and Other Kernel Methods*; PHI Learning Pvt. Ltd.: New Delhi, India, 2009; ISBN 978-81-203-3435-9.
21. Singh, A.; Thakur, N.; Sharma, A. A Review of Supervised Machine Learning Algorithms. In Proceedings of the 2016 3rd International Conference on Computing for Sustainable Global Development (INDIACom), New Delhi, India, 16–18 March 2016; pp. 1310–1315.

Article

Experimental In Vitro Microfluidic Calorimetric Chip Data towards the Early Detection of Infection on Implant Surfaces

Signe L. K. Vehusheia [1,*], Cosmin I. Roman [1], Markus Arnoldini [2] and Christofer Hierold [1]

[1] Department of Mechanical and Process Engineering, ETH Zurich, 8092 Zurich, Switzerland; cosmin.roman@micro.mavt.ethz.ch (C.I.R.); christofer.hierold@micro.mavt.ethz.ch (C.H.)

[2] Department of Health Sciences and Technology, ETH Zurich, 8093 Zurich, Switzerland; markus.arnoldini@hest.ethz.ch

* Correspondence: signe.vehusheia@micro.mavt.ethz.ch; Tel.: +41-446-32-7115

Abstract: Heat flux measurement shows potential for the early detection of infectious growth. Our research is motivated by the possibility of using heat flux sensors for the early detection of infection on aortic vascular grafts by measuring the onset of bacterial growth. Applying heat flux measurement as an infectious marker on implant surfaces is yet to be experimentally explored. We have previously shown the measurement of the exponential growth curve of a bacterial population in a thermally stabilized laboratory environment. In this work, we further explore the limits of the microcalorimetric measurements via heat flux sensors in a microfluidic chip in a thermally fluctuating environment.

Keywords: heat flux measurement; microfluidics; microbiology

Citation: Vehusheia, S.L.K.; Roman, C.I.; Arnoldini, M.; Hierold, C. Experimental In Vitro Microfluidic Calorimetric Chip Data towards the Early Detection of Infection on Implant Surfaces. *Sensors* **2024**, *24*, 1019. https://doi.org/10.3390/s24031019

Academic Editors: Bruno Ando, Luca Francioso and Pietro Siciliano

Received: 20 December 2023
Revised: 24 January 2024
Accepted: 25 January 2024
Published: 5 February 2024

1. Introduction

Heat flux measurements are very interesting for measuring the metabolic heat output of chemical and biological systems [1]. For example, heat produced by living organisms reflects metabolic activity and metabolic changes [2,3]. Large instruments such as microcalorimeters have been used for studying the heat produced during the growth of different bacterial strains in medicine [4] and microbiology [5,6]. On-chip calorimetric measurements such as micro- [7,8], nano- [9–11] and pico-calorimetric chips [12] have also been utilized for investigating metabolic changes through exothermic heat production. Microfluidic systems offer a number of benefits, such as optical access, controllable fluid mixing and lab-on-chip/in vitro systems capabilities, as summarized in the literature [13]. Many different microfluidic chips have been developed to emulate the in vivo environment in the body for diagnostic [14,15] or research purposes of different regions in the body such as the kidney [16], lung [17] or tumors [18]. These systems focus on the detection of biochemical markers or on optical investigations of the in vitro system. We wish to adapt the concept of in vitro microfluidic systems to emulate the in vivo environment of an infection and use physical sensors as opposed to biochemical sensors to avoid drift and promote long-term stability of the sensor in vivo. Having heat flux measurements accessible in microfluidic chips would expand the range of tools for bioengineering with tabletop access to microcalorimetric data. This work is an extension of work presented at the XXXV Eurosensors Conference 2023 in Lecce (10–13 September).

Microcalorimetric microfluidic chips have been proposed to mimic the thermal environment of a growing infection on a vascular graft implant surface [19,20]. For example, Figure 1a shows a 2 + 1 channel microfluidic system, where a growing infection is emulated by an exothermic chemical reaction. The heat produced is measured differentially by two heat flux sensors. Heat transfer to blood flowing in an aorta is accounted for by a top channel. Its heat transfer coefficient, h, is designed to match that of a physiological aorta [19].

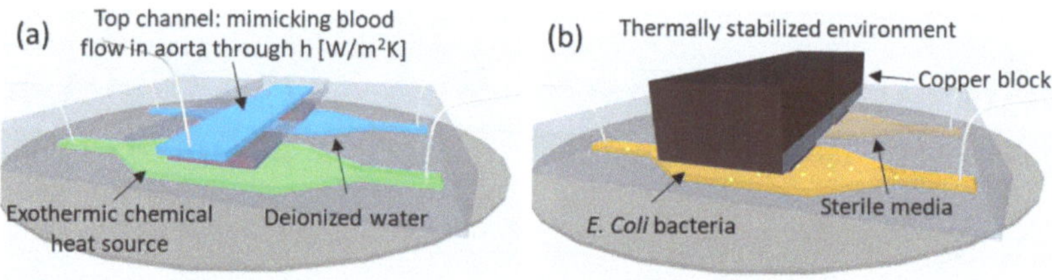

Figure 1. (**a**) Sketch of a 2 + 1 channel microfluidic chip [19] reproduced with permission. The heat transfer coefficient of the top channel matches that of aortic heat transfer. (**b**) Microfluidic chip in a thermally stabilized environment able to resolve bacterial growth though heat flux measurement [1]. Schematics adapted from [1,19].

In another system, shown in Figure 1b, the exponential growth curve of a bacterial population was measured using a similar concept involving differential heat flux sensors [1] in a microfluidic chip. The individual bacterial thermal power could be determined thanks to a thermally stabilized environment. This system was developed to facilitate calorimetric measurements on microfluidic chips in laboratory environments. The thermally stabilized environment consists of an incubator at 37 °C, a PMMA (Poly(methyl methacrylate)) thermally stabilizing box and a block of copper placed on top of the two channels to keep the thermal fluctuation to a minimum.

These in vitro experimental systems have shown the potential application of heat flux sensing towards the early detection of infection. However, some aspects still need to be investigated before considering in vivo applications. On implant surfaces, to detect an infection early, one has to resolve the onset of bacterial growth prior to biofilm formation, after which bacteria stop responding to antibiotic treatment. Such early detection could significantly decrease the patient mortality rate of, for example, vascular graft implant infections, which is currently between 17 and 40% of cases [21]. In the system shown in Figure 1b, the resolution with respect to the number of bacteria is supported by a thermally stabilized environment [1]. In contrast, in vivo, the thermal environment of the implant would undergo temperature fluctuations (sleep cycles, outside temperature, level of patient activity, and fever, for example). To determine the feasibility of the early detection of infection in vivo, it is thus important to investigate the influence of higher temperature fluctuations on the measurement capabilities of microcalorimetric chip systems. Furthermore, smaller heat flux sensors than those used in [1,19] would be needed for integration of the heat flux sensor in an aortic vascular graft, for example in a mesh structure [20]. Since smaller sensors might be more limited by noise, their impact on the resolution of measuring bacterial growth needs to be studied.

Here, we present a microfluidic chip measuring bacterial growth through heat flux in a thermally not-stabilized environment. The system uses smaller heat flux sensors and a smaller microfluidic channel (smaller thermal mass). We investigate the system in a fluctuating temperature range of 1 K, which matches the standard body temperature fluctuations in the human body (36.5–37.5 °C) [22]. This study gives further information about the limit of detection with respect to bacterial growth possible on implant surfaces, a step closer towards in vivo conditions, and extends the study presented at the Eurosensors Conference.

2. Materials and Methods

2.1. Experimental System and Setup

Two different experimental systems, as illustrated in Figure 2, will be compared. To the left, Figure 2a,c,e show the microfluidic chip in the thermally not-stabilized environment, and to the right, Figure 2b,d,f show the thermally stabilized system as introduced in [1]. Pictures of both experimental systems are shown in the Supplementary Information in Figures S1 and S2.

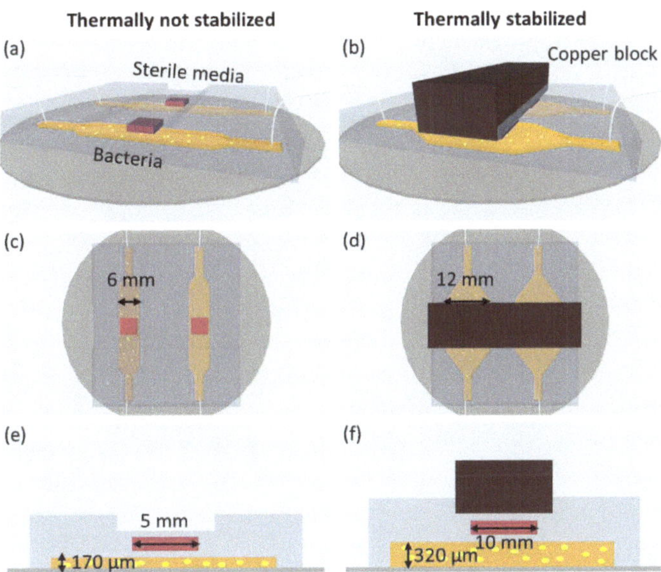

Figure 2. Left column (**a**,**c**,**e**) show the channels of the microfluidic chip in the thermally not-stabilized environment. Right column (**b**,**d**,**f**) show the channels of the microfluidic chip in the thermally stabilized environment [1] with the additional thermally stabilizing copper block. (**e**,**f**) show the cross-section of the chip. The groove is to ensure a given thickness of the PDMS above the heat flux sensor. Sketches not to scale.

The two channels shown in Figure 2a,c,e are 170 µm tall, 6×6 mm wide and long with a volume of 6 µL. The heat flux sensor is a gSKIN XM (greenTEG, Zurich, Switzerland, resolution: 0.41 W/m^2) with dimensions measuring 5 mm $\times$ 5 mm $\times$ 0.5 mm, whereas the two channels shown in Figure 2b,d,f are 320 µm tall and 12×12 mm wide and long (volume = 46 µL) with a heat flux sensor of 10 mm $\times$ 10 mm and 0.5 thickness [1]. The heat flux sensor used is the gSKIN XP (resolution: 0.06 W/m^2). In both systems, one of the channels is filled with bacteria and the other one is filled with sterile lysogeny broth (LB) media for differentially compensated heat flux measurements. The heat flux sensors are embedded in polydimethylsiloxane (PDMS) and placed 150 µm above the channels.

Both the thermally not-stabilized setup and the thermally stabilized setup are shown comparatively in Figure 3. In the former, the microfluidic chip is placed in a temperature-controlled room at 37 ± 0.5 °C without any further temperature stabilization. A PT1000 temperature sensor measures the temperature in close proximity to the microfluidic chip. In the thermally stabilized system [1], the microfluidic chip is placed within a double-walled PMMA box with an additional thermally insulating air layer, as illustrated in Figure 3. The PMMA box is placed inside an incubator at 37 ± 0.15 °C. As also shown in Figure 2b,d,f, a copper block is placed directly above the microfluidic chip for further thermal stability of the heat flux sensors.

Escherichia coli (*E. coli*) MG 1655 in lysogeny broth (LB) is utilized for bacterial experiments. The LB medium consists of 10 g of Tryptone, 5 g of yeast extract, and 5 g of NaCl per 1 L of media. To prevent bacterial cells sticking to the tubing walls or in the channel, a 1:100 ratio of Tween20 is added to the LB medium. Each experiment is performed in two steps, with an initial calibration phase in a sterile environment. First, the flow is controlled using a peristaltic pump connected to both channels of the microfluidic chip. LB without bacteria is pumped at a flow rate of 98.75 µL/min. Second, bacteria are added to one of the channels, into the flask connected to the respective channel. PTFE tubing of an inner diameter of 0.02

in is inserted in the microfluidic chip inlet and outlet and connected to the peristaltic pump. The growth of the bacteria is measured via optical density (OD) analysis of the bacterial population at the outlet of the tubing from the microfluidic chip. The samples collected at the outlet are put on ice and measured subsequently using an optical density (OD) meter. This experimental process is based on a previously published protocol of the thermally stabilized system [1]. During the setup and the initial filling of the channels with liquid, it is important to ensure that no air bubbles are present in the microfluidic channels prior to the start of the measurement.

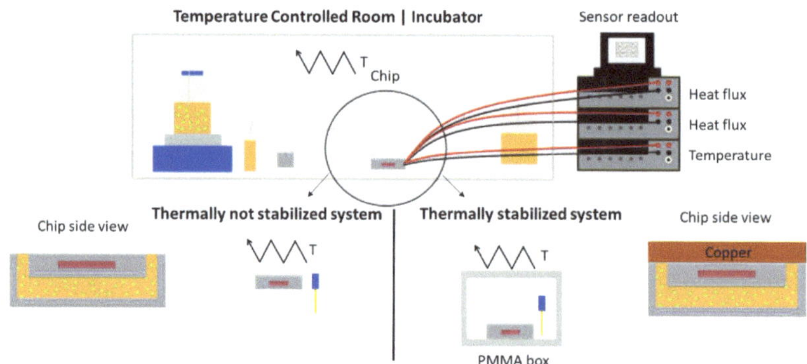

Figure 3. A schematic of the measurement setups of the thermally not-stabilized and the thermally stabilized systems. In the thermally not-stabilized system, the microfluidic chip is placed directly on an arbitrary surface in a temperature-controlled room without further thermal stabilization measures. The thermally stabilized system is a previously published system which shows the data of a microfluidic chip placed inside a thermally stabilizing PMMA box and with a block of copper on the top side of the microfluidic chip [1].

2.2. Microfluidic Chip Fabrication Steps

The lithography and soft lithography microfluidic chip fabrication steps are illustrated step-by-step in Figure 4. The microfluidic chip is fabricated by patterning SU8 on a silicon wafer. A 4 mL quantity of SU8-100 is spin coated on the wafer, with parameters as described in the Supplementary Information (SI), for a total thickness of 170 μm. The wafer is soft baked at 95 °C for 70 min and subsequently exposed with 400 mW/cm^2 on a mask aligner (MA6 Karl Suss, Suss Microtec SE, Garching, Germany). The mask used for the exposure contains the dimensions of the two channels, as shown in Figure 2. Following the exposure, the SU8 is baked at 95 °C for 16 min and subsequently developed in Mr Dev600 (micro resist technology GmbH, Berlin, Germany) for 16 min. After cleaning the wafer with isopropyl alcohol and DI water, it is ready for dicing and the soft lithography steps with PDMS.

PDMS (Sylgard 184, Suter-Kunststoffe AG, Fraubrunnen, Switzerland) is mixed in a 1:10 ratio between the elastomer and crosslinker and poured on the prepared and diced silicon wafer with the desired channel diameters fabricated on top, as shown in Figure 4. Following 40 min of degassing in a vacuum desiccator, the PDMS is cured at 80 °C for 3 h. Spacers are used to define the 150 μm PDMS thickness above the microfluidic channels, as shown in the sketch. Following the first layer of PDMS, the two heat flux sensors are placed on top of the cured PDMS and above the SU8 structures. More uncured PDMS is poured on top of the sensors and degassed for 40 min. The PDMS is again cured at 80 °C for 3 h.

The PDMS with the embedded heat flux sensors is cut out from the SU8 patterned wafer mold. Inlet and outlets are punched out of the PDMS using a 20 G needle. The PDMS is then fused with a glass slide using oxygen plasma ashing for 1 min and 30 s at 100 W in an O$_2$ environment. For further adhesion improvement, the fused chip is directly placed in an oven at 80 °C for 15 min.

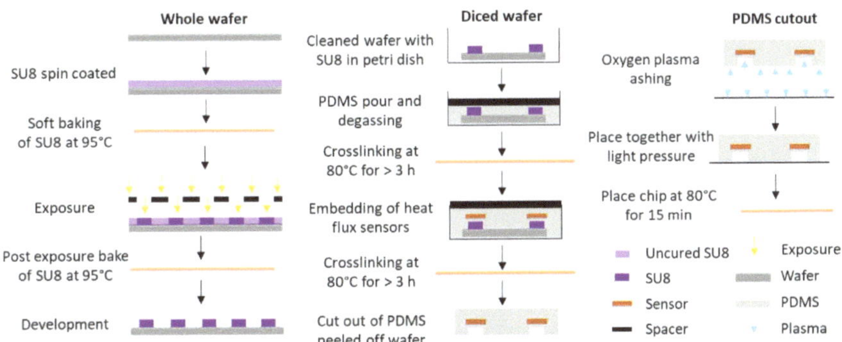

Figure 4. Fabrication steps of the microfluidic chip with two integrated heat flux sensors. A patterned silicon wafer with SU8 is used as the basis of the microfluidic chip channel fabrication. Multiple steps of PDMS curing are applied to create and control different layer thicknesses around the heat flux sensors. After removing the cured PDMS from the wafer, the inlets and outlets are punched out, and it is fused with a glass slide using oxygen plasma.

The fabrication steps are different to the fabrication steps of the microfluidic chip used in the thermally stabilized environment. The channel height for that system is 320 µm, and the specific fabrication steps are described in detail elsewhere [1].

3. Results

Temperature Measurement

Temperature fluctuations of the environment of both the thermally not-stabilized and thermally stabilized environment, as measured by the PT1000 temperature sensors, are shown in Figure 5. The thermally not-stabilized system shows a temperature fluctuation range of about 1 K for 2.3 h, and the thermally stabilized system shows a temperature fluctuation range of 0.3 K for over 4.2 h. The different temperature fluctuation ranges are due to differences in thermal shielding of the microfluidic chip from the environment, as described in the previous sections.

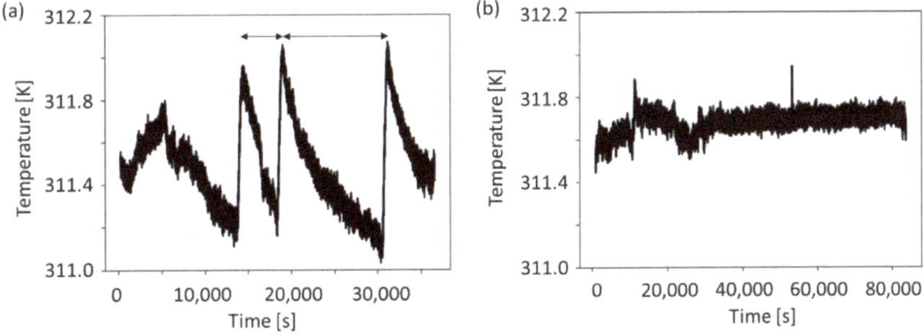

Figure 5. Temperature measurements in both the (**a**) thermally not-stabilized and (**b**) thermally stable environments. The fluctuations in (**a**) are in the range of 1 K from 1.0 h to 3.3 h, whereas in (**b**) the range of temperature is 0.3 K over 4.2 h. The data shown is for the duration of the whole experiment—both calibration and bacterial phase. The temperature data shown in (**b**) is part of the previously published heat flux data [1].

As shown in Figure 2a, two microfluidic channels with different functions are used. One contains only LB media for control purposes, and the other contains LB media together with growing *E. coli*. In the calibration phase, parameters for the differential compensation scheme are determined. These are thereafter applied to heat flux signals in the sensing phase when bacteria are added. The differential compensation scheme allows for the extraction of heat produced by the bacterial growth via a common-mode rejection scheme, as described previously [1]. Figure 6 shows the heat flux signal of the two sensors in the calibration phase (Figure 6a,b) and upon the addition of *E. coli*.

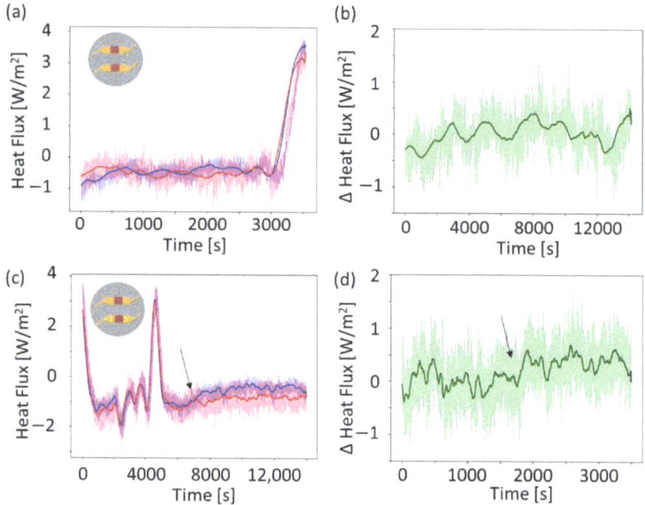

Figure 6. Raw and differentially compensated heat flux values. (**a**) Data during the calibration phase and (**b**) data after the differential compensation is applied. The peaks in the heat flux correspond to peaks in the temperature of the temperature-controlled room, as visible in Figure 5a. (**a,c**) Both have a blue heat flux signal and a red heat flux signal where the different heat fluxes represent the different channels. The blue data belong to the channel in which the bacteria is subsequently added, and the red data belong to the calibration channel. (**c,d**) Heat flux data upon addition of *E. coli*. Peaks in the heat flux correspond to peaks in the temperature, as visible in Figure 5a (shifted by time). The raw data are shown as opaque, and the data averaged over 200 datapoints are shown in the darker color and with a thicker line.

Figure 6a shows raw and averaged data for the calibration phase where both channels are filled with LB media only. The differentially compensated heat flux signal is shown in Figure 6b and is calculated according to the scheme introduced in [1]:

$$q = q_1 - q_2^*$$

(1)

With q as the differentially compensated heat flux signal, q_1 as the heat flux in the channel in which bacteria is to be added, q_2 as the heat flux in the control channel, $\bar{q}$ indicates the average over 200 datapoints, L_c is the ratio of the standard deviations σ_{c1}/σ_{c2}, and q_{c1} and q_{c2} indicate the calibration phase [1]. q_2^* is defined as

$$q_2^* = L_c(q_2 - \bar{q}_{c2}) + \bar{q}_{c1}$$

(2)

The differential compensation scheme manages to cancel out temperature peaks in the heat flux upon thermal fluctuations [1]. Figure 6c shows the data of the different channels upon addition of *E. coli* in one of the channels. The blue signal is from the heat flux sensor above the channel with bacteria, and the red signal is from the heat flux sensor above the calibration channel. Equations (1) and (2) yield the differentially compensated heat flux value, which are also used upon the addition of bacteria, as shown in Figure 6d. At around 8000 s, the heat flux signal increases like a step-function, as indicated by the arrow in Figure 6c,d.

4. Discussion

Figure 7 compares the differentially compensated heat flux signals in the thermally not-stabilized and the thermally stabilized system. Figure 7a,d show the measured heat flux data in parallel with the optical density measurement of the bacterial population. Data for the thermally not-stabilized system shown in Figure 7a is only able to reveal a step-function-like increase in the heat flux. On the other hand, heat flux measurements in the thermally stabilized setup in Figure 7d is able to show a clear exponential growth of the bacteria with the heat flux measurement. In our previous work [1], we have shown that the exponential region of the heat flux measurement matches the exponential region as measured in the optical density measurement. The larger noise in the heat flux values around the averaged signal visible in Figure 7a,c, reflects the two different models of heat flux sensors used.

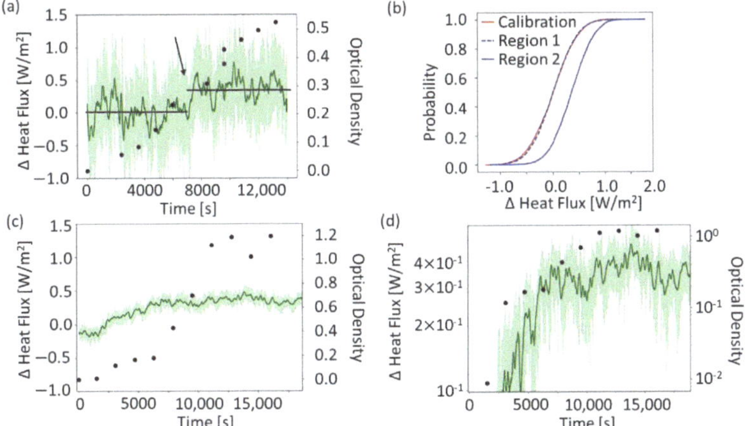

Figure 7. Differentially compensated heat flux measurements in the two systems. (**a**) Data in the thermally not-stabilized system. A change in the heat flux is distinguishable in the exponential growth phase (as indicated by the arrow at 7500 s). Region 1 is indicated by a line in the first part before the arrow, and Region 2 is indicated by the line starting around 7500 s. (**b**) Comparison of the cumulative distribution functions of the calibration phase with raw data with those of Regions 1 and 2. (**c**) Data for the thermally stabilized system, shown previously in [1]. The data shown are the points at which the bacteria was added to the system ($t = 0$ s is the addition of bacteria). In all figures, the opaque data is the raw heat flux signal, and the dark line shows a 200-point moving average. (**d**) Semilogarithmic plot of the data. An exponential increase in both the measured heat flux and also the bacterial population growth is identifiable in the same region.

To distinguish the regions in which the heat flux produced by the bacteria is below or above the detection limit for the thermally not-stabilized system, empirical cumulative distribution functions (eCDF) are utilized in different regions. The first is the calibration phase shown in Figure 6b. The second is Region 1, as indicated in Figure 7a by a horizontal line from $t = 0$ to 7500 s. The third is Region 2, as indicated by the horizontal line from $t = 7500$ to 13,000 s. Comparing eCDFs of these three regions in Figure 7b, the calibration phase and Region 1 are very close. Region 2 is significantly different from both the calibration and Region 1. There is a clear shift in the eCDF at the point where the heat produced by the bacteria surpasses the threshold of detection of the heat flux sensor. The median of the heat flux difference in the calibration phase is 0.01 W/m^2; in Region 1 it is 0.00 W/m^2, and in Region 2 it is 0.35 W/m^2, as indicated by the horizontal lines in Figure 7a. The detection of bacterial growth in the thermally not-stabilized system occurs at an optical density of 0.3. In comparison, the system with the larger, more sensitive sensor [1] is able to resolve the exponential growth of the bacteria in correlation to the optical density measurement.

Table 1 compares the performance of the thermally not-stabilized and the thermally stabilized systems. The differential measurement scheme present in both systems measures metabolic heat flux and suppresses the influence of large temperature peaks and fluctuations. The spread in heat flux values are mainly related to the resolution limitations of the heat flux sensors themselves. The different detection limits are calculated via an optical density conversion, where the optical density unit corresponds to 10^9 cells/mL. The detection limit of the cells is the calculated number of cells at the determined heat flux limit of detection converted using the given channel volume (6 µL or 46 µL). We show in Figure 7b that it is still possible to detect bacterial growth with smaller sensors and smaller thermal volumes.

Table 1. Table overview over system performance and design properties of both the thermally not-stabilized and the thermally stabilized system [1].

Property	Thermally Not-Stabilized	Thermally Stabilized [1]
Growth detection	Yes	Yes
Channel size (underneath sensor)	6 mm × 6 mm × 170 µm	12 mm × 12 mm × 320 µm
Channel volume	6 µL	46 µL
Sensor	gSKIN XM	gSKIN XP
Sensor resolution	0.41 W/m^2	0.06 W/m^2
Temperature fluctuation	1 K	0.3 K
Standard deviation heat flux	0.32 W/m^2	0.05 W/m^2
Standard deviation averaged heat flux	0.20 W/m^2	0.02 W/m^2
OD limit of detection *	3×10^8 cells/mL	2×10^7 cells/mL
Cell population limit of detection *	1.8×10^6 cells	9.2×10^5 cells

* These values are given for completeness; however, they are determined by different methods, and therefore direct comparison between the systems should be carefully considered.

5. Conclusions

This work extends our study presented at the Eurosensors conference. Here, we are able to resolve bacterial growth via the metabolic heat produced in a microfluidic channel with an integrated differential heat flux sensing system by measuring the bacterial growth via metabolic heat. In contrast to the previous system involving more controlled thermal environments and larger sensors and microfluidic channels [1], we expose a smaller microfluidic chip with a smaller size and worse-resolution sensors to a thermally fluctuating environment without additional thermal isolation around the microfluidic chip. The 1 K range of temperature fluctuations matches the in vivo temperature fluctuation range in the human body [22]. However, the rate of change of temperature is within minutes, which is faster than in physiological conditions. These higher rates of temperature change perturb the heat flux measurement of the bacterial growth more than what is expected in physiological conditions. We demonstrate that, leveraging a differential sensing strategy, our system can detect bacterial growth in the exponential growth phase (OD = 0.3), despite the large environmental temperature fluctuations, smaller sensor sizes, and different sensor performances.

Further, assuming that each *E. coli* bacteria produces 3.5 pW [23] of thermal power, we estimate the limit of detection to be 1707 W/m^3 for the thermally stabilized system [1]. The limit of detection is determined by extrapolation based on sensitivity and noise. This method cannot be applied for the thermally not-stabilized system, where the limit of detection in relation to the bacterial population is determined from the OD level that causes a detecable increase in the heat flux measurement using the cumulative distribution function (Figure 7b). Therefore, we refrain from calculating the heat flux limit of detection in this case.

The bacterial concentration which constitutes the early onset of infectious growth on implant surfaces is at present unknown. Therefore, we compare the experimentally determined heat flux limit of detection to the thermal density of a biofilm of 350,000 W/m^3 [1]. The limit of detection in our system corresponds to 0.3% of the thermal density of a biofilm. Whether or not 0.3% of the heat of a biofilm corresponds to the early stages of biofilm formation is yet to be investigated in a clinically-relevant environment. In the literature, there are papers investigating implant infections which use implant bacterial seeding amounts between 2.0×10^6 and 6.2×10^6 [24–26], which is comparable to the determined detection limit in this study. From this point of view, using bacterial growth in LB media only reflects the total heat produced by a given bacterial population, which we, due to the low bacterial population at the determined limit of detection, consider being at the onset of infection.

We thus demonstrate the feasibility of bacterial growth detection using the cumulative distribution function. The detection of bacterial growth is statistically confirmed by comparing the cumulative distribution functions of the calibration phase to the point at which the bacterial growth is detected. Admittedly, this scheme is only able to detect whether a bacterial colony has grown beyond a threshold and is not able to follow the actual exponential growth curve of the bacterial colony.

These setups prove the possibility of introducing microcalorimetric measurements on microfluidic chips in thermally fluctuating environments. They may find use as tabletop microfluidic systems for more accessible microcalorimetric measurements of samples in microfluidic chips. Furthermore, these results give an additional insight in the possibility of using heat flux sensors in thermally not-stabilized environments such as implant surfaces. Further investigations should emulate in vivo environments and investigate the feasibility of surface-covering infection detection on implants, for example via a mesh of sensors on the implant [20]. We show that by applying the differential compensation scheme, the heat flux signal removes the influence of the thermal background fluctuations in the environment around the area of interest. The smaller size of the sensor could allow for integration, for example, in the vascular graft implant wall material, and could open the way to bacterial growth detection directly at the vascular graft implant surface.

Supplementary Materials: The following supporting information can be downloaded at: https://www.mdpi.com/article/10.3390/s24031019/s1.

Author Contributions: Conceptualization, S.L.K.V., C.I.R., M.A. and C.H.; methodology, S.L.K.V., M.A. and C.I.R.; formal analysis, S.L.K.V., C.I.R. and M.A.; investigation, S.L.K.V.; writing—original draft preparation, S.L.K.V., C.I.R., M.A. and C.H.; writing—review and editing, S.L.K.V., C.I.R., M.A. and C.H.; visualization, S.L.K.V. All authors have read and agreed to the published version of the manuscript.

Funding: M.A. was supported as a part of NCCR Microbiomes, a National Centre of Competence in Research, funded by the Swiss National Science Foundation (grant number 180575).

Institutional Review Board Statement: Not applicable.

Informed Consent Statement: Not applicable.

Data Availability Statement: The data is available upon request to the corresponding author.

Acknowledgments: The project was supported by the ETHeart Initiative. We acknowledge valuable discussions and support by Volkmar Falk, Viola Vogel and Emma Slack, and Nikola Cesarovic.

Conflicts of Interest: The authors declare no conflicts of interest.

References

1. Vehusheia, S.L.K.; Roman, C.; Braissant, O.; Arnoldini, M.; Hierold, C. Enabling direct microcalorimetric measurement of metabolic activity and exothermic reactions onto microfluidic platforms via heat flux sensor integration. *Microsystems Nanoeng.* **2023**, *9*, 56. [CrossRef]
2. Al-Hallak, M.H.D.K.; Sarfraz, M.K.; Azarmi, S.; Kohan, M.H.G.; Roa, W.H.; Löbenberg, R. Microcalorimetric method to assess phagocytosis: Macrophage-nanoparticle interactions. *AAPS J.* **2011**, *13*, 1. [CrossRef]
3. Hansen, L.D.; Macfarlane, C.; McKinnon, N.; Smith, B.N.; Criddle, R.S. Use of calorespirometric ratios, heat per CO2 and heat per O2, to quantify metabolic paths and energetics of growing cells. *Thermochim. Acta* **2004**, *422*, 55–61. [CrossRef]

4. Trampuz, A.; Salzmann, S.; Antheaume, J.; Daniels, A.U. Microcalorimetry: A novel method for detection of microbial contamination in platelet products. *Transfusion* **2007**, *47*, 1643–1650. [CrossRef]

5. Braissant, O.; Wirz, D.; Göpfert, B.; Daniels, A.U. Use of isothermal microcalorimetry to monitor microbial activities. *BMC Microbiol.* **2010**, *13*, 171. [CrossRef]

6. Butini, M.E.; Abbandonato, G.; Di Rienzo, C.; Trampuz, A.; Di Luca, M. Isothermal microcalorimetry detects the presence of persister cells in a *Staphylococcus aureus* biofilm after vancomycin treatment. *Front. Microbiol.* **2019**, *10*, 332. [CrossRef]

7. Wang, Y.; Zhu, H.; Feng, J.; Neuzil, P. Recent advances of microcalorimetry for studying cellular metabolic heat. *TrAC Trends Anal. Chem.* **2021**, *143*, 115353. [CrossRef]

8. Maskow, T.; Lerchner, J.; Peitzsch, M.; Harms, H.; Wolf, G. Chip calorimetry for the monitoring of whole cell biotransformation. *J. Biotechnol.* **2006**, *122*, 431–442. [CrossRef] [PubMed]

9. Wang, S.; Sha, X.; Yu, S.; Zhao, Y. Nanocalorimeters for biomolecular analysis and cell metabolism monitoring. *Biomicrofluidics* **2020**, *14*, 011503. [CrossRef]

10. Xu, J.; Reiserer, R.; Tellinghuisen, J.; Wikswo, J.P.; Baudenbacher, F.J. A microfabricated nanocalorimeter: Design, characterization, and chemical calibration. *Anal. Chem.* **2008**, *80*, 2728–2733. [CrossRef]

11. Johannessen, E.A.; Weaver, J.M.R.; Cobbold, P.H.; Cooper, J.M. Heat conduction nanocalorimeter for pl-scale single cell measurements. *Appl. Phys. Lett.* **2002**, *80*, 2029–2031. [CrossRef]

12. Bae, J.; Zheng, J.; Zhang, H.; Foster, P.J.; Needleman, D.J.; Vlassak, J.J. A micromachined picocalorimeter sensor for liquid samples with application to chemical reactions and biochemistry. *Adv. Sci.* **2021**, *8*, 1–11. [CrossRef]

13. Yang, Y.; Chen, Y.; Tang, H.; Zong, N.; Jiang, X. Microfluidics for Biomedical Analysis. *Small Methods* **2020**, *4*, 1900451. [CrossRef]

14. Soleimany, A.P.; Bhatia, S.N. Activity-Based Diagnostics: An Emerging Paradigm for Disease Detection and Monitoring. *Trends Mol. Med.* **2020**, *26*, 450–468. [CrossRef] [PubMed]

15. Zelenin, S.; Hansson, J.; Ardabili, S.; Ramachandraiah, H.; Brismar, H.; Russom, A. Microfluidic-based isolation of bacteria from whole blood for sepsis diagnostics. *Biotechnol. Lett.* **2015**, *37*, 825–830. [CrossRef] [PubMed]

16. Jang, K.-J.; Mehr, A.P.; Hamilton, G.A.; McPartlin, L.A.; Chung, S.; Suh, K.-Y.; Ingber, D.E. Human kidney proximal tubule-on-a-chip for drug transport and nephrotoxicity assessment. *Integr. Biol.* **2013**, *5*, 1119–1129. [CrossRef] [PubMed]

17. Postek, W.; Garstecki, P. Droplet Microfluidics for High-Throughput Analysis of Antibiotic Susceptibility in Bacterial Cells and Populations. *Acc. Chem. Res.* **2021**, *55*, 605–615. [CrossRef] [PubMed]

18. Shirure, V.S.; Bi, Y.; Curtis, M.B.; Lezia, A.; Goedegebuure, M.M.; Goedegebuure, S.P.; Aft, R.; Fields, R.C.; George, S.C. Tumor-on-a-chip platform to investigate progression and drug sensitivity in cell lines and patient-derived organoids. *Lab Chip* **2018**, *18*, 3687–3702. [CrossRef] [PubMed]

19. Vehusheia, S.L.K.; Roman, C.; Cesarovic, N.; Hierold, C. Microfluidic thermal model for early detection of infection on aortic grafts. In Proceedings of the Technical Digest of 22nd International Conference on Solid State Sensors, Actuators and Microsystems (Transducers), Kyoto, Japan, 25–29 June 2023; pp. 951–954.

20. Vehusheia, S.L.K.; Roman, C.; Sonderegger, R.; Cesarovic, N.; Hierold, C. Finite element-based feasibility study on utilizing heat flux sensors for early detection of vascular graft infections. *Sci. Rep.* **2023**, *13*, 16198. [CrossRef]

21. Nagpal, V.; Sohail, M.R. Prosthetic Vascular Graft Infections: A Contemporary Approach to Diagnosis and Management. *Curr. Infect. Dis. Rep.* **2011**, *13*, 317–323. [CrossRef] [PubMed]

22. Hutchison, J.S.; Ward, R.E.; Lacroix, J.; Hébert, P.C.; Barnes, M.A.; Bohn, D.J.; Dirks, P.B.; Doucette, S.; Fergusson, D.; Gottesman, R.; et al. Hypothermia Therapy after Traumatic Brain Injury in Children. *N. Engl. J. Med.* **2008**, *358*, 2447–2456. [CrossRef] [PubMed]

23. Higuera-Guisset, J.; Rodríguez-Viejo, J.; Chacón, M.; Muñoz, F.; Vigués, N.; Mas, J. Calorimetry of microbial growth using a thermopile based microreactor. *Thermochim. Acta* **2005**, *427*, 187–191. [CrossRef]

24. Arens, S.; Kraft, C.; Schlegel, U.; Printzen, G.; Perren, S.M.; Hansis, M. Susceptibility to local infection in biological internal fixation. *Arch. Orthop. Trauma Surg.* **1999**, *119*, 82–85. [CrossRef]

25. Harrasser, N.; Gorkotte, J.; Obermeier, A.; Feihl, S.; Straub, M.; Slotta-Huspenina, J.; von Eisenhart-Rothe, R.; Moser, W.; Gruner, P.; de Wild, M.; et al. A new model of implant-related osteomyelitis in the metaphysis of rat tibiae. *BMC Musculoskelet. Disord.* **2016**, *17*, 152. [CrossRef]

26. Monzon, M.; García-Álvare, F.; Lacleriga, A.; Gracia, E.; Leiva, J.; Oteiza, C.; Amorena, B. A simple infection model using pre-colonized implants to reproduce rat chronic *Staphylococcus aureus* osteomyelitis and study antibiotic treatment. *J. Orthop. Res.* **2001**, *19*, 820–826. [CrossRef] [PubMed]

Article

Enhancing the Deposition Rate and Uniformity in 3D Gold Microelectrode Arrays via Ultrasonic-Enhanced Template-Assisted Electrodeposition

Neeraj Yadav [1,2,*], **Flavio Giacomozzi** [2], **Alessandro Cian** [2], **Damiano Giubertoni** [2] and **Leandro Lorenzelli** [2]

1   Department of Industrial Engineering, University of Trento, 38123 Trento, Italy
2   Center for Sensors & Devices (SD), FBK—Foundation Bruno Kessler, 38123 Trento, Italy; giaco@fbk.eu (F.G.); acian@fbk.eu (A.C.); lorenzel@fbk.eu (L.L.)
*   Correspondence: nyadav@fbk.eu; Tel.: +39-3294716563

Abstract: In the pursuit of refining the fabrication of three-dimensional (3D) microelectrode arrays (MEAs), this study investigates the application of ultrasonic vibrations in template-assisted electrode-position. This was driven by the need to overcome limitations in the deposition rate and the height uniformity of microstructures developed using conventional electrodeposition methods, particularly in the field of in vitro electrophysiological investigations. This study employs a template-assisted electrodeposition approach coupled with ultrasonic vibrations to enhance the deposition process. The method involves utilizing a polymeric hard mask to define the shape of electrodeposited microstructures (i.e., micro-pillars). The results show that the integration of ultrasonic vibrations significantly increases the deposition rate by up to 5 times and substantially improves the uniformity in 3D MEAs. The key conclusion drawn is that ultrasonic-enhanced template-assisted electrodeposition emerges as a powerful technique and enables the development of 3D MEAs at a higher rate and with a superior uniformity. This advancement holds promising implications for the precision of selective electrodeposition applications and signifies a significant stride in developing micro- and nanofabrication methodologies for biomedical applications.

Keywords: electrodeposition; MEA; uniformity; three-dimensional; template assisted; deposition rate

Citation: Yadav, N.; Giacomozzi, F.; Cian, A.; Giubertoni, D.; Lorenzelli, L. Enhancing the Deposition Rate and Uniformity in 3D Gold Microelectrode Arrays via Ultrasonic-Enhanced Template-Assisted Electrodeposition. *Sensors* **2024**, *24*, 1251. https://doi.org/10.3390/s24041251

Academic Editor: Nigel T. Maidment

Received: 4 January 2024
Revised: 22 January 2024
Accepted: 12 February 2024
Published: 15 February 2024

1. Introduction

Microelectrode arrays (MEAs) are powerful tools for studying the electrical activity in cells and tissues and consist of electrodes that record or stimulate electrical activity in vitro. Three-dimensional (3D) MEAs have become increasingly popular, as they provide a larger surface area for cell–electrode interactions and can capture more physiologically relevant data [1–4]. These MEAs are essential in various fields, particularly neuroscience and neuro-engineering, and serve as the foundation for groundbreaking technologies like brain–computer interfaces (BCIs). The efficacy of an MEA hinges on its sampling capabilities, which are determined by the electrode density and its capacity to precisely target specific regions of interest [5]. Although current fabrication techniques have made significant strides in enhancing the recording density, primarily through advancements in micro-electromechanical system (MEMS) fabrication methods [6–8], there are inherent limitations to existing technologies. Silicon-based arrays, for instance, have a limited volu-metric electrode density and lack customization options. Similarly, alternative fabrication approaches such as bead stacking, 3D printing, and direct laser writing techniques provide options for individual shank customization and allow reproducibility; however, these techniques lack scalability in production and cost-effectiveness [9–13].

The forthcoming generation of electrophysiological recording tools must transcend these limitations and allow for the customization of probes to the specific study [14]. Historically, MEA fabrication methods have mirrored trends in the semiconductor industry, transitioning from micro-wires to lithography [15–17]. However, the emergence of template-assisted electrodeposition techniques presents a promising new avenue for the development of 3D MEAs with customizable shank heights and array topographies. Template-assisted electrodeposition utilizes a polymeric hard mask as a template to define the shape of the electrodeposited microstructures [18–21]. More recently, the development of multi-depth probing 3D MEAs with customizable microelectrode heights has been reported using the template-assisted electrodeposition technique [22]. Electrodeposited gold electrodes have a higher surface roughness compared to the electrodes developed using physical vapor deposition methods; this surface roughness translates to a larger electrochemically active area at the cell–electrode interface, promoting interactions between the cell and the electrode [4]. The cell–electrode adhesion could be improved further by functionalizing the electrode surface with adhesion promoters as a post-processing step [23].

The development of advanced 3D MEAs requires high electrodeposition rates to enable rapid prototyping and remain cost-effective, while the uniformity of the thickness of the electrodeposited microstructures plays a critical role in defining the array topography; however, conventional processes, particularly template-assisted electrode processes, suffer from low deposition rates due to localized depletion of ions in the electroplating solution, leading to a non-uniform deposition. The use of ultrasonic baths, which accelerate chemical reactions within liquid media using high-frequency pressure waves, appears to be very promising for overcoming the limitations of the conventional electrodeposition process. These baths are widely used in various industrial and laboratory applications, such as cleaning, degreasing, and electroplating [24,25]. In recent years, ultrasonic baths have also been proposed as a potential solution for accelerating the electrodeposition process, which is often limited by its low deposition rate. When an ultrasonic bath is used for electrodeposition, the sound waves create microscopic cavitation bubbles in the liquid medium. These bubbles are subjected to intense pressure and temperature changes, forming highly reactive sites on their surface. These sites can then act as catalysts, accelerating the electrodeposition process [26,27]. The potential applications of ultrasonic baths in electrodeposition are numerous. Ultrasonic baths could be used to produce gold microstructures with a high accuracy and uniformity. Additionally, using ultrasonic baths can reduce the cost of the electrodeposition process by increasing the deposition rate without needing an additional catalytic agent. While the effect of ultrasonic agitation on electrodeposition has been extensively studied for two-dimensional (2D) structures, its influence on template-assisted electrodeposition of 3D microstructures remains largely unexplored [28–33].

In this work, we investigate the influence of ultrasonic vibrations on the deposition rate and uniformity of microelectrode arrays developed using template-assisted electrodeposition. Various characterization techniques, including scanning electron microscopy, optical profilometry, and X-ray diffraction analysis (XRD), are used to assess the quality of the fabricated structures. A mechanical shear strength testing tool is utilized to test the adhesion strength of the electrodeposited micro-pillars with the planar substrate. The results of this study provide a deeper insight into the optimization of ultrasonic parameters for template-assisted electrodeposition, leading to an improved 3D MEA prototyping process. This study could also have broader implications for using ultrasonic agitation in other electrodeposition techniques for micro- and nanofabrication.

2. Materials and Methods

2.1. Materials

An additive-free electroplating solution (AUROLYTE CN200, Atotech Deutschland GmbH & Co. KG, Berlin, Germany) was chosen for the electrodeposition experiments. Three types of planar MEAs arranged in a hexagonal pattern (as described in [22]) were used for the experiments: (S1) consisting of 60 electrodes with a diameter of 65 μm and a pitch of 265 μm, of which 21 were connected to a custom routing for electrodeposition; (S2) consisting of 60 electrodes with a diameter of 65 μm and a pitch of 265 μm, of which 41 were connected to a custom routing for electrodeposition; and (S3) consisting of 60 electrodes with a diameter of 35 µm and a pitch of 195 µm, of which 44 were connected to a custom routing for electrodeposition. A thick negative-type photoresist (KMPR-1035, Kayaku Advanced Materials, Inc., Westborough, MA, USA) was utilized for template fabrication. An ultrasonic bath (Sonorex Digitec DT 514 BH-RC, BANDELIN electronic GmbH & Co. KG, Berlin, Germany) operating at a peak power of 640 W at 35 kHz was utilized for the experiments. The galvanostat used for these experiments was assembled in-house. Remover PG (Kayaku Advanced Materials, Inc., Westborough, MA, USA) was used to strip the photoresist template after completion of the electrodeposition process.

2.2. Template Development

Custom planar MEAs were designed and developed for this study, and the design and fabrication protocols are described in detail in Appendix B. The planar MEAs were first coated with a 110 µm-thick layer of KMPR photoresist. This was achieved by using a spin coater operating at 1000 rpm. This was followed by a 30 min soft-bake at a temperature of 100 °C. UV exposure was performed using an i-line mask aligner (KARL SÜSS MA6, SÜSS MICROTEC SE, Garching, Germany) to define the custom pattern on the photoresist. The photoresist was subjected to a post-exposure bake at 100 °C for a span of 6 min. This step was crucial to complete the curing reaction of the exposed regions. The final step was to develop the photoresist using the SU-8 developer solution supplied by MicroChemicals GmbH, Ulm, Germany. This 20 min process required a shaker plate for mild agitation for assistance. This process resulted in an array of 110 µm deep, 65 µm diameter cylindrical holes strategically aligned over the planar electrodes for S1- and S2-type MEAs, while the S3 MEA had cylindrical cavities with an internal diameter of 35 µm and a height of 110 µm.

2.3. Experimental Setup

All of the electrodeposition experiments were set up inside an ultrasonic bath with a temperature maintained at 55 °C, as illustrated in Figure 1. The electroplating solution bath was conditioned for 1 h before initiating gold electrodeposition for each experiment. The experiments were conducted in four phases. Experiment 1: Electrodeposition of gold MEAs with a high current density (i.e., 8 mA/cm^2) for 1 h (H1) and various operational modes of the ultrasonic bath (i.e., no sonication mode (NS), pulsed sonication (PS) mode with a duty cycle of 50%, and continuous sonication (CS) mode) to investigate the influence of the ultrasonic bath operation mode on the deposition rate and electrode height uniformity across the array, as illustrated in Figure 1.

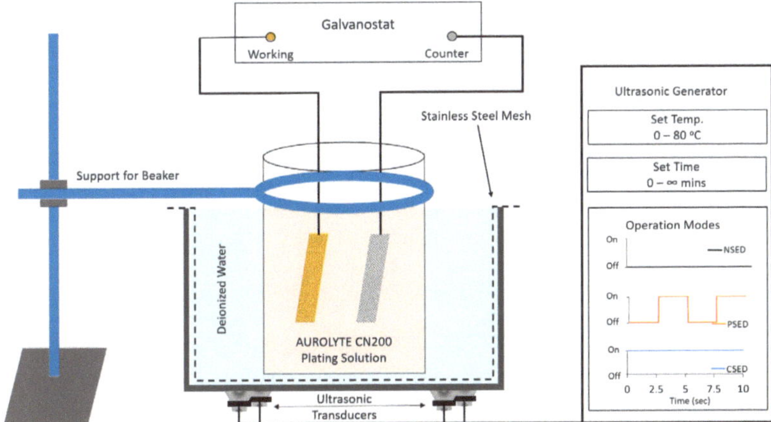

Figure 1. Schematic representation of the experimental setup including the various operational modes of the ultrasonic bath, i.e., NSED (electrodeposition without the ultrasonic vibrations), PSED (electrodeposition with ultrasonic vibrations in pulsed mode with a duty cycle of 50%), and CSED (electrodeposition with continuous ultrasonic vibrations).

Experiment 2 (L1): This was a repeat of Experiment 1 with a lower current density (4 mA/cm^2) to determine the influence of the deposition current density on the deposition rate and the uniformity of the electrode height. Experiment 3: The best electrodeposition parameters were selected from the previous experiments (i.e., CS-L1, based on the optimal deposition rate and the highest uniformity). In this case, the tests were repeated for longer durations of 2 and 3 h (i.e., CS-L2 and CS-L3, respectively) to verify the consistency of the process. Finally, a fourth experiment was conducted to verify the scalability of the process for the deposition of high-density (HD) MEAs for long durations. For the first two experiments, the S1 MEAs with 21 active electrodes were subjected to template-assisted electrodeposition for 1 h. For the third experiment, the S2 MEAs with 41 active electrodes were subjected to template-assisted electrodeposition for 2 and 3 h, and the experiment was repeated to ensure reproducibility. S3 MEAs were used in the fourth experiment by subjecting them to template-assisted electrodeposition for 4 h. The experimental details are tabulated in Table 1.

Table 1. Design of experiments. List of experiments performed under various conditions.

	Sample Name	MEA Type	No. of Active Electrodes	Electrode Diameter (μm)	Deposition Current Density (mA/cm^2)	Electrodeposition Duration (Minutes)	Ultrasonic Bath Mode *
Experiment 1	NS-H1	S1	21	65	8	60	NS
	PS-H1	S1	21	65	8	60	PS
	CS-H1	S1	21	65	8	60	CS
Experiment 2	NS-L1	S1	21	65	4	60	NS
	PS-L1	S1	21	65	4	60	PS
	CS-L1	S1	21	65	4	60	CS
Experiment 3	CS-L2	S2	41	65	4	120	CS
	CS-L3	S2	41	65	4	180	CS
	CS-L2 r **	S2	41	65	4	120	CS
	CS-L3 r	S2	41	65	4	180	CS
Experiment 4	CS-L4 (HD)	S3	44	65	4	240	CS

* Ultrasonic bath modes: NS—ultrasonic vibrations OFF, PS—ultrasonic vibrations ON in pulsed mode, and CS—ultrasonic vibrations ON continuously. ** Repeated.

2.4. Analysis

A scanning electron microscope (SEM) was utilized to assess the process yield (defined as the number of electrodeposited micro-pillars divided by the number of electrodes subjected to electrodeposition for each MEA), and in order to determine the electrodeposition rate and uniformity, a two-step process was employed. First, an optical profilometer was used to measure the height of the electrodeposited micro-pillars, which were subjected to electrodeposition for one hour. To evaluate the height uniformity of the electrodeposited micro-pillars, the heights of multiple micro-pillars sourced from different regions of the MEA were analyzed comparatively.

To evaluate the mechanical, or more specifically, the shear strength, of the electrodeposited micro-pillars, a destructive die shear strength test was performed using the Condor Ez Pull&Shear test tool from XYZtec, as shown in Figure 2a.

For each experimental condition from the first experiment, five micro-pillars randomly selected from each MEA were subjected to a shear test at room temperature. A shear tool, moving at a constant velocity of 2 μm/s, was placed in contact with the side of the micro-pillar until fracture or split occurred, as illustrated in Figure 2b. To obtain a valid measure of the adhesion strength at the pillar–substrate interface, it was ensured that failure occurred at the pillar–substrate interface by placing the shear tool 1 μm above the substrate surface. The maximum force applied by the tool at the point of failure was recorded for all five measurements. From the mean maximum shear force and its standard deviation, the maximum shear strength τ_{max} was calculated from the maximum shear force F_{max} using the following equation:

$$F_{\text{max}} = \tau_{\text{max}} \cdot A \tag{1}$$

where A is the cross-sectional area of the pillar, obtained from the measured pillar diameter.

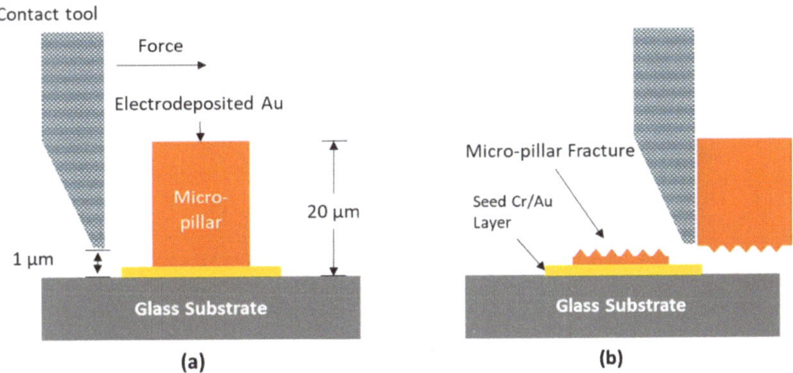

Figure 2. Illustration of the shear test setup. (**a**) Initial position of the contact tool, (**b**) final position of the contact tool.

In order to evaluate the influence of ultrasonic vibrations on the morphological characteristics of the electrodeposited gold microstructures, three separate samples were prepared. Square (17×17 mm^2) glass wafer pieces coated with 5 nm chromium (Cr) and 200 nm gold (Au) deposited via the thermal evaporation technique were utilized as substrates for this experiment. The substrates were masked with 70 µm-thick Kapton tape consisting of a circular opening with a diameter of 6 mm. The masked substrates were subjected to electrodeposition using the parameters from the first experiment. The first substrate was subjected to electrodeposition for 1 h in the absence of ultrasonic vibrations, i.e., NSED. The second substrate was subjected to electrodeposition with an ultrasonic bath operating in pulsed sonication mode for one hour, i.e., PSED. Finally, the third substrate was subjected to 1 h of electrodeposition under continuous sonication mode, i.e., CSED, as shown in Figure 3. The influence of ultrasonic vibrations on the morphology of the electrodeposited gold films was determined by measuring the surface roughness using atomic force microscopy (AFM; PX, NT-MDT SI, Moscow, Russia).

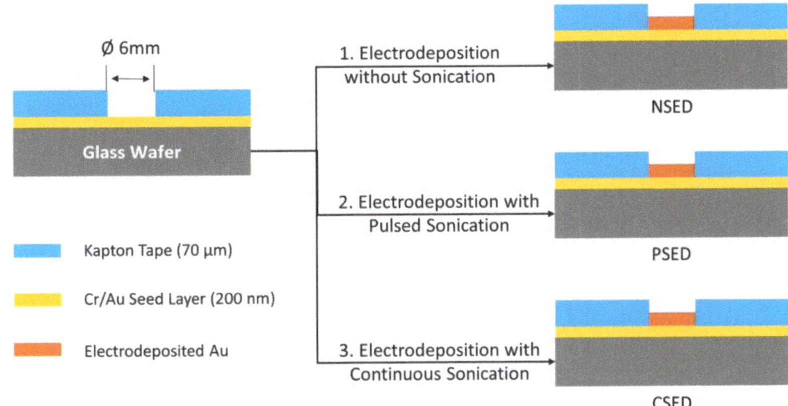

Figure 3. Schematic representation for preparation of the samples for the structural and morphological characterizations.

The evaluation of the influence of ultrasonic vibrations on the structural (crystal phase) characteristics of the electrodeposited structures was carried out via X-ray diffraction (XRD) on the NSED, PSED, and CSED samples and the substrate with the seed layer using a high-resolution XRD instrument (ITALSTRUCTURES APD2000, Austin AI, Austin, TX, USA). The average grain size D was determined via XRD for multiple crystallographic planes, including (111), (311), (220), and (200), via Scherrer's equation:

$$D = K \cdot \lambda / (\beta \cdot cos(\theta))$$ (2)

where the Scherrer constant K is typically taken as 0.94 [34], λ = 1.5418 Å is the wavelength of the Cu Kalpha radiation employed for the XRD, β is the full-width at half-maximum (FWHM) of the measured peak in radians, and theta is the Bragg angle at which the peak occurs.

3. Results and Discussions

3.1. Deposition Rate and Uniformity

The deposition rate and uniformity are critical attributes in the template-assisted electrodeposition process, and they play a pivotal role in determining the quality of the micro-pillars. An optical profilometer was utilized to measure the height of the electrodeposited micro-pillars across each array to derive the deposition rate and assess the uniformity of the height of the electrodeposited micro-pillars. In the first experiment, the S1 MEAs were subjected to electrodeposition for 1 h with a high current density of 8 mA/cm^2 (H1) under different operational modes of the ultrasonic bath. The MEA subjected to electrodeposition without ultrasonic vibrations (i.e., NS-H1) showed a deposition rate of 0.10 µm/min and a percentage standard deviation of 35.8%. This indicates a low deposition rate compared to standard electrodeposition without the use of a template under similar experimental conditions (i.e., ~0.45 µm/min) and a considerable variation in micro-pillar heights across the array, revealing a lack of uniformity. The MEA subjected to electrodeposition with pulsed ultrasonic vibrations (i.e., PS-H1) showed an improved deposition rate of 0.24 µm/min. The percentage standard deviation decreased to 16.19%, suggesting an improved uniformity compared to NS-H1. Finally, the MEA subjected to electrodeposition with continuous ultrasonic vibrations (i.e., CS-H1) demonstrated the highest deposition rate of 0.55 µm/min. It also showed a further decrease in the percentage standard deviation to 13.52%, implying more consistency in micro-pillar heights and an enhanced uniformity. The normalized height distributions of electrodeposited microstructures (micro-pillars) for each sample from individual experiments are plotted in the bar graphs in Appendix A.

In the second experiment, the S1 MEAs were subjected to electrodeposition for 1 h with a lower current density of 4 mA/cm^2 (L1) under different operational modes of the ultrasonic bath. The MEA subjected to electrodeposition without ultrasonic vibrations (i.e., NS-L1) showed a deposition rate of 0.06 µm/min proportional to NS-H1 and a percentage standard deviation of 14.77%, indicating a considerable improvement in the uniformity. The MEA subjected to electrodeposition with pulsed ultrasonic vibrations (i.e., PS-L1) showed an improved deposition rate of 0.13 µm/min compared to NS-L1. However, the experiment revealed a slight increase in the percentage standard deviation (i.e., 16.15%) with respect to NS-L1. Finally, the MEA subjected to electrodeposition with continuous ultrasonic vibrations and lower deposition current (i.e., CS-L1) showed a deposition rate of 0.24 µm/min and also a further decrease in the percentage standard deviation to 9.63%, indicating more consistency in the micro-pillar heights, and therefore an improved uniformity.

The first experiment clearly indicated that the use of ultrasonic vibrations significantly increases the deposition rate and improves uniformity. A higher deposition rate is always a desirable parameter; however, uniformity is an essential parameter for developing MEAs. The second experiment demonstrated a reduction in the deposition rate compared to the first experiment but this was proportional to the applied deposition current density in all three cases, as shown in Figure 4a. The CS-L1 showed the lowest percentage standard deviation, indicating the highest uniformity compared to all the other samples from the first and second experiments, as shown in Figure 4b. Based on these results, continuous sonication with a low current density was chosen as the optimal combination for further investigations.

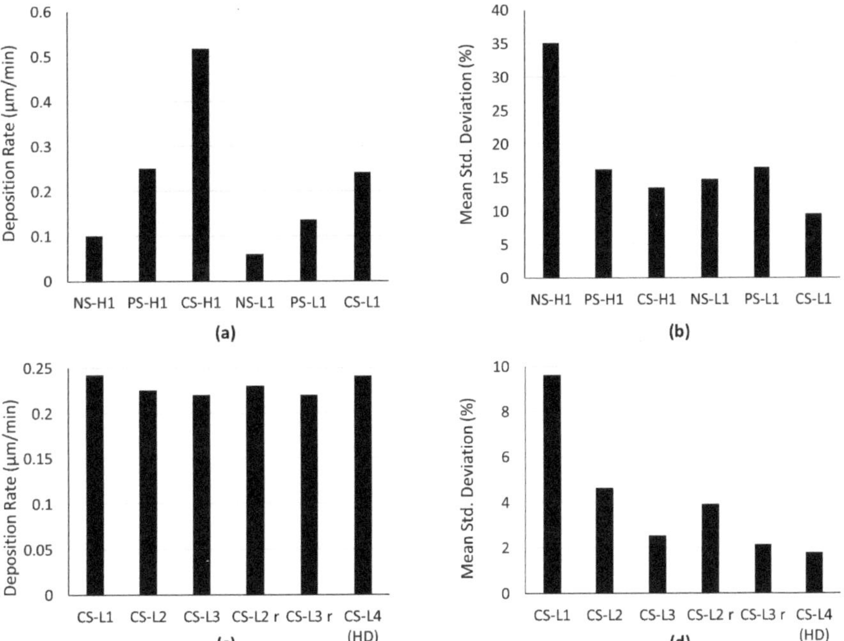

Figure 4. Bar graphs comparing the deposition rate and the percentage mean standard deviation across the height of the electrodeposited micro-pillar array (as a measure of uniformity) for the various experiments. (**a**) Comparison of the deposition rate of the different microelectrode arrays (MEAs) from the first (H1) and the second (L1) experiment. (**b**) Comparison of the percentage mean standard deviation in the thickness of 21 electrodeposited micro-pillars for each MEA in the first and second experiment. (**c**) Comparison of the deposition rate of the various microelectrode arrays (MEAs) from the second (CS-L1), third (CS-L2, CS-L3, CS-L2 r, and CS-L3 r), and fourth (CS-L4 (HD)) experiments. (**d**) Percentage mean standard deviation pertaining to each sample from the second (CS-L1), third (CS-L2, CS-L3, CS-L2 r, and CS-L3 r), and fourth (CS-L4 (HD)) experiments.

In the third experiment, S2 MEAs consisting of 41 active electrodes were subject to template-assisted electrodeposition for 2 and 3 h using continuous sonication with a low current density (i.e., CS-L2 and C3-L3, respectively). The third experiment was repeated to ensure reproducibility (i.e., CS-L2 r and C3-L3 r). Both CS-L2 and the CS-L3 have comparable but slightly lower deposition rates (CS-L3 having the lowest deposition rate) compared to the CS-L1 experiment, and the trend was confirmed when the experiment was repeated, i.e., CS-L2 r and C3-L3 r, as shown in Figure 4c. This decrease in the deposition rate during the longer deposition is likely due to the depletion in gold ions in the electroplating solution, However, this is mere speculation based on the experimental setup; further investigations are warranted to fully understand this observation. On the other hand, the percentage standard deviation decreased to as low as 2.13% for the CS-L3 experiment as the duration of the electrodeposition increased, as shown in Figure 4d.

Finally, the fourth experiment was conducted by subjecting an HD MEA (S3) to template-assisted electrodeposition for 4 h (i.e., CS-L4 (HD)) with continuous ultrasonic vibrations and a current density of 4 mA/cm^2. The experiment led to a deposition rate of 0.24 μm/min, consistent with the previous experiments, as shown in Figure 4c. The experiment resulted in the lowest standard deviation of 1.76%, indicating a very high uniformity, as shown in Figure 4d.

These observations suggest that employing ultrasonic assistance, especially in continuous mode, substantially boosts the deposition rates in template-assisted electrodeposition. Moreover, the lower deposition current density significantly enhances the uniformity in the electrode height across the array, as shown in Figure 5. The empty circular sites within the images in Figure 5 are the planar electrodes that were not subjected to electrodeposition. All the MEAs subjected to electrodeposition demonstrated a 100% deposition yield (i.e., the number of electrodeposited micro-pillars divided by the number of electrodes subjected to electrodeposition). The data for the micro-pillar heights are available in Spreadsheet S2 in the Supplementary Materials. This finding may provide a crucial strategy for optimizing the deposition parameters when precise control over micro-pillar dimensions and uniformity is desired.

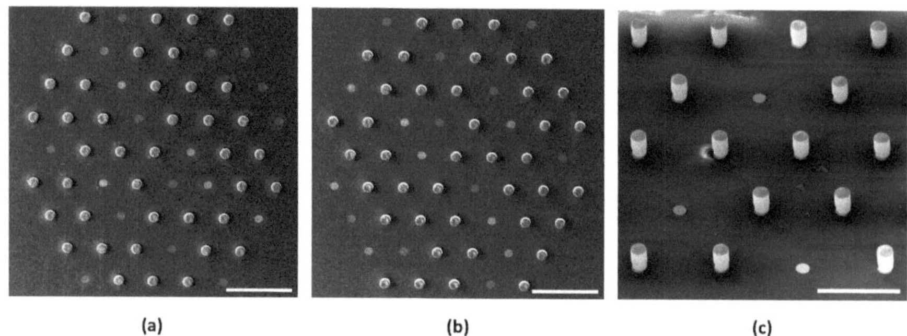

(a)   (b)   (c)

Figure 5. SEM images of the MEAs. (**a**) CS-L2 with a tilt of 10 degrees (scale: 500 μm). (**b**) CS-L3 with a tilt of 10 degrees (scale: 500 μm). (**c**) CS-L4 (HD) with a tilt of 25 degrees (scale: 200 μm).

3.2. Mechanical Strength

The mechanical strength of the electrodeposited micro-pillars was a crucial parameter evaluated in this study to investigate the influence of ultrasonic vibrations on the shear strength of the electrodeposited micro-pillars. To this end, a destructive shear stress test was employed to determine the maximum shear force the micro-pillars could endure before failure, as shown in Figure 2. The three MEAs (five micro-pillars from each MEA) from the first experiment and the HD MEA from the fourth experiment were subjected to this test.

For the NS-H1 MEA, the maximum shear force was recorded at 0.610 N with a percentage standard deviation of 6.58%. The PS-H1 MEA displayed a slightly higher resistance, having a maximum shear force of 0.707 N and a lower percentage standard deviation of 3.57%, suggesting a marginally superior and more consistent shear strength compared to NS-H1. The CS-H1 MEA documented a mean maximum shear force similar to NS-H1, at 0.636 N, but with a decreased percentage standard deviation of 3.38%, hinting at a more uniform shear strength under this process.

The CS-L4 (HD) MEA registered a substantially lower shear force resistance at 0.158 N with a percentage standard deviation of 3.62%. However, it is imperative to note that the cross-sectional area of the micro-pillars from the CS-L4 (HD) MEA was 3.5 times smaller than the micro-pillars from the other three methods. As shear strength is directly proportional to the cross-sectional area, the seemingly weaker shear strength in the CS-L4 (HD) MEA is not an indication of a poor shear resistance but a consequence of a smaller cross-sectional area. Adjusting for the reduced cross-sectional area, the actual shear strength (τ_{max}) of the CS-L4 (HD) MEA is similar to the CS-H1 MEA, as shown in Figure 6a. This observation suggests that the reduced current density in continuous ultrasonic bath-assisted electrodeposition with a lower deposition rate does not necessarily undermine the shear strength. The findings from this research highlight the impact of ultrasonic agitation and the current density on shear strength, having potential implications for the design of electrodeposition processes for gold microstructures that require strong resistance to shear forces on the substrate. The data pertaining to the shear strength measurements are presented in Table S1 of the Supplementary Materials. Figure 6b shows an optical image of the shear strength measurement setup with the contact tool at the initial measurement position for the CS-L4 (HD) MEA.

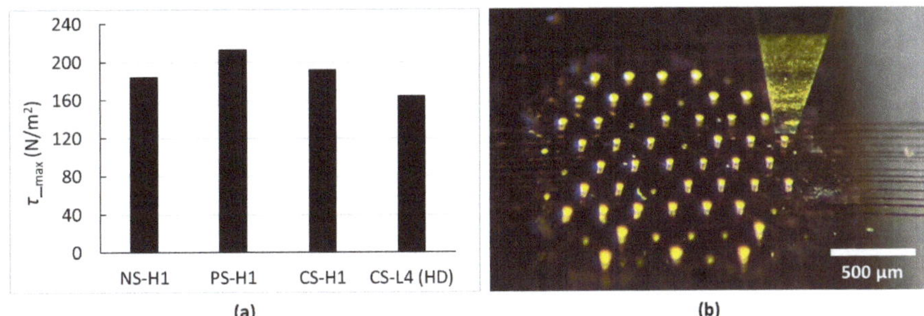

Figure 6. Shear strength analysis. (**a**) Bar chart presenting the average (n = 5) of the maximum shear strength τ_{max} faced by the micro-pillars before failure for each MEA subjected to the test. (**b**) A microscopic image of the CS-L4 (HD) MEA undergoing the shear test with the contact tool at the initial position.

3.3. Structural and Morphological Analysis

Figure 7 shows the Y-stacked plot of the XRD spectra for the three electrodeposited gold samples (i.e., NSED, PSED, and CSED) and the substrate with the gold seed layer (substrate) deposited via the thermal evaporation technique. As expected, the seed layer has a crystalline structure with a single peak at 38.2° (i.e., the (111) plane). In the case of the electrodeposited samples, strong peaks were observed in the (111) plane, and weaker peaks were observed in the (311), (220), and (200) planes, indicating a polycrystalline structure. No significant shift in the 2theta locations in the XRD spectra was observed for the various electrodeposited gold samples. As shown in Table 2, the full width at half maximum (FWHM) of the (111), (311), (220), and (200) planes is highest for the NSED sample and lowest for the PSED sample, indicating a lower surface roughness and average gain size compared to the PSED and CSED samples, which could also be observed in the AFM images in Figure 8.

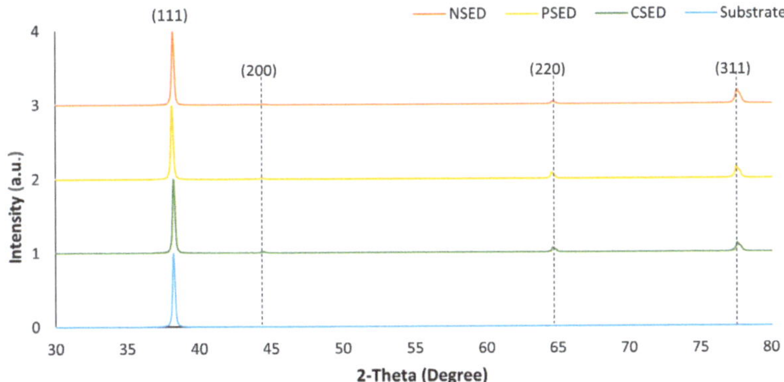

Figure 7. Measured XRD patterns of the electrodeposited gold (without template) under different ultrasonic vibration modes and the thermally evaporated gold seed layer (i.e., substrate).

Table 2. Full width at half maximum (FWHM) of 111, 311, 220, and 200 peaks of electrodeposited gold.

| Sample | (111) [1] | (311) | (220) | (200) | Average Grain |
	FWHM	FWHM	FWHM	FWHM	Size (nm)
Substrate	0.20524	-	-	-	40.96
NSED	0.15433	0.25735	0.21134	0.23835	43.63
PSED	0.14565	0.2244	0.1736	0.23286	48.51
CSED	0.14751	0.23751	0.18359	0.2303	47.01

[1] Crystallographic plane.

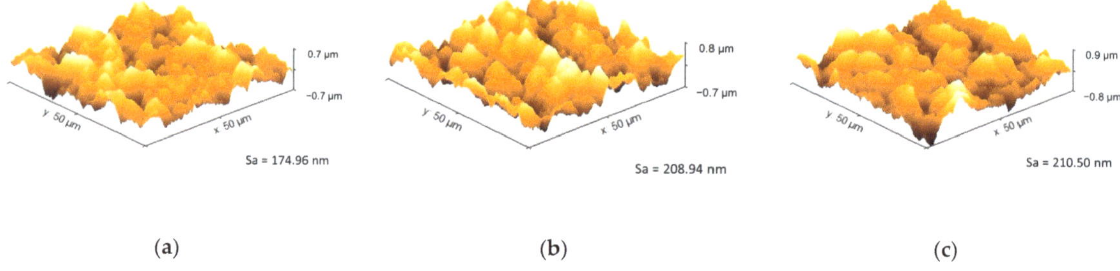

Figure 8. AFM 3D image and average surface roughness (Sa) of (**a**) NSED, (**b**) PSED, and (**c**) CSED.

3.4. Implications for 3D MEAs

These results demonstrate that ultrasonic-bath-assisted electrodeposition, especially under continuous operation (CSED and LC-CSED), significantly improves the performance of template-assisted electrodeposition. This could allow for the fabrication of higher quality 3D MEAs, as it leads to a higher deposition rate, improved uniformity, and an enhanced adhesion strength of gold micro-pillars. The findings could also guide the optimization of ultrasonic parameters in electrodeposition processes, advancing the field of micro- and nanofabrication. Our results highlight the promising potential of ultrasonic-bath-assisted electrodeposition for fabricating 3D MEAs. Future research can further validate these findings by exploring other factors that influence the electrodeposition process and optimizing them for superior MEA performance.

4. Conclusions

This research explored the impact of ultrasonic vibrations on template-assisted electrodeposition of gold micro-pillars for the development of 3D MEAs intended for in vitro electrophysiological investigations. This study found that continuous ultrasonic-bath-assisted electrodeposition, at both high and low deposition current densities, significantly enhanced the deposition rate and improved the thickness/height uniformity of the micro-pillars across the array, specifically for template-assisted electrodeposition processes. Along with the application of continuous ultrasonic vibrations, the deposition time and current density also play a crucial role in improving uniformity. The direct relationship between the current density and the grain size is well established. However, the observed relationship between the electrodeposition duration and uniformity presents an interesting opportunity for further investigation. The investigation outcomes are crucial in understanding how to optimize the use of ultrasonic baths in template-assisted electrodeposition, thereby improving the fabrication of 3D MEAs. These results therefore significantly contribute to the advancement of micro- and nanofabrication. These findings could serve as a valuable guide for optimizing the parameters of ultrasonic-bath-assisted electrodeposition.

While the study's results offer exciting possibilities for ultrasonic-assisted electrodeposition, further investigations are warranted. These could include investigations of different materials for electrodeposition, alternative ultrasonic setups, and a wider range of ultrasonic parameters. Also, further investigations of how these findings translate into real-world applications in biomedical devices and MEMSs would be beneficial. The research outcomes, nevertheless, offer promising avenues to explore and optimize ultrasonic-assisted electrodeposition methods for the fabrication of 3D MEAs, bringing us a step closer to achieving high-performance biomedical interfaces.

Supplementary Materials: The following supporting information can be downloaded at: https://www.mdpi.com/article/10.3390/s24041251/s1, Spreadsheet S1: Optical profilometer data—electrode heights; Spreadsheet S2: XRD data; Table S1: Shear strength analysis; Layout_Substrate_S1: Layout of the substrate S1; Layout_Substrate_S2: Layout of the substrate S2; Layout_Substrate_S3: Layout of the substrate S3.

Author Contributions: Conceptualization, N.Y. and L.L.; methodology, N.Y., F.G. and D.G.; investigation, N.Y.; resources, L.L. and A.C.; data curation, N.Y.; writing—original draft preparation, N.Y. and L.L.; writing—review and editing, N.Y., F.G. and L.L.; visualization, N.Y. and A.C.; supervision, L.L.; project administration, N.Y. and L.L.; funding acquisition, L.L. All authors have read and agreed to the published version of the manuscript.

Funding: This work was partially funded by the European Union (NextGeneration EU) through the MUR-PNRR project SAMOTHRACE (ECS00000022).

Institutional Review Board Statement: Not applicable.

Data Availability Statement: The original contributions presented in the study are included in the article/supplementary material, further inquiries can be directed to the corresponding author/s.

Acknowledgments: We would like to thank Victor Micheli of the Functional Materials and Photonics Structures (FMAPS) research group, Fondazione Bruno Kessler (FBK), Trento (Italy), for carrying out XRD measurements. We would like acknowledge the MUR-PNRR project SAMOTHRACE (ECS00000022) for supporting the study financially.

Conflicts of Interest: The authors declare no conflicts of interest.

Appendix A. Normalized Height Distributions of Electrodeposited Micro-Pillars for Each Experiment

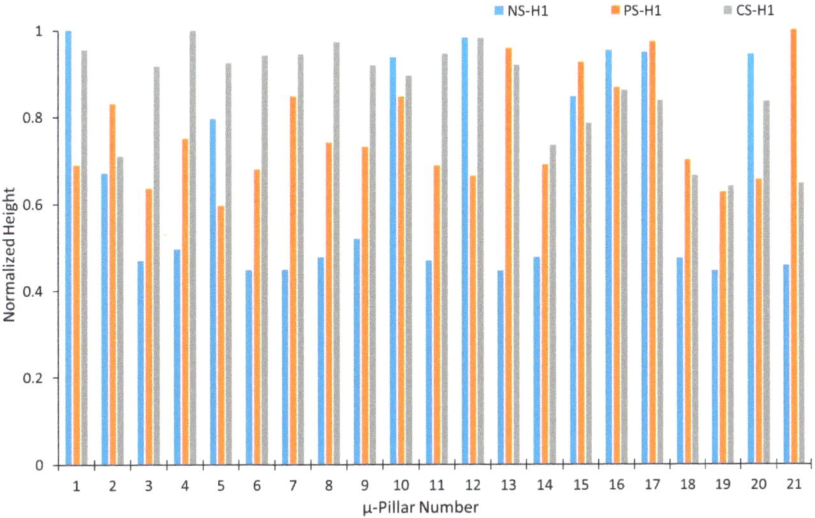

Figure A1. Normalized height distribution of electrodes/micro-pillars for each MEA from Experiment 1. The bar graph clearly indicates that the use of ultrasonic vibrations during electrodeposition (i.e., PS-H1 and CS-H1) significantly improves the uniformity in the height distribution of the micro-pillars compared to the MEA subjected to electrodeposition without ultrasonic vibrations (i.e., NS-H1).

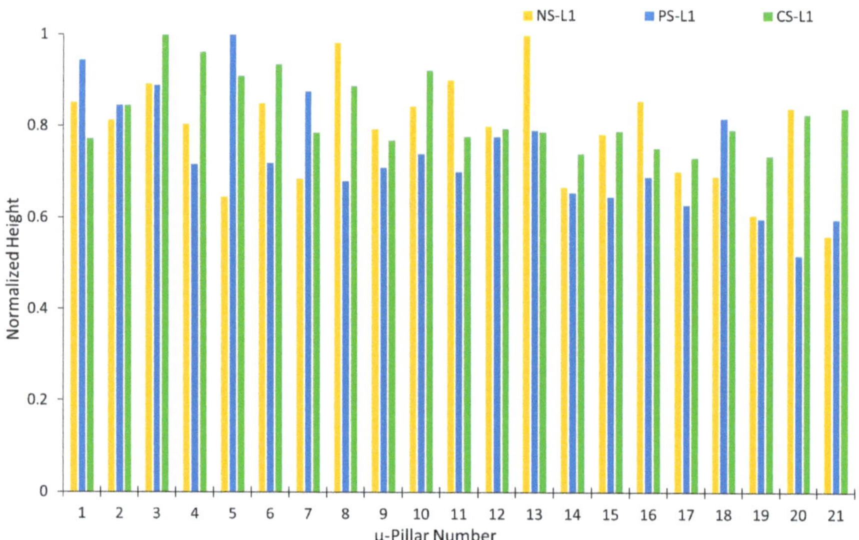

Figure A2. Normalized height distribution of electrodes/micro-pillars for each MEA from Experiment 2. The bar graph indicates that using the lower current density for electrodeposition leads to a slight improvement in the uniformity even without the use of ultrasonic vibrations (i.e., NS-L1), and the use of ultrasonic vibrations further enhances the uniformity (i.e., PS-L1 and CS-L1).

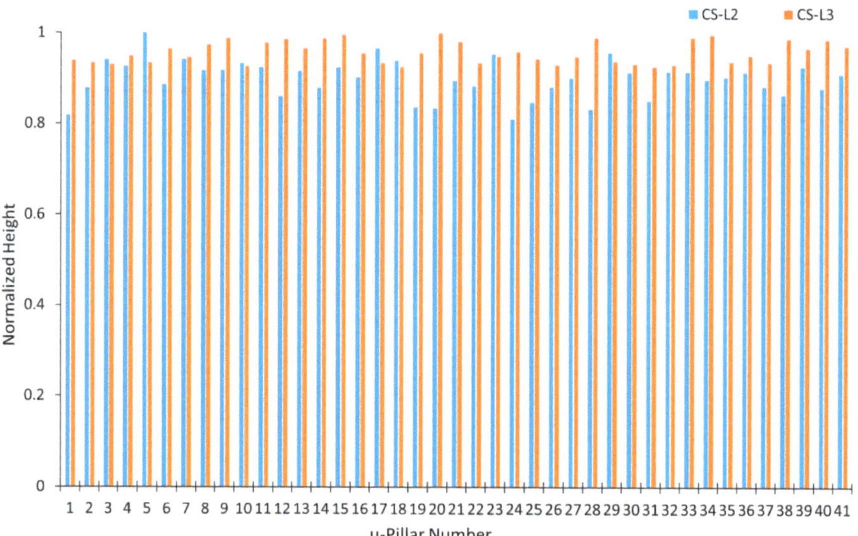

Figure A3. Normalized height distribution of electrodes/micro-pillars for each MEA from Experiment 3. The bar graph indicates that the longer electrodeposition durations with the use of continuous ultrasonic vibrations also contribute to an enhancement in uniformity.

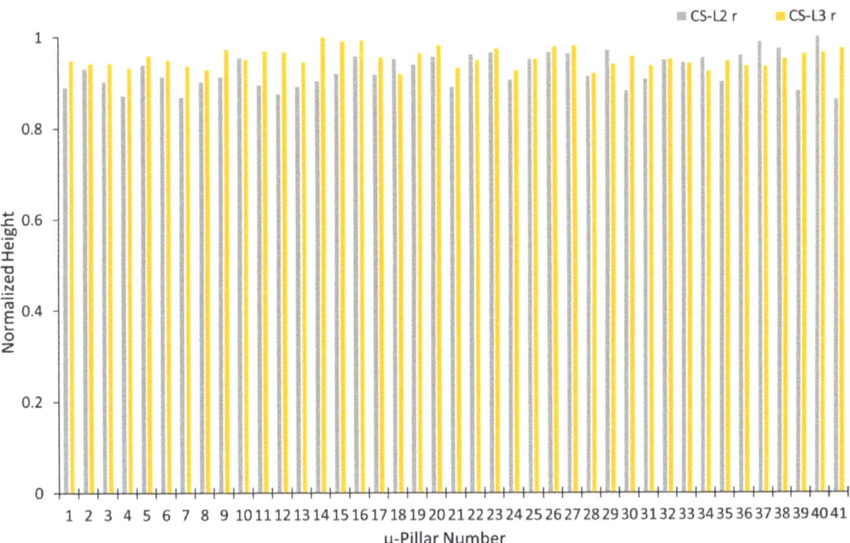

Figure A4. Normalized height distribution of electrodes/micro-pillars for each MEA from Experiment 3 (repeated). The bar graph confirms the trends observed in Figure A3 when the experiment was repeated, demonstrating the reliability and reproducibility of the process.

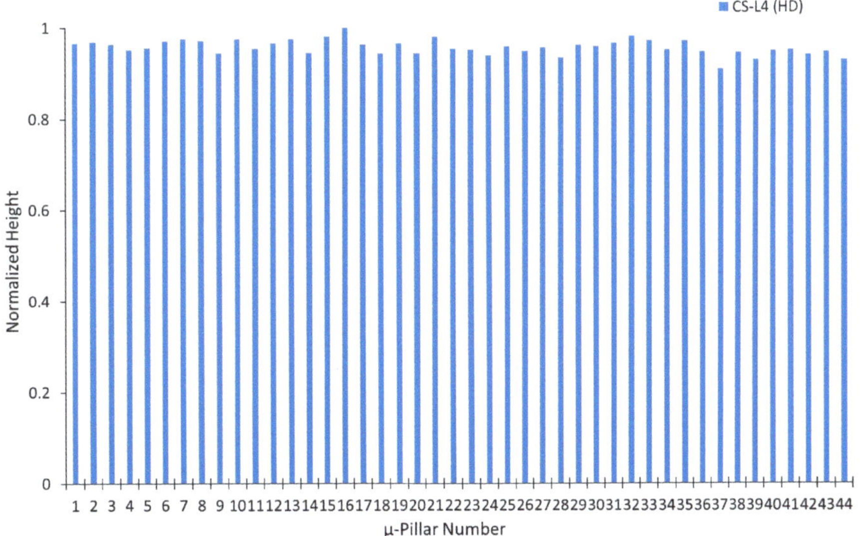

Figure A5. Normalized height distribution of electrodes/micro-pillars for each MEA from Experiment 4. The experiment was conducted to check the scalability (in terms of higher aspect ratios of the micro-pillars/template) of the process using a high-density MEA subjected to ultrasonic-vibration-assisted electrodeposition for a long duration (i.e., CS-L4 (HD)). The bar graph demonstrates the scalability of the process.

Appendix B. Design, Layout, and Fabrication of MEA Substrates

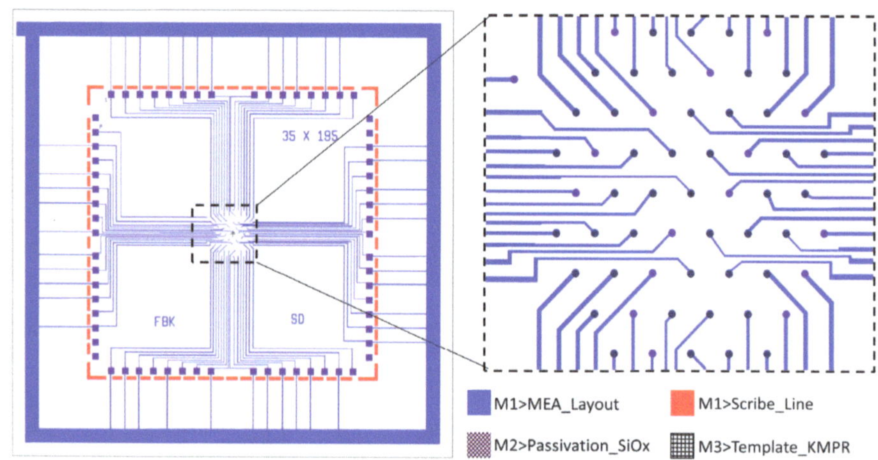

Figure A6. Design and layout of the S3 MEA substrate. The layout consists of three layers: M1 defines the layout of the MEA consisting of 60 channels (inside the scribe line), the scribe line (red dashed), and the external routing (outside the periphery of the scribe line) to support electrodeposition. The second layer (M2) defines the device passivation, and the third layer (M3) is designed to define the template for electrodeposition.

In Figure A6, the layout for the S3 MEA substrate is presented. The initial mask (M1), designed as a light-field mask, transfers the chip layout onto the wafer's metal layer. This chip layout encompasses a planar 60-electrode MEA (10×10 mm^2, including the dashed scribe line) and a tailored routing configuration along the MEA chip's periphery to facilitate electrodeposition. The planar electrodes, each with a 40 µm diameter, are arranged in a hexagonal layout with a pitch of 195 µm, optimizing internal routing between electrodes and contact pads. Notably, 44 of the 60 contact pads are connected to the external metal ring through custom routing, which is the working electrode during electrodeposition experiments.

The second mask (M2), designed as a dark-field mask, is required for creating openings in the passivation layer over the electrodes (circular openings with a 35 µm diameter) and contact pads. On the other hand, the third mask (M3), designed as a light-field mask, is tailored to establish a mold for the electrodeposition of gold micro-pillars on the top of the planar electrodes. This is achieved by creating cylindrical holes in the photoresist.

Figure A7 presents a detailed process for the fabrication of the MEA substrates. The planar MEA substrates were fabricated in a class 1000 cleanroom, employing polished borosilicate glass wafers sourced from MicroChemicals GmbH, Ulm, Germany, as the substrate (Figure A7a). To begin the process, a Cr/Au layer, with a 10/200 nm thickness, was deposited onto the pristine glass wafer using an ultralow vacuum (ULVAC) metal evaporator (Figure A7b). The wafer underwent priming through an HMDS (Hexamethyl-disilazane, MicroChemicals GmbH, Ulm, Germany) process at 150 °C, succeeded by the application of a positive-tone photoresist (AZ® 1518 from MicroChemicals GmbH, Ulm, Germany) through spin coating (Figure A7c), followed by a pre-exposure bake at 100 °C.

The pattern from mask M1 (Figure A7d) was transferred to the photoresist using an i-line mask aligner setup, followed by a standard post-exposure bake and a subsequent photoresist development step (Figure A7e). Further, the pattern transfer from the photoresist to the metal layers involved wet chemical etching of gold and chrome. Subsequently, the photoresist was removed using acetone (Figure A7f), followed by a deionized (DI) rinse and a drying cycle. The Cr/Au-patterned wafer was sintered at 200 °C for 60 min.

Following gold sintering, a 200 nm-thick passivation layer of SiO$_2$ was deposited using the plasma-enhanced chemical vapor deposition (PECVD) technique (Figure A7g).

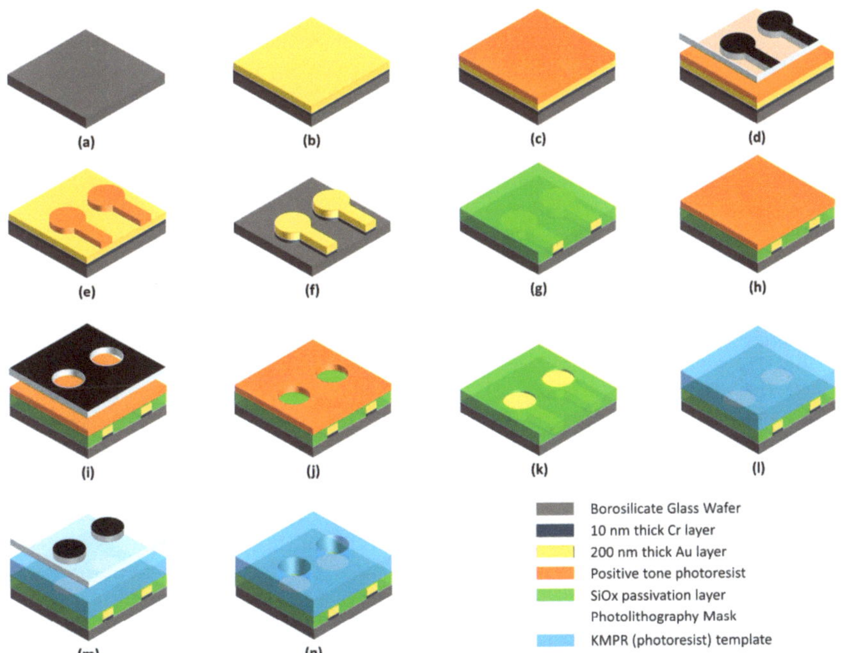

Figure A7. Fabrication of MEA substrates utilized for the study. (**a**) Substrate, (**b**) Cr/Au layer deposited thermally over the substrate. (**c**) Application of photoresist over the Cr/Au layer. (**d**) A bright-field photolithography mask is aligned over the wafer stack to pattern the photoresist through UV radiation. (**e**) Patterned photoresist. (**f**) Photoresist pattern transferred to the Cr/Au layer through wet chemical etching. (**g**) Deposition of a passivation layer over the patterned substrate. (**h**) Photoresist layer coated over the passivated substrate stack. (**i**) A dark-field photolithography mask aligned over the substrate stack to pattern the photoresist. (**j**) Patterned photoresist. (**k**) Photoresist pattern transferred to the passivation layer through reactive ion etching, yielding partially passivated planar device architecture. (**l**) A negative tone photoresist (KMPR) coated over the MEA substrate. (**m**) A bright-field photolithography mask aligned over the substrate stack to pattern the photoresist through UV radiation. (**n**) Planar MEA consisting of the photoresist template.

The wafer was subject to another lithography process using the second mask (M2) to define the pattern on the photoresist, following a similar process as described earlier (Figure A7h–i). The pattern transfer from the photoresist to the passivation layer was executed through a dry etching technique (AW-903ER Plasma Etch RIE, Allwin21 Corp., Morgan Hill, CA, USA), followed by a resist stripping process. The photoresist was stripped (Figure A7j–k), and the wafer was prepared for the final lithography process.

Finally, to fabricate the template for electrodeposition, the wafer with planar MEAs was first coated with a 110 μm-thick layer of chemically amplified negative photoresist (KMPR-1035, Kayaku Advanced Materials, Inc., Westborough, MA, USA), followed by a 30 min soft-bake at 100 °C (Figure A7l). The photoresist was then exposed to UV light through the third mask (M3) using an i-line mask aligner setup (Figure A7m). Further post-exposure baking was performed at 100 °C for 6 min to cure the exposed area of the photoresist completely. The photoresist was then developed using SU-8 developer solution for 20 min, assisted by a shaker plate and mild agitation to obtain a template with 110 μm-deep cylindrical holes on top of the planar electrodes (Figure A7n).

Following the electrodeposition process, it is imperative to highlight that the photoresist template, serving as a foundational structure for the 3D MEA, underwent a meticulous stripping procedure. This is crucial for removing the residual photoresist, leaving behind the fully realized 3D microelectrode array. Subsequently, the completed 3D MEA was precisely diced along the predefined scribe line to enhance the practical utility and functionality, as shown in Figure A6. This dicing procedure generated 60 independent channels by eliminating the traces to the external routing, each equipped to record or stimulate electrophysiological activity autonomously. It is essential to note that this detailed process ensures the creation of individualized and functional microelectrode channels, optimizing the overall performance and versatility of the 3D MEA. Utilizing this precise dicing method further enhances the adaptability of the 3D MEA for various experimental setups and applications.

References

1. Choi, J.S.; Lee, H.J.; Rajaraman, S.; Kim, D.H. Recent Advances in Three-Dimensional Microelectrode Array Technologies for in Vitro and in Vivo Cardiac and Neuronal Interfaces. *Biosens. Bioelectron.* **2021**, *171*, 112687. [CrossRef] [PubMed]
2. Lam, D.; Fischer, N.O.; Enright, H.A. Probing Function in 3D Neuronal Cultures: A Survey of 3D Multielectrode Array Advances. *Curr. Opin. Pharmacol.* **2021**, *60*, 255–260. [CrossRef] [PubMed]
3. Jones, I.L.; Livi, P.; Lewandowska, M.K.; Fiscella, M.; Roscic, B.; Hierlemann, A. The Potential of Microelectrode Arrays and Microelectronics for Biomedical Research and Diagnostics. *Anal. Bioanal. Chem.* **2011**, *399*, 2313–2329. [CrossRef] [PubMed]
4. Decker, D.; Hempelmann, R.; Natter, H.; Pirrung, M.; Rabe, H.; Schäfer, K.H.; Saumer, M. 3D Nanostructured Multielectrode Arrays: Fabrication, Electrochemical Characterization, and Evaluation of Cell–Electrode Adhesion. *Adv. Mater. Technol.* **2019**, *4*, 1800436. [CrossRef]
5. Saleh, M.S.; Ritchie, S.M.; Nicholas, M.A.; Gordon, H.L.; Hu, C.; Jahan, S.; Yuan, B.; Bezbaruah, R.; Reddy, J.W.; Ahmed, Z.; et al. CMU Array: A 3D Nanoprinted, Fully Customizable High-Density Microelectrode Array Platform. *Sci. Adv.* **2022**, *8*, eabj4853. [CrossRef]
6. Berényi, A.; Somogyvári, Z.; Nagy, A.J.; Roux, L.; Long, J.D.; Fujisawa, S.; Stark, E.; Leonardo, A.; Harris, T.D.; Buzsáki, G. Large-Scale, High-Density (up to 512 Channels) Recording of Local Circuits in Behaving Animals. *J. Neurophysiol.* **2014**, *111*, 1132–1149. [CrossRef] [PubMed]
7. Campbell, P.K.; Jones, K.E.; Huber, R.J.; Horch, K.W.; Normann, R.A. A Silicon-Based, Three-Dimensional Neural Interface: Manufacturing Processes for an Intracortical Electrode Array. *IEEE Trans. Biomed. Eng.* **1991**, *38*, 758–768. [CrossRef] [PubMed]
8. Goncalves, S.B.; Peixoto, A.C.; Silva, A.F.; Correia, J.H. Fabrication and Mechanical Characterization of Long and Different Penetrating Length Neural Microelectrode Arrays. *J. Micromechanics Microengineering* **2015**, *25*, 055014. [CrossRef]
9. Soscia, D.A.; Lam, D.; Tooker, A.C.; Enright, H.A.; Triplett, M.; Karande, P.; Peters, S.K.G.; Sales, A.P.; Wheeler, E.K.; Fischer, N.O. A Flexible 3-Dimensional Microelectrode Array for: In Vitro Brain Models. *Lab Chip* **2020**, *20*, 901–911. [CrossRef]
10. Hartmann, J.; Lauria, I.; Bendt, F.; Rütten, S.; Koch, K.; Blaeser, A.; Fritsche, E. Alginate-Laminin Hydrogel Supports Long-Term Neuronal Activity in 3D Human Induced Pluripotent Stem Cell-Derived Neuronal Networks. *Adv. Mater. Interfaces* **2023**, *10*, 2200580. [CrossRef]
11. Bartsch, H.; Baca, M.; Fernekorn, U.; Müller, J.; Schober, A.; Witte, H. Functionalized Thick Film Impedance Sensors for Use in In Vitro Cell Culture. *Biosensors* **2018**, *8*, 37. [CrossRef]
12. Lorenzelli, L.; Spanu, A.; Pedrotti, S.; Tedesco, M.; Martinoia, S. Three-Dimensional Microelectrodes Array Based on Vertically Stacked Beads for Mapping Neurons' Electrophysiological Activity. In Proceedings of the 2019 20th International Conference on Solid-State Sensors, Actuators Microsystems Eurosensors XXXIII, TRANSDUCERS 2019 EUROSENSORS XXXIII, Berlin, Germany, 23–27 June 2019; pp. 987–990. [CrossRef]
13. Wang, P.; Wu, E.G.; Uluşan, H.; Phillips, A.J.; Hays, M.R.; Kling, A.; Zhao, E.T.; Madugula, S.; Vilkhu, R.S.; Vasireddy, P.K.; et al. Direct-Print Three-Dimensional Electrodes for Large-Scale, High-Density, and Customizable Neural Interfaces. *bioRxiv* **2023**. [CrossRef]
14. Acarón Ledesma, H.; Li, X.; Carvalho-de-Souza, J.L.; Wei, W.; Bezanilla, F.; Tian, B. An Atlas of Nano-Enabled Neural Interfaces. *Nat. Nanotechnol.* **2019**, *14*, 645–657. [CrossRef]
15. Ghane-Motlagh, B.; Sawan, M. A Review of Microelectrode Array Technologies: Design and Implementation Challenges. In Proceedings of the 2013 2nd International Conference on Advances in Biomedical Engineering, Tripoli, Lebanon, 11–13 September 2013. [CrossRef]
16. Zeck, G.; Jetter, F.; Channappa, L.; Bertotti, G.; Thewes, R. Electrical Imaging: Investigating Cellular Function at High Resolution. *Adv. Biosyst.* **2017**, *1*, 1700107. [CrossRef] [PubMed]
17. Gross, G.W. Multielectrode Arrays. *Scholarpedia* **2011**, *6*, 5749. [CrossRef]
18. Spanu, A.; Colistra, N.; Farisello, P.; Friz, A.; Arellano, N.; Rettner, C.T.; Bonfiglio, A.; Bozano, L.; Martinoia, S. A Three-Dimensional Micro-Electrode Array for in-Vitro Neuronal Interfacing. *J. Neural Eng.* **2020**, *17*, 036033. [CrossRef] [PubMed]

19. Yadav, N.; Lorenzelli, L.; Giacomozzi, F. A Novel Additive Manufacturing Approach towards Fabrication of Multi-Level Three-Dimensional Microelectrode Array for Electrophysiological Investigations. In Proceedings of the 2021 23rd European Microelectronics and Packaging Conference & Exhibition (EMPC), Gothenburg, Sweden, 13–16 September 2021; IEEE: Piscataway, NJ, USA, 2021; pp. 1–5.

20. Teixeira, H.; Dias, C.; Aguiar, P.; Ventura, J. Gold-Mushroom Microelectrode Arrays and the Quest for Intracellular-Like Recordings: Perspectives and Outlooks. *Adv. Mater. Technol.* **2021**, *6*, 2000770. [CrossRef]

21. Steins, H.; Mierzejewski, M.; Brauns, L.; Stumpf, A.; Kohler, A.; Heusel, G.; Corna, A.; Herrmann, T.; Jones, P.D.; Zeck, G.; et al. A Flexible Protruding Microelectrode Array for Neural Interfacing in Bioelectronic Medicine. *Microsyst. Nanoeng.* **2022**, *8*, 131. [CrossRef] [PubMed]

22. Yadav, N.; Di Lisa, D.; Giacomozzi, F.; Cian, A.; Giubertoni, D.; Martinoia, S.; Lorenzelli, L. Development of Multi-Depth Probing 3D Microelectrode Array to Record Electrophysiological Activity within Neural Cultures. *J. Micromechanics Microengineering* **2023**, *33*, 115002. [CrossRef]

23. Mescola, A.; Canale, C.; Prato, M.; Diaspro, A.; Berdondini, L.; MacCione, A.; Dante, S. Specific Neuron Placement on Gold and Silicon Nitride-Patterned Substrates through a Two-Step Functionalization Method. *Langmuir* **2016**, *32*, 6319–6327. [CrossRef]

24. Ning, F.; Cong, W. Ultrasonic Vibration-Assisted (UV-A) Manufacturing Processes: State of the Art and Future Perspectives. *J. Manuf. Process.* **2020**, *51*, 174–190. [CrossRef]

25. Wei, C.; Shixing, W.; Libo, Z.; Jinhui, P.; Gengwei, Z. The Application of Ultrasound Technology in the Field of the Precious Metal. *Russ. J. Non-Ferrous Met.* **2015**, *56*, 417–427. [CrossRef]

26. Hyde, M.E.; Compton, R.G. How Ultrasound Influences the Electrodeposition of Metals. *J. Electroanal. Chem.* **2002**, *531*, 19–24. [CrossRef]

27. Gadkari, S.A.; Nayfeh, T.H. Micro Fabrication Using Electro Deposition and Ultrasonic Acoustic Liquid Manipulation. *Int. J. Adv. Manuf. Technol.* **2008**, *39*, 107–117. [CrossRef]

28. Viventi, J.; Kim, D.H.; Vigeland, L.; Frechette, E.S.; Blanco, J.A.; Kim, Y.S.; Avrin, A.E.; Tiruvadi, V.R.; Hwang, S.W.; Vanleer, A.C.; et al. Flexible, Foldable, Actively Multiplexed, High-Density Electrode Array for Mapping Brain Activity in Vivo. *Nat. Neurosci.* **2011**, *14*, 1599–1605. [CrossRef]

29. Peng, Q.; Xiong, W.; Tan, X.; Venkataraman, M.; Mahendran, A.R.; Lammer, H.; Kejzlar, P.; Militky, J. Effects of Ultrasonic-Assisted Nickel Pretreatment Method on Electroless Copper Plating over Graphene. *Sci. Rep.* **2022**, *12*, 21159. [CrossRef]

30. Tudela, I.; Zhang, Y.; Pal, M.; Kerr, I.; Mason, T.J.; Cobley, A.J. Ultrasound-Assisted Electrodeposition of Nickel: Effect of Ultrasonic Power on the Characteristics of Thin Coatings. *Surf. Coatings Technol.* **2015**, *264*, 49–59. [CrossRef]

31. Costa, J.M.; Almeida Neto, A.F. de Ultrasound-Assisted Electrodeposition and Synthesis of Alloys and Composite Materials: A Review. *Ultrason. Sonochemistry* **2020**, *68*, 105193. [CrossRef]

32. Bonin, L.; Bains, N.; Vitry, V.; Cobley, A.J. Electroless Deposition of Nickel-Boron Coatings Using Low Frequency Ultrasonic Agitation: Effect of Ultrasonic Frequency on the Coatings. *Ultrasonics* **2017**, *77*, 61–68. [CrossRef] [PubMed]

33. Coleman, S.J.; Roy, S. Electrodeposition of Copper Patterns Using EnFACE Technique under Ultrasonic Agitation. *Chem. Eng.* **2014**, *41*, 37–42. [CrossRef]

34. Scherrer, P. Bestimmung Der Größe Und Der Inneren Struktur von Kolloidteilchen Mittels Röntgenstrahlen. In *Nachrichten von der Gesellschaft der Wissenschaften zu Göttingen, Mathematisch-Physikalische Klasse*; Weidmannsche Buchhandlung: Berlin, Germany, 1918; pp. 98–100.

Article

Classification in Early Fire Detection Using Multi-Sensor Nodes—A Transfer Learning Approach

Pascal Vorwerk [1,*], Jörg Kelleter [2], Steffen Müller [2] and Ulrich Krause [1]

[1] Faculty of Process- and Systems Engineering, Institute of Apparatus and Environmental Technology, Otto von Guericke University of Magdeburg, Universitätsplatz 2, 39106 Magdeburg, Germany; ulrich.krause@ovgu.de

[2] GTE Industrieelektronik GmbH, Helmholtzstr. 21, 38-40, 41747 Viersen, Germany; joerg.kelleter@gte.de (J.K.); steffen.mueller@gte.de (S.M.)

* Correspondence: pascal.vorwerk@ovgu.de

Abstract: Effective early fire detection is crucial for preventing damage to people and buildings, especially in fire-prone historic structures. However, due to the infrequent occurrence of fire events throughout a building's lifespan, real-world data for training models are often sparse. In this study, we applied feature representation transfer and instance transfer in the context of early fire detection using multi-sensor nodes. The goal was to investigate whether training data from a small-scale setup (source domain) can be used to identify various incipient fire scenarios in their early stages within a full-scale test room (target domain). In a first step, we employed Linear Discriminant Analysis (LDA) to create a new feature space solely based on the source domain data and predicted four different fire types (smoldering wood, smoldering cotton, smoldering cable and candle fire) in the target domain with a classification rate up to 69% and a Cohen's Kappa of 0.58. Notably, lower classification performance was observed for sensor node positions close to the wall in the full-scale test room. In a second experiment, we applied the TrAdaBoost algorithm as a common instance transfer technique to adapt the model to the target domain, assuming that sparse information from the target domain is available. Boosting the data from 1% to 30% was utilized for individual sensor node positions in the target domain to adapt the model to the target domain. We found that additional boosting improved the classification performance (average classification rate of 73% and an average Cohen's Kappa of 0.63). However, it was noted that excessively boosting the data could lead to overfitting to a specific sensor node position in the target domain, resulting in a reduction in the overall classification performance.

Keywords: multi-sensor nodes; early fire detection; gas sensors; transfer learning; electronic nose; feature fusion; linear discriminant analysis (LDA); classification

Citation: Vorwerk, P.; Kelleter, J.; Müller, S.; Krause, U. Classification in Early Fire Detection Using Multi-Sensor Nodes—A Transfer Learning Approach. *Sensors* **2024**, *24*, 1428. https://doi.org/10.3390/s24051428

Academic Editor: Eduard Llobet

Received: 17 January 2024
Revised: 2 February 2024
Accepted: 21 February 2024
Published: 22 February 2024

1. Introduction

The advantages of multi-sensor approaches to early fire detection over traditional smoke detectors have been extensively discussed in the previous literature [1–3]. The main advantages are improved coverage of the detection area [4], a shorter detection time [5–7], more accurate detection (improved sensitivity to real fires) [8–11] and a reduction in the false alarm rate [12,13].

In addition to the temporal and robustness aspects of early fire detection, the ability to differentiate between different types of fire scenarios can provide additional information to laypersons or first responders during alarms [14]. This can support effective identification and intervention, especially in the early stages of ongoing incipient fires where combustion products are barely visible [15].

Previous research has demonstrated the effectiveness of employing multi-sensor approaches to distinguish various fire materials based on their unique "odor prints" [16–18]. However, these studies faced limitations in their training and validation datasets. Some were confined to a single room setting [19], while others were constrained to a binary output (fire/no fire) when utilizing data from different environments [20,21].

Generally, fire events are infrequent occurrences throughout a building's lifespan. The scarcity of real event data poses challenges and necessitates reliance on data obtained from experimental setups or simulations [22]. However, conducting such (large-scale) experiments is expensive, and the availability of large-scale test rooms is very limited [21]. Given these constrains, there is an urgent need to investigate the effective transfer of data from small-scale laboratory setups to real room applications.

In this work, we address the research question (RQ) of whether multi-sensor data generated in a small-scale laboratory setup can be used to identify various incipient fire scenarios in a large-scale room setup.

To our knowledge, existing transfer learning methodologies have not been employed in the field of early fire detection using multi-sensor nodes. Furthermore, it remains uncertain whether, in general, the differentiation of various incipient fire scenarios during their initial stages is achievable based on multi-sensor data.

In this study, we employed two primary methodologies from the transfer learning research domain. We leveraged both feature representation transfer and instance transfer in order to identify different incipient fire scenarios in a real EN54 standard test room, relying solely on training data generated in a small-scale laboratory setup. Subsequently, we assessed the classifier's performance at various sensor node positions within a large-scale test room.

The novelty of this work lies in its approach to distinguish between various incipient fire scenarios in their initial phases using solely training data from a small-scale setup. Prior research has typically been confined to a single experimental setup for both model construction and testing, or it has been restricted to binary model prediction (fire/no fire), simplifying the classification problem and incurring high data generation costs. This study addresses two primary limitations in the existing literature. Firstly, we present a comprehensive workflow for cost-effective data acquisition and model development in the field of early fire detection employing multi-sensor nodes. Secondly, we apply this workflow to a multi-classification problem, for which we differentiate between four distinct fire scenarios in their earliest stages. Previous work has predominantly focused on simpler binary classification problems and more advanced fire scenarios where detection is generally more straightforward. The proposed approach provides valuable additional information about the nature of an ongoing incipient fire event, enabling first responders or firefighters to make more informed decisions, such as formulating intervention recommendations or enhancing situational awareness.

2. Related Work

Prior research has explored various methodologies for fire detection and identification using multi-sensor data.

Solórzano et al. [21] achieved a classification rate of approximately 68% using training and test data from normative test fires conducted in a standard EN-54 test room. The authors stated that the classification rate could be increased to 96% by incorporating additional training and test data from laboratory experiments. In their recent publication [20], Solórzano et al. corroborated these findings, reporting a classification rate ranging from 52% to 70% (or 88% with additional training and test data generated in a small-scale setup).

However, in both studies, the model output was confined to a binary prediction (fire/no fire), leading to a significantly simpler classification problem compared to our study. Additionally, the test data consistently encompassed data from the same room environment that had already been utilized for training the model.

Other studies, as summarized in [3], were also primarily constrained to a binary decision problem (fire/no fire) and/or confined to a single experimental environment.

Milke et al. [23] defined hard rules utilizing a sensor array comprising temperature, light obscuration, CO_2, MOX and O_2 sensors in order to distinguish between "flaming fire", "smoldering fire" and "nuisance". The authors attained a classification rate of 90% and could enhance the classification rate up to 97% by employing a three-layer neural network as the model instead of hard rules. However, the training and test data were derived from experiments conducted in the same test room.

Ni et al. [24] constructed a classification model to categorize various wire insulation materials (PVC, Teflon, Kapton and silicone rubber) based on the volatiles released during electrical overload. The authors employed dimension reduction (PCA) and a K-NN classifier as the classification model and achieved a classification rate of up to 82% for four different classes. However, the training and test data were derived from the same experimental setup using the leave-one-out method.

Experiments in prior studies primarily utilized standard test fires, resulting in considerably higher emissions and, consequently, clearer sensor signals. In contrast, our study encompasses the initial phases of ongoing incipient fires within the experimental setup. Moreover, previous studies often focused on binary or ternary classification problems, with Ni et al. [24] being a notable exception. Another limitation in previous research is the generation of training and test data within the same experimental environment, which poses a constraint for real-world applications. The novelty of our work lies in utilizing data from two distinct experimental environments.

2.1. Early Fire Indicators

Previous studies have employed various combinations of multi-sensor measurements for early fire detection. Solórzano et al. [20] utilized hydrogen (H_2), methane (CH_4), nitrogen oxides (NO_x) and volatile organic compounds (VOCs) in a multi-sensor array. The authors emphasized the significance of CO and VOCs as early fire indicators due to their substantial emissions during incipient fire scenarios such as smoldering fires. Nazir et al. [25] corroborated these findings by including air temperature, humidity, CO_2 and ammonia (NH_3) in their study.

Krüger et al. [26] and Hayashi et al. [27] identified substantial releases of H_2 during the smoldering process of various polymeric materials commonly present in households such as wood, PUR foam and PE. The authors concluded that H_2 can serve as an early fire indicator that precedes the substantial emissions of CO and smoke.

Gutmacher et al. [28] corroborated these findings, emphasizing that CO and H_2 are the most crucial gases for detecting the early stages of smoldering fires.

In our previous study [29], we validated these observations. We examined particulate matter (PM), VOCs, CO, CO_2, H_2, ultraviolet radiation (UV), air temperature and humidity as early fire indicators during different incipient fires conducted in a standard EN 54 test room. By varying the distance between the sensor node and the fire source, we identified five significant early fire indicators: H_2, CO, PM0.5 (PM < 0.5 μm), PM1.0 (0.5 μm < PM < 1.0 μm) and VOC.

2.2. Transfer Learning

Weiss et al. [30] emphasized the challenges in obtaining training and test data from the same domain for real-world machine learning applications, particularly in cases where data collection is impractical due to high costs or difficulty. This challenge is particularly relevant in the context of (early) fire detection using multi-sensor nodes, where generating data in real room setups is prohibitively expensive and the availability of fire test rooms is extremely limited. The authors emphasize the importance of employing less expensive training data from a different domain for model building. This concept is known as transfer learning.

Zhuang et al. [31] defined transfer learning as the enhancement of a target learner using knowledge from a "[...] different but related" [31] source domain. The primary objective is to decrease reliance on (expensive) data from the target domain.

According to Kim et al. [32], transfer learning aims to learn a target predictive function $f_T(\cdot)$ from pairs $\{x_i, y_i\}$ generated in a source domain D_S, where $x_i \in \mathcal{X}$ and $y_i \in \mathcal{Y}$. In the subsequent work, the notation provided by Kim et al. [32] given in Table 1 is adopted, with the index S representing the source domain D_S and the index T representing the target domain D_T.

Table 1. Transfer learning definitions and notations for source domain D_S and target domain D_T according to Kim et al. [32].

Notation Source Domain D_S	Notation Target Domain D_T	Description
$D_S = \{\mathcal{X}_S, P(X_S)\}$	$D_T = \{\mathcal{X}_T, P(X_T)\}$	domain
$\mathcal{X}_S = x_{S1}, \ldots, x_{Sn}$	$\mathcal{X}_T = x_{T1}, \ldots, x_{Tn}$	feature space
$P(\mathcal{X}_S)$	$P(\mathcal{X}_T)$	marginal probability distribution
$\mathcal{Y}_S = y_{S1}, \ldots, y_{Sn}$	$\mathcal{Y}_T = y_{T1}, \ldots, y_{Tn}$	label space
$f_S(\cdot) = P(y_{Si}\|x_{Si})$	$f_T(\cdot) = P(y_{Ti}\|x_{Ti})$	objective predictive function
$\mathcal{T}_S = \{\mathcal{Y}_S, f_S(\cdot)\}$	$\mathcal{T}_T = \{\mathcal{Y}_T, f_T(\cdot)\}$	task

According to Cook et al. [33], a certain relationship must exist between D_S and D_T in order to be able to transfer knowledge from D_S to D_T. In our case, the feature space in both D_S and D_T is essentially the same (sensors, and selected sensor measurements are identical), thus satisfying Equation (1).

$$\mathcal{X}_S = \mathcal{X}_T \tag{1}$$

However, the scaling and rotation of the feature spaces $\mathcal{X}_S$ and $\mathcal{X}_T$ differs slightly due to the distinct room settings.

In these feature spaces $\mathcal{X}_S$ and $\mathcal{X}_T$, the marginal probability distribution $P(\mathcal{X})$ is not equal because the "activity" in D_S and D_T, respectively, is not exactly the same (the experiments in D_S are downscaled; see Section 3.2). This assumption is given in the following Equation (2).

$$P(\mathcal{X}_S) \neq P(\mathcal{X}_T) \tag{2}$$

In this work, the label space $\mathcal{Y}$ in D_S and D_T is identical, as we conducted the same types of fire experiments in both domains (see Section 3.2), as given in Equation (3).

$$\mathcal{Y}_S = \mathcal{Y}_T \tag{3}$$

As the objective prediction function $f(\cdot)$ is defined as $f(\cdot) = P(y|x)$ and $P(\mathcal{X})$ varies between D_S with respect to D_T (see Equation (2)), $f(\cdot)$ differs for D_S and D_T, as shown in Equation (4).

$$f_S(\cdot) \neq f_T(\cdot) \tag{4}$$

This finally results in a different task $\mathcal{T}$ to learn, so that

$$\mathcal{T}_S \neq \mathcal{T}_T \tag{5}$$

Cook et al. [33] defined two primary types of transfer learning approaches to address disparities between D_S and D_T.

The first approach is feature representation transfer, which aims to mitigate the differences between the feature spaces $\mathcal{X}_S$ and $\mathcal{X}_T$. According to Cook et al. [33], feature representation transfer is typically achieved by mapping both $\mathcal{X}_S$ and $\mathcal{X}_T$ to a new feature space $\mathcal{X}$ through functions $g : \mathcal{X}_S \rightarrow \mathcal{X}$ and $f : \mathcal{X}_T \rightarrow \mathcal{X}$. Dimension reduction is a commonly employed technique in this context [33].

The second transfer learning approach is instance transfer, where a small amount of data from the target domain is utilized to weight instances from the source domain. Since this approach works particularly well under the condition of equivalent feature spaces $\mathcal{X}_S$ and $\mathcal{X}_T$, instance transfer is typically applied after feature representation transfer [33].

A common method for instance transfer is the TrAdaBoost algorithm proposed by Dai [34], which has already been employed in combination with an SVM classifier to categorize atmospheric dust aerosol particles in a transfer learning application [30].

3. Materials and Methods

3.1. Sensor Nodes

We employed muti-sensor nodes for data collection, as illustrated in Figure 1. Each sensor node was equipped with sensors, including an SPS30, SGP40, SHT4x, CO/MF-1000, UST6xxx and SCD40, that measured parameters such as PM, VOC, relative air temperature, air humidity, CO, H_2 and CO_2.

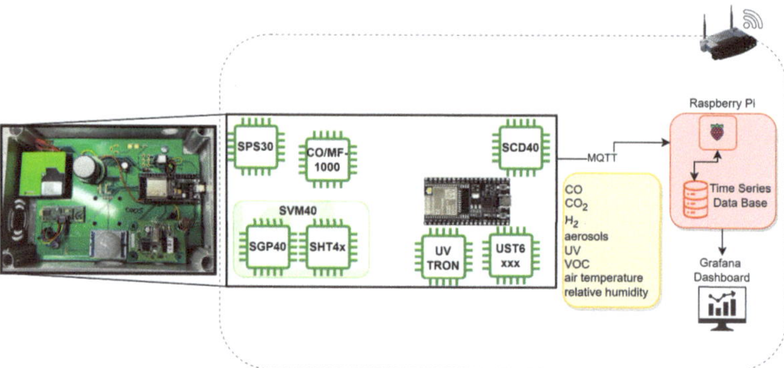

Figure 1. Sensor node with multiple sensors and data transfer via MQTT to Raspberry Pi.

The sensors on each sensor node were controlled by a microcontroller (ESP32). Communication between the microcontroller and the broker/server (Raspberry Pi) was via WiFi using the MQTT protocol. The microcontroller sent sensor data in JSON format to the Raspberry Pi, where a Python script decoded the information and recorded it in an Influx time series database. The database automatically assigned an unique UTC timestamp to each measurement vector.

For real-time monitoring during the experiments, a Grafana dashboard was utilized. Data were exported from the Influx time series database as a CSV file using a Python script. Each sensor node in the network was equipped with the sensors listed in Table 2.

A consistent sampling rate of one sample per 10 s was maintained throughout all experiments. This decision was influenced by the characteristics of the sensors in use. Specifically, the CO/MF-1000 sensor had a T90 response time of approximately 25 s: capturing 90% of the gas concentration within this time frame [35]. Likewise, the UST6xxx sensor relied on internal temperature cycles with a 10 s interval for H_2 detection [36]. Hence, opting for a sampling rate exceeding one sample per 10 s would not yield any additional information.

To minimize cross-sensitivity between CH_4, CO and alcohol, we selected the UST6xxx sensor containing the GGS 6530 T gas sensing element. The UST6xxx exhibits nearly no response to CH_4 exposure up to 1000 vppm, and it sustains this characteristic at a heating temperature of 475 °C [36].

Table 2. Overview of sensors in each sensor node.

Sensor	Manufacturer	Measurand	Unit
SPS30	Sensirion (Stäfa, Switzerland)	PM	cm^{-3}
SGP40	Sensirion	VOC	A.U.
CO/MF-1000	MEMBRAPOR (Wallisellen, Switzerland)	CO	ppm
UST6xxx	UST (Sydney, Australia)	H_2	ppm
SCD40	Sensirion	CO_2	ppm
UVTRON	HAMAMATSU (Shizuoka, Japan)	UV photon	#
SHT4x	Sensirion	Temperature, Relative air humidity	°C, %

3.2. Experiments and Datasets

Following the idea of transfer learning discussed in Section 2.2, we used two experimental setups in order to represent the source domain D_S and the target domain D_T. The two setups are exemplarily shown in Figure 2.

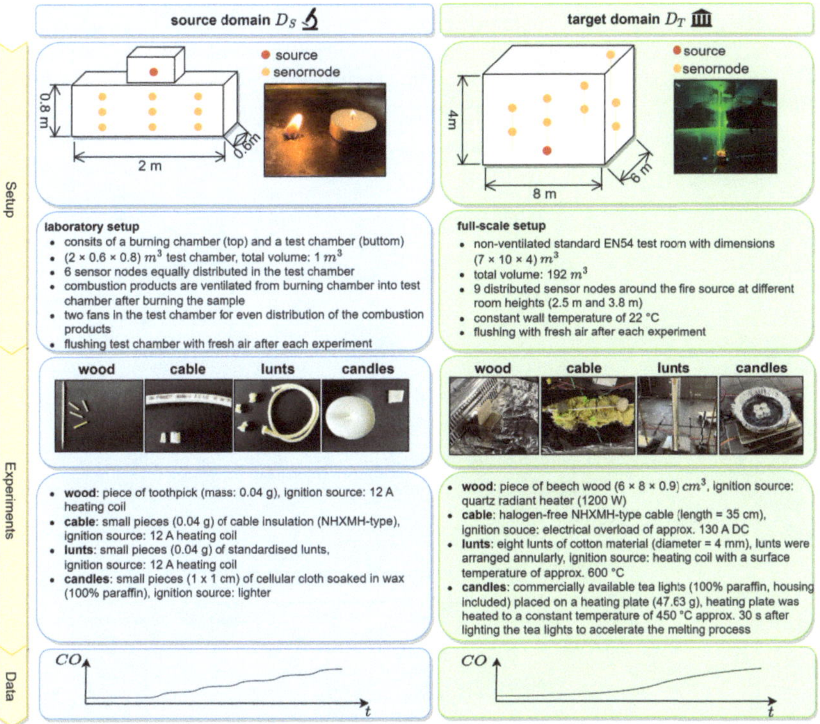

Figure 2. Experimental setup in D_S (**left**) and D_T (**right**); 4 different incipient fire experiments: smoldering wood, smoldering cable, glowing lunts and candle fire.

A $(2 \times 0.6 \times 0.8)$ m^3 test chamber served as the small-scale setup (source domain D_S), and we exposed six sensor nodes to various fire loads using cotton, cable insulation, candle wax and wood (see Figure 2, left). This experimental setup was used to generate the source domain dataset (ds_dataset).

An unventilated standard EN54 test room with dimensions $(7 \times 10 \times 4)$ m^3 was used as the large-scale setup (target domain D_T) to generate the target domain dataset (dt_dataset). The fire source was positioned in the center of the room. Nine distributed sensor nodes were placed around the source as shown in Figure 3.

In both domains, four distinct fire types—wood, cable, lunt and candle fires—were executed. Table 3 provides a summary of the burning material mass, repetitions, stages and ignition source type. A more comprehensive description of the experiments conducted in the target domain D_T is given in [29].

The experiments conducted in D_S represent scaled-down setups of the experiments performed in D_T. For equivalence, we employed identical materials in both domains but adjusted the mass of the burning material and the combustion process as follows.

To represent the smoldering wood fire, we used small pieces of toothpick. The toothpicks were standardized, and the mass of one piece of toothpick was 0.04 g. A DC heating coil (12 A) was used as the ignition source in order to ensure non-flaming combustion. The heating coil was a 1 mm-thick constantan wire twisted into a spiral consisting of 15 windings and an inner diameter of 100 mm.

The cable fire was simulated using small pieces (0.04 g) of the same cable insulation material used in D_T. As with the wood scenario, the 12 A DC heating coil was used as the ignition source.

The lunt fire was scaled down equivalently by using small pieces (0.04 g) of the lunts used in D_T. The ignition source was again the 12 A DC heating coil.

Downscaling of the candle fire was not trivial, as the wax fire produces high flames even with smaller amounts of wax material. To control the size of the flame, we used small pieces of cotton that were soaked in wax. The cotton acted as a wick. Its surface size served as the controlling parameter for the size of the flame.

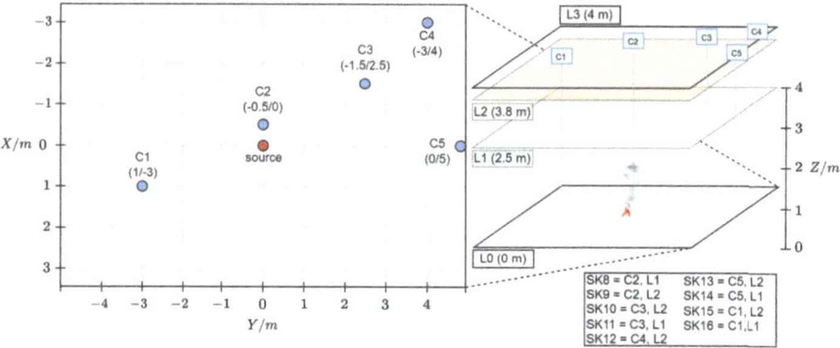

Figure 3. Sensor node positions in D_T; L0 = ground layer, L1 = height at 2.5 m, L2 = height at 3.8 m, C1–C5 = chains containing two sensor nodes (one at L1 and one at L2), and SK8–SK16: unique sensor node IDs.

Table 3. Overview of the experiments carried out in D_S and D_T.

Domain	Scenario	Mass	Stages	Repetitions	Material	Ignition Source
source	wood	0.04 g	5	4	beech wood	heating coil (12A)
	cable	0.04 g	5	6	cable isolation [1]	heating coil (12A)
	lunts	0.04 g	6	4	cotton [2]	heating coil (12A)
	candles	0.90 g	1	6	cellular cloth [3]	lighter
target	wood	30.8 g	1	3	beech wood	quartz radiant heater [4]
	cable	29.99 g	1	3	cable isolation [1]	electrical overload [7]
	lunts	100.20 g	1	3	cotton [2]	heating coil [5]
	candles	47.63 g	1	3	tea lights	heating plate [6]

[1] NHXMH-type; [2] manufacturer: "FehrErlen"; [3] soaked with candle wax (100% paraffin); [4] 1200 W, distance between heater and source = 5 cm; [5] surface temperature = 600 °C; [6] surface temperature = 450 °C; [7] 130 A DC.

As depicted in Figure 2, variations were observed in the temporal increase of sensor measurements in D_S and D_T. This aligns with the findings reported by Solórzano et al. [21].

This contrast can be attributed to two primary factors. Firstly, there is a significant difference in the propagation behavior in D_T with respect to D_S due to the size of the room and the ventilation conditions. In D_S, the combustion products exhibit nearly uniform distribution due to static ventilation and the small room size. In contrast, the propagation behavior in the non-ventilated D_T is predominantly influenced by agglomeration and gravitational settling [29].

Secondly, the combustion undergoes variations over time as a consequence of the downscaling of the sample size in D_S. The sub-processes of the combustion process, including heating, release of pyrolysis gases, smoldering and glowing, take place at considerably shorter time intervals in D_S due to the small sample sizes.

We simulated various intensity levels that may occur in D_T by accumulating the combustion products from multiple experimental stages in the test chamber in D_S. Consequently, we excluded the temporal component from our data and focused on the absolute values of the sensor measurements in the transfer learning approach, as described in more detail in Section 3.3 following.

The resulting datasets, ds_dataset and dt_dataset, underwent a data pre-processing step to achieve balance by randomly down-sampling to the minority class in order to avoid implicit class weights. After data balancing, the ds_dataset (training dataset) contained 770 datapoints per class and the dt_dataset (validation and boost dataset) contained 432 datapoints per class and sensor node position.

3.3. Methodology

As proposed by Cook et al. [33], we applied both feature representation transfer and instance transfer in our study. The aim was to investigate the suitability of these two methods for classification in early fire detection considering the challenge of limited or no access to extensive data from large-scale experiments during model development. The overall workflow of data generation and processing is illustrated in Figure 4.

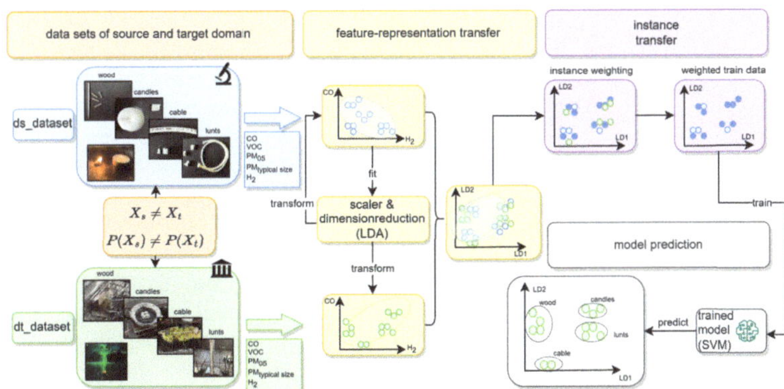

Figure 4. Methodology workflow; (1) data generation, (2) feature representation transfer using LDA, (3) instance transfer (weighting of training data) and (4) model building.

3.3.1. Feature Representation Transfer

Linear Discriminant Analysis (LDA) was employed as a supervised dimension reduction method in the feature representation transfer step. LDA aimed to extract crucial information (reduced features) that are most relevant for distinguishing between fire scenarios based on data from D_S. As outlined in Section 2.1, the original input features for the LDA comprised CO, H_2, VOC and PM (PM0.5 and PM1.0).

Both LDA and the scaler (min–max scaler with bounds [0, 1]) were applied to the data from D_S. The resulting transformation parameters were then utilized to transform the data in both D_S and D_T into the new feature space. Subsequently, the transformed data were employed to train a support vector machine (SVM) classifier using the transformed data from D_S, and its performance was validated at various sensor node positions in D_T.

3.3.2. Instance Transfer

In addition to feature representation transfer, we implemented instance transfer using the TrAdaBoost algorithm presented in [34]. TrAdaBoost is a supervised domain adaptation method that utilizes limited data from D_T to adjust a pre-trained model to new data: specifically, the target domain D_T in our case [34]. The fundamental concept of TrAdaBoost is to adapt the knowledge learned from D_S and apply it to a slightly different D_T, assuming that labeled data from D_T are generally rare.

By definition, this approach requires the availability of limited instances from the target domain, which are employed to re-weight the training instances from D_S.

In practical terms, the target domain data for TrAdaBoost could be sourced from an actual fire event occurring in D_T during the operation of the fire detection system or from a small number of large-scale experiments. Consequently, this method serves as a means to adapt the fundamental model trained on laboratory data to real-world application environments.

The objective of this study was to investigate how the performance of a classifier trained solely on laboratory data (ds_dataset) can be enhanced by incorporating small amounts of available data from D_T. To achieve this, we utilized from 1% up to 30% of the dt_dataset to re-weight the source domain instances using TrAdaBoost. Higher proportions of D_T instances were employed to identify overfitting boundaries during the instance transfer step.

4. Results

This section is structured as follows. First, Section 4.1 presents the performance of the boosted model independent of the sensor node position in D_T. This means that instance transfer (boosting) was executed using data from the same sensor node position in D_T as utilized for validation.

In Section 4.2, the boosting data were selected from a fixed sensor node position, and the model's performance was subsequently validated across all sensor node positions in D_T to identify potential overfitting effects based on the amount of boosting data taken from a specific sensor node position.

Beyond the performance assessments using various boosting strategies, the model was validated without any boosting; this served as the baseline for performance. This implies that only the feature representation transfer described in Section 3.3.1 was performed before applying the model to the D_T data. This baseline performance facilitates the assessment of performance improvement when employing additional boosting strategies.

The Manhattan distance between the sensor node and the fire source in D_T was employed to arrange different sensor node positions along the x-axis.

To enable performance comparisons across different models, the classification rate (average accuracy) was used as our primary performance metric. According to [37], the average accuracy for a multi-class classification problem is defined as shown in Equation (6).

$$\frac{\sum_{i=1}^{l} \frac{tp_i + tn_i}{tp_i + fn_i + fp_i + tn_i}}{l} \tag{6}$$

Since we considered a balanced dataset for model validation, the classification rate is a suitable performance measure [37].

To compare the performance of the baseline model (non-boosted, only trained on ds_dataset) with a model that randomly assigns labels based on the given class distribution, we utilized Cohen's κ as an additional performance metric, as suggested by Artstein et al. [38]. Cohen's κ is a scaled value in the range of $[-1, 1]$ that evaluates the model's classification accuracy against the accuracy achieved by random label assignment according to a specified class distribution [38].

4.1. Classification Performance Independent of the Node Position

Table 4 shows the results of the baseline model only trained on the ds_dataset in terms of precision, recall, F1 score, classification rate and Cohen's κ.

Table 4. Precision, recall, F1 score, classification rate and Cohen's κ for non-boosted model based on test sensor node position.

Test Sensor Node	Manhattan Distance [m]	Boost	Precision	Recall	F1 Score	Classific. Rate (Accuracy)	Cohen's κ
08	3.0	no boost	0.74	0.68	0.69	0.69	0.58
09	4.3	no boost	0.69	0.64	0.65	0.66	0.53
10	7.8	no boost	0.69	0.64	0.65	0.66	0.54
11	6.5	no boost	0.67	0.56	0.58	0.58	0.48
12	10.5	no boost	0.68	0.60	0.62	0.61	0.47
13	8.5	no boost	0.60	0.51	0.52	0.53	0.36
14	7.5	no boost	0.66	0.56	0.58	0.58	0.42
15	7.8	no boost	0.71	0.64	0.66	0.66	0.53
16	6.5	no boost	0.69	0.63	0.64	0.64	0.51

The baseline model exhibits its lowest performance at sensor node positions close to the wall—specifically, at sensor node 13 (global minimum, classification rate of 53%) and sensor node 14 (local minimum, classification rate of 58%)—in the dt_dataset, as shown in Table 4. This implies that the most significant difference between our laboratory setup (D_S) and D_T occurs at positions close to the wall in D_T.

The Cohen's κ ranges from 0.36 (minimum at test sensor node 13) up to 0.58 (maximum at test sensor node position 08). According to Landis et al. [39], this can be categorized as "fair" ($0.2 \leq \kappa \leq 0.4$) to "moderate" ($0.4 \leq \kappa \leq 0.6$) model performance.

To adapt the model derived from D_S, additional model boosting was performed. Initially, boosting was performed assuming knowledge about the distance between the sensor node and the fire source in D_T.

Figure 5 shows the classification rate as a function of the sensor node position used for boosting and testing. The different lines represent the amount of data used for boosting (1% to 30%) from the test position in D_T. The "no_boost" line represents the classification performance of the baseline model.

It can be seen from Figure 5 that the classification rate of the non-boosted baseline model ranges from a global minimum (classification rate of 53% at sensor node position 13) up to a global maximum at sensor node position 08 (classification rate of 69%; see also Table 4). There is a continuous decrease in the classification rate from the lowest Manhattan distance of 3.0 m (sensor node position 08) to a Manhattan distance of 7.5 m (sensor node position 14). The classification rate then reaches a local minimum of 58% at sensor node position 14. A local maximum with a 66% classification rate can be observed at sensor node positions 10 and 15. Moving to the next-higher Manhattan distance (sensor node position 13), the classification rate reaches a global minimum of 53%. Sensor node position 12 (highest Manhattan distance to the source) shows a classification rate of 61%, which is 3% more than the local minimum at sensor node position 14.

Looking at the different boosting curves (0.01 to 0.3) in Figure 5, it is evident that additional information from D_T used for model boosting cannot completely offset the local minima in the classification rate at sensor node positions 13 and 14 close to the wall. The previously described trend in the classification rate remains essentially the same. However, differences in the classification rate between different sensor node positions (except for sensor nodes 13 and 14 close to the wall) can be mitigated by using additional boosting, particularly for boost amounts up to 5%. Although higher boost amounts (from 10% to 30%) lead to a global maximum of the classification rate (sensor node position 09, 20% boosting data), the differences in the classification rates between different sensor node positions increase compared to boost amounts of around 5%.

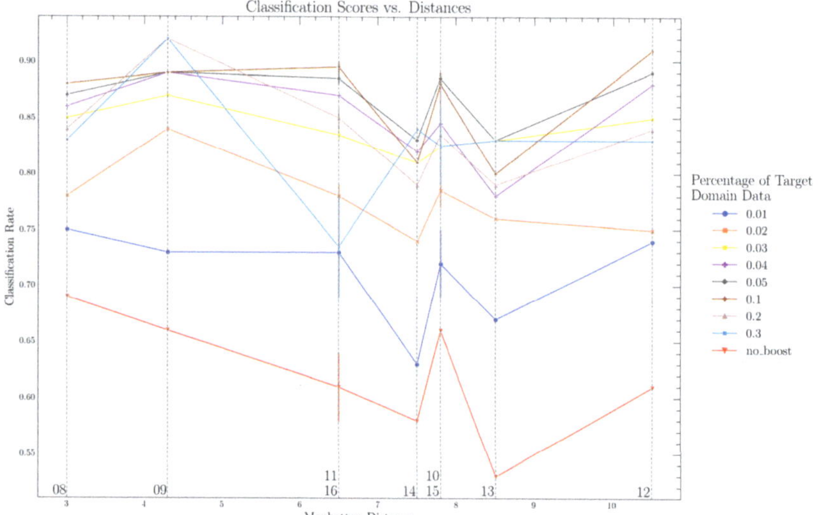

Figure 5. Classification rate using instance weighting (boosting) based on test position.

Nevertheless, the differences between the sensor node positions (except for the positions close to the wall) are increasingly compensated for by additional boosting. This implies that the model trained only based on D_S can be adapted to a new environment with small amounts of available data from D_T.

Figure 6 following illustrates the model's performance for sensor node position 8 in D_T for the non-boosted model (left) and the boosted model (boosted with 5% of the data from sensor node 8).

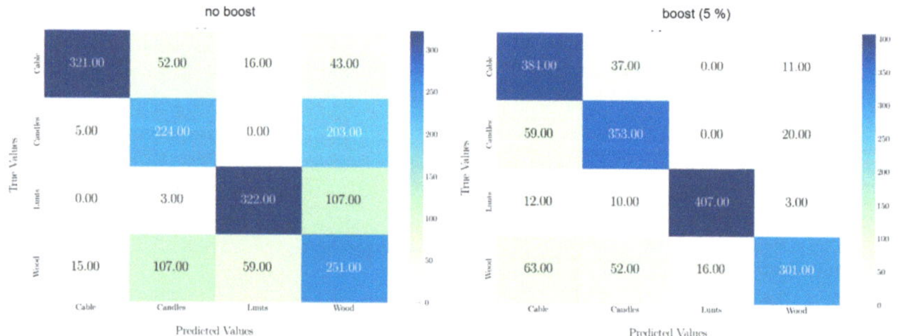

Figure 6. Comparison of confusion matrices for non-boosted case (**left**) and boosted case (**right**) for sensor node position 8 in D_T.

It can be seen from Figure 6 that the baseline model only trained on data from D_S primarily misclassifies between the candle scenario and the wood scenario. This misclassification can be attributed to the experimental procedure. In D_S, we employed small pieces of cellular cloth soaked with candle wax to represent the candle wax fire in a small-scale test. However, when the wax was fully burned, the cellular cloth (wick) started to glow and smolder at the end of each experiment. Since this combustion process closely resembles the glowing process of wood, it likely led lead to misclassification between the wood and candle fires.

Comparing the wick size to the mass of wax in D_S and D_T, the ratio is considerably higher in D_S than in D_T. As discussed in Section 3.2, scaling down a wax fire is challenging. To regulate the flame size of the wax fire, we had to use a much higher ratio of wick volume to wax volume. The volume of the wick compared to the volume of the burning wax was negligible in D_T. Consequently, fewer smoldering or glowing artifacts were observed in the dt_dataset than in the ds_dataset, resulting in the aforementioned misclassification.

This misclassification was evident at other test sensor node positions in D_T. Figure 6 (right) illustrates that the misclassification can be minimized by employing additional boosting.

Table 5 provides an overview of the average model performance across all sensor node positions in D_T, represented by the mean classification rate and the mean Cohen's κ for different boosting scenarios (ranging from no boosting to 30% boosting data).

Table 5. Mean classification rate and mean Cohen's κ for different boosting strategies (dynamic boost).

Boost	Mean Classification Rate	Mean Cohen's κ
no_boost	0.62	0.49
0.01	0.71	0.61
0.02	0.78	0.70
0.03	0.84	0.78
0.04	0.85	0.80
0.05	**0.87**	**0.83**
0.10	**0.87**	**0.83**
0.20	0.84	0.78
0.30	0.82	0.75

bold: maximum mean classification rate and Cohen's κ for dynamic boosting.

It can be seen from Table 5 that the mean model performance (mean classification rate and mean Cohen's κ) generally improves with model boosting. The model performance increases with an increasing amount of boosting data and reaches its maximum (87% mean classification rate and a mean Cohen's κ of 0.83) at 5% (up to 10%) of boosting data. The Cohen's κ ranges from 0.49 ("moderate") up to 0.83 ("perfect") according to Landis et al. [39].

As the amount of boosting data increases, the average model performance decreases, although it remains higher than the model performance without boosting. This phenomenon has already been discussed based on Figure 5. Even though higher amounts of boosting data lead to global maxima for the classification rate, the difference in the classification rate increases, causing the mean classification rate to decrease.

In summary, we found higher classification rates with boosting compared to the no-boost baseline model. The sensor node positions close to the wall show a local minimum of the classification rate regardless of the model used (no boost vs. different amounts of boosting data).

4.2. Classification Performance Dependent of the Sensor Node Position

The results presented in Section 4.1 represent an optimal boosting scenario with respect to the distance between the sensor node and the fire source. The data used for boosting were derived from the same sensor node position used for testing without utilizing the validation data already employed for boosting from the dt_dataset. However, in a real-world application, the distance between the sensor node and the fire source will be unknown. To investigate this scenario, we utilized boosting data from one fixed sensor node position and evaluated the model performance across all sensor node positions. The results are shown in Figure 7.

The red line in Figure 7 represents the baseline model performance without model boosting. The sub-figures are labeled based on the sensor node position used for boosting. We utilized the same amount of boosting data as in Section 4.1 (1% to 30%).

We observed that the global maximum of the classification rate was reached when the test sensor node position and the sensor node position used for boosting were the same (e.g., see sub-figure "sensornode0009" at the Manhattan distance of sensor node position 09 in Figure 7). However, it can be seen from Figure 7 that the model performance at test sensor node positions different from the boosting sensor node achieves higher classification rates compared to the baseline model (no boosting). This holds true for boosting data amounts up to 5%, while higher amounts of boosting data from a particular sensor node position lead to overfitting to the boosting sensor node position. This effect is visible in Figure 7 when the classification rate of the boosted model falls below the baseline classification rate.

Another observation from Figure 7 is that there is still a local minimum in the classification rate at sensor node positions 13 and 14 (positions close to the wall). However the difference between D_S and D_T can be compensated for (see sub-figures "sensornode0013" and "sensornode0014" in Figure 7) if data from these sensor node positions are used for model boosting.

Table 6 shows the mean classification rates and the mean Cohen's κ values over all sensor node positions used for testing and boosting as a function of the amount of boosting data (0–30%).

Table 6. Mean classification rate and Cohen's κ for different boosting strategies (static boost).

Boost	Classification Rate	Cohen's κ
no_boost	0.62	0.49
0.01	0.67	0.56
0.02	0.69	0.58
0.03	0.71	0.61
0.04	0.72	0.62
0.05	**0.73**	**0.63**
0.10	0.71	0.61
0.20	0.67	0.56
0.30	0.64	0.51

bold: maximum mean classification rate and Cohen's κ for static boosting.

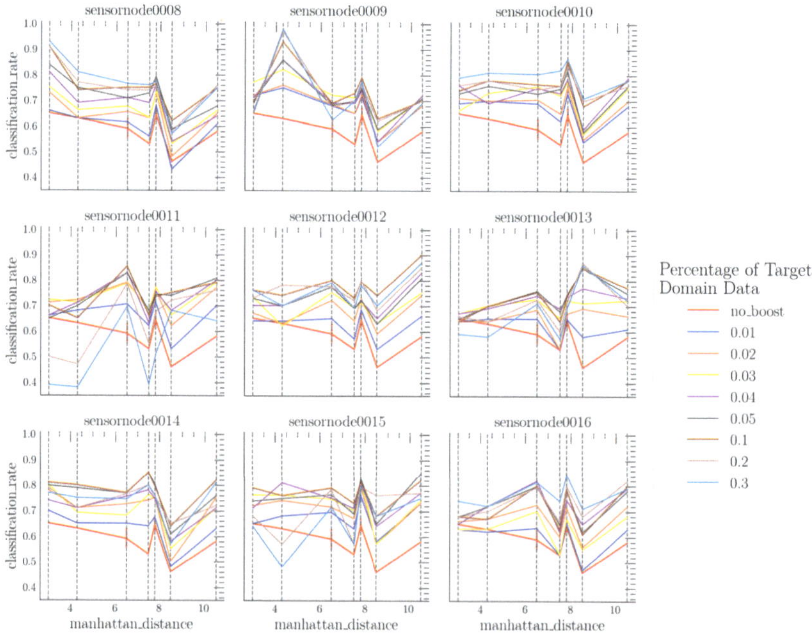

Figure 7. Classification rate using instance weighting (boosting) and random sensor node positions.

It is essential to emphasize that the mean performance measure represents the static boost scenario (boosting data were taken from only one sensor node position in D_T, and the model was then tested on all sensor node positions in D_T).

Table 6 indicates that the mean classification rate, as well as the mean Cohen's κ, is significantly higher when using additional boosting compared to the cases without any boosting (no_boost). Furthermore, it can be observed that the performance increases with the amount of boosting data used, up to the maximum performance at 5%.

At higher amounts of boosting data, the average performance decreases again due to increased overfitting to individual sensor node positions. At 30% boosting, the average performance in terms of mean classification rate and mean Cohen's κ is comparable to the average performance without boosting.

5. Discussion

As highlighted by Burgués et al. [40], a common challenge in machine-learning-based prediction lies in the limitations of examples available in the training data.

In this study, we considered four different incipient fire scenarios that have been identified as the main initial fire sources in historic and cultural buildings in Germany [41]. However, the model's predictive accuracy may be compromised in the presence of different or additional burning materials (or superpositions of different materials) that have not been accounted for in this study.

Nevertheless, our research demonstrates the feasibility of classifying various incipient fire scenarios using multi-sensor training data from a small-scale setup. This opens up the possibility of generating cost-effective and extensive data for other burning materials that encompass different combustion conditions and/or superpositions with different nuisance scenarios (such as deodorant, dust, etc.).

Burgués et al. [40] also highlighted the model's limitation to a specific range of tested (in their case, odor) concentrations. In our study, we focused on early phases of incipient fires, which are primarily characterized by the combustion process and the masses of burning material relative to the room volume. From our current results, we cannot extrapolate the model performance to more advanced stages of the conducted fire scenarios. Different combustion conditions result in the distinct release of combustion products over time. However, the experimental setup presented for D_S enables the generation of data for these diverse combustion conditions, including those of more advanced courses of various fire scenarios.

Another consideration is that we did not include test positions where the sensor node is positioned very close to the fire source in D_T. This might lead to significantly different sensor signals due to sensor override. In such cases, deterioration in model performance would be expected.

We found that the baseline model trained only on D_S data tends to misclassify between the candle and the wood fire scenarios. As discussed in Section 4.1, this misclassification can be attributed to the experimental setup used for the candle fire in D_S. In further experiments, it would be advisable to alter the wick material to a non-combustible substance to minimize glow and smolder effects. Alternatively, stopping the experiment before complete wax combustion could prevent glow and smolder artifacts in the data. In the small-scale setup (D_S), we used a fan to transport combustion products from the combustion chamber into the test chamber where the sensor nodes were located. The uncertainty regarding when combustion products from the smoldering wick entered the test chamber makes it challenging to remove these artifacts from the ds_dataset afterward.

When comparing the classification performance of our study with previous research, it is noteworthy that our non-boosted model already achieves comparable results (classification rate up to 69%) compared to studies such as Solórzano et al. [20] (52% to 70%). With additional boosting, the model performance can be further increased up to 87%: yielding results comparable to [20] with additional laboratory data (88%) or only slightly lower performance than in [23] (90%). It is essential to recognize that, unlike previous research, we addressed a four-class classification problem and employed two distinct experimental settings to generate the test and training data, thereby limiting direct comparisons.

In comparison with a similar classification problem ([24]), our boosted model achieves an average classification rate that surpasses Ni et al.'s [24] result (82% classification rate) by 5%. It is important to note that Ni et al. utilized a single experimental setup to generate the training and test data.

Another limitation to comparing our model's performance with previous studies is our consideration of performance across various sensor node positions. It is evident that positions with lower classification rates will adversely affect the average model performance. The previous literature did not account for position-dependent performance measures, which hold great relevance in practical applications. Hence, it can be assumed that the performance comparison of our model with the previous work leans towards the conservative side.

Drawing from the outcomes presented in this study, we posit that the introduced approach, which combines transfer learning methods with multi-sensor data, is promising and highly relevant for the practical application of data-driven models relying on multi-sensor data. For instance, cost-effective generation of data for various fire materials or combinations can be accomplished on a small laboratory scale, including overlays with nuisance variables, to facilitate the early detection of fires in real room environments.

The demonstrated approach can be expanded to diverse application domains. For instance, investigating outdoor applications such as forest fire detection or air monitoring in industrial plants is a plausible direction for future investigations. However, outdoor environments exhibit distinct ventilation conditions, characterized by the formation of plumes, and a reduced tendency for the accumulation of combustion products. In scenarios like forest fire detection, combustion products tend to accumulate beneath the canopy or due to atmospheric inversion, leading to substantial influence of environmental conditions on the propagation behavior of combustion products.

The misclassification between candles and wood highlights that similar combustion processes lead to lower classification performance, particularly in the early detection phase. The classification rate of the non-boosted model (53% to 69%) indicates potential uncertainty in the classification, which should not be underestimated, especially during the initial stages of incipient fires. We presume that the classification rate might improve with more advanced fires that give clearer sensor signals. However, in an application scenario, an anomaly detector would be connected before the classifier to act as a trigger. One approach could involve using the time interval between the current classification and the triggering of the anomaly detector as a measure of the expected information quality of the classifier.

It is essential to acknowledge that, despite the approach presented in this work, the individual propagation behavior in the application room significantly influences the model's performance. Notably, the model performance experienced a significant reduction at sensor node positions close to the wall in our study. A classification rate of 53% (sensor node position 13, non-boosted model) is relatively low even in a four-class classification problem and may result in misjudgment of the situation in a real application scenario. Since distance effects have not been considered in the previous literature when calculating performance measures, there is a pressing need for further research in this area. In general, a model can only recognize scenarios reliably if the sensor generates reliable input data. Future work should pay more attention to limitations associated with sensor positioning and incorporate these limitations into the evaluation process.

6. Conclusions and Outlook

This paper presents the results of employing two transfer learning methodologies—namely, feature representation transfer and instance transfer—within the context of early fire detection through multi-sensor nodes. The primary objective (RQ) of this study was to investigate whether multi-sensor data from a small-scale setup (D_S) can be used to classify various incipient fires in their early stages within an authentic room setting without the need to generate time- and cost-intensive data in large-scale setups.

In conclusion, we successfully generated multi-sensor data for four distinct types of incipient fires in a time- and cost-efficient manner within the small-scale experimental setup (D_S) outlined in this study. The D_S data facilitated the extraction of crucial information to differentiate between various types of incipient fires. Based on this new feature space, a state-of-the-art classifier (SVM) was trained to classify unseen data from a large-scale setup.

We observed that the baseline model, trained exclusively on the D_S data, consistently demonstrated the ability to classify four different incipient fire scenarios within D_T: achieving a classification rate of up to 69% and a Cohen's κ of 0.58. However, the model's performance is notably influenced by the distance between the sensor node and the fire source. In particular, we found that sensor node positions close to the wall exhibited lower classification performance (minimum classification rate of 53% and minimum Cohen's κ of 0.36).

We identified that the decrease in performance primarily resulted from misclassification between the candle and wood scenarios. This misclassification was attributed to the experimental setup of the candle (wax) fire in D_S. In further investigations, we recommend optimizing the experimental setup to prevent the ds_dataset from acquiring glowing or smoldering artifacts. Based on our findings, we anticipate that such optimization will indeed enhance the performance of the baseline model.

Another finding of this study is that the model's performance can be enhanced through additional model boosting (instance transfer), which is applicable when there is access to (small) amounts of real room data. However, it is crucial to keep the amount of boosting data low to avoid overfitting the model to a particular room situation or sensor node position. In our study, we determined the optimal amount of boosting data to be approximately 5% of the training instances in D_T.

In further research, we aim to extend the ds_dataset to include a broader range of combustible materials. Additionally, we plan to investigate superpositions of different combustible materials in D_S and D_T. This is crucial to investigate, as real-world combustible objects often consist of mixtures of various materials. Another noteworthy aspect is the examination of superposition of nuisance scenarios with different fire scenarios, which will enhance the model's robustness against side effects such as dust, humidity changes, etc.

To ensure a wider range of applications, future research should involve generating test data in diverse full-scale environments. This could encompass test rooms with varying geometries beyond the standard fire test room. Additionally, conducting full-scale outdoor tests would be valuable for extending the application of this concept to areas such as wildland fire detection or industrial facilities.

Another aspect to consider is that data processing, including the classification model presented in this study, is currently executed using the resources of the server (Raspberry Pi). In future research, we aim to explore the feasibility of conducting data processing directly on the ESP32. This would enhance the autonomy of the multi-sensor node, potentially reducing the notification time.

Author Contributions: Conceptualization, P.V. and J.K.; methodology, P.V.; software, P.V.; validation, P.V.; formal analysis, P.V.; investigation, P.V. and J.K.; resources, P.V., J.K. and S.M.; data curation, P.V.; writing—original draft preparation, P.V.; writing—review and editing, J.K. and S.M.; visualization, P.V.; supervision, U.K.; project administration, U.K.; funding acquisition, U.K. All authors have read and agreed to the published version of the manuscript.

Funding: This research was funded by the German Federal Ministry of Education and Research as part of the "Research for Civil Security" program (funding codes: 13N15415 to 13N15420 and 13N15565).

Institutional Review Board Statement: Not applicable.

Informed Consent Statement: Not applicable.

Data Availability Statement: The used datasets can be provided on request.

Conflicts of Interest: Authors Jörg Kelleter and Seffen Müller were employed by the company Industrieelektronik GmbH. The remaining authors declare that the research was conducted in the absence of any commercial or financial relationships that could be construed as a potential conflict of interest.

References

1. McAvoy, T.J.; Milke, J.; Kunt, T.A. Using multivariate statistical methods to detect fires. *Fire Technol.* **1996**, *32*, 6–24. [CrossRef]
2. Chen, S.J.; Hovde, D.C.; Peterson, K.A.; Marshall, A.W. Fire detection using smoke and gas sensors. *Fire Saf. J.* **2007**, *42*, 507–515. [CrossRef]
3. Fonollosa, J.; Solórzano, A.; Marco, S. Chemical Sensor Systems and Associated Algorithms for Fire Detection: A Review. *Sensors* **2018**, *18*, 553. [CrossRef] [PubMed]
4. Rachman, F.Z.; Hendrantoro, G.; Wirawan. A Fire Detection System Using Multi-Sensor Networks Based on Fuzzy Logic in Indoor Scenarios. In Proceedings of the 2020 8th International Conference on Information and Communication Technology (ICoICT), Yogyakarta, Indonesia, 24–26 June 2020; pp. 1–6. [CrossRef]
5. Wu, L.; Chen, L.; Hao, X. Multi-Sensor Data Fusion Algorithm for Indoor Fire Early Warning Based on BP Neural Network. *Information* **2021**, *12*, 59. [CrossRef]
6. Liang, Y.H.; Tian, W.M. Multi-sensor Fusion Approach for Fire Alarm Using BP Neural Network. In Proceedings of the 2016 International Conference on Intelligent Networking and Collaborative Systems (INCoS), Ostrava, Czech Republic, 7–9 September 2016; pp. 99–102. [CrossRef]
7. Nakıp, M.; Güzeliş, C. Multi-Sensor Fire Detector based on Trend Predictive Neural Network. In Proceedings of the 2019 11th International Conference on Electrical and Electronics Engineering (ELECO), Bursa, Turkey, 28–30 November 2019; pp. 600–604. [CrossRef]
8. Jana, S.; Shome, S.K. Hybrid Ensemble Based Machine Learning for Smart Building Fire Detection Using Multi Modal Sensor Data. *Fire Technol.* **2023**, *59*, 473–496. [CrossRef]
9. Yu, M.; Yuan, H.; Li, K.; Wang, J. Research on multi-detector real-time fire alarm technology based on signal similarity. *Fire Saf. J.* **2023**, *136*, 103724. [CrossRef]
10. Gottuk, D.T.; Peatross, M.J.; Roby, R.J.; Beyler, C.L. Advanced fire detection using multi-signature alarm algorithms. *Fire Saf. J.* **2002**, *37*, 381–394. [CrossRef]
11. Nakip, M.; Güzeliş, C.; Yildiz, O. Recurrent Trend Predictive Neural Network for Multi-Sensor Fire Detection. *IEEE Access* **2021**, *9*, 84204–84216. [CrossRef]
12. Milke, J.A.; Hulcher, M.E.; Worrell, C.L.; Gottuk, D.T.; Williams, F.W. Investigation of Multi-Sensor Algorithms for Fire Detection. *Fire Technol.* **2003**, *39*, 363–382. [CrossRef]
13. Conrad, T.; Reimann, P.; Schutze, A. A hierarchical strategy for under-ground early fire detection based on a T-cycled semiconductor gas sensor. In Proceedings of the 2007 IEEE SENSORS, Atlanta, GA, USA, 28–31 October 2007; pp. 1221–1224, ISSN 1930-0395. [CrossRef]
14. von der Linde, M.; Herbster, C.; Dobel, C.; Festag, S.; Thielsch, M.T. Creating safe environments: Optimal acoustic alarming of laypeople in fire prevention. *Ergonomics* **2023**, *66*, 2193–2211. [CrossRef]
15. Gutmacher, D.; Hoefer, U.; Wöllenstein, J. Gas sensor technologies for fire detection. *Sens. Actuators A Chem.* **2012**, *175*, 40–45. [CrossRef]
16. Scorsone, E.; Pisanelli, A.M.; Persaud, K.C. Development of an electronic nose for fire detection. *Sens. Actuators B Chem.* **2006**, *116*, 55–61. [CrossRef]
17. Fujinaka, T.; Yoshioka, M.; Omatu, S.; Kosaka, T. Intelligent Electronic Nose Systems for Fire Detection Systems Based on Neural Networks. In Proceedings of the 2008 The Second International Conference on Advanced Engineering Computing and Applications in Sciences, Valencia, Spain, 29 September–4 October 2008; pp. 73–76. [CrossRef]
18. Joseph, P.; Bakirtzis, D.; Vieille, A. An "electronic nose" as a potential device for fire detection of forest product fire loads in enclosures. *Wood Mater. Sci. Eng.* **2015**, *10*, 130–144. [CrossRef]
19. Andrew, A.M.; Shakaff, A.; Zakaria, A.; Gunasagaran, R.; Kanagaraj, E.; Saad, S.M. Early Stage Fire Source Classification in Building using Artificial Intelligence. In Proceedings of the 2018 IEEE Conference on Systems, Process and Control (ICSPC), Melaka, Malaysia, 14–15 December 2018; pp. 165–169. [CrossRef]
20. Solórzano, A.; Eichmann, J.; Fernández, L.; Ziems, B.; Jiménez-Soto, J.M.; Marco, S.; Fonollosa, J. Early fire detection based on gas sensor arrays: Multivariate calibration and validation. *Sens. Actuators B Chem.* **2022**, *352*, 130961. [CrossRef]
21. Solórzano, A.; Fonollosa, J.; Marco, S. Improving Calibration of Chemical Gas Sensors for Fire Detection Using Small Scale Setups. *Proceedings* **2017**, *1*, 453. [CrossRef]
22. Kim, Y.J.; Kim, H.; Lee, S.; Kim, W.T. Trustworthy Building Fire Detection Framework With Simulation-Based Learning. *IEEE Access* **2021**, *9*, 55777–55789. [CrossRef]
23. Milke, J.A. Monitoring Multiple Aspects of Fire Signatures for Discriminating Fire Detection. *Fire Technol.* **1999**, *35*, 195–209. [CrossRef]
24. Ni, M.; Stetter, J.R.; Buttner, W.J. Orthogonal gas sensor arrays with intelligent algorithms for early warning of electrical fires. *Sens. Actuators B Chem.* **2008**, *130*, 889–899. [CrossRef]

25. Nazir, A.; Mosleh, H.; Takruri, M.; Jallad, A.H.; Alhebsi, H. Early Fire Detection: A New Indoor Laboratory Dataset and Data Distribution Analysis. *Fire* **2022**, *5*, 11. [CrossRef]
26. Krüger, S.; Despinasse, M.C.; Raspe, T.; Nörthemann, K.; Moritz, W. Early fire detection: Are hydrogen sensors able to detect pyrolysis of house hold materials? *Fire Saf. J.* **2017**, *91*, 1059–1067. [CrossRef]
27. Hayashi, Y.; Akimoto, Y.; Hiramatsu, N.; Masunishi, K.; Saito, T.; Yamazaki, H.; Nakamura, N.; Kojima, A. Smoldering Fire Detection Using Low-Power Capacitive MEMS Hydrogen Sensor for Future Fire Alarm. In Proceedings of the 2021 21st International Conference on Solid-State Sensors, Actuators and Microsystems (Transducers), Orlando, FL, USA, 20–24 June 2021; pp. 267–270, ISSN 2167-0021. [CrossRef]
28. Gutmacher, D.; Foelmli, C.; Vollenweider, W.; Hoefer, U.; Wöllenstein, J. Comparison of gas sensor technologies for fire gas detection. *Procedia Eng.* **2011**, *25*, 1121–1124. [CrossRef]
29. Vorwerk, P.; Kelleter, J.; Müller, S.; Krause, U. Distance-Based Analysis of Early Fire Indicators on a New Indoor Laboratory Dataset with Distributed Multi-Sensor Nodes. *Fire* **2023**, *6*, 323. [CrossRef]
30. Weiss, K.; Khoshgoftaar, T.M.; Wang, D. A survey of transfer learning. *J. Big Data* **2016**, *3*, 9. [CrossRef]
31. Zhuang, F.; Qi, Z.; Duan, K.; Xi, D.; Zhu, Y.; Zhu, H.; Xiong, H.; He, Q. A Comprehensive Survey on Transfer Learning. *Proc. IEEE* **2021**, *109*, 43–76. [CrossRef]
32. Kim, H.E.; Cosa-Linan, A.; Santhanam, N.; Jannesari, M.; Maros, M.E.; Ganslandt, T. Transfer learning for medical image classification: A literature review. *BMC Med. Imaging* **2022**, *22*, 69. [CrossRef] [PubMed]
33. Cook, D.; Feuz, K.D.; Krishnan, N.C. Transfer learning for activity recognition: A survey. *Knowl. Inf. Syst.* **2013**, *36*, 537–556. [CrossRef] [PubMed]
34. Dai, W.; Yang, Q.; Xue, G.R.; Yu, Y. Boosting for transfer learning. In Proceedings of the 24th International Conference on Machine Learning, ICML '07, New York, NY, USA, 20–24 June 2007; pp. 193–200. [CrossRef]
35. Carbon-Monoxide-Gas-Sensor_Datasheet. 2023. Available online: https://www.membrapor.ch/sheet/Carbon-Monoxide-Gas-Sensor-CO-MF-1000.pdf (accessed on 22 May 2023).
36. DataSheet-GGS-6530-T_Rev2203. 2023. pp. 1–2. Available online: https://www.umweltsensortechnik.de/fileadmin/assets/downloads/gassensoren/single/DataSheet-GGS-6530-T_Rev2203.pdf (accessed on 12 June 2023).
37. Sokolova, M.; Lapalme, G. A systematic analysis of performance measures for classification tasks. *Inf. Process. Manag.* **2009**, *45*, 427–437. [CrossRef]
38. Artstein, R.; Poesio, M. Inter-Coder Agreement for Computational Linguistics. *Comput. Linguist.* **2008**, *34*, 555–596. [CrossRef]
39. Landis, J.R.; Koch, G.G. The Measurement of Observer Agreement for Categorical Data. *Biometrics* **1977**, *33*, 159–174. [CrossRef] [PubMed]
40. Burgués, J.; Doñate, S.; Esclapez, M.D.; Saúco, L.; Marco, S. Characterization of odour emissions in a wastewater treatment plant using a drone-based chemical sensor system. *Sci. Total Environ.* **2022**, *846*, 157290. [CrossRef]
41. Kabat, S. *Brandschutz in Kirchen und Klöstern*; Springer Fachmedien Wiesbaden: Wiesbaden, Germany, 2021. [CrossRef]

Article

Quantification of UV Light-Induced Spectral Response Degradation of CMOS-Based Photodetectors

Pablo F. Siles * and Daniel Gäbler

X-FAB Global Services GmbH, 99097 Erfurt, Germany; daniel.gaebler@xfab.com
* Correspondence: pablo.siles@xfab.com

Abstract: High-energy radiation is known to potentially impact the optical performance of silicon-based sensors adversely. Nevertheless, a proper characterization and quantification of possible spectral response degradation effects due to UV stress is technically challenging. On one hand, typical illumination methods via UV lamps provide a poorly defined energy spectrum. On the other hand, a standardized measurement methodology is also missing. This work provides an approach where well-defined energy spectrum UV stress conditions are guaranteed via a customized optical set up, including a laser driven light source, a monochromator, and a non-solarizing optical fiber. The test methodology proposed here allows performing a controlled UV stress between 200 nm and 400 nm with well-defined energy conditions and offers a quantitative overview of the impact on the optical performance in CMOS-based photodiodes, along a wavelength range from 200 to 1100 nm and 1 nm step. This is of great importance for the characterization and development of new sensors with a high and stable UV spectral response, as well as for implementation of practical applications such as UV light sensing and UV-based sterilization.

Keywords: UV sensing; photodiodes; UV degradation; spectral response; UVC; sterilization

Citation: Siles, P.F.; Gäbler, D. Quantification of UV Light-Induced Spectral Response Degradation of CMOS-Based Photodetectors. *Sensors* **2024**, *24*, 1535. https://doi.org/10.3390/s24051535

Academic Editor: Clement Yuen

Received: 31 January 2024
Revised: 23 February 2024
Accepted: 26 February 2024
Published: 27 February 2024

1. Introduction

A wide number of the CMOS-based photodetectors in use today have no (or very poor) sensitivity in the UV spectral range, mainly because the backend layers are absorbing the UV light. The junction design is also an important factor, but more likely for degradation effects. Nevertheless, UV light sensing has become in recent years a topic of increasing interest, due to the surge of a plethora of new technological applications, such as sterilization, UV spectroscopy, biological analysis, space imaging, UV-based cure processes, and more [1–8]. Although, there are still a few major challenges in the field of UV light sensing. One of these challenges is regarding the typical low sensitivity to UV light due to the short penetration depth in Si. This means that most of the photo-generated carriers within the upper atomic layers in Si cannot be detected because of the strong recombination through the interface states. Several approaches dealing, for example, with the photodiode dopant profile [9,10] or the engineering of photodetectors optimizing design and manufacturing processes have made it possible to reach not only high UV spectral response [11,12] but also to address the poor stability or spectral response degradation due to exposure to UV light conditions. Achieving such optical robustness under high-energy UV-light illumination conditions is certainly the second major challenge in UV light sensing. Such degradation mechanisms may be explained due to trap generation [13] or changes in the fixed charges and the interface states at the Si/SiO_2 interface above the photodetector, originating from the very high photon energy (6.2–4.1 eV for wavelengths between 200 and 300 nm). There exists also a plethora of approaches for the development of UV photodetectors, which include the usage of compound semiconductors with a wide bandgap such as SiC or ZnO [14,15], usage of Silicon-in-Insulator (SOI) structures with a shallow surface detection layer [8], the proper tunning of doping levels to reach high concentration surface layers near the Si surface [16], back side illumination structure approaches, or other thin-film or nanostructure-based approaches [17]. Our more recent research on the development of new UV photodiodes, using X-FAB's Semiconductors Foundries XS018 technology, shows a significant spectral response improvement especially for wavelengths around 260 nm, which is relevant for applications in sterilization. It also shows a remarkable robustness (<5% degradation) under UV light stress conditions.

The UV instability of Si-based photodetectors is certainly a very complex phenomenon, which may depend on several parameters such as irradiance, duration of UV light exposure, radiant exposure, wavelength of the UV radiation, type of photodetector, etc. [18]. Therefore, the robustness of photodetectors must be investigated systematically, with a reliable measurement methodology which covers a wide range of experimental parameters. It is important to precisely quantify the spectral response degradation due to UV light exposure, at any stress wavelength of interest. In addition, it is also very important that the impact on the optical performance of photodetectors is determined not only at the wavelength of exposure but also along the entire spectral range where such photodetectors are expected to be used. It then becomes necessary to count with a reliable, fast, and standardized measurement methodology which allows us to perform such systematic characterizations. Due to the lack of a common standard, the present work intends to propose a systematic measurement methodology for the investigation and quantification of spectral response degradation due to UV light exposure in CMOS-based photodetectors. It is also to determine the overall impact on the optical performance of such photodetectors along the entire spectral range of operation, while maintaining always well-defined energy conditions for the UV exposure.

2. Materials and Methods

All the photodetectors characterized and presented in this work were fabricated with X-FAB's Semiconductors Foundries XS018 technology [19]. Data are shown for photodiodes belonging to three different main modules: module A, which offers a good spectral sensitivity over a broad wavelength range; module B, which provides specially outstanding spectral response in the human eye response range, with its maximum point of sensitivity close to the green region; and module C, which is specially dedicated to performing with a high and stable spectral response in the UV range. Table 1 briefly summarizes the different XS018-based photodetectors for which experimental UV stress and spectral response degradation investigations are shown in this work.

Table 1. X-FAB's XS018 technology-based photodetectors investigated in this work.

Photodetector	Basic Structure	Sensitivity Application
doa	Module A, DNWell/p-Sub	Wide spectral range
dob	Module A, DNWell (pinched to PWell_1)/p-Sub	Red and NIR spectral range
doe	Module A, NWell/p-Sub	Wide spectral range
doeher	Module B, NWell/p-Sub	Enhanced Visible and NIR
UVC [1]	Module C, PWell_2/DNWell/p-Sub	Enhanced for UV range

[1] This new photodetector is soon to be released as dosuv.

Figure 1 shows a conceptual diagram of most of the devices listed in Table 1. All photodetectors are fabricated following X-FAB's XS018 process flow technology. All devices can be fabricated scalable in size, as well as in arrays of photodiodes. Nevertheless, for the investigations performed here, all measurements are performed on single devices. For simplification, all front-end layers which comprise different metallization levels (M2, M3, etc.) and the passivation layer are not shown in the diagrams. Some of these metallization levels may be used as a light shield, for example, to avoid photons reaching other regions instead of the photodiode's pn-junction active region. All devices are built over a bulk p-type substrate, where the electrical connection is realized via a p+ implant and a metal contact at the M1 level. In the case of the doa device, the pn-junction proper depth is realized via an NWell plus a DNWell (deep NWell), which can be contacted via an n+ implant and a metal contact at the M1 level. The dob device conforms its pn-junction also via a DNWell. Although, in this case, the electrical connection is realized via a p+ implant, once the DNWell is pinched to a PWell_1. On the other hand, the doe and doeher devices realize the pn-junction only via an NWell, where electrical contact is possible thanks to an n+ implant and a metal contact at the M1 level. Different from the doe device, the doeher photodetector possesses a special layer at the back end with special optical properties which allows precise optimization of the optical performance, especially in the Visible and NIR spectral ranges. As described in Table 1, the UVC photodetector is realized by a special structure comprised by two pn-junctions, the PWell_2 to DNWell upper junction and the DNWell to p-Sub lower junction.

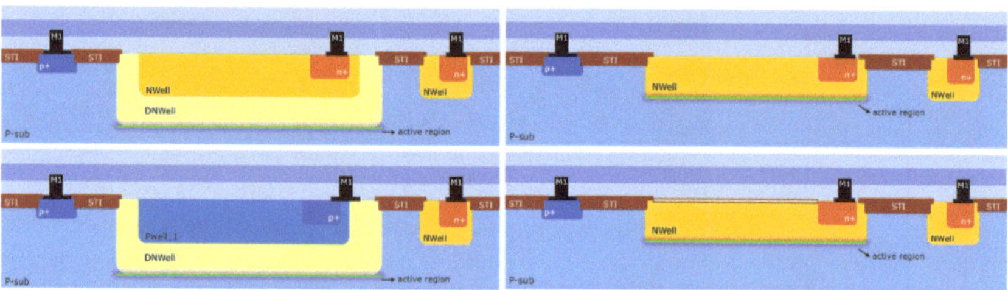

Figure 1. Conceptual cross-section diagram of the different CMOS-based photodiodes investigated in this work. In accordance with the naming shown in Table 1, top-left corresponds to doa, bottom-left corresponds to dob, top-right corresponds to doe, and bottom-right corresponds to doeher.

2.1. Test Setup

The photocurrent measurements were performed in situ, during illumination UV conditions, using a Keithley 4200A DC Parameter Analizer (Tektronix, Beaverton, OR, USA). All measurements have been performed at 27 °C (slightly elevated room temperature, to ensure precise temperature control). Wafers were placed on the chuck of a Summit 200-FA-AP probe station (Form Factor GmbH, Thiendorf, Germany). All photodetectors were operated with standard specification conditions, with $V_{Cathode} = 0.9$ V, $V_{Guard} = 3.3$ V and $V_{Anode} = 0.0$ V (substrate/bulk/chuck).

2.2. UV Stress Methodology

Conventional artificial UV light sources such as UV lamps were found to be inconvenient for UV degradation wafer-level investigations of photodetectors. On the one hand, the global illumination may affect other light-sensitive test structures and the close electrical contact via probe-wedge may induce undesired light reflections, when the illumination is performed in situ. On the other hand, such light sources provide a poorly defined energy spectrum, making it challenging to define the specific wavelength of illumination and consequently quantifying at which wavelength (and intensity) the UV stress is applied. Despite the points mentioned here, in the case of UV lamps, it is always possible to devise a way to normalize the illumination area and dose onto the photodetector under test. In fact, in most research investigations about UV degradation, often an intensity value per area is given, but without a complete spectrum. So, it remains unknown at which wavelength (and intensity) the UV stress is provided. Due to the typical broadband nature of a UV lamp as well as variations on the spectral output, a UV lamp does not provide stress at all wavelengths of the UV spectrum. Other light sources such as UV LEDs could also be used, but such LEDs provide non-uniform light, which brings other technical challenges to improve the illumination uniformity. Nevertheless, in this case, the illumination is also global, which may also induce undesired light reflections.

To perform the optical characterization of the photodetectors and to quantify the impact on their optical performance due to UV stress, the spectral response is measured before and after a specially devised UV stress step. Here, a laser-driven light source (Model EQ-99X LDLS, Energetiq (Wilmington, MA, USA), a monochromator (Hyperchromator, Mountain Photonics GmbH, Landsberg am Lech, Germany), and a non-solarizing optical fiber (Multimode Solarization Resistant Optical Fiber, 0.22 NA, Ø105 µm core, Thorlabs, Newton, NJ, USA) are used in a customized manner. A reference detector is also used before or after every wavelength sweep, as indicated in Figure 2. This allows for a precise estimation of the light power intensity brought via the optical fiber onto the photodetector. In addition, a second control detector continuously monitors the output at the light source to counter for possible fluctuations or variations due to aging and other factors. For every conventional wavelength sweep, a dark current correction is always performed. The monochromator receives the white light output of the laser-driven light source and splits the light into every wavelength between 200 nm and 1100 nm (with a smaller step possible of 1 nm), thanks to a special optical set up which comprises the usage of proper filters, mirrors, and gratings. This allows bringing, via the non-solarizing fiber, any desired wavelength only onto the photodetector of interest, avoiding the illumination (or stress) of other devices as well as undesired light reflections.

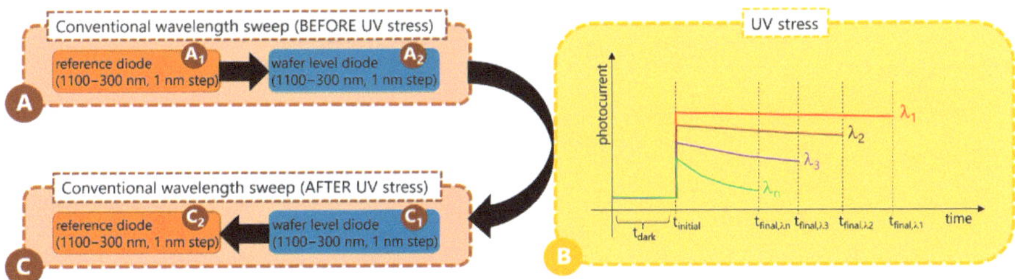

Figure 2. General scheme for photocurrent measurements before (**A**) and after (**C**) the UV stress methodology (**B**). Differently colored lines indicate different stress wavelengths.

The spectral response measurements before (step A) and after (step C) the UV stress step (step B) correspond to standard monitoring of the photocurrent level as the wavelength is swept between 1100 nm and 300 nm, with 1 nm steps (2 nm FWHM). Such measurement before the UV stress (step A) includes a wavelength sweep on a reference calibrated diode (which allows the determination of the light power intensity) and a subsequent wavelength sweep on the photodetector of interest (wafer-level).

For the UV stress step (step B), it is ensured that the UV stress methodology is realized under well-defined energy conditions. The wavelength is fixed at different values between 400 nm and 200 nm, with a step as small as 1 nm (2 nm FWHM). The stress time at each wavelength is defined in such a way that the exact illumination conditions (fluence rate) are maintained. It is important to mention that the UV stress is applied locally and is delimited only by the optical fiber diameter (~105 nm), which means that not the entire photodetector has been stressed during the illumination. Also, to perform a reliable UV stress and subsequent quantification of the photodetector's degradation, the spectral response measurement after the UV stress (step C) must be performed in a reversed order with respect to the initial measurement (step A). This means that first, the wavelength sweep is performed on the photodiode of interest (wafer level), and subsequently, the wavelength sweep on the reference diode is performed. In this way, it is ensured that the electrical contact always remains the same during the entire process and that the optical fiber also does not change its position over the photodetector. These in situ pre/post optical photocurrent measurements, as well as the UV stress step, are summarized in detail in Figure 2.

2.3. UV Stress Methodology Applied on CMOS-Based Photodetectors

As mentioned previously, the UV stress is carried out at pre-defined fixed stressing wavelengths, between 400 nm and 200 nm (see Figure 3 for some wavelength examples between 200 nm and 350 nm). At such wavelength values, the photocurrent is monitored over a specific stress time ($t_{stress} = t_{final} - t_{initial}$). A dark current correction is also performed by subtracting the average dark current which is measured for about 10 s (t_{dark}) before the shutter is opened (without light exposure). The degradation of the spectral response (shown in % in Figure 3b) is calculated from the variation in photocurrent measured at the final measurement time (t_{final}) and the initial measurement time ($t_{initial}$). For all the photodiodes tested in this work, it is observed that shorter wavelengths (therefore more photon energetic) have a higher degradation potential. In the device case shown in Figure 3, such degradation reaches some level of saturation at around 90% close to 200 nm. Also, for all the photodiodes tested in this work, for wavelengths > 300 nm, there is no evident spectral response degradation noticed due to UV light stress.

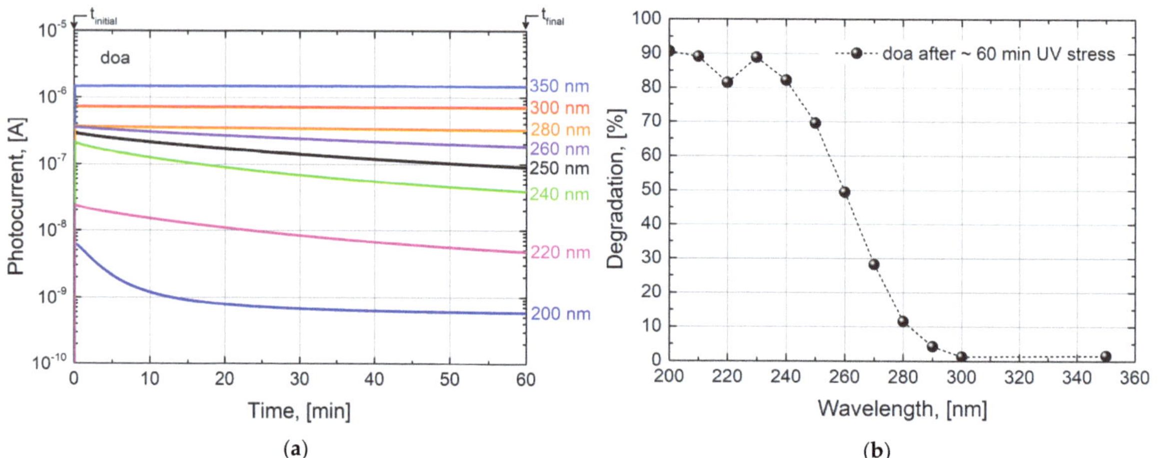

Figure 3. UV stress methodology for CMOS photodetectors. Data are shown for the case of module A, doa photodiode: (**a**) over-time photocurrent monitoring under UV stress, shown for some pre-defined wavelength values; (**b**) estimation of the absolute spectral response degradation suffered at each stressing wavelength. Dotted line serves merely as a guide-to-the-eye to follow the tendency of the UV degradation behavior.

The light power per area reaching the photodetector varies with the stress wavelength used, as shown in Figure 4. Also, as shown in Figure 3, different stress levels (and therefore UV degradation) can be induced at different stress wavelengths. Therefore, to ensure a fair comparison and evaluation in terms of UV degradation of different photodetectors, as well as the stress induced at every stressing wavelength, it is necessary to properly control the dosage of UV light reaching the photodetector at every stress wavelength. The approach in this work is to expose the photodetectors to a wide range of stress wavelengths while always maintaining the same light illumination conditions. In this way, no UV wavelength is favored, and each stress wavelength will induce the same stress over the photodetector.

Therefore, we employ a constant light fluence over the light-exposed surface of the photodetector. This is known as the fluence rate or radiant exposure and is defined as the product of the power of the UV electromagnetic radiation incident on a surface per unit surface area and the duration of the light exposure or simply the product of the effective light irradiance and the UV stress time [20]. For simplicity, and in accordance with some experimental measurement requirements, such as reasonable UV stress and measurement times, in this work, we defined a fluence rate of 3550.24 W·min/m². Consequently, the exposure times at every stress wavelength must be adjusted accordingly. Table 2 summarizes the stress wavelengths used during the UV stress procedure (step B in Figure 2), as well as their corresponding exposure times. This is also displayed in Figure 4.

Table 2. UV light exposure stress times applied at each stress wavelength.

Stress Wavelength (nm)	Exposure Time (s) [1]
200	3151.8
210	1197.0
220	637.3
230	322.9
240	170.0
250	112.8
260	93.2
270	84.6
280	84.8
290	82.8
300	74.1
350	48.0

[1] This is precisely controlled by an automated optical shutter mechanism.

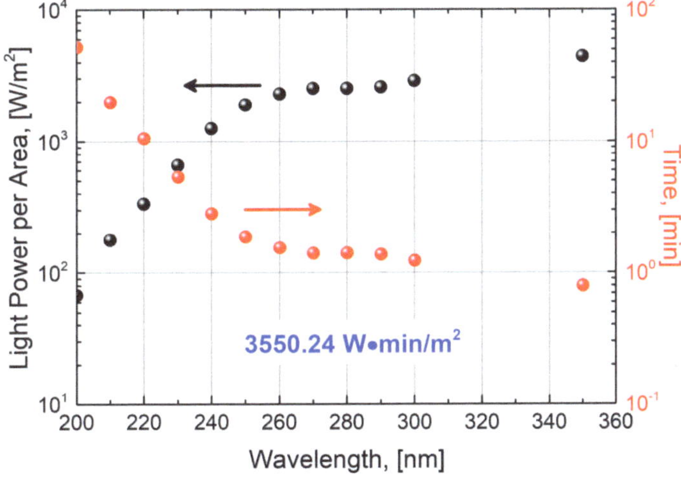

Figure 4. Light power per area dependency on wavelength and determination of UV stress exposure times for a constant fluence rate of 3550.24 W·min/m². The colored arrows indicate the corresponding proper axis of light power per area (black) and time (red).

3. Results and Discussions

3.1. Quantification of Optical Degradation Due to UV Light Exposure

Once an adequate fluence rate is defined, it is then required to re-normalize the estimations of UV degradation shown in Figure 3 with precise UV exposure times to guarantee the same UV light exposure conditions at each stress wavelength. The results of such normalization are shown in Figure 5, considering the case of the doa device (previously shown in Figure 3), as well as other examples of X-FAB's photodetectors (doe, doeher, and UVC) designed for operation in a variety of spectral ranges (see Table 1 for further details). In general, these results confirm that non-UV photodetectors should not be exposed to stress wavelengths below 300 nm. Degradation effects become more evident around 260 nm, growing exponentially as the photo energy increases and reaching almost a total loss of optical sensitivity at 200 nm. Due to some structural characteristics, some non-UV photodetectors (see doeher type in Figure 5) appear to be slightly more resistant to UV light exposure. Nevertheless, these devices also suffer almost a total loss of optical sensitivity as the stress wavelength approaches 200 nm. A remarkable highlight is the robustness of the UV photodetectors under UV stress conditions, where a spectral response degradation below 5% is observed for the more energetic wavelength applied in this work. Due to technical limitations, and to ensure good reliability of the experimental results presented here, 200 nm is the lower limit investigated in this work. For wavelengths below 200 nm, not only the stability of the light source system may be compromised, but also strong solarization effects are observed in the optical fiber. This is caused by the formation of absorbing centers due to the intense UV light flux. Therefore, it is no longer possible to properly bring the light into the photodetectors without avoiding undesired significant light loss.

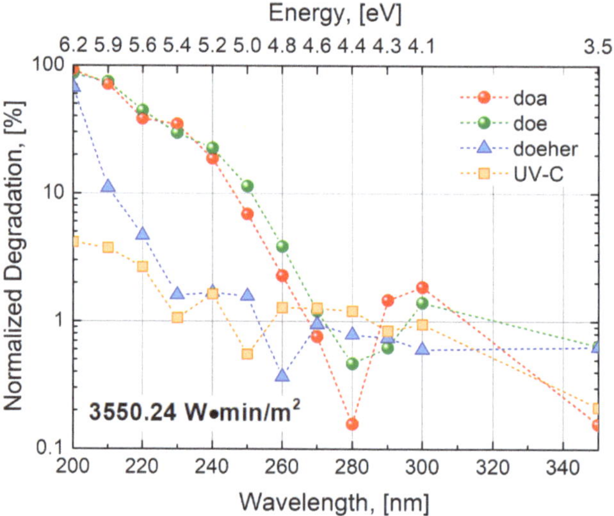

Figure 5. Quantification of the degradation induced due to UV light exposure, considering stress wavelength between 400 nm and 200 nm. Data are shown for some representative examples including photodetectors with a wide range of spectral applications.

3.2. Overall Impact of Optical Performance for an Extended Spectral Range

Once a robust methodology has been laid down to quantify the degradation of sensitivity suffered by CMOS-based photodetectors due to UV stress, it is of great interest to determine the impact that such degradation has on the overall optical performance of such devices. Particularly, when such an impact may possibly be irreversible (to the best of the author's knowledge, there is no clear evidence that a UV degradation mechanism can be reversed). The undesired or unproper exposure to UV light may adversely compromise the safe and successful outcome of the specific technological application in which a silicon-based photodetector may be implemented. Under UV light exposure, the optical performance of a photodetector is impacted, not only in the UV wavelength range but certainly in the entire spectral range. This would turn the device unresponsive, even in specific spectral ranges where such a device is expected to offer its maximum light sensing capabilities.

Figure 6 shows the example of three photodetectors which were submitted to the UV stress methodology proposed in Section 2.3 and following the test sequence shown in Figure 2. Solid data represent conventional wavelength sweep measurements before UV stress (step A, Figure 2); meanwhile, dotted data represent conventional wavelength sweep measurements after UV stress (step C, Figure 2). Dashed back data indicate the ideal spectral response of a photodetector, which would be the ideal case when its quantum efficiency is equal to one. Each photodetector in Figure 6 was selected to showcase three main particular cases. The first case corresponds to a non-UV photodiode (doe), which is severely affected under UV stress illumination conditions (especially for wavelengths below 260 nm), as shown also in Figure 5. After UV stress, the optical performance of the device is severely impacted along the entire spectral range measured (1100 nm to 300 nm). The strongest impact is clearly in the UV range, where a degradation of about 88% is observed. Note that the UV range for this device corresponds to wavelengths only from 400 nm to 300 nm, once such non-UV photodiodes are not specified to be operated below 300 nm. Nevertheless, the responsivity of the device is also affected beyond the UV range, about 23% in the Visible range and about 5% in the NIR range. The second case corresponds also to a non-UV photodiode, which is especially designed to be responsive for the Red and NIR spectral range (dob). This device is designed to be non-responsive in the UV spectral range. Therefore, no impact on its optical performance is expected to appear due to UV stress light illumination, as clearly shown in Figure 6. The third case considered here is a newly developed UV photodetector (UVC), which shows high UV response down to 200 nm. This is the lower wavelength operation limit specified for these kinds of photodetectors. Such devices also show a strong robustness under UV stress conditions, with a maximum responsivity degradation of 5–6% in the UV spectral range. Such degradation is almost unnoticeable in the Visible and NIR spectral ranges (below 3% in both cases). For a fair comparison, it should be noted that the result shown in Figure 6b does not imply that a photodetector like the dob is more robust to UV light conditions compared to the UVC photodetector. The low degradation for dob (in %) is because such a device is specially designed to be unresponsive for wavelengths below ~450 nm. Therefore, the photocurrent levels and variations before and after UV stress illumination conditions are extremely small in this case.

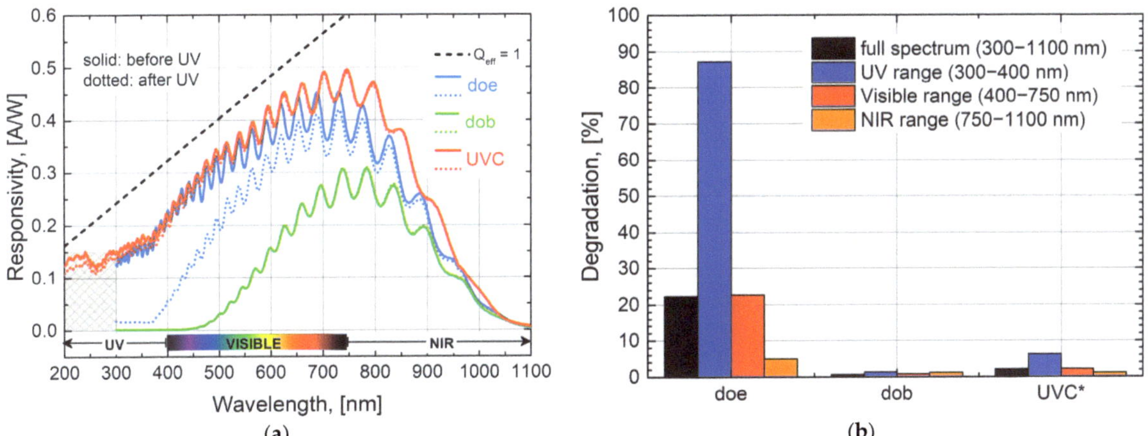

Figure 6. Overall impact on the optical performance of CMOS-based photodetectors due to UV stress light exposure: (**a**) Photodetectors responsivity before (solid data) and after (dotted data) UV stress conditions. Dashed lines correspond to the ideal responsivity ($Q_{eff} = 1$). (**b**) Estimation of the spectral response degradation suffered along different spectral ranges of interest. (*) For the case of the UVC photodetector the labels for spectral ranges should be read as full spectrum (200–1100 nm) and UV range (200–400 nm), once the experimental data collection starts at 200 nm instead of 300 nm.

As a further comparison of the improved optical performance shown by the UVC photodetector, a comparison is made with other previous X-FAB's technologies such as the XH018 technology-based UV detector. Figure 7 shows the clear improvement in optical performance in the UV range for the UVC detector (red solid data), in comparison to the XH018 UV detector (blue solid data). Such improvement starts to be more evident below 300 nm, and especially of interest is the performance around ~260 nm, which opens up important possibilities for applications of UV sterilization. As a comparison, the case of a commercially available discrete back side illuminated UV photodetector (gray dotted data) is also shown. Nevertheless, it must be pointed out that such discrete devices are based on a back side illumination technology, which is very different from the technology approach for the case of the UVC photodetector, based on X-FAB's CMOS XS018 technology. Also, the data for this case (at 25 °C) are obtained from available public documentation and such a device has not been directly measured with the measurement methodology approach proposed in this work. Therefore, this serves solely as a first general performance comparison.

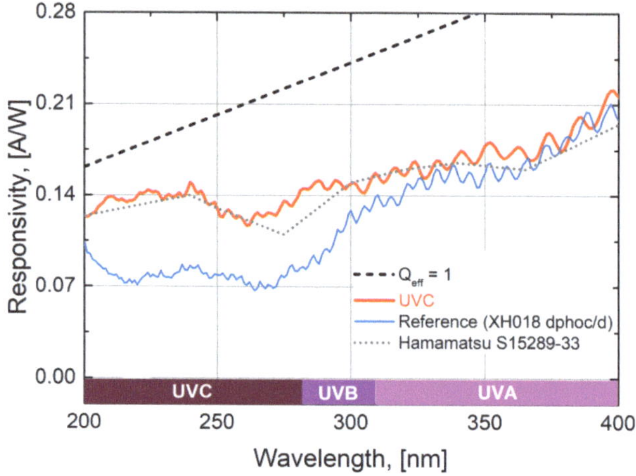

Figure 7. Optical performance in the UV spectral range for the XS018 technology-based UVC photodetector, in comparison to other X-FAB's technologies (XH018) as well as other discrete back side illuminated commercially available photodetectors.

4. Conclusions

As expected, different photodetectors are impacted differently under UV stress illumination conditions, depending on the specific technical application for which they are designed. This also defines aspects such as photodiode structure, design or layout characteristics and even adequate operation conditions. Therefore, it becomes very important to develop a robust measurement methodology which allows a reliable characterization of CMOS-based photodetectors in terms of UV stress and degradation. The work presented here is the result of several optimization loops in terms of electrical and optical measurements methodologies which are easily applied to any kind of front-side illuminated photodetector in a fully automated manner. In the case of the UV photodetectors shown here, designs and process flavors have been optimized to achieve high spectral responsivity and robustness under UV light illumination conditions. Compared with X-FAB's technologies (XH018) and even other discrete backside illuminated devices, we have observed a significant spectral response improvement, especially for wavelengths around 260 nm, which is of great interest for UV-based sterilization applications. Our planned research on UV light sensing applications continues and will also include the influence of operation temperature on spectral response degradation performance due to UV stress conditions, as well as the impact on other important photodetector parameters such as dark current (leakage current) and capacitance.

5. Patents

Four patents related to UV light sensing are pending to protect the IP provided by X-FAB regarding the development and realization of silicon-based photodetectors with an exceptionally high UV spectral response.

Author Contributions: Conceptualization, D.G. and P.F.S.; methodology, D.G.; software, P.F.S.; validation, D.G. and P.F.S.; formal analysis, P.F.S.; investigation, P.F.S.; data curation, P.F.S.; writing—original draft preparation, P.F.S.; writing—review and editing, D.G. and P.F.S.; visualization, P.F.S. All authors have read and agreed to the published version of the manuscript.

Funding: This research received no external funding.

Institutional Review Board Statement: The work was internally discussed and approved at X-FAB for external publication.

Informed Consent Statement: Not applicable.

Data Availability Statement: Raw data related to the measurements of the photodetectors shown here are available. Specimens of all different fabricated photodetectors as well as demonstration kits can be provided to interested parties under individual agreements.

Conflicts of Interest: Both authors were employed by the company X-FAB Global Services GmbH.

References

1. Haraguchi, T.; Ding, D.Q.; Yamamoto, A.; Kaneda, T.; Koujin, T.; Hiraoka, Y. Multiple-color fluorescence Imagining of Chromosomes and Microtubules in Living Cells. *Cell Struct. Funct.* **1999**, *24*, 291–298. [CrossRef] [PubMed]
2. Sipauba Carvalho da Silva, Y.R.; Kuroda, R.; Sugawa, S. A Shigly Robust Silicon Ultraviolet Selective Radiation Sensor Using Differential Spectral Response Method. *Sensors* **2019**, *19*, 2755. [CrossRef] [PubMed]
3. Malinowski, P.E.; Duboz, J.Y.; Moor, P.D.; John, J.; Minoglou, K.; Srivastava, P.; Creten, Y.; Torfs, T.; Putzeys, J.; Semond, F.; et al. 10 µm Pixel-to-Pixel Pitch Hybrid Backside Illuminated AlGaN-in-Si Imagers for Solar Blind EUV Radiation Detection. In Proceedings of the Electron Devices Meeting (IEDM), 2010 IEEE International, San Francisco, CA, USA, 6–8 December 2010; pp. 348–351.
4. Biasin, M.; Bianco, A.; Pareschi, G.; Cavalleri, A.; Cavatorta, C.; Fenizia, C.; Galli, P.; Lessio, L.; Lualdi, M.; Tombetti, E.; et al. UV-C irradiation is highly effective in inactivating SARS-CoV-2 replication. *Sci. Rep.* **2021**, *11*, 6260. [CrossRef] [PubMed]
5. Raeiszadeh, M.; Adeli, B. A Critical Review on Ultraviolet Disinfection Systems against COVID-19 Outbreak: Applicability, Validation, and Safety Considerations. *ACS Photonics* **2020**, *7*, 2941–2951. [CrossRef] [PubMed]
6. Heilinglog, C.S.; Aufderhorst, U.W.; Schipper, L.; Dittmer, U.; Witzke, O.; Yang, D.; Zheng, X.; Sutter, K.; Trilling, M.; Alt, M.; et al. Susceptibility of SARS-CoV-2 to UV Irradiation. *Am. J. Infect. Control* **2020**, *48*, 1273–1275. [CrossRef] [PubMed]
7. Hoenk, M. Stability and Photometric Accuracy of CMOS Image Sensors in Space: Radiation Damage, Surface Charge and Quantum Confinement in Silicon Detectors. In Proceedings of the IISW, Scotland, UK, 21–25 May 2023.
8. Yamada, H.; Miura, N.; Okihara, M.; Hinohara, K. A UV Sensor IC based on SOI Technology for UV Care Applications. In Proceedings of the SICE Annual Conference, UEC, Tokyo, Japan, 20–22 August 2008; pp. 317–320.
9. Kuroda, R.; Nakazawa, T.; Hanzawa, K.; Sugawa, S. Highly Ultraviolet Light Sensitive and Highly Reliable Photodiode with Atomically Flat Si Surface. In Proceedings of the IISW, Hokkaido, Japan, 8–11 June 2011.
10. Nakazawa, T.; Kuroda, R.; Koda, Y.; Sugawa, S. Photodiode Dopant Structure with Atomically Flat Si Surface for High-Sensitivity to UV Light. *Proc. SPIE* **2012**, *8298*, 82980M.
11. Gaebler, D.; Henkel, C.; Thiele, S. CMOS Integrated UV Photodiodes. *Procedia Eng.* **2016**, *168*, 1208–1213. [CrossRef]
12. Yampolsky, M.; Pikhay, E.; Roizin, Y. Embedded UV Sensors in CMOS SOI Technology. *Sensors* **2022**, *22*, 712. [CrossRef] [PubMed]
13. Shi, L.; Nihitianov, S. Comparative Study of Silicon-Based Ultraviolet Photodetectors. *IEEE Sens. J.* **2012**, *12*, 2453–2459. [CrossRef]
14. Wright, N.; Horsfall, A. SiC Sensors: A Review. *J. Phys. D Appl. Phys.* **2007**, *40*, 6345–6354. [CrossRef]
15. Kewei, L.; Sakurai, M.; Aono, M. ZnO-Based Ultraviolet Photodetectors. *Sensors* **2010**, *10*, 8604–8634.
16. Aiyagi, Y.; Fujihara, Y.; Murata, M.; Shike, H.; Kuroda, R.; Sugawa, S. A CMOS image sensor with dual pixel reset voltage for high accuracy ultraviolet light absorption spectral imaging. *Jpn. J. Appl. Phys.* **2019**, *58*, SBBL03. [CrossRef]
17. Liwen, S.; Liao, M.; Sumiya, M. A Comprehensive Review of Semiconductor Ultraviolet Photodetectors: From Thin Film to One-Dimensional Nanostructures. *Sensors* **2013**, *13*, 10482–10518.
18. Werner, L. Ultraviolet Stability of Silicon Photodiodes. *Metrologia* **1998**, *35*, 407–411. [CrossRef]
19. XS018 Photodiodes Application Note V2.0.0. Available online: https://my.xfab.com (accessed on 16 January 2024).
20. *ISO 15858*; UV-C Devices—Safety Information—Permissible Human Exposure. ISO: Geneva, Switzerland, 2016.

Article

Highly Selective Tilted Triangular Springs with Constant Force Reaction [†]

Lisa Schmitt *, Philip Schmitt and Martin Hoffmann

Microsystems Technology, Faculty of Electrical Engineering and Information Technology, Ruhr University Bochum, 44801 Bochum, Germany; philip.schmitt@rub.de (P.S.); martin.hoffmann-mst@rub.de (M.H.)
* Correspondence: lisa.schmitt-mst@rub.de; Tel.: +49-(0)-234-32-24830
[†] This is an expanded research article based on the conference paper "Tilted Triangular Springs with Constant Force Reaction" that was presented at EUROSENSORS2023 conference, 10–13 September 2023 in Lecce, Italy.

Abstract: Guiding mechanisms are among the most elementary components of MEMS. Usually, a spring is required to be compliant in only one direction and stiff in all other directions. We introduce triangular springs with a preset tilting angle. The tilting angle lowers the reaction force and implements a constant reaction force. We show the influence of the tilting angle on the reaction force, on the spring stiffness and spring selectivity. Furthermore, we investigate the influence of the different spring geometry parameters on the spring reaction force. We experimentally show tilted triangular springs exhibiting constant force reactions in a large deflection range and a comb-drive actuator guided by tilted triangular springs.

Keywords: micromechanical spring; constant force mechanism; MEMS; electrostatic actuator

1. Introduction

Guiding mechanisms enable translational in-plane displacement of Micro-Electro-Mechanical Systems (MEMS) and are, therefore, among the most important components of microsystems technology. In MEMS, linear guiding mechanisms are used in a wide variety of applications such as in electrostatic actuators [1], or in inertial MEMS to guide proof masses. Usually, these applications require a guiding mechanism that is compliant in the displacement direction and stiff in all other directions. For a solid spring, this means that it should have a low stiffness, k_x, in the deflection direction, x (i.e., the slope of the force–displacement characteristic (FDC)), but high stiffnesses in the orthogonal y-direction (k_y). The ratio of the stiffness can be defined as the selectivity, S, with

$$S = \left. \frac{k_y}{k_x} \right|_{y \to 0} \tag{1}$$

Guiding mechanisms have many different forms of appearance and geometries. We give a comprehensive overview of the guiding mechanisms in [2]. The clamped–clamped beams are among the most common guiding mechanisms [1,3] that fit the requirements for guiding springs quite well, as their stiffness in the displacement direction, k_x, is much smaller than their stiffness in the direction orthogonal to the displacement, k_y. Figure 1a shows a clamped–clamped beam and its typical non-linear FDC [4]. Consequently, the mechanical reaction force, F_x, of the clamped beams increases strongly with progressing displacement. This characteristic is quite unattractive for, e.g., electrostatic actuators. Serpentine springs have a linear FDC, as shown in Figure 1b; hence the mechanical force, F_x, increases only slightly during deflection, making the serpentine springs attractive for electrostatic comb-drive actuation [1,5]. However, serpentine springs only realize a unidirectional displacement and suffer from a strongly decreasing selectivity with progressing displacement that results, e.g., in the side instability of comb-drive actuators [1].

Citation: Schmitt, L.; Schmitt, P.; Hoffmann, M. Highly Selective Tilted Triangular Springs with Constant Force Reaction. *Sensors* **2024**, *24*, 1677. https://doi.org/10.3390/s24051677

Academic Editors: Bruno Ando, Pietro Siciliano and Luca Francioso

Received: 29 January 2024
Revised: 22 February 2024
Accepted: 29 February 2024
Published: 5 March 2024

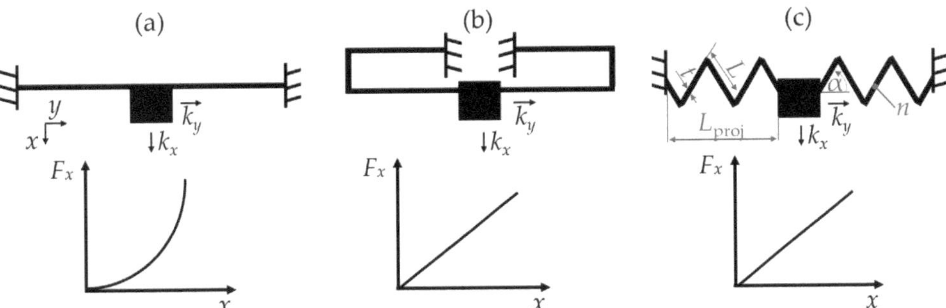

Figure 1. A sketch of the FDC of a (**a**) clamped–clamped beam, (**b**) serpentine spring, (**c**) triangular spring [2,3].

Springs with an M-shaped geometry have a non-linear and asymmetric FDC [6]. Both triangular and sinusoidal springs, presented in [2], have a geometry similar to the M-shaped springs. The triangular springs (Figure 1c) are characterized by a linear FDC, high selectivity, and a large displacement range. They achieve large deflections without rupture and can be deflected in both the positive and negative x-directions. The appearance of these springs is highly variable, since the number of spring segments n, the angle of inclination α as well as the thickness, t, and the length, L, can be varied. The triangular springs can be migrated into sinusoidal shaped springs, which reduces the stress peaks that occur in the spring kinks and limits the maximum deflection of the spring [2].

There are also springs with a constant FDC [7]. The springs realize a constant force feedback when applying an external force, e.g., a voltage [8,9]. Consequently, the resulting force is not proportional to the deflection [9]. In the constant force range, the stiffness, k_x, of such springs is ideally 0 N/m. Constant-force springs are either used to maintain the functionality of sensitive MEMS, which could be affected by fluctuations of the spring force [10] or for specific applications that require a constant force [11]. A constant force can be accomplished by complex force feedback systems that control the force by means of an actuator [12,13]. To generate constant forces in microsystems, electrostatic actuators with closed-loop feedback control are often used to control the force in MEMS [14]. Alternatively, the spring itself can be used to generate a constant and displacement-independent force. Such systems include, for example, buckling beams combined with a linear spring [8,11,15]. Here, the FDC of the negative buckling spring is superimposed with the FDC of the linear spring. The superposition of both springs results in a displacement independent constant force reaction [11,16–19].

In this contribution, we present a tilted triangular shaped spring achieving constant force reactions. Unlike other solutions, the tilted triangular spring does not require a superimposing of multiple individual reaction forces making it a simple and compact solution for a linear guidance with a constant force. In Section 2, we present the spring design and discuss the simulation results. Here, we take a close look at the influence of the design parameters on the constant force range. In Section 3, we present the experimental setup and discuss the results obtained by the experiments. In Section 4, we show a comb-drive actuator displaced by a tilted triangular spring. Section 5 gives a short summary of this article.

2. Materials and Methods

2.1. Spring Design

The tilted triangular spring is characterized by a tilting angle, β, as shown in Figure 2a. The properties of this spring are defined by the number of straight elements, n, their length, L, the thickness, t, depth, d, of the device layer, and the inclination angle, α. Figure 2 also shows the projected length, L_{proj}. Based on the results that we achieved in [2], we aim to achieve symmetrical force-displacement curves. Therefore, we only used springs with an even number of beams, i.e., $n = 2, 4, 6, \ldots$, and the length of the first, L_1, and the last beam, L_n, is half the length of the other beams, i.e., $L/2$. The triangular spring achieves a constant force reaction with a preset initial tilting angle, β, as an additional degree of freedom. To realize a translational displacement, we use a symmetric design (Figure 2b). As shown in Figure 2b, the constant force range (CFR) arises with the progressing displacement of the spring; consequently, the tilted triangular spring achieves a partially constant spring force.

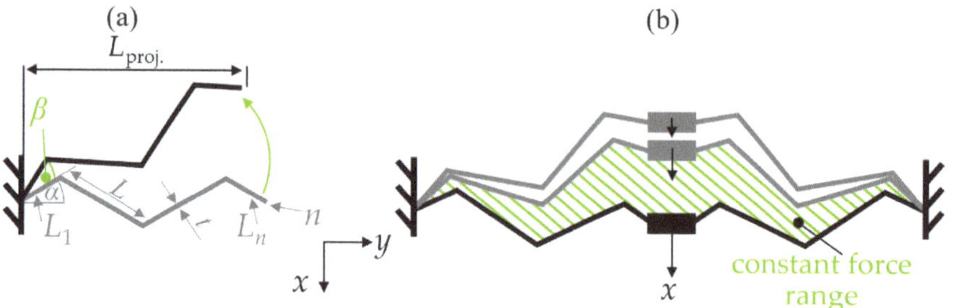

Figure 2. (**a**) A sketch of the tilted triangular spring with a preset tilting angle, β, (**b**) displacement of the tilted triangular spring with a constant force region.

2.2. Simulation Procedure

The springs are simulated by FEM using COMSOL *Multiphysics*. We simulate a system with four identical springs with identical design parameters (Figure 3) to guarantee a translational displacement and to maintain the system stability. The structures are simulated with respect to the geometric non-linearity.

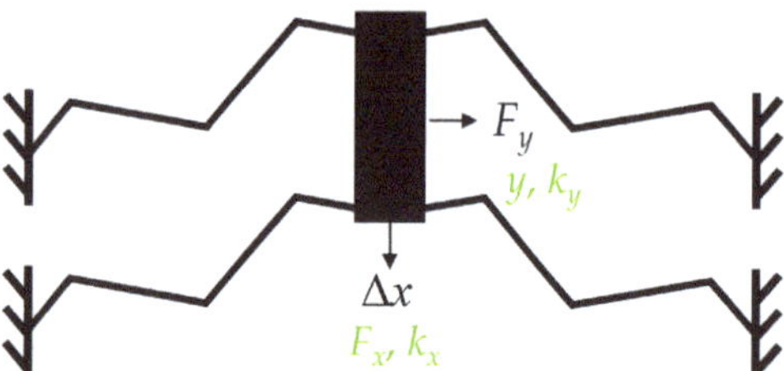

Figure 3. Simulation setup; applied force or displacement (black); resulting values (green).

The springs are displaced in defined steps, Δx, resulting in a change in the mechanical reaction force of the spring, ΔF_x. With

$$k_x = \frac{\Delta F_x}{\Delta x} \tag{2}$$

we determine the stiffness, k_x, in the displacement direction. In a second simulation, a constant force, F_y, also acts in the y-direction, resulting in a displacement, y. With

$$k_y = \frac{F_y}{y} \tag{3}$$

we determine the stiffness, k_y, of the spring when displaced in the x-direction. The selectivity, S, of the spring is given by (1) as the quotient of k_y and k_x.

The design parameters are given in Table 1. To analyze the influence of the tilting angle on the reaction force, stiffness, and selectivity, we designed triangular springs with the ability to vary the tilting angle, β, from $0°$ to $8°$ as shown in Figures 4 and 5. To analyze the correlation between the position of the maximum selectivity, $x(S_{\text{max}})$, the projected length, $L_{\text{proj.}}$, and the tilting angle, the tilting and inclination angle, as well as the number of beams were varied in Figure 5b. In Figure 6, the influence of the design parameters was investigated. Therefore, in each diagram, one single parameter was varied and the influence on the constant force range was analyzed.

Table 1. Parameters of simulated triangular-shaped springs; the varied parameters are in bold.

Analyzed in	β	α	n	L	t	d	Analysis of
Figure 4a–c	**0–8°**	20°	6	400 μm	5 μm	20 μm	reaction force F_x, stiffness k_x and k_y
Figure 5a	**0–8°**	20°	6	400 μm	5 μm	20 μm	selectivity S correlation
Figure 5b	**0°–8°**	**10°, 20°, 30°**	**4, 6**	400 μm	5 μm	20 μm	$S_{max.}$, $L_{proj.}$ and β
Figure 6a standard spring	**7°**	**20°**	**6**	**400 μm**	**5 μm**	**20 μm**	**constant force**
Figure 6b	**0–30°**	20°	6	400 μm	5 μm	20 μm	Influence of β
Figure 6c	7°	**5–30°**	6	400 μm	5 μm	20 μm	Influence of α
Figure 6d	7°	20°	**4–14**	400 μm	5 μm	20 μm	Influence of n
Figure 6e	7°	20°	6	**100…500 μm**	5 μm	20 μm	Influence of L
Figure 6f	7°	20°	6	400 μm	**3…10 μm**	20 μm	Influence of t

2.3. Simulation Results

2.3.1. Constant Force Range (CFR) and Selectivity

Figure 4a shows the simulated force reaction of a triangular spring for different preset tilting angles β. For tilting angles from 0° to 5°, the reaction force increases continuously. For tilting angles of 6° and 7°, the spring exhibits an approximately constant force range. For the selected geometry, the force reaction shows local maxima and minima for a tilting angle $\beta = 8°$ or higher.

Figure 4b shows the resulting spring stiffness, k_x, in the displacement direction, x. For the springs with constant force reaction ($\beta = 6°$ and 7°), the stiffness is approx. 0 N/m in the CFR. As shown in Figure 4c, the stiffness orthogonal to the displacement, k_y, slightly increases for springs with no or small tilting angles, whereas springs with larger tilting angles have a slightly decreasing stiffness, k_y.

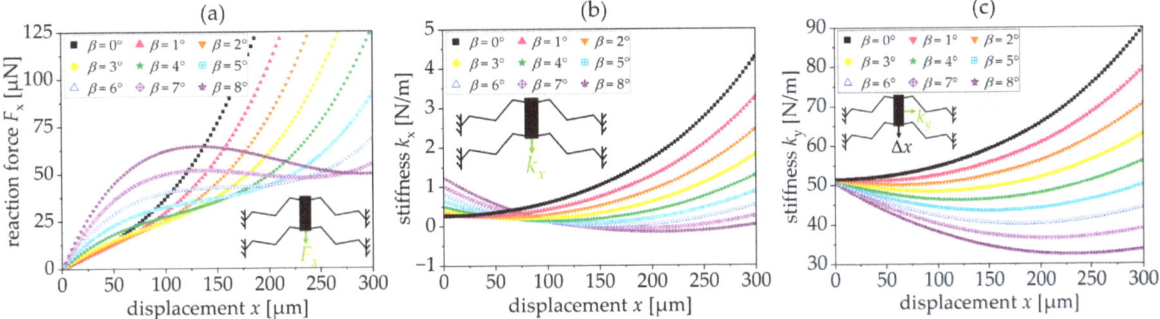

Figure 4. (a) Reaction force, F_x, as a function of β, (b) stiffness, k_x, as a function of β, (c) stiffness, k_y, as a function of β; simulation results for an exemplary spring with $L = 400$ μm, $t = 5$ μm, $n = 6$ and $\alpha = 20°$.

Figure 5a shows the resulting selectivity, S. Apart from generating a constant force, tilted triangular springs exhibit a high level of selectivity, as defined by the stiffness ratio between the y- and x-directions (cf. (1)). With the increase in the tilting angle, the maximum selectivity, $S_{\max}$, increases, too. Consequently, the tilted triangular spring exhibits a selectivity, S, as a function of the tilting angle. Thereby, the position of the maximum selectivity, $x(S_{\max})$, shifts towards larger displacements. Springs with a reaction force featuring a local maximum and minimum do not have a defined maximum of selectivity. The increase in selectivity is important, e.g., for the lateral instability of the comb-drive actuators. A higher selectivity allows a higher voltage, U_{SI}, to be applied, and thus allows larger displacements before lateral-side instability occurs. This correlation results from the analytical model for the displacement of comb-drive actuators that is highly dependent on the selectivity, S, [1].

$$U_{\mathrm{SI}}^2 = \frac{b^2 k_x}{2\varepsilon_0 \varepsilon_r d n_e}\left(\sqrt{2\frac{k_y}{k_x} + \frac{x_0^2}{b^2}} - \frac{x_0}{b}\right) = \frac{b^2 k_x}{2\varepsilon_0 \varepsilon_r t n_e}\left(\sqrt{2S + \frac{x_0^2}{b^2}} - \frac{x_0}{b}\right) \tag{4}$$

Here, U_{SI} is the voltage when side instability occurs, b is the distance between two electrodes, n_e is the number of electrodes, d is the thickness of the device layer, and x_0 is the initial overlap of electrodes.

Figure 5b shows that the position of maximum selectivity normalized to the projected spring length, $L_{\mathrm{proj.}}$, depends only on the tilting angle, β, and increases almost linearly with it. Consequently, a spring with a tilting angle of, e.g., $\beta = 6°$ reaches its maximum selectivity at a deflection of about 4.5% of its projected length.

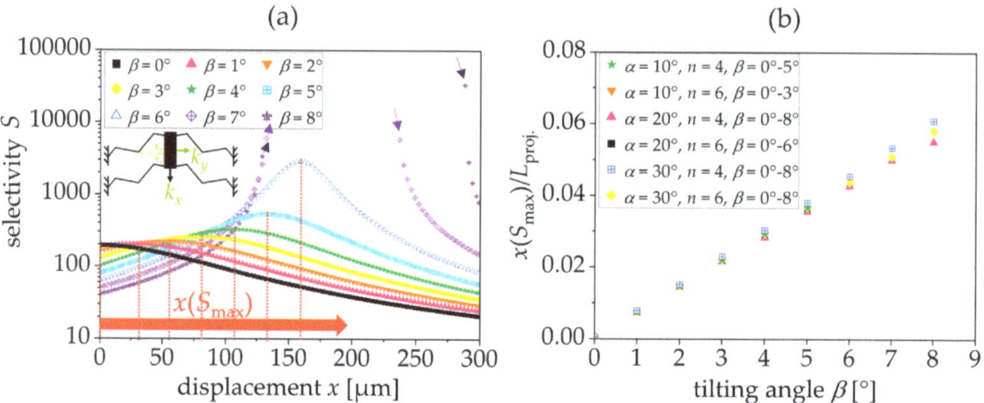

Figure 5. (**a**) Selectivity, S, as a function of β; (**b**) position of maximum selectivity, $x(S_{\max})$, normalized to $L_{\mathrm{proj.}}$ depending on the tilting angle, β; simulation results for an exemplary spring with $L = 400$ µm and $t = 5$ µm.

2.3.2. Influence of the Spring Parameter on the Constant Force Range

To better analyze the influence of the different geometry parameters of the tilted triangular spring on the constant force reaction, we use a standard spring with a force reaction, as shown in Figure 6a, with a constant force of 50.1 ± 1.5 µN in the range of 100 µm to 250 µm. In Figure 6b–f, we vary a single geometry parameter.

Figure 6b shows the influence of the tilting angle, β, on the constant force. At small tilting angles (in this case, $\beta = 5°$), no constant force range is achieved, while at larger tilting angles, the force response decreases slightly (negative stiffness) after reaching a maximum.

Figure 6c shows the influence of the inclination angle, α, which in this case defines the constant force range at 20°. If the inclination angle decreases, the reaction force increases sharply before it drops sharply. For a larger α (here, $\alpha = 30°$), the reaction force increases with increasing deflection without establishing an extended constant force range.

Figure 6d shows the influence of the number of beam segments, n. If the spring consists of few beam segments, in this case $n = 4$, there is no constant force; if there are many beam segments, the reaction force decreases after reaching a maximum.

Figure 6e shows that springs with long segments have a low reaction force and a large linear force range. If the segments of the tilted triangular spring are short, in this case 100 μm and 200 μm, respectively, the spring force is higher and the constant force range is non-existent or very small.

The segment thickness, t, is used to set the level of the reaction force, so that as the thickness decreases, the reaction force decreases and the constant force range increases, as shown in Figure 6f.

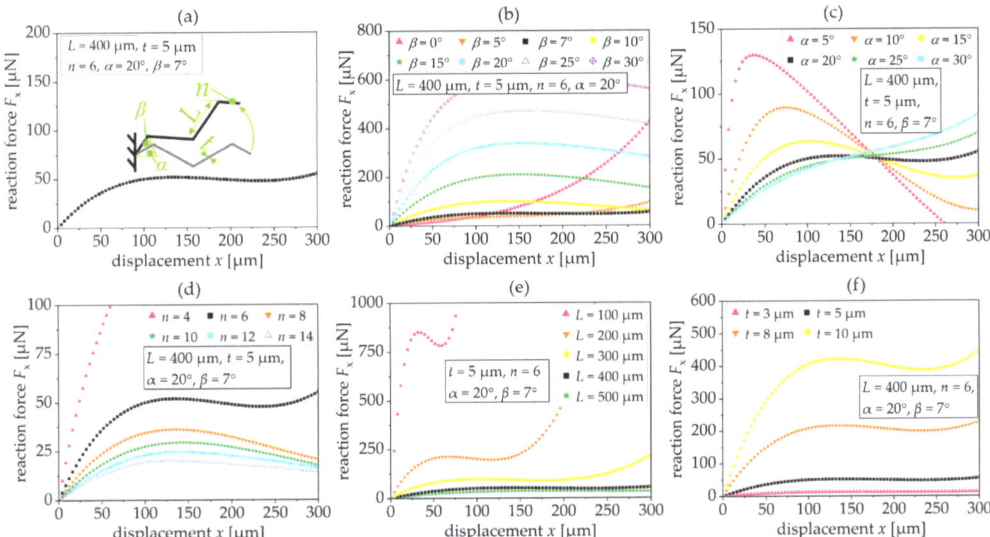

Figure 6. Simulated force–displacement characteristic of (**a**) a standard spring (L = 400 μm, t = 5 μm, n = 6, β=7° and α = 20°), when varying (**b**) the tilting angle, β, (**c**) the inclination angle, α, (**d**) the number of beams, n, (**e**) the beam length, L, (**f**) the beam thickness, t.

The optimal design depends on the application and the design of interacting components on the chip. The design can be optimized to, e.g., a large constant force range, a very high or very low constant force, or a spring with a small length or height. An example of this is as follows: (1) If a large constant force range is required, it is useful to have a long beam length, L, or thin beam thickness, t. However, large beams require a lot of space, and the beam thickness can be limited by the quality of the etching process or the thickness of the device layer. (2) If a constant and high reaction force is required, the tilting angle or the thickness can be increased. However, a spring with a high tilting angle requires lots of space. Consequently, the optimal design always depends on the application, other chips' design properties or the etching processes, and we cannot derive a single expression for an optimal spring design.

2.4. Fabrication

Based on the simulations conducted, demonstrator springs were designed. In addition to that, an electrostatic comb-drive actuator was designed to provide the tilted springs with a constant force.

The chips were manufactured in a dicing-free process for silicon-on-insulator (SOI) substrates known from [20] using a device layer of 20 μm and a handle layer of 300 μm; this is shown in Figure 7. First, a 100 nm aluminum layer was sputtered on the device layer. The aluminum was used for the electrostatic activation of the comb-drive electrodes. Afterwards, the electrical contacts were patterned by lithography and wet chemical etching, so that the mechanical structures were not covered with aluminum (Figure 7a). In the next step, we etched the handle layer by deep reactive ion etching (DRIE) using a SiO₂-hard mask (Figure 7b). To structure the device layer, we used a hard mask of aluminum nitride (AlN) and DRIE (Figure 7c). The chips were released from the substrate by hydrofluoric vapor etching (Figure 7d). Afterwards, the chips were fixed on PCBs. The simulated thickness of the springs was 5 µm; however, the fabricated thickness was 4.8 µm. Figure 7e shows a stacking photo of a fabricated chip with the spring demonstrator. Each spring system consists of a slider with two tilted guiding springs on each side. We provided the springs with additional force sensors to experimentally determine the force reaction on-chip. The chips also contain reference force sensors to determine the stiffness of the force sensors.

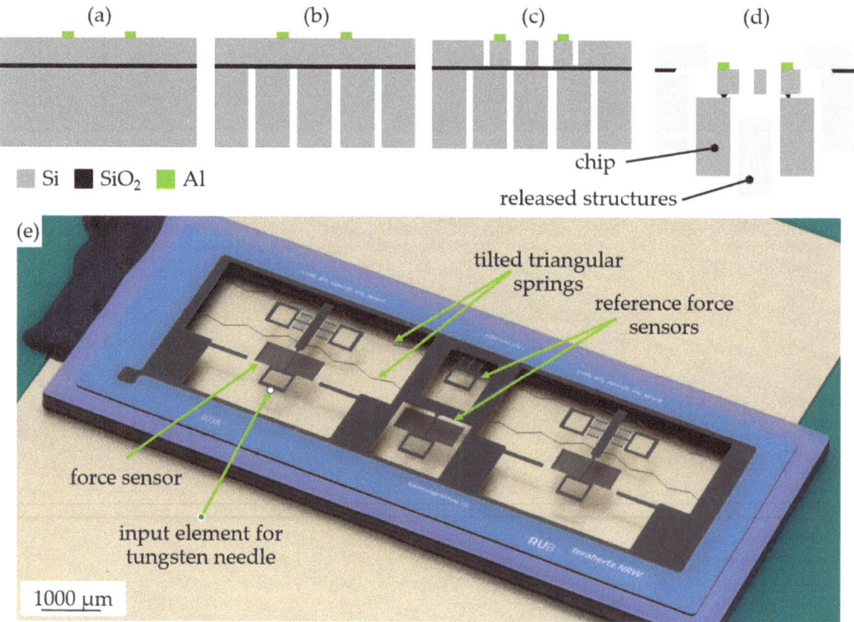

Figure 7. (**a–d**) Fabrication flowchart, (**e**) stacking photo of the fabricated chip.

2.5. Experimental Charcterization Procedure

First, the integrated force sensors were calibrated. Therefore, we could determine the spring constant, k_{FS}, of the force sensor. Instead of using the springs that were already connected to the tested device, we used the reference force sensors shown in Figure 7e. The calibration procedure included the simultaneous measurement of force and displacement of the force sensor, as illustrated in Figure 8a. The chips were mounted on a piezoelectric stage (PI Q-545). A tungsten needle was inserted into the input element of the reference force sensor. This needle was connected to a load cell (Sartorius WZA224-ND) with a resolution of 0.1 µN. The piezo stage was shifted upwards stepwise in defined increments. This resulted in an expansion of the force sensor and changed the force reaction measured by the load cell. Using (2), the spring constants of the reference force sensors were measured to be 0.2 N/m for force sensor 1 and 0.55 N/m, 1.5 N/m, and 4.6 N/m for force sensors 2, 3, and 4, respectively.

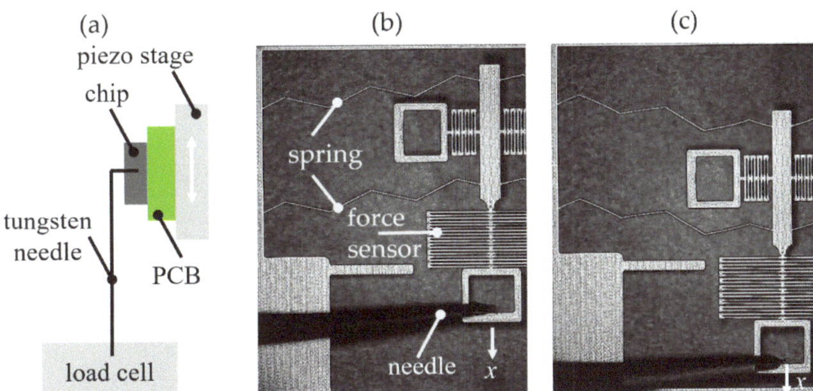

Figure 8. (**a**) sketch of the setup for calibrating the force sensor; left-hand-side of the tilted triangular spring in (**b**) the initial position, and (**c**) the displaced spring.

Knowing the spring constant of the force sensor, the triangular springs could be characterized using the integrated force sensors on the device under test. Therefore, again, a needle was connected to the input element of the force sensor and a displacement controlled by a piezo stage was imposed at the spring, as shown in Figure 8b. Both the deflection of the slider and the extension of the force sensor were measured visually with a microscope camera. With

$$F_{\mathrm{DUT}}(x) = F_{\mathrm{FS}}(x) = k_{\mathrm{FS}}x \tag{5}$$

the force was measured as a function of the spring displacement, which yielded the FDC. A deflected spring is shown in Figure 8c. Each experiment was repeated three times and the mean values and a standard deviation were calculated.

3. Experimental Results and Discussion

3.1. Overview of the Results

Table 2 summarizes the parameters of the selected spring designs.

Table 2. Parameters of selected spring designs.

Analyzed in	β [°]	α [°]	n	L [µm]	t [µm]	d [µm]	$F_{x,\mathrm{konst.}}$ (exp.)	$F_{\mathrm{konst., range}}$ (exp.)
Figure 9	0, 2, 4, 6, 7, 8, 10	20	6	400	4.8	20	6.8 ± 0.3 µN ($\beta = 7°$)	97…267 µm ($\beta = 7°$)
Figure 10a	0, 8, 9, 10, 11	30	6	400	4.8	20	6.4 ± 1.1 µN ($\beta = 10°$)	105…355 µm ($\beta = 10°$)
Figure 10b	7	20	6	400	4.8, 9.4, 14.8, 19.7	20	147.4 ± 1.1 µN ($t = 9.4$ µm)	103…228 µm ($t = 9.4$ µm)
Figure 10c	7	20	6	400, 600	4.8	20	4.1 ± 0.3 µN ($L = 600$ µm)	74…324 µm ($L = 600$ µm)

3.2. Characterization of the Tilted Triangular Springs

We designed triangular springs where the tilting angle varied from 0° to 8°. The design parameters of the spring are given in Table 2.

Figure 9a shows the experimentally derived reaction force depending on the tilting angle β. As shown in the simulation results, a constant force appears within the span of $\beta = 6°$ to $\beta = 8°$. For $\beta = 10°$, and the force reaction curve has a local maximum. The experimentally derived reaction force (6.8 ± 0.3 µN for $\beta = 7°$) is smaller than the simulated reaction force, which correlates to the smaller thickness of the fabricated springs (see Section 2.3). However, the region with the constant force is also larger, e.g., for $\beta = 7°$, the

CFR starts with a displacement of 97 μm and ends at 267 μm. Figure 9b shows the resulting spring stiffness for exemplary springs that is close to 0 N/m for the constant force range of the springs with a tilting angle of 7° and 8°.

Comparing the experimental results with the simulation presented in Figure 4a, the reaction force is lower, which is attributed to the reduced thickness of the manufactured structures (t = 4.8 μm (fabrication) instead of t = 5.0 μm (simulation)). This is because the thickness, t, of the springs has a cubic influence on the reaction force. The manufactured springs show a lower increase in the reaction force and a smaller influence of the tilting angle on the constant force range. In Figure 4a, the simulated springs with a tilting angle of 4° and 8° do not show a constant force range, while the fabricated springs with these tilting angles show a constant force behavior in Figure 9a.

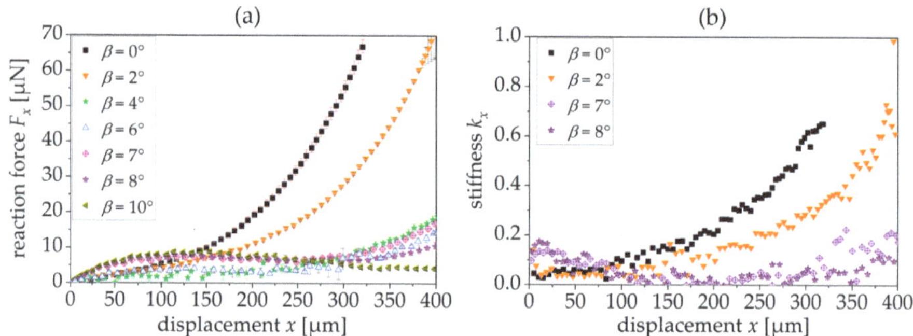

Figure 9. (**a**) Experimentally derived FDC of triangular springs when varying the tilting angle β; (**b**) resulting stiffness k_x (L = 400 μm, t = 4.8 μm, n = 6 and α = 20°).

As shown in Figure 6c, the inclination angle, α, also influences the reaction force. In Figure 10a, we show the corresponding experimental results for a spring with an inclination angle α = 30°. To generate a constant force, we need a larger tilting angle when the inclination angle is higher. Figure 10b shows the influence of the spring thickness, t, on the reaction force. The reaction force increases with increasing thickness. Thereby, the CFR is becoming smaller. Figure 8c shows that the larger beam segments also achieve larger constant force regions with a slightly smaller reaction force.

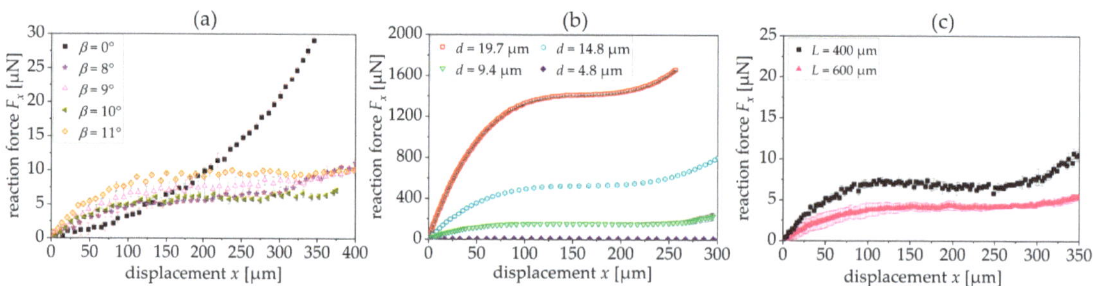

Figure 10. (**a**) Experimentally derived FDC of triangular springs when varying the tilting angle, β, (α = 30°); (**b**) experimentally shown influence of the spring thickness, t, on the reaction force; (**c**) experimentally shown influence of the beam length, L, on the reaction force.

4. Application in Comb Drive Actuators

An application for the presented springs are comb-drive actuators that struggle with side instability, which means that electrodes guided by the spring escape to the mounted electrodes, limiting the displacement of the comb-drive actuator (cf. Section 2.3). This happens as soon as the first derivate of the electrostatic force with respect to y becomes larger than the restoring spring constant in the y-direction [1].

We use the spring characterized in Figure 9 with a tilting angle of $\beta = 7°$ to guide an electrostatic comb-drive actuator. Figure 11a shows a schematic of the setup of the actuator; Figure 11b shows the voltage-dependent displacement with a relatively large displacement at low voltage. The maximum displacement is 71 µm at 28 V, and at larger voltages, side instability occurs. Figure 11c,d show the displacing electrodes.

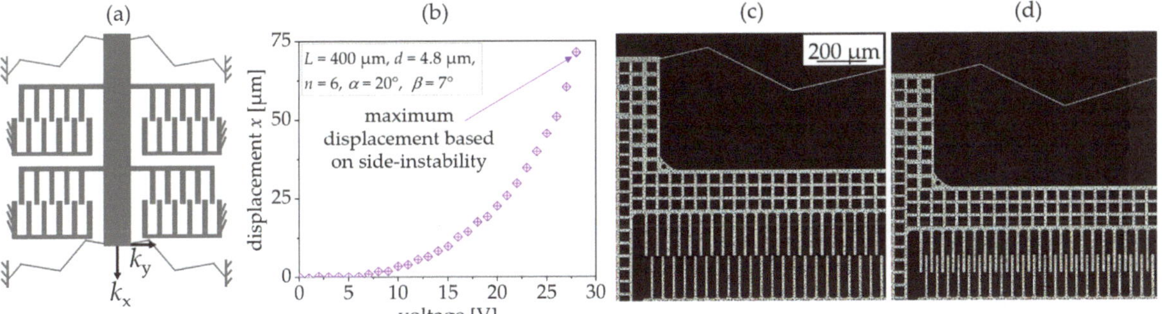

Figure 11. (**a**) sketch of the actuator setup; (**b**) experimentally derived voltage-displacement characteristics of a comb-drive actuator guided by the tilted triangular spring presented in Figure 7 ($\beta = 7°$); comb-drive actuator at (**c**) 0 V and (**d**) 28 V.

5. Conclusions

In this contribution, we present tilted triangular springs that achieve a partially constant reaction force using a preset tilting angle, β. Constant force springs are suitable for mechanisms which control an object with a defined force, as well as for all MEMS in which a mechanical overload could lead to a functional restriction or destruction of the sensitive components. Here, we focus on the influence of the different geometry parameters on the spring reaction force and analyze the spring selectivity based on the tilting angle. An exemplary spring achieves a constant force of 6.8 ± 0.3 µN in the range of 97 µm to 267 µm. We show that tilted triangular springs are very well-suited for guiding large displacement comb-drive actuators due to their high selectivity.

Author Contributions: Conceptualization, L.S. and P.S.; methodology, L.S. and P.S.; software, L.S. and P.S.; validation, L.S. and P.S.; formal analysis, L.S.; investigation, L.S.; resources, M.H.; data curation, L.S.; writing—original draft preparation, L.S.; writing—review and editing, P.S. and M.H.; visualization, L.S.; supervision, P.S. and M.H.; project administration, M.H.; funding acquisition, M.H. All authors have read and agreed to the published version of the manuscript.

Funding: This research was partly funded by the project terahertz.NRW. The project "terahertz.NRW" is receiving funding from the programme "Netzwerke 2021" (NW21-068D), an initiative of the Ministry of Culture and Science of the State of Northrhine Westphalia. The sole responsibility for the content of this publication lies with the authors. This research was also partly funded by the Deutsche Forschungsgemeinschaft (DFG, German Research Foundation)—Project-ID 287022738—TRR 196, Project C12.

Institutional Review Board Statement: Not applicable.

Informed Consent Statement: Not applicable.

Data Availability Statement: The data can be provided by L.S. upon reasonable request.

Acknowledgments: The chips were partly fabricated at the ZGH, Ruhr-Universität Bochum. The authors thank the technical staff of the Microsystems Technology (MST) for the support in the cleanroom.

Conflicts of Interest: The authors declare no conflicts of interest.

References

1. Legtenberg, R.; Groeneveld, A.W.; Elwenspoek, M. Comb-drive actuators for large displacements. *J. Micromech. Microeng.* **1996**, *6*, 320. [CrossRef]
2. Schmitt, P.; Schmitt, L.; Tsivin, N.; Hoffmann, M. Highly Selective Guiding Springs for Large Displacements in Surface MEMS. *J. Microelectromech. Syst.* **2021**, *30*, 597–611. [CrossRef]
3. Schomburg, W.K. *Introduction to Microsystem Design*; Springer: Berlin, Germany, 2011.
4. Gao, Y.; You, Z.; Zhao, J. lectrostatic comb-drive actuator for MEMS relays/switches with double-tilt comb fingers and tilted parallelogram beams. *J. Micromech. Microeng.* **2015**, *25*, 45003. [CrossRef]
5. Schmitt, L.; Conrad, P.; Kopp, A.; Ament, C.; Hoffmann, M. Non-Inchworm Electrostatic Cooperative Micro-Stepper-Actuator Systems with Long Stroke. *Actuators* **2023**, *12*, 150. [CrossRef]
6. Leadenham, S.; Erturk, A. M-shaped asymmetric nonlinear oscillator for broadband vibration energy harvesting: Harmonic balance analysis and experimental validation. *J. Sound Vib.* **2014**, *333*, 6209–6223. [CrossRef]
7. IMorkvenaite-Vilkonciene; Bucinskas, V.; Subaciute-Zemaitiene, J.; Sutinys, E.; Virzonis, D.; Dzedzickis, A. Development of Electrostatic Microactuators: 5-Year Progress in Modeling, Design, and Applications. *Micromachines* **2022**, *13*, 1256. [CrossRef] [PubMed]
8. Wang, P.; Xu, Q. Design and modeling of constant-force mechanisms: A survey. *Mech. Mach. Theory* **2018**, *119*, 1–21. [CrossRef]
9. Yang, S.; Xu, Q. Design and simulation of a passive-type constant-force MEMS microgripper. In Proceedings of the 2017 IEEE International Conference on Robotics and Biomimetics (ROBIO), Macau, Macao, 5–8 December 2017; pp. 1100–1105.
10. Ma, C.; Du, J.; Liu, Y.; Chu, Y. Overview of micro-force sensing methods. *Appl. Mech. Mater.* **2014**, *462–463*, 25–31. [CrossRef]
11. Thewes, A.C.; Schmitt, P.; Löhler, P.; Hoffmann, M. Design and characterization of an electrostatic constant-force actuator based on a non-linear spring system. *Actuators* **2021**, *10*, 192. [CrossRef]
12. Li, B.; Li, G.; Lin, W.; Xu, P. Design and constant force control of a parallel polishing machine. In Proceedings of the 2014 4th IEEE International Conference on Information Science and Technology, Shenzhen, China, 26–28 April 2014; pp. 324–328.
13. Erlbacher, E. Method for Applying Constant Force with Nonlinear Feedback Control and Constant Force Device Using Same. U.S. Patent 5,448,146, 5 September 1995.
14. Boudaoud, M.; Haddab, Y.; Le Gor, Y. Modeling and optimal force control of a nonlinear electrostatic microgripper. *IEEE/ASME Trans. Mechatron.* **2012**, *18*, 1130–1139. [CrossRef]
15. Shahan, D.; Fulcher, B.; Seepersad, C.C. Robust design of negative stiffness elements fabricated by selective laser sintering. In Proceedings of the 2011 International Solid Freeform Fabrication Symposium, Austin, TX, USA, 11–14 August 2024; University of Texas at Austin: Austin, TX, USA, 2011.
16. Qiu, J.; Lang, J.H.; Slocum, A.H. A curved-beam bistable mechanism. *J. Microelectromech. Syst.* **2004**, *13*, 137–146. [CrossRef]
17. Qiu, J.; Lang, J.H.; Slocum, A.H. A centrally-clamped parallel-beam bistable MEMS mechanism. Technical Digest. In Proceedings of the MEMS 2001. 14th IEEE International Conference on Micro Electro Mechanical Systems (MEMS), Interlaken, Switzerland, 25 January 2001; pp. 353–356.
18. Vysotskyi, B.; Parrain, F.; Aubry, D.; Gaucher, P.; Le Roux, X.; Lefeuvre, E. Engineering the structural nonlinearity using multimodal-shaped springs in MEMS. *J. Microelectromech. Syst.* **2018**, *27*, 40–46. [CrossRef]
19. Vysotskyi, B.; Parrain, F.; Aubry, D.; Gaucher, P.; Lefeuvre, E. Innovative energy harvester design using bistable mechanism with compensational springs in gravity field. *Proc. J. Phys. Conf.* **2016**, *773*, 12064. [CrossRef]
20. Sari, I.; Zeimpekis, I.; Kraft, M. A dicing free SOI process for MEMS devices. *Microelectron. Eng.* **2012**, *95*, 121–129. [CrossRef]

sensors

MDPI

Communication

Correction of 2π Phase Jumps for Silicon Photonic Sensors Based on Mach Zehnder Interferometers with Application in Gas and Biosensing

Loic Laplatine [1,*], Sonia Messaoudene [1], Nicolas Gaignebet [1], Cyril Herrier [2] and Thierry Livache [3]

[1] Univ. Grenoble Alpes, CEA, LETI, 38054 Grenoble, France; sonia.messaoudene@cea.fr (S.M.); gaignebet@insa-toulouse.fr (N.G.)
[2] Aryballe, 38000 Grenoble, France; c.herrier@aryballe.com
[3] Univ. Grenoble Alpes, CEA, CNRS, Grenoble INP, IRIG, SyMMES, 38000 Grenoble, France; thierry.livache@cea.fr
[*] Correspondence: loic.laplatine@cea.fr; Tel.: +33-438783014

Abstract: Silicon photonic sensors based on Mach Zehnder Interferometers (MZIs) have applications spanning from biological and olfactory sensors to temperature and ultrasound sensors. Although a coherent detection scheme can solve the issues of sensitivity fading and ambiguity in phase direction, the measured phase remains 2π periodic. This implies that the acquisition frequency should ensure a phase shift lower than π between each measurement point to prevent 2π phase jumps. Here, we describe and experimentally characterize two methods based on reference MZIs with lower sensitivities to alleviate this drawback. These solutions improve the measurement robustness and allow the lowering of the acquisition frequency. The first method is based on the phase derivative sign comparison. When a discrepancy is detected, the reference MZI is used to choose whether 2π should be added or removed from the nominal MZI. It can correct 2π phase jumps regardless of the sensitivity ratio, so that a single reference MZI can be used to correct multiple nominal MZIs. This first method relaxes the acquisition frequency requirement by a factor of almost two. However, it cannot correct phase jumps of 4π, 6π or higher between two measurement points. The second method is based on the comparison between the measured phase from the nominal MZI and the phase expected from the reference MZI. It can correct multiple 2π phase jumps but requires at least one reference MZI per biofunctionalization. It will also constrain the corrected phase to lie in a limited interval of $[-\pi, +\pi]$ around the expected value, and might fail to correct phase shifts above a few tens of radians depending on the disparity of the nominal sensors responses. Nonetheless, for phase shift lower than typically 20 radians, this method allows the lowering of the acquisition frequency almost arbitrarily.

Keywords: silicon photonics; Mach Zehnder interferometers; phase unwrapping; biosensors; olfactory sensors

Citation: Laplatine, L.; Messaoudene, S.; Gaignebet, N.; Herrier, C.; Livache, T. Correction of 2π Phase Jumps for Silicon Photonic Sensors Based on Mach Zehnder Interferometers with Application in Gas and Biosensing. *Sensors* **2024**, *24*, 1712. https://doi.org/10.3390/s24051712

Academic Editors: Bruno Ando, Luca Francioso and Pietro Siciliano

Received: 24 January 2024
Revised: 21 February 2024
Accepted: 26 February 2024
Published: 6 March 2024

1. Introduction

Silicon photonics make it possible to engineer highly sensitive and miniaturized sensors with high yield and high reproducibility. Among the various transduction components available, such as micro-ring resonators, photonic crystals or Bragg gratings, Mach Zehnder Interferometers (MZIs) have been widely explored and optimized [1]. MZIs are probably the easiest to design and manufacture, as they do not necessarily require a feature size below 300 nm, which is accessible using 248-UV dry lithography. They also exhibit state-of-the-art performances while only requiring commercially available optoelectronics parts, such as VCSELs for light source and CMOS imagers for light detection.

The basis of refractive-index-based sensing with integrated MZIs was introduced in the late 1980s [2] with proofs of concept for gas and biosensors in the early 1990s [3,4]. Since then, one of the major improvements was to use a coherent detection scheme to solve the issues of absolute intensity noise, sensitivity fading near extrema and ambiguity in phase direction [5,6]. Nowadays, integrated MZI sensors have been applied to a wide range of applications such as biosensing [7], temperature monitoring [8], gas detection in the ppb range [9], odor identification in the ppm range [10], ultrasound recording [11], refractive index sensing [12] down the the 10^{-8} RIU range and wavelength tracking [13], just to name a few.

Typically, the measured light intensity is converted into a phase φ expressed in radians. φ corresponds to the phase delay difference between the sensing and reference arm and its value lies in the interval $[0, 2\pi]$ or $[-\pi, \pi]$. Thus, the signal measured by interferometric methods is always intrinsically 2π periodic. To track the physical value of interest over time, also called the sensorgram, φ needs first to be unwrapped in order to measure the accumulated phase $\Phi(t)$ which is not bonded to any interval anymore. This phase is finally converted into another unit through a calibration.

This 2π periodicity imposes a lower limit on the acquisition frequency (Nyquist frequency) to ensure that less than a π phase shift has happened between each measurement point. Otherwise, the unwrapping function will output an erroneous $\Phi(t)$, offset by multiples of 2π (referred to as "2π phase jumps" later in the text). Unfortunately, the minimum acquisition frequency is proportional to the sensor sensitivity and can reach tens of Hz for gas sensing, especially with highly concentrated samples. This implies relatively high-speed imagers and electronics, as well as powerful light sources with high optical coupling to reduce the exposure time.

Phase unwrapping error is a well-known issue in many interferometry applications, such as 3D-imaging [14], displacement sensors [15] or distributed fiber-based strain sensors [16]. However, for silicon photonic sensors, this drawback is rarely mentioned in the literature. In 2017, Milvich et al. proposed an elegant solution based on coupling a coherent detection with a wavelength modulation in order to access the absolute interference-order [17]. However, the wavelength modulation range should exceed the MZI free spectral range (FSR). Since VCSEL wavelength spanning by current modulation does not exceed a few nm, this technique requires short FSR MZIs which are more sensitive to laser wavelength drift, jitter and line-width, whereas large FSR MZIs are compatible with lower-cost lasers [18], on-chip non-monochromatic organic lasers [19] or even LEDs [20]. Moreover, low-pass filters or lock-in amplifiers are required to access the precise phase shift since several modulation periods should be averaged. The acquisition frequency should thus be tens of times faster than the typical time constant of the physical phenomenon to be tracked.

Here, we describe two different methods to detect and correct 2π phase jumps by using reference sensors with lower sensitivities. Our solution makes it possible to lower the acquisition frequency without affecting the optoelectronic hardware or the nominal MZI design. Our experimental demonstrations are based on MZI sensors but can be easily extrapolated to micro-ring resonator sensors, which suffer from the same periodicity issue with respect to their FSR.

2. Materials and Methods

Figure 1 show pictures of two silicon photonic dies. The left die is a typical MZI die with a 60-MZI matrix and two lower sensitivity MZIs. Each nominal MZI has a sensing arm length of 9.5 mm. The right die has been designed to assess the presented correction methods and contains 5 groups of 12 identical MZIs. The MZI arm lengths, and thus their sensitivities, are increased across the 5 groups. The sensing arm of the less sensitive group is 250 μm long. The arm lengths of the following groups are increased by factors of 7, 11, 23 and 31 to reach a maximum length of 7.75 mm. A wavelength and a temperature variation tracker are also visible in both dies but were not used in this study. The silicon photonic dies were designed and produced at CEA-Leti on a 200 mm CMOS platform as described in Ref. [10].

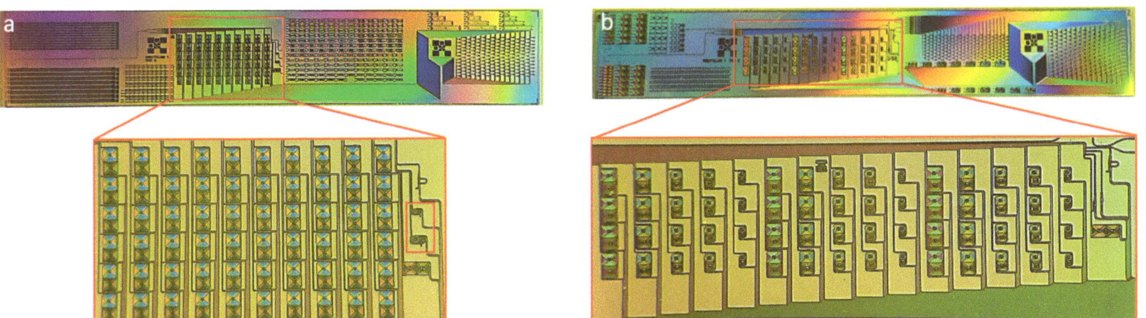

Figure 1. Dies with lower sensitivity reference Mach Zehnder Interferometers (MZIs). (**a**) Standard die with only two lower sensitivity reference MZIs highlighted in the red rectangle. (**b**) Special die designed to evaluate both methods with 5 sensitivity groups, each composed of 12 identical MZIs .

Measurements were performed on a custom optical bench with a single mode VCSEL emitting at $\lambda = 850$ nm aligned to the input grating coupler (GC) and a CMOS imager aligned to the output GC array. Data acquisition and processing were performed by custom MATLAB and Python scripts. Liquid samples were injected by a pressure-regulated system (Fluigent, Paris, France) at a constant flow rate. Gas samples were manually approached from the sensor input tubing while the output tubing was connected to a vacuum pump. A fluidic restriction was used to limit the flow rate. Note that the absolute values of flow rates or sample concentrations have no particular importance here since performance comparisons are relative (i.e., how much the frame rate can be reduced without signal integrity loss). Phase monitoring was performed at 20 Hz for liquid sample injection and 60 Hz for gas sample injection. The frame rate was then artificially lower in post-processing by skipping measurement points until 2π phase jumps were observed. The experiment with a liquid sample consisted in acquiring a baseline with deionized water, then injecting a Phosphate Buffer Saline solution at a concentration of $0.5\times$ (Thermo Scientific Chemicals (Waltham, MA, USA)), and finally rinsing with deionized water. The experiment with a gas sample consisted in acquiring a baseline with air, then injecting the head space with a pure β-pinene solution (1 mL) (Sigma-Aldrich (St. Louis, MO, USA)) contained in a 10 mL flask, and finally rinsing with air.

3. Results

3.1. Phase Extraction and Phase Unwrapping Errors

Figure 2 presents the main steps to compute the phase shift $\Delta\Phi(t)$ from the measured intensity variations P_1, P_2 and P_3 from a three-port coherent MZI.

Each P_i can be written as:

$$P_i \propto 1 + V \times cos[\varphi + (i-1)\frac{2\pi}{3}].\qquad(1)$$

V is the visibility of the interference which varies from 1 to 0. It is a function of the lengths L and propagation losses α of both arms:

$$V = \frac{2 \times e^{-\frac{1}{2}(\alpha_s L_s + \alpha_r L_r)}}{e^{-\alpha_s L_s} + e^{-\alpha_r L_r}}.\qquad(2)$$

In Equation (1), the MZI phase φ is a function of the wavelength, the waveguide lengths and the effective indices:

$$\varphi = \frac{2\pi}{\lambda}(L_s \times n_{eff-s} - L_r \times n_{eff-r}).\qquad(3)$$

From the three P_i which are phase shifted by $2\pi/3$, it is possible to calculate two values phase shifted by $\pi/2$ called the in-phase (I) and quadrature (Q) components:

$$I = P_2 - \frac{P_1 + P_3}{2}, \tag{4}$$

$$Q = \sqrt{3} \times \frac{P_1 - P_3}{2}. \tag{5}$$

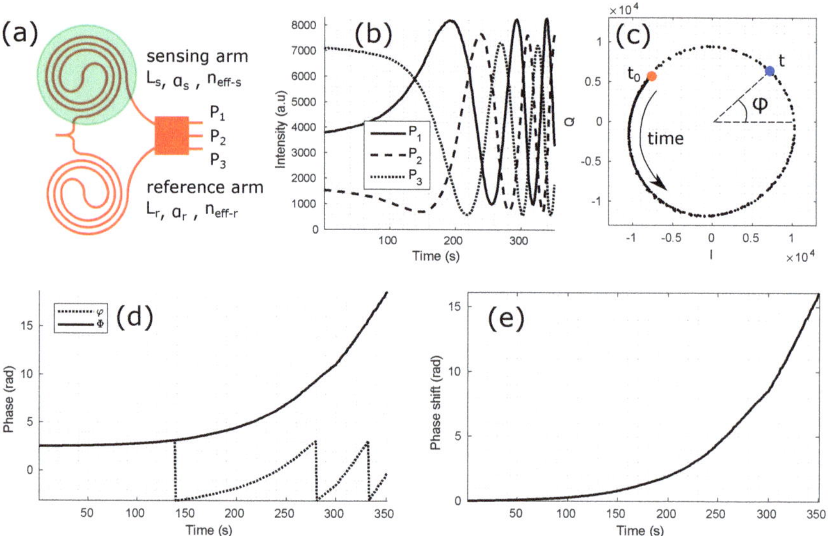

Figure 2. Main steps for phase shift calculation: (**a**) Schematic of a three-port MZI. (**b**) Acquisition of the three optical outputs. (**c**) I and Q diagram. (**d**) Phase unwrapping. (**e**) Phase shift.

I and Q can be represented as the x and y coordinates of a point circularly turning in the IQ diagram. Alternatively, they correspond to the real and imaginary part of a complex number. In that case, the IQ diagram is equivalent to the complex plane. The phase φ corresponds to the angle of each point with respect to the x-axis:

$$\varphi = arctan(I, Q) = Arg(I + iQ) \tag{6}$$

It is obvious from the 4-quadrant inverse tangent function that φ can only have values from $-\pi$ to $+\pi$. To access a meaningful measurement, φ needs to be unwrapped to obtain Φ as shown in Figure 2d. The phase unwrapping algorithm finds the smallest angle between two successive times t in the IQ diagram and adds it to the accumulated phase Φ with a positive sign if the smallest angle is found in the counter clockwise direction or a negative sign if it is found in the clockwise direction. The measured phase φ and the unwrapped phase Φ are thus always equal modulo 2π:

$$\Phi = U(\varphi) = \varphi + 2\pi n \ ; \ \ n \in Z, \tag{7}$$

where U is the unwrapping function. For gas and biosensors based on long sensing waveguides, the initial phases of a matrix of MZIs are often randomly distributed because of small fabrication variations, surface biofunctionalization differences (e.g., different probe molecules on each MZI) as well as differences in surface interaction from previous experiments. Therefore, only the phase shift $\Delta\Phi$ with respect to the initial phase of a measurement is taken into account:

$$\Delta\Phi(t) = \Phi(t) - \Phi(t = 0). \tag{8}$$

From the phase measurement and a calibration, it is finally possible to extract the physical value of interest (temperature, wavelength, mechanical deformation or constraint, sample refractive index, surface adsorption, etc. ...).

A phase unwrapping error is illustrated in Figure 3 by comparing two MZIs with different sensitivities. The phase φ of the less sensitive MZI, in blue, can be correctly unwrapped since the phase steps between two successive times t remain well below π. In contrast, the unwrapping function fails to output the real phase shift evolution for the most sensitive MZI, in red, as the phase step reaches a value higher than π near 9 s. This condition can be written as:

$$max\{|\frac{\partial \Delta\Phi(t)}{\partial t}|\} \leq \pi. \tag{9}$$

We can derive two different methods to detect and correct phase unwrapping errors. Both methods are based on at least one reference MZI sensor with a lower sensitivity than the nominal MZI to be corrected. It is assumed that no 2π phase jump has happened in the reference MZI.

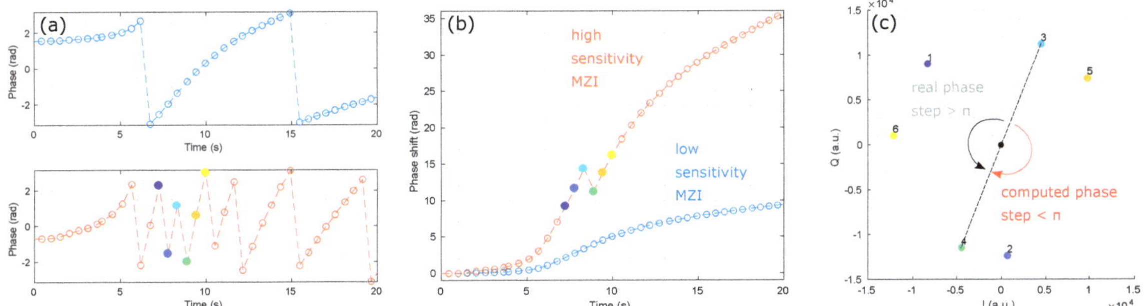

Figure 3. Phase unwrapping error. (**a**) Measured phase φ of two MZIs with different sensitivities. (**b**) Unwrapped phase shift $\Delta\Phi$. (**c**) IQ diagram highlighting the origin of the phase unwrapping error.

3.2. First Method: Phase Shift Derivative Sign Comparison

As illustrated in Figure 4, in the first method, 2π phase jumps are detected for every time t when two conditions are met:

1. The nominal and reference MZI phase shift derivatives have opposite signs:

$$sgn(\frac{\partial \Delta\Phi_{nom}(t)}{\partial t}) \neq sgn(\frac{\partial \Delta\Phi_{ref}(t)}{\partial t}). \tag{10}$$

2. The phase shift derivative of the nominal MZI exceeds a threshold value ε (for instance 0.3 rad). This aims at avoiding random sign noise from the flat parts of the sensorgram.

$$|\frac{\partial \Delta\Phi_{nom}(t)}{\partial t}| \geq \varepsilon. \tag{11}$$

To obtain the corrected phase shift $\widehat{\Delta\Phi}_{nom}(t)$, a 2π offset is added to or removed from any erroneous point according to the sign of the reference MZI derivative:

$$\widehat{\Delta\Phi}_{nom}(t) = \Delta\Phi_{nom}(t) + sgn(\frac{\partial \Delta\Phi_{ref}(t)}{\partial t}) \times 2\pi. \tag{12}$$

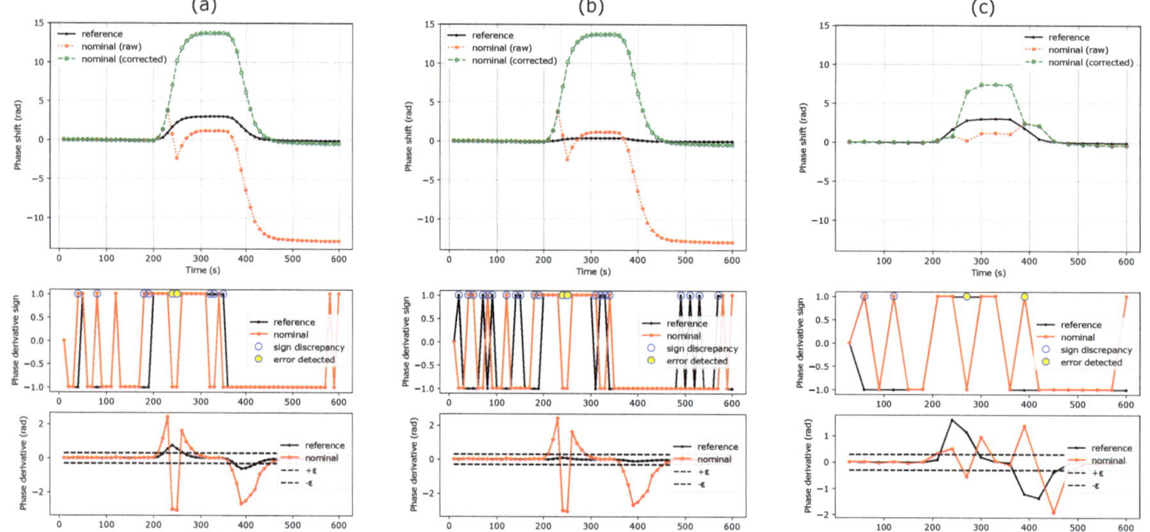

Figure 4. Original and corrected phase shifts using the phase shift derivative sign comparison method for a refractive index step from water to PBS 0.5×. (**a**) Sensitivity ratio of 4.5, threshold of 0.3 rad. (**b**) Sensitivity ratio of 31, threshold of 0.3 rad. (**c**) Sensitivity ratio of 4.5, threshold of 0.3 rad at a frame rate divided by 3. This method fails to retrieve the real phase shift.

3.3. Second Method: Phase Shift Ratio Comparison

As shown in Figure 5, in the second method, the phase shift ratio R between the nominal and reference MZIs is used to detect erroneous points. In the absence of a phase unwrapping error, this ratio should be equal to the sensitivity ratio of the nominal (S_{nom}) to the reference (S_{ref}) MZI. R can be set by design, for instance, using a sensing waveguide length difference between the nominal and reference MZI, or measured experimentally during a calibration protocol as the ratio of maximum phase shifts. In the latter case, the protocol should ensure that no 2π phase jump can happen. This second method assumes that R is known and remains constant:

$$R = \frac{S_{nom}}{S_{ref}} \quad or \quad R = \frac{max(\Delta\Phi_{nom})}{max(\Delta\Phi_{ref})}. \tag{13}$$

The algorithm minimizes the discrepancy $\delta(t)$ between an expected and a potential phase shift. The former is equal to the reference MZI phase shift times the sensitivity ratio:

$$\Delta\Phi_{nom-expected}(t) = R \times \Delta\Phi_{ref}(t). \tag{14}$$

The later corresponds to the nominal MZI phase shift plus a given number of 2π phase jumps:

$$\Delta\Phi_{nom-k}(t) = \Delta\Phi_{nom}(t) + 2k\pi \; ; \; k \in Z. \tag{15}$$

In practice, only a limited number of 2π phase jumps can be tested at every time t, for instance $-K \leq k \leq K$ with $K = 8$. The smallest discrepancy δ can be written as:

$$\delta(t) = \min_{-K \leq k \leq K} \left\{ \Delta\Phi_{nom-k}(t) - \Delta\Phi_{nom-expected}(t) \right\}. \tag{16}$$

Note that $|\delta| \leq \pi$. At every time t, there is a corresponding discrepancy $\delta(t)$ and a best 2π phase jump number $\hat{k}(t)$ so that the corrected phase shift can be written as:

$$\widehat{\Delta\Phi}_{nom}(t) = \Delta\Phi_{nom}(t) + 2\hat{k}(t)\pi = R \times \Delta\Phi_{ref}(t) + \delta(t). \tag{17}$$

Alternatively, this method aims at finding the k value that makes the measured phase shift ratio $R_k(t) = \Delta\Phi_{nom-k}(t)/\Delta\Phi_{ref}(t)$ approach the expected one R.

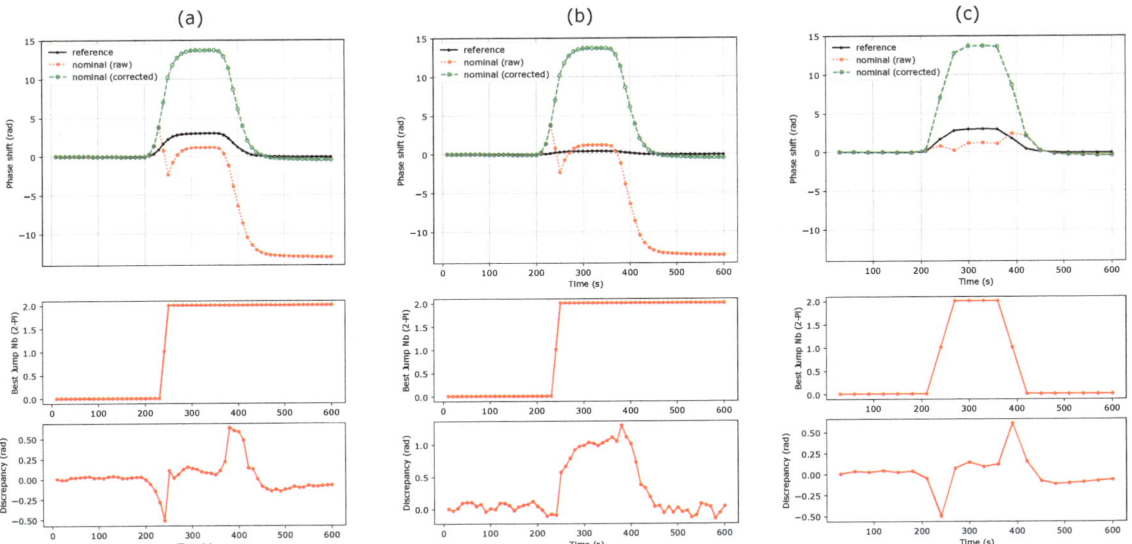

Figure 5. Original and corrected phase shifts using the phase shift ratio comparison method. (**a**) $R = 4.5$. (**b**) $R = 31$. (**c**) $R = 4.5$ at a frame rate divided by 3. The method remains efficient regardless the frame rate reduction.

4. Discussion

As seen in Figure 4a,b, the first method can efficiently correct successive phase unwrapping errors. Note that the sign discrepancy count increases as the sensitivity of the reference MZI decreases. This can be attributed to the lower signal-to-noise ratio of the reference MZI. Depending on the measurement noise floor and sensitivity ratio between the nominal and reference MZIs, a threshold can also be set on the reference MZI to avoid false error detection. This method cannot correct more than one 2π phase jump at a given time (Figure 4c). The threshold condition slightly reduces the maximum phase shift derivative that can be efficiently corrected to:

$$max\left\{|\frac{\partial\Delta\Phi_{nom}(t)}{\partial t}|\right\} \leq 2\pi - \varepsilon. \tag{18}$$

This improves the condition stated in Equation (9) by a factor of almost 2. Since no assumption is needed on the sensitivity ratio, a single reference MZI can be used for the correction of a full MZI matrix (Figure 1a), which makes this method efficient in terms of silicon real estate.

The second method is not restricted in terms of phase shift derivative and is even able to correct multiple 2π phase jumps at once, as shown in Figure 6. However, it relies on a constant sensitivity ratio, which, in practice, might slightly evolve during an experiment or across several experiments. This can be caused by uncompensated drifts, biofunctionalization aging and variability, poor synchronicity between the reference and nominal MZIs as well as sample inhomogeneity inside the fluidic chamber. Thus, the higher the ratio, the higher the uncertainty on the correction truthiness. However, as for the first method, the higher the ratio, the lower the risk of phase unwrapping errors on the reference MZI. A trade off must thus be found. As mentioned in Section 2, using sensitivity ratios corresponding to prime numbers slightly limits the risk of missing 2π phase jumps, as one jump on a reference MZI should not result exactly in two or more jumps on another reference or nominal MZI.

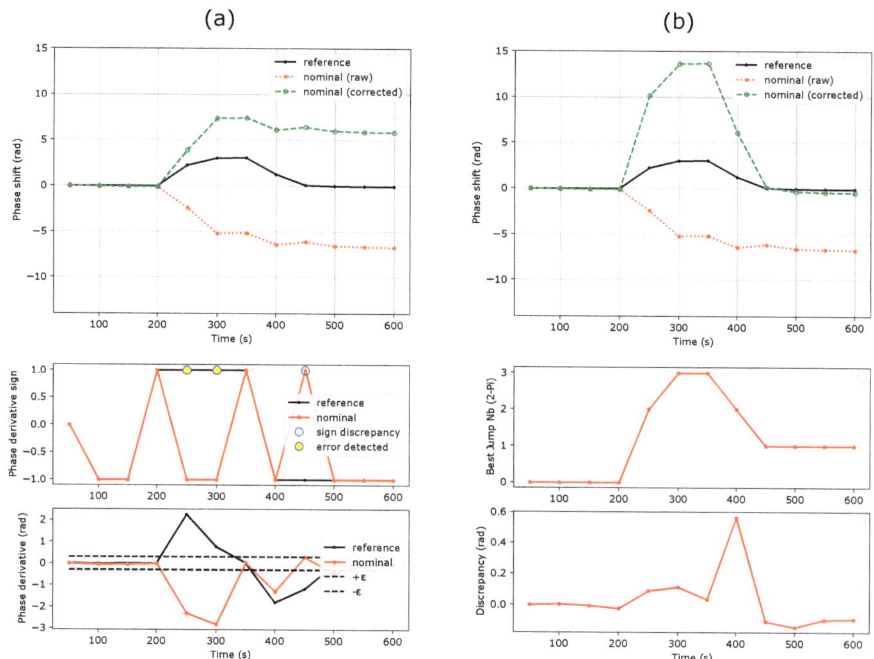

Figure 6. Original and corrected phase shifts using both methods in the case of multiple 2π phase jumps (frame rate further reduced by a factor of 1.7 compared to Figures 4 and 5). (**a**) Method 1 fails. (**b**) Method 2 succeeds.

In addition, as opposed to the first method, for biosensing experiments, each biofunctionalization group must have its own reference MZI since this would affect the sensitivity ratio. In addition, if a single reference MZI is used to correct multiple nominal MZIs, it will force all their phase shifts to lie a in restricted interval of $[+\pi, -\pi]$, while true values might exhibit a higher disparity, especially for the high absolute phase shifts. Figure 7 illustrates the aforementioned limitations and compares methods 1 and 2 for a β-pinene vapour injection. Here, the phase shifts of the 12 MZIs of the less sensitive group (in black) were averaged in order to lower the noise of the reference signal.

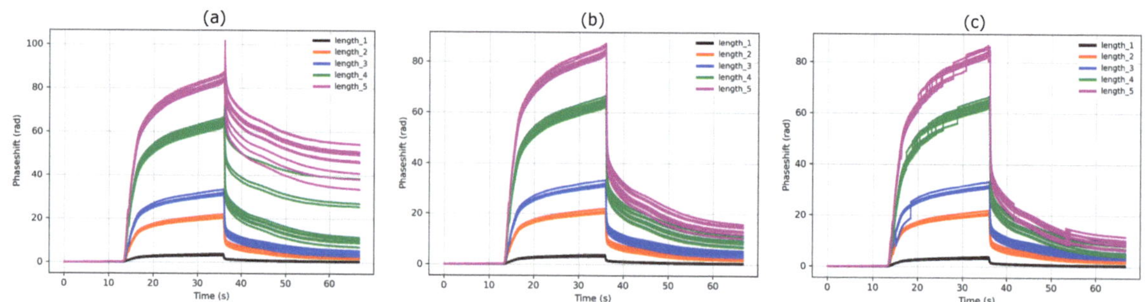

Figure 7. β-pinene vapor injection on die b. The 5 different sensitivity groups are clearly visible, each containing 12 identical MZIs. The mean of the less sensitive group is used as reference. (**a**) Phase unwrapping errors happen for the two most sensitive groups during desorption in the original phase shift. (**b**) Method 1 succeeds in correcting these errors. (**c**) Method 2 fails at perfectly correcting the two most sensitive groups and generates new errors in the three most sensitive groups.

Figure 8 puts both methods in perspective by defining regions of efficacy with respect to the nominal MZI absolute phase shift and absolute phase shift derivative, each method being primarily limited by one of these two aspects. To robustify the correction algorithm, both methods can be combined and several reference MZIs with different sensitivity ratios can be used.

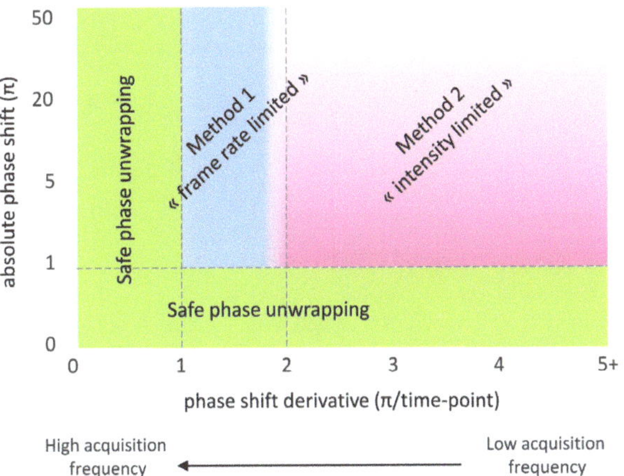

Figure 8. Region of efficacy of the presented methods in terms of absolute phase shift (i.e., intensity) and absolute phase shift derivative (i.e., frame rate).

5. Conclusions

This study propose two new methods to relax the acquisition frequency requirement for silicon photonics sensors based on Mach Zehnder interferometers. Both methods are based on reference MZI sensors with lower sensitivities than the nominal sensors. A dedicated photonic MZI sensor die was designed, fabricated and used to acquire representative experimental data. Both methods are theoretically detailed and applied on datasets to illustrate their respective strengths and weaknesses. From these examples, we can conclude that the acquisition frequency can be lowered by a factor of $\sim$2 in most cases. We can also predict that if the maximum achievable phase shift can be known in advance and remains below a few tens of radians, then the acquisition frequency can be almost arbitrarily lowered. This work should help photonic designers to make more reliable and power-efficient sensors for a wide range of applications beyond gas and biosensing.

6. Patents

Patent referenced FR3129480 results from the work reported in this manuscript.

Author Contributions: Conceptualization, L.L., C.H. and T.L.; Formal analysis, L.L.; Investigation, L.L. and S.M.; Methodology, L.L. and S.M.; Project administration, T.L.; Software, N.G.; Validation, L.L. and N.G.; Visualization, N.G.; Writing—original draft, L.L.; Writing—review and editing, L.L., N.G., C.H. and T.L. All authors have read and agreed to the published version of the manuscript.

Funding: This research was funded by Bpifrance, 2020-PSPC-5 and IPCEI Nano 2022.

Institutional Review Board Statement: Not applicable.

Informed Consent Statement: Not applicable.

Data Availability Statement: Dataset available on request from the authors.

Acknowledgments: The authors sincerely acknowledge CEA-Leti silicon fabrication division, especially Maryse Fournier.

Conflicts of Interest: L.L. and T.L. hold a patent on this work. T.L. was CSO of Aryballe. C.H. is an employee of Aryballe. The remaining authors have no conflicts of interest to declare.

References

1. Luan, E.; Shoman, H.; Ratner, D.M.; Cheung, K.C.; Chrostowski, L. Silicon photonic biosensors using label-free detection. *Sensors* **2018**, *18*, 3519. [CrossRef] [PubMed]
2. Hollenbach, U.; Efstathiou, C.; Fabricious, N.; Oeste, H.; Gotz, H. Integrated Optical Refractive Index Sensor by Ion-Exchange in Glass. In *Micro-Optics*; SPIE: Bellingham, WA, USA, 1989; Volume 1014, pp. 77–80. [CrossRef]
3. Fabricius, N.; Gauglitz, G.; Ingenhoff, J. A gas sensor based on an integrated optical Mach-Zehnder interferometer. *Sens. Actuators B Chem.* **1992**, *7*, 672–676. [CrossRef]
4. Ingenhoff, J.; Drapp, B.; Gauglitz, G. Biosensors using integrated optical devices. *Fresenius' J. Anal. Chem.* **1993**, *346*, 580–583. [CrossRef]
5. Luff, B.J.; Wilkinson, J.S.; Piehler, J.; Hollenbach, U.; Ingenhoff, J.; Fabricius, N. Integrated optical Mach-Zehnder biosensor. *J. Light. Technol.* **1998**, *16*, 583–592. [CrossRef]
6. Halir, R.; Vivien, L.; Le Roux, X.; Xu, D.X.; Cheben, P. Direct and sensitive phase readout for integrated waveguide sensors. *IEEE Photonics J.* **2013**, *5*, 6800906. [CrossRef]
7. Angelopoulou, M.; Petrou, P.; Misiakos, K.; Raptis, I.; Kakabakos, S. Simultaneous Detection of Salmonella typhimurium and Escherichia coli O157:H7 in Drinking Water and Milk with Mach–Zehnder Interferometers Monolithically Integrated on Silicon Chips. *Biosensors* **2022**, *12*, 507. [CrossRef] [PubMed]
8. Guan, X.; Wang, X.; Frandsen, L.H. Optical temperature sensor with enhanced sensitivity by employing hybrid waveguides in a silicon Mach-Zehnder interferometer. *Opt. Express* **2016**, *24*, 16349. [CrossRef] [PubMed]
9. Antonacci, G.; Goyvaerts, J.; Zhao, H.; Baumgartner, B.; Lendl, B.; Baets, R. Ultra-sensitive refractive index gas sensor with functionalized silicon nitride photonic circuits. *APL Photonics* **2020**, *5*, 081301. [CrossRef]
10. Laplatine, L.; Fournier, M.; Gaignebet, N.; Hou, Y.; Mathey, R.; Herrier, C.; Liu, J.; Descloux, D.; Gautheron, B.; Livache, T. A silicon photonic olfactory sensor based on an array of 64 biofunctionalized Mach-Zehnder interferometers. *Opt. Express* **2022**, *30*, 33955–33968. [CrossRef] [PubMed]
11. Ouyang, B.; Li, Y.; Kruidhof, M.; Horsten, R.; van Dongen, K.W.A.; Caro, J. On-chip silicon Mach–Zehnder interferometer sensor for ultrasound detection. *Opt. Lett.* **2019**, *44*, 1928. [CrossRef] [PubMed]
12. Leuermann, J.; Fernández-Gavela, A.; Torres-Cubillo, A.; Postigo, S.; Sánchez-Postigo, A.; Lechuga, L.M.; Halir, R.; Molina-Fernández, Í. Optimizing the limit of detection of waveguide-based interferometric biosensor devices. *Sensors* **2019**, *19*, 3671. [CrossRef] [PubMed]
13. Ba, R.; Mu, P. Opto-electronic wavelength tracker in PDs. In Proceedings of the ECIO Conference, P004, Berlin, Germany, 23–26 April 2014; pp. 25–26.
14. Lv, Z.; Zhu, K.; He, X.; Zhang, L.; He, J.; Mu, Z.; Wang, J.; Zhang, X.; Hao, R. Phase Unwrapping Error Correction Based on Multiple Linear Regression Analysis. *Sensors* **2023**, *23*, 2743. [CrossRef] [PubMed]
15. Ellis, J.D.; Baas, M.; Joo, K.N.; Spronck, J.W. Theoretical analysis of errors in correction algorithms for periodic nonlinearity in displacement measuring interferometers. *Precis. Eng.* **2012**, *36*, 261–269. [CrossRef]
16. Lu, X.; Krebber, K. Phase error analysis and unwrapping error suppression in phase-sensitive optical time domain reflectometry. *Opt. Express* **2022**, *30*, 6934. [CrossRef] [PubMed]
17. Milvich, J.; Kohler, D.; Freude, W.; Koos, C. Mach-Zehnder interferometer readout for instantaneous sensor calibration and extraction of endlessly unwrapped phase. In Proceedings of the 2017 IEEE Photonics Conference (IPC), Orlando, FL, USA, 1–5 October 2017; pp. 567–568. [CrossRef]
18. Leuermann, J.; Fernandez-Gavela, A.; Halir, R.; Ortega-Monux, A.; Wanguemert-Perez, J.G.; Lechuga, L.M.; Molina-Fernandez, I. Silicon photonic label free biosensors with coherent readout. In Proceedings of the International Conference on Transparent Optical Networks, Bari, Italy, 19–23 July 2020; Volume 2020. [CrossRef]
19. Kohler, D.; Schindler, G.; Hahn, L.; Milvich, J.; Hofmann, A.; Länge, K.; Freude, W.; Koos, C. Biophotonic sensors with integrated Si3N4-organic hybrid (SiNOH) lasers for point-of-care diagnostics. *Light. Sci. Appl.* **2021**, *10*, 64. [CrossRef] [PubMed]
20. Deng, Q.; Li, X.; Chen, R.; Zhou, Z. Low-cost silicon photonic temperature sensor using broadband light source. In Proceedings of the 11th International Conference on Group IV Photonics (GFP), Paris, France, 27–29 August 2014; Volume 23, pp. 85–86.

sensors

Article

Exploring Aesthetic Perception in Impaired Aging: A Multimodal Brain—Computer Interface Study

Livio Clemente [1], Marianna La Rocca [2,3], Giulia Paparella [1], Marianna Delussi [1], Giusy Tancredi [1], Katia Ricci [1], Giuseppe Procida [1], Alessandro Introna [1], Antonio Brunetti [4], Paolo Taurisano [1], Vitoantonio Bevilacqua [4] and Marina de Tommaso [1,*]

[1] Translational Biomedicine and Neuroscience (DiBraiN) Department, University of Bari, 70124 Bari, Italy; livio.clemente@uniba.it (L.C.); giulia.paparella@uniba.it (G.P.); marianna.delussi@uniba.it (M.D.); tancredi.giusy.96@gmail.com (G.T.); katiari86@gmail.com (K.R.); beppepine@gmail.com (G.P.); ale.introna@gmail.com (A.I.); paolo.taurisano@uniba.it (P.T.)

[2] Interateneo Department of Fisica 'M. Merlin', University of Bari, 70125 Bari, Italy; marianna.larocca@uniba.it

[3] Laboratory of Neuroimaging, USC Stevens Neuroimaging and Informatics Institute, Keck School of Medicine of USC, University of Southern California, Los Angeles, CA 90033, USA

[4] Electrical and Information Engineering Department, Polytechnic of Bari, 70125 Bari, Italy; antonio.brunetti@poliba.it (A.B.); vitoantonio.bevilacqua@poliba.it (V.B.)

* Correspondence: marina.detommaso@uniba.it; Tel.: +39-08-0559-6739 (ext. 6870)

Abstract: In the field of neuroscience, brain–computer interfaces (BCIs) are used to connect the human brain with external devices, providing insights into the neural mechanisms underlying cognitive processes, including aesthetic perception. Non-invasive BCIs, such as EEG and fNIRS, are critical for studying central nervous system activity and understanding how individuals with cognitive deficits process and respond to aesthetic stimuli. This study assessed twenty participants who were divided into control and impaired aging (AI) groups based on MMSE scores. EEG and fNIRS were used to measure their neurophysiological responses to aesthetic stimuli that varied in pleasantness and dynamism. Significant differences were identified between the groups in P300 amplitude and late positive potential (LPP), with controls showing greater reactivity. AI subjects showed an increase in oxyhemoglobin in response to pleasurable stimuli, suggesting hemodynamic compensation. This study highlights the effectiveness of multimodal BCIs in identifying the neural basis of aesthetic appreciation and impaired aging. Despite its limitations, such as sample size and the subjective nature of aesthetic appreciation, this research lays the groundwork for cognitive rehabilitation tailored to aesthetic perception, improving the comprehension of cognitive disorders through integrated BCI methodologies.

Keywords: BCI; aesthetic; fNIRS; EEG; impaired aging

Citation: Clemente, L.; La Rocca, M.; Paparella, G.; Delussi, M.; Tancredi, G.; Ricci, K.; Procida, G.; Introna, A.; Brunetti, A.; Taurisano, P.; et al. Exploring Aesthetic Perception in Impaired Aging: A Multimodal Brain—Computer Interface Study. *Sensors* **2024**, *24*, 2329. https://doi.org/10.3390/s24072329

Academic Editor: Chang-Hwan Im

Received: 15 March 2024
Revised: 3 April 2024
Accepted: 3 April 2024
Published: 6 April 2024

1. Introduction

The field of neuroscience is constantly evolving in both research and healthcare, and a major role in this process is assumed by the brain–computer interface (BCI) [1,2]. BCIs are advanced systems that enable direct communication between the human brain and external devices [3], and their efficacy has been demonstrated in multiple settings, from detection to intervention. The main goal of BCI research is to collect the most accurate real-time data on brain activity in the simplest and least invasive way, with the shortest calibration and set-up time [4]. Studies have shown significant improvements in motor function through the use of BCIs, exceeding the performance of traditional control therapies and highlighting its potential in post-stroke motor rehabilitation [5,6]. Additionally, it has been noted that adding local network features to BCIs based on steady-state visual evoked potentials (SSVEPs) can greatly increase the information transfer rate and classification accuracy, providing new opportunities for BCI optimization of performance across a broad spectrum of application domains [7]. The integration of BCIs with other technologies, such as neural signal analysis and data processing, is improving the accuracy and effectiveness of these systems. For example, the use of advanced machine learning algorithms [8] and brain connectivity [9] has refined the decoding of neural signals, improving the interface between the brain and external devices.

To classify and understand the wide range of BCI technologies available, an important distinction is made between invasive and non-invasive systems, each with distinct characteristics and application areas [10]. Invasive BCI systems, which require surgical implantation of electrodes in the brain, including intracortical electrodes [11] and electrocorticography (ECoG) [12], offer high-quality signals with high spatial and temporal resolution, making them particularly useful in research settings and for applications that require very precise control [10]; however, these systems carry greater risks and complications related to the surgical procedure [13].

Non-invasive BCIs capture neural signals directly from the surface of the skull using devices such as electroencephalography (EEG), functional near-infrared spectroscopy (fNIRS), functional magnetic resonance imaging (fMRI), magnetoencephalography (MEG), and positron emission tomography (PET), without the need for invasive surgery [14]. These methods provide unprecedented access to brain activity and enable advanced communication between the brain and external devices in a safe and non-invasive manner. However, MEG, PET, and fMRI are still technically demanding and expensive [15].

In the context of neurosciences, these non-invasive techniques are fundamental for the study of central nervous system activity, and among them, EEG emerges as the most widely used method for measuring electrical brain activity [16]. EEG is particularly prized for its portability, non-invasiveness, and, most importantly, its ability to monitor brain activity with high temporal resolution, making it one of the most effective techniques in this field [17]. In EEG signals, it is possible to identify so-called evoked potentials, which are specific neurophysiological responses measured in reaction to standardized and known sensory stimuli. These evoked potentials provide a functional assessment of sensory systems, including auditory, visual, and somatosensory [18]. The various categories of evoked potentials, which include event-related evoked potentials (ERPs), visual evoked potentials (VEPs) [19], acoustic evoked potentials (AEPs), motor evoked potentials (MRPs), and steady-state visual evoked responses (SSVEPs) [20], are fundamental tools for reflecting brain activity in response to specific sensory stimuli and find application in the monitoring and diagnosis of various neurological disorders in clinical and domestic settings.

Similarly, fNIRS is praised for its ability to monitor cognitive activity by measuring task-related hemodynamic responses in the prefrontal cortex, providing a unique window into brain function during mental tasks [21,22]. Recent studies have shown that fNIRS can be effectively used to record signals from cognitive, visual, and auditory functions [20,23]. This significantly expands the field of application of BCIs, allowing not only the improvement of motor functions as demonstrated in post-stroke rehabilitation [5,6] but also the exploration of new frontiers in the monitoring and intervention of complex cognitive functions. The ability to detect responses related to problem-solving tasks and activities requiring concentration and memory underscores the adaptability of fNIRS to a variety of BCI applications, making the signals not only easily monitored by users, but also closely related to the underlying cognitive activity [24].

Neuroscience research has investigated the impact of visual stimuli, both static and dynamic, on the motor areas of the brain during aesthetic perception [25]. Studies have shown that viewing dynamic works of art, such as those by van Gogh, not only enriches the perceptual experience through complex details but also activates movement-related brain areas, such as the M+ area [26]. These findings suggest a link between aesthetic perception and motor processing [27]. Research suggests that observing static human actions involved in motion can increase cortical activation. The brain areas involved in this process include V5/MT, EBA, and motor cortices [28]. Therefore, it is important to consider the motor area when discussing aesthetic appreciation.

The combination of EEG and fNIRS offers new directions for a comprehensive analysis of brain function by integrating electrical and hemodynamic data [29]. This integrated approach, EEG-fNIRS, has demonstrated its value in various areas of neuroscience, providing a holistic perspective on brain functioning [30].

With this in mind, the purpose of this study is to explore aesthetic perception in individuals with mild impairment using a multimodal approach that combines EEG and fNIRS in the context of non-invasive BCI technologies. The study investigates how changes in brain activity, as measured through these two methodologies, may reflect differences in response to aesthetic stimuli. Our study aims to provide new insights into the underlying neural dynamics and offer suggestions for the development of more effective diagnostic and therapeutic tools by analyzing responses evoked by aesthetic stimuli. We will contribute to exploring the potential of BCI technologies in clinical and research contexts.

2. Materials and Methods

2.1. Study Participants

Twenty participants were recruited from the Psychology and Clinical Neuropsychology Clinic and the U.O.C. University Neurophysiopathology Unit of Bari General Hospital. All participants had a Mini-Mental State Examination (MMSE) score greater than 18 to exclude moderate to severe cognitive impairment [31]. The sample was then divided according to their MMSE score (Figure 1), creating two groups of ten subjects each: a clinical group with MMSE scores from 18 to 23, indicating impaired aging (IA), and a control (CT) group with MMSE scores from 24 to 30 [31]. Healthy and impaired aging groups did not differ in terms of age ($t(18) = -0.67$, $p = 0.510$).

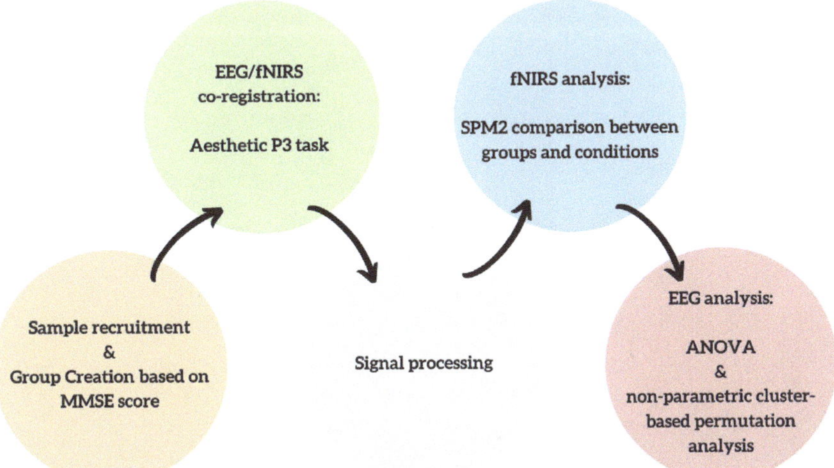

Figure 1. Flow chart of the study design illustrating the stages of recruitment, recording, data processing, and group comparison.

The study was conducted from May 2022 to February 2023. All participants were right-handed and older than 65 years of age. The experimental procedures were approved by the ethics committee of the General Polyclinic of Bari and were performed in accordance with the Declaration of Helsinki. Participants were informed of the purpose of the study before the experiment and provided written informed consent. None of the subjects in this study had any prior experience with the recording devices or the experimental task, and they were all able to independently follow the study's instructions.

2.2. Recording Techniques

The cerebral hemodynamic and bioelectrical activity was recorded using an EEG-fNIRS co-recording cap with an additional black cap placed over it to block out any potential ambient light interference (Figure 1).

The EEG cap consisted of 61 encephalic EEG channels positioned according to the enlarged international standard 10–20 system, as shown in Figure 1 (Fp1, FPZ, Fp2, F7, F3, Fz, F4, F8, T3, C3, Cz, C4, T4, T5, P3, Pz, P4, T6, O1, Oz, O2, AF7, AF3, AFz, AF4, AF8, F5, F1, F2, F6, FT7, FC5, FC3, FC1, FCz, FC2, FC4, FC6, FT8, C5, C1, C2, C6, TP7, CP5, CP3, CP1, CPz, CP2, CP4, CP6, TP8, P5, P1, P2, P6, PO7, PO3, POz, PO4, and PO8). A biauricular reference electrode was used. In addition, to remove any ocular blink artifacts, two electrooculogram detection electrodes were applied at the level of the outer canthus of the right and left eye, while the ground electrode was placed on the right forearm. The impedance was kept below 5 KW. During the EEG recording, we used a digital filter in the 0.1–70 Hz range and a 50 Hz notch filter to allow signal inspection.

fNIRS sensors were placed on the EEG headset, allowing the simultaneous acquisition of both parameters (Figure 2). The continuous wave fNIRS system (NIRSport 8 8, Nirx Medical Technologies LLC, Berlin, Germany) was used to conduct the current investigation. The fNIRS device consists of a multi-channel system able to measure hemodynamic activity fluctuations. For data acquisition, NIRStar 14.2 software was adopted (Version 14, Revision 2, Release Build, 15 April 2016 NIRx Medizintechnik GmbH, Berlin, Germany). The easy-to-use device involved light-emitting diode (LED) sources and photosensitive detectors (sensitivity: >1 pW, dynamic range: >50 dB). More exactly, the data were recorded by eight light sources, from which lights were emitted from each source at two different wavelengths (760 nm and 850 nm), and eight detectors. The transmitter–receiver distance was 30 mm, as recommended in the scientific literature [32]. Sources, detectors, and the layout of the 20 fNIRS channels were located as illustrated in Figure 1. For these specific registrations, the probes were placed on the cap over the primary motor cortex (M1) and supplementary motor cortex. The sampling rate was 7.81 Hz. The optical density was converted to variations in oxyhemoglobin (HbO_2) and deoxyhemoglobin (HbR), using the modified Beer–Lambert law (mBLL) [33]. Before each fNIRS recording, a calibration procedure was employed to determine the signal amplification required for every source–detector pattern.

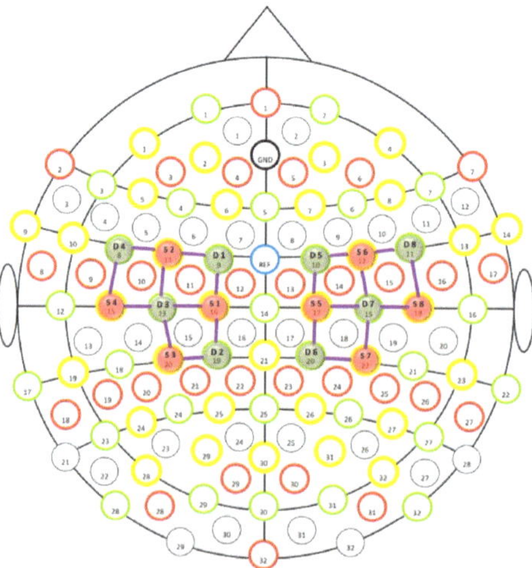

Figure 2. Combined EEG/fNIRS system with 10/20 64 electrodes and 20 NIR channels resulting from 16 optodes (8 sensors in red and 8 detectors in green).

2.2.1. Paradigm

Participants were examined in a well-ventilated room and positioned in front of a screen in an ergonomic position. The tasks included an initial two-minute resting state baseline, in which the subject was asked to fix a cross in the center of the black screen. The methodology adopted consisted of presenting an odd-ball paradigm to the subjects in which frequent and rare stimuli were alternated. A total of 165 visual stimuli were selected, divided into 115 single-color frequent images and 50 target images, consisting of 25 static and 25 dynamic aesthetic stimuli (Figure 3). Each visual stimulus had a duration of 5 s, and the inter-stimulus interval was 9 s. The frequency of occurrence of the standard stimulus was 70%, while for the target stimulus, it was 30%. The motor task at the appearance of the target stimulus (static or dynamic images) consisted of pressing, as quickly as possible, the space bar of a PC keyboard provided to the participant. This keyboard was triggered on the EEG and NIRS trace, and each time it was pressed, an indicator marker appeared on the trace. At the end of the task, participants were asked to rate each image on a Likert scale from 1 to 10 based on their aesthetic appreciation.

Figure 3. Experimental paradigm, showing (**a**) common stimuli; (**b**) target dynamic stimuli on the left side and target static stimuli on the right; (**c**) Likert scale for evaluation of aesthetic appreciation.

2.2.2. fNIRS Data Processing

Data preprocessing was performed using NirsLAB v.2019.04 software. Before data preprocessing, the images were separated into different files according to the level of attractiveness of the images for each subject, dividing into separate files the images that each subject had considered attractive (scores of 7 or higher), neutral (scores of 5 and 6), and unattractive (scores of 4 or lower). Markers were then set to indicate the onset of each condition (common stimulus, target static image, target dynamic image, and resting state period), with a stimulus duration of 14–14–14–120 s, respectively. For fNIRS data preprocessing, discontinuities were removed (STD threshold = 5) and channels were interpolated (maximum frames = 4). Subsequently, the signals recorded in each channel were inspected by removing channels containing sufficient noise (gain setting > 8; coefficient of variation > 7.5), and a bandpass filter of 0.01–0.2 Hz was applied. The hemodynamic states were then calculated for each channel, and the Beer–Lambert law was used to convert the processed signals into variations of oxygenated and deoxygenated hemoglobin concentrations (baseline in frames = 39–195, corresponding in 5–25 s of baseline).

During our fNIRS analysis, we specifically observed the changes in oxyhemoglobin (HbO) triggered by the P300 test. Our analysis focused on stimulus-related changes within a 14-s time window after stimulus onset. This enabled us to capture the immediate hemodynamic dynamics evoked by the different levels of image attractiveness assessed by the subjects.

2.3. EEG Data Processing

The EEG data analysis was performed using an automated workflow based on EEGLAB (version 2022) in MATLAB. We applied an FIR filter to limit the frequency range between 1 and 30 Hz. Next, we implemented the Artifact Subspace Reconstruction (ASR) method to correct the continuous data by discarding unsuitable channels and data segments. We then interpolated the bad channels and recalibrated the entire dataset to the mean. For each 0.5-s time window, we set a maximum deviation in the standard deviation of 20. In addition, we eliminated channels that were inactive for more than 5 s, those with a high-frequency noise standard deviation of less than 4, and those with a correlation greater than 0.8 with adjacent channels. Next, an independent component analysis (ICA) was performed, and artifactual components were automatically excluded through the use of a machine learning algorithm called the Multiple Artifact Rejection Algorithm (MARA). Components with a greater than 50% probability of being artifacts were removed. Finally, the data were divided into epochs in the time interval from -0.3 s to 1 s, and the baseline was corrected.

2.4. Statistical Methods

2.4.1. fNIRS

To conduct an in-depth investigation of the brain areas involved, we performed a topographical analysis using the Generalized Linear Model (GLM) based on Statistical Parametric Mapping for fNIRS (NIRS-SPM), implemented via SPM 8 software in NirsLAB. This approach enabled us to accurately identify the brain regions that were active during task execution for each case. The study employed the Hemodynamic Response Function (HRF) to represent the hemodynamic response to the experimental tasks in the Statistical Parametric Mapping analysis (SPM1, intra-subject). The degree of activation of each channel relative to the baseline was calculated via the beta value.

The study analyzed the differences between subjects using SPM 2, a function implemented in NirsLab, to identify fNIRS channels that showed significant changes in HbO during the P300 task between the different groups.

2.4.2. EEG

In our study, we used two approaches to analyze the EEG data, focusing on the P300 evoked potential. Using Letswave7 software, we first performed a permutation-based ANOVA to identify significant differences in ERP responses between subject groups. This method allowed us to assess statistical differences while maintaining strict control of the type I error rate by generating null distributions from permuted data.

We then implemented a non-parametric cluster-based permutation analysis [34], a robust technique for comparing ERP distributions between control groups and subjects with impaired aging in all conditions (static vs. dynamic * pleasant vs. unpleasant vs.

neutral). This approach, which uses cluster statistics to maximize statistical power while maintaining adequate control over multiple comparisons, is based on the construction of clusters of adjacent electrodes that show statistically significant differences.

For the analysis, electrode neighbors were defined based on the physical distance to construct the neighbor matrix, and ERP averages for the static and dynamic conditions were calculated separately, aggregating data from all subjects for each condition.

A cluster-based analysis was performed on a time interval between 300 and 500 ms, focusing on differences between ERP averages. We identified clusters of electrodes that showed statistically significant differences using a critical value derived from the distribution of the maximum test statistics observed in the permutations.

3. Results

3.1. fNIRS Results

A Student's t-test with a significance level of $p < 0.05$ was used for this analysis. The results revealed significant differences in the hemodynamic response to the P300 test, providing valuable insights into the neural dynamics underlying stimulus processing.

3.1.1. Between Groups Comparison

The analysis showed that patients had significantly higher HbO levels than controls in response to visual stimuli perceived as aesthetically pleasing, in both static (ch1, t = 2.90, $p < 0.05$) and dynamic (ch5, t = 2.61, $p < 0.05$) conditions, indicating greater brain activation (Figure 4). Conversely, controls showed higher HbO levels than patients in the static condition when presented with visual stimuli perceived as unpleasant. For neutral dynamic stimuli (ch10, t = 3, $p < 0.05$), the patients exhibited higher HbO levels than the controls, indicating differential responsiveness depending on the aesthetic value of the stimulus and condition.

3.1.2. ANOVA

An ANOVA provided additional information on the interactions between group, condition, and aesthetic valence of the stimuli. Specifically, on channel 10, the interaction between group and aesthetic valence was significant (F(2) = 3.28, p = 0.042), indicating that the response to different types of aesthetic stimuli varies significantly between patients and controls (Table 1). Channel 20 showed a significant interaction between group (Table 2), condition, and aesthetic valence (F(2) = 3.45, p = 0.036). This suggests that the two groups' brain activation is affected differently by the condition (static or dynamic) and their aesthetic valence.

Table 1. fNIRS Channel 10: ANOVA on group, condition, and aesthetic appreciation.

ch 10	Sum of Squares	df	Mean Square	F	p
group	4.37×10^{-7}	1	4.37×10^{-7}	1.63	0.205
condition	4.76×10^{-7}	1	4.76×10^{-7}	1.77	0.186
aesthetic	1.02×10^{-6}	2	5.08×10^{-7}	1.89	0.156
group x condition	6.08×10^{-7}	1	6.08×10^{-7}	2.27	0.135
group x aesthetic	1.76×10^{-6}	2	8.80×10^{-7}	3.28	0.042 *
condition x aesthetic	1.07×10^{-6}	2	5.36×10^{-7}	2.00	0.141
group x condition x aesthetic	1.20×10^{-6}	2	5.98×10^{-7}	2.23	0.113
Residuals	2.63×10^{-5}	98	2.68×10^{-7}		

Note: * $p < 0.05$.

Table 2. fNIRS Channel 20: ANOVA on Group, Condition, and Aesthetic appreciation.

ch 20	Sum of Squares	df	Mean Square	F	p
group	2.30×10^{-8}	1	2.30×10^{-8}	0.06	0.809
condition	9.68×10^{-7}	1	9.68×10^{-7}	2.47	0.119
aesthetic	3.99×10^{-8}	2	2.00×10^{-8}	0.05	0.950
group x condition	1.48×10^{-6}	1	1.48×10^{-6}	3.78	0.055
group x aesthetic	9.25×10^{-8}	2	4.62×10^{-8}	0.12	0.889
condition x aesthetic	2.22×10^{-6}	2	1.11×10^{-6}	2.84	0.063
group x condition x aesthetic	2.70×10^{-6}	2	1.35×10^{-6}	3.45	0.036 *
Residuals	3.83×10^{-5}	98	3.91×10^{-7}		

Note: * $p < 0.05$.

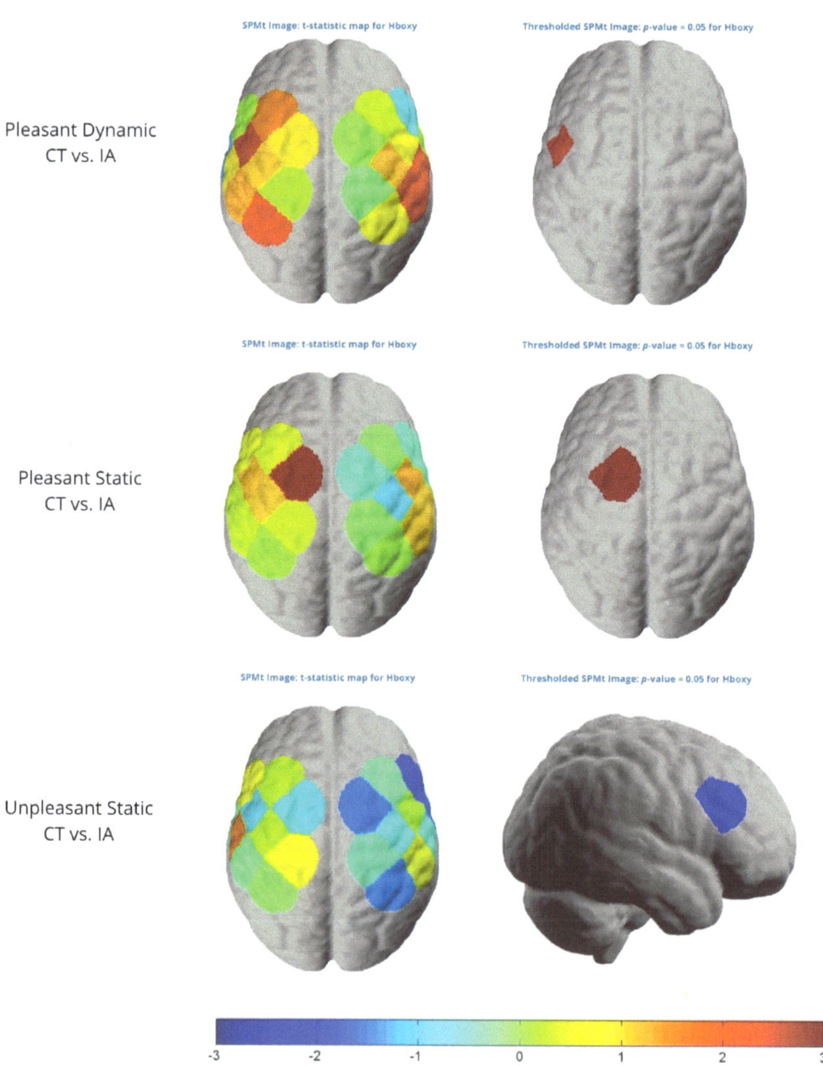

Figure 4. The analysis of oxyhemoglobin (HbO) is presented in two parts: the left side displays the raw beta value, while the right side shows the same values filtered to display only areas where brain activity is statistically significant ($p < 0.05$).

3.2. EEG Results

In this study, we conducted an ANOVA to examine the effects of the interaction between condition, aesthetic appreciation, and group on ERP components, specifically on the late positive potential (LPP) and P300. The results indicated that the interaction between condition and aesthetic appreciation significantly influenced the LPP in the parietal region (POZ) (F = 3.29, p = 0.04; Figure 5). Additionally, the LPP also reached significance (F = 4.09, p = 0.019) in the prefrontal region (FPZ). Regarding image dynamism, significant modulation of P300 (F = 8.1, p = 0.005) and LPP (F = 5.24, p = 0.02) was observed in the POZ (Figure 5). This indicates that the experimental condition has a significant impact on brain activity measured by these metrics. The analysis also revealed that aesthetic appreciation has a significant effect on the LPP in the POZ (F = 3.67, p = 0.029; Figure 5), suggesting that aesthetic evaluations modulate brain activity in this area. The analysis revealed a significant association between the group factor and brain activity in the POZ (F = 14.1, p < 0.001; Figure 5), indicating substantial differences between groups. Additionally, a significant interaction effect was found between the group, experimental condition, and aesthetic appreciation in the FPZ and PO4 (F = 4.08, p = 0.019; Figure 5), suggesting that the relationship between brain activity and aesthetic appreciation is influenced by these factors.

Non-Parametric Cluster-Based Permutation Analysis

To evaluate the robustness of the results of the ANOVA, a non-parametric cluster-based permutation analysis was performed (Table 3).

Table 3. Summary of cluster-based permutation analysis: Cohen's d, cluster Mass, and raw effect differences of clusters.

	Cohen's D	Max Positive Cluster			Max Negative Cluster		
		Mass	ERP (Mean)	ERP (sd)	Mass	ERP (Mean)	ERP (sd)
Pleasant dynamic CT vs. *IA*	2.10	118.10	3.37	3.24	102.92	0.20	1.55
Pleasant static CT vs. *IA*	2.27	128.42	2.80	4.83	88.80	−0.69	2.76
Unpleasant dynamic CT vs. *IA*	1.96	38.21	0.57	5.06	16.06	1.30	8.12
Unpleasant static CT vs. *IA*	2.33	34.67	−1.31	1.57	73.25	1.52	4.18
Neutral dynamic CT vs. *IA*	2.05	53.02	4.44	4.45	14.24	1.37	6.17
Neutral static CT vs. *IA*	1.76	21.31	−0.65	2.26	27.98	0.95	3.15
Control dynamic pleasant vs. *unpleasant*	2.18	11.40	−0.48	2.09	34.99	0.06	2.36
Control dynamic pleasant vs. *neutral*	-	-	-	-	-	-	-
Control dynamic unpleasant vs. *neutral*	1.95	16.85	0.06	2.36	7.13	−3.48	1.07
Patient dynamic pleasant vs. *unpleasant*	2.12	130.55	1.37	1.31	75.97	0.28	6.43
Patient dynamic pleasant vs. *neutral*	1.81	41.33	0.46	1.64	19.79	4.17	7.08
Patient dynamic unpleasant vs. *neutral*	2.04	11.98	−0.28	4.73	30.83	1.98	6.07
Control static pleasant vs. *unpleasant*	2.15	20.29	0.21	1.94	7.66	−0.76	2.41
Control static pleasant vs. *neutral*	2.10	15.41	2.25	3.14	9.44	0.56	1.03
Control static unpleasant vs. *neutral*	2.82	20.15	0.69	3.99	17.51	−1.70	0.94
Patient static pleasant vs. *unpleasant*	2.32	24.60	−1.47	2.62	46.32	0.40	3.31
Patient static pleasant vs. *neutral*	1.75	26.75	−1.17	2.65	12.06	0.42	3.66
Patient static unpleasant vs. *neutral*	2.16	27.86	0.10	5.33	22.55	0.42	3.66
Control pleasant dynamic vs. *static*	1.62	15.38	−3.57	2.98	14.80	−0.85	2.60
Control unpleasant dynamic vs. *static*	2.45	61.84	1.62	4.66	21.37	−1.60	2.18
Control neutral dynamic vs. *static*	2.23	16.94	0.37	1.60	24.52	−1.85	1.75
Patient pleasant dynamic vs. *static*	1.57	31.34	0.71	1.04	14.01	−1.86	2.42
Patient unpleasant dynamic vs. *static*	2.51	60.60	0.22	5.44	57.08	0.93	4.19
Patient neutral dynamic vs. *static*	1.49	27.86	4.36	5.26	33.90	0.32	3.33

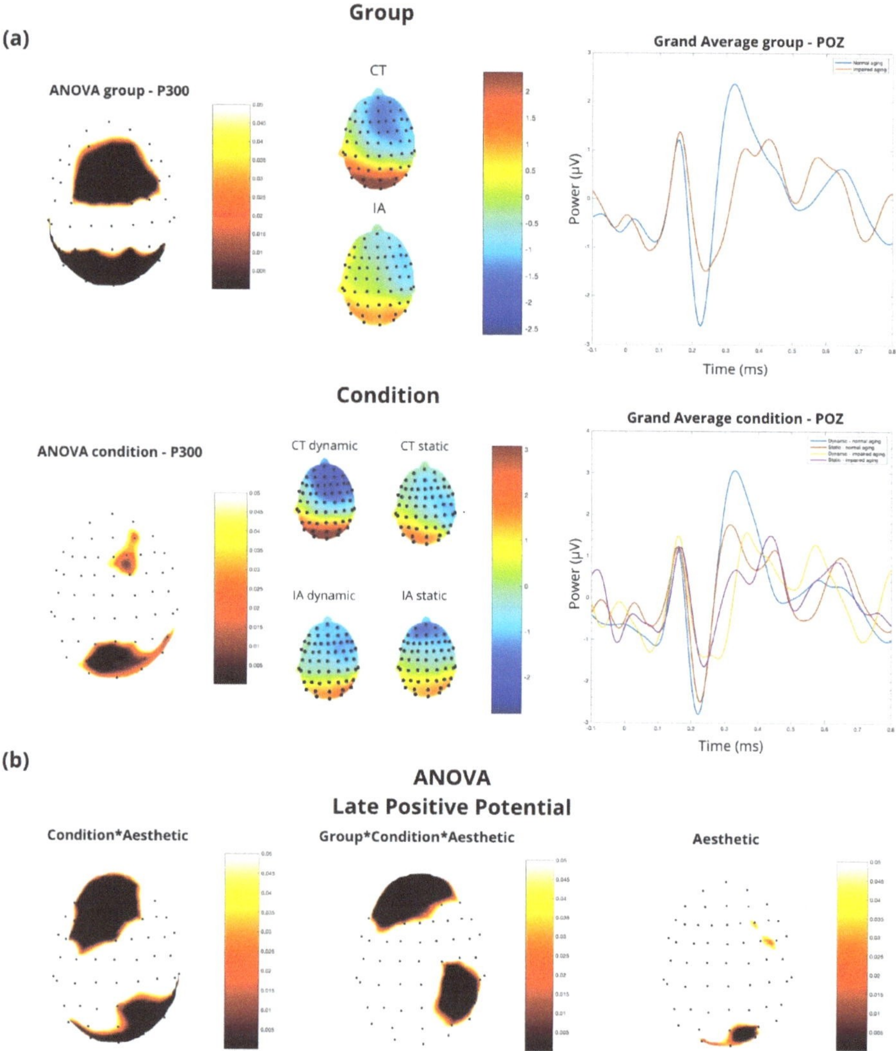

Figure 5. Statistical analysis and electrophysiological responses were used to evaluate aesthetic stimuli. (**a**) P300 ANOVA p-value in group and condition, showing the topographical map of raw activity and the wavelet; (**b**) LPP ANOVA p-value in condition*aesthetic (left), group*condition*aesthetic (central), and aesthetic (right).

When comparing the healthy control (CT) and impaired aging (IA) groups, significant clusters reflecting differences in P300 amplitude were identified. Positive clusters were observed for both pleasant (d = 2.10, Mass = 118.10) and unpleasant (d = 1.96, Mass = 18.21) dynamic stimuli, indicating that the CT group had a more pronounced P300 amplitude than the IA group (Figure 6). This suggests a more pronounced neurophysiological reactivity in controls, indicative of a greater ability to process dynamic stimuli, regardless of their aesthetic valence.

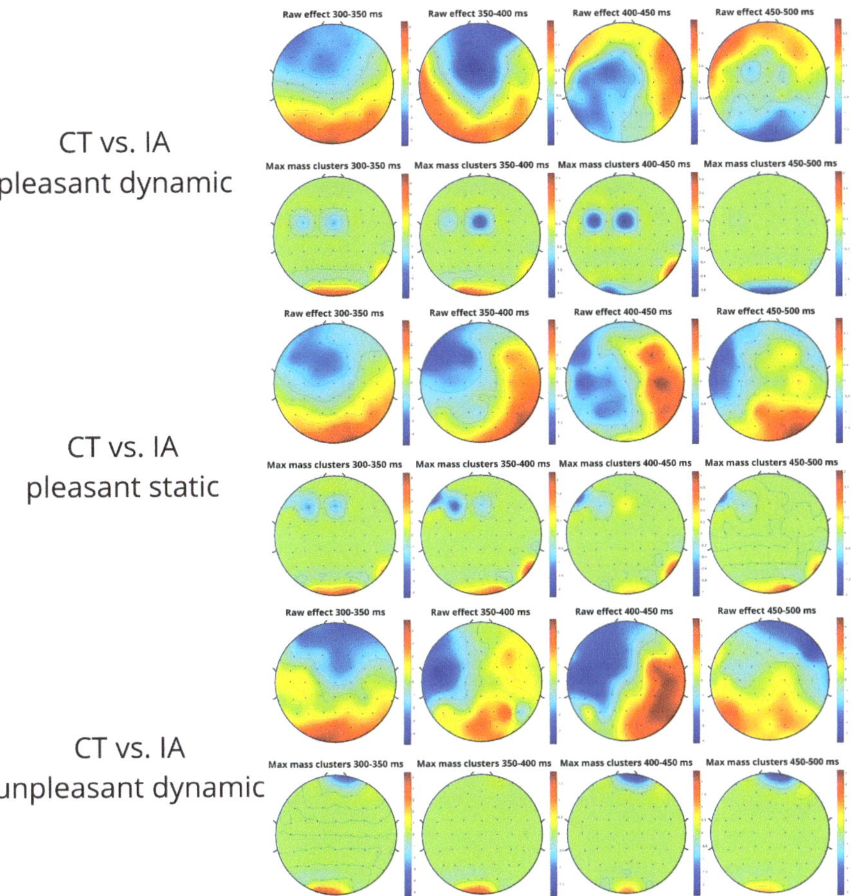

Figure 6. Non-parametric cluster-based permutation analysis showing topographical raw difference (top row) and raw difference filtered by cluster (down row) between groups in pleasant dynamic, pleasant static, and unpleasant dynamic stimuli.

In the dynamic neutral stimuli, both positive (d = 2.05, Mass = 53.02) and negative (d = 2.05, Mass = 14.24) clusters were identified, revealing differences in P300 amplitude in the posterior area of the cortex (Figure 6). Between 300 and 350 ms, the positive cluster indicates that the CTs have a more pronounced P300 amplitude than the IA group. However, thereafter, the cluster becomes negative, suggesting a delay in latency response for the IA group.

4. Discussion

Our pilot study highlights the importance of multimodal neurophysiological assessment, particularly through the combined use of EEG and fNIRS, to investigate cognitive changes. The integration of these techniques provides a comprehensive view of neural dynamics, combining the strengths of each method to provide a more detailed and accurate understanding of brain function. This new BCI technology has gained popularity because it can more accurately decode brain activity in specific cortical regions while producing less electrical noise [35].

Recent literature highlights how the multimodal approach allows for a more holistic and detailed understanding of neural dynamics. This approach is particularly useful when studying complex neurological conditions [36,37], providing a richer and more multifaceted perspective that improves our ability to analyze and intervene in specific neurological changes.

To the best of our knowledge, this is the first study evaluating the EEG response to paintings in subjects with impaired aging.

4.1. fNIRS

Our investigation using functional near-infrared spectroscopy (fNIRS) revealed significant differences in brain activation between individuals with impaired aging and healthy controls. The study highlights how the nature of visual stimuli, both static and dynamic, affects brain activation in distinct ways.

Specifically, we observed that the group with impaired aging exhibited significantly higher levels of oxyhemoglobin (HbO) than controls in response to aesthetically pleasing stimuli, indicating greater brain activation. This increase in activation in patients can be explained by the neural inefficiency theory [38], which suggests that individuals with cognitive impairment may require more brain activation to perform the same cognitive tasks as healthy controls. According to this theory, individuals with neurological conditions that result in impaired cortical functioning, such as multiple sclerosis, Parkinson's disease, and mild cognitive impairment, must compensate for their neural inefficiencies by increasing activation in specific brain areas to maintain a comparable level of behavioral performance [39–41]. The heightened activation observed in response to aesthetically pleasing stimuli may indicate the brain's effort to compensate for challenges in processing such stimuli, which necessitate more intricate interpretation and appreciation.

The evaluation of beauty, indeed, involves complex cognitive processing that engages multiple brain areas [42]. This process is known as the 'aesthetic triad' and comprises "neural systems that contribute to emergent aesthetic experience. Aesthetic experiences are emergent states resulting from interactions between sensory-motor, emotional-evaluative, and meaning-knowledge neural systems" [42]. Furthermore, our findings align with previous studies that have explored the processing of complex stimuli in individuals with mild cognitive impairment (MCI). For instance, Niu et al. [43] and Yeung et al. [44] have demonstrated that patients with MCI display modified brain activation patterns when performing tasks that require a greater working memory load.

4.2. EEG

In our EEG results, we observed significant differences in P300 amplitude between groups of subjects with impaired aging and healthy controls, particularly in response to aesthetic stimuli.

Neurophysiological markers, such as EEG and event-related potential (ERP) components, have been identified as significant indicators of cognitive decline in the scientific literature [45]. They offer valuable tools for determining pathological stages [46]. P300, in particular, emerges as a key ERP component for the study of attention, working memory, and cognitive impairment. P300 is a positive peak that is prominent in the centro-parietal areas of the scalp. It usually occurs between 250 and 500 ms after stimulus presentation. The first subcomponent of P300, P3a, is related to attention mechanisms and the handling of novel stimuli [47]. Research has shown that changes in P300 latency and amplitude may indicate the progression of neurodegeneration [48], showing a longer latency and a more attenuated amplitude [49]. Other studies have confirmed significant differences in P300 parameters between healthy individuals and those with MCI or Alzheimer's Disease (AD) [50,51].

Based on the evidence discussed above, these differences appear to be consistent with studies that have reported reduced P300 amplitude in patients with MCI compared to healthy controls in both auditory [52] and visual [53] tasks, suggesting an alteration in cognitive processing ability in this population. The study's findings suggest that the processing of aesthetically pleasing stimuli requires a complex set of cognitive functions, which may be impaired in individuals with impaired aging.

Indeed, aesthetic appreciation is a complex process that involves perception, integration of implicit memory, explicit classification, cognitive mastery, and evaluation [54]. This process is also influenced by additional factors, such as the environment and social interaction. According to our data, healthy controls show greater amplitude in response to stimuli perceived as pleasant compared to the impaired aging group. This suggests a greater sensitivity or ability to process aesthetic stimuli and supports the integrity of the previously discussed processes. In contrast, the group of patients exhibited greater activation in response to neutral stimuli than to aesthetically pleasing stimuli. This could indicate a reduced ability to discriminate aesthetics and process the aesthetic qualities of stimuli, and/or deficits in one or more of those processes [55].

In our study, the LPP, which is associated with emotional and evaluative processing of stimuli, exhibited a more right-lateralized distribution when comparing the CT and IA groups, consistent with previous studies that have shown that aesthetic judgments can elicit a stronger distribution in the right hemisphere of the brain [56]. This statement is in line with the literature that links the LPP to a heightened response to objects with high affordance and aesthetic levels [57]. This suggests that objects that are both highly attractive and perceived as functional elicit a privileged neural response.

4.3. Limitations

Although this pilot study provides significant insights into the combined use of EEG and fNIRS to explore neurophysiological differences associated with aesthetic appreciation in individuals with impaired aging, it has some limitations. The small sample size limits the generalizability of our results, suggesting the need for future studies with larger samples to confirm and expand our findings. The assessment of aesthetic appreciation, which is inherently subjective, may have been influenced by uncontrolled factors, such as cultural and personal preferences. Additionally, the interpretation of ERP data, particularly the P300 data, despite the extensive literature present, requires caution, as alterations may be attributable to a variety of factors. Finally, this study highlights the potential of BCIs in the clinical setting, but further research is needed to refine interfaces and analysis techniques for targeted clinical applications.

5. Conclusions

This pilot study highlights the relevance of a multimodal approach combining EEG and fNIRS to investigate the neural dynamics associated with aesthetic appreciation and impaired aging, demonstrating significant differences in brain activation between the two groups in response to aesthetic stimuli. Although the group with impaired aging showed difficulty in the P300-related task compared to the control group, the results suggest compensation in terms of hemodynamic recruitment. This finding could provide a basis for cognitive rehabilitation by adjusting the environment according to the subjects' aesthetic perception to improve outcomes.

The combined use of non-invasive BCI technologies is a valuable tool for neurophysiological assessment, contributing to the emerging literature on their use in clinical settings. The results highlight the potential of these technologies in improving the diagnosis and understanding of cognitive disorders. The ability to detect these dynamics only through the multimodal use of BCIs emphasizes the effectiveness of integrating different methodologies for a more comprehensive understanding of the neural processes involved in aesthetic appreciation and cognitive impairment.

Author Contributions: Conceptualization, M.d.T.; methodology, M.d.T.; software, V.B. and A.B.; formal analysis, L.C. and M.L.R.; investigation, L.C. and M.D.; data curation, M.d.T., L.C., G.P. (Giuseppe Procida) and M.D.; writing—original draft preparation, M.d.T. and L.C.; writing—review and editing, L.C., M.L.R. and M.D.; visualization, A.I., P.T., K.R., G.T., G.P. (Giulia Paparella) and G.P. (Giuseppe Procida).; supervision, M.d.T.; project administration, M.d.T.; funding acquisition, M.d.T. All authors have read and agreed to the published version of the manuscript.

Funding: This study was supported by BRIEF—Biorobotics Research and Innovation Engineering Facilities—Missione 4; 'Istruzione e Ricerca'—Componente 2, 'Dalla ricerca all'impresa'—Linea di investimento 3.1; 'Fondo per la realizzazione di un sistema integrato di infrastrutture di ricerca e innovazione', funded by the European Union—NextGenerationEU, CUP: J13C22000400007.

Institutional Review Board Statement: The study was conducted in accordance with the Declaration of Helsinki and approved by the Ethics Committee of the "Azienda Policlinico" of Bari and was conducted according to the principles outlined in the Declaration of Helsinki (No. ET-20-01). The participants provided their written informed consent to contribute to this study.

Informed Consent Statement: Informed consent was obtained from all subjects involved in the study.

Data Availability Statement: The raw data supporting the conclusions of this article will be made available by the authors, without undue reservation.

Acknowledgments: We express our gratitude to all study participants for their valuable contribution, as well as to the research assistants and trainees involved in patient evaluation and recruitment.

Conflicts of Interest: The authors declare that the research was conducted in the absence of any commercial or financial relationships that could be construed as potential conflicts of interest.

References

1. Mudgal, S.K.; Sharma, S.K.; Chaturvedi, J.; Sharma, A. Brain computer interface advancement in neurosciences: Applications and issues. *Interdiscip. Neurosurg.* **2020**, *20*, 100694. [CrossRef]
2. Rosenfeld, J.V.; Wong, Y.T. Neurobionics and the brain–computer interface: Current applications and future horizons. *Med. J. Aust.* **2017**, *206*, 363–368. [CrossRef] [PubMed]
3. Nicolas-Alonso, L.F.; Gomez-Gil, J. Brain Computer Interfaces, a Review. *Sensors* **2012**, *12*, 1211–1279. [CrossRef] [PubMed]
4. Buccino, A.P.; Keles, H.O.; Omurtag, A. Hybrid EEG-fNIRS Asynchronous Brain-Computer Interface for Multiple Motor Tasks. *PLoS ONE* **2016**, *11*, e0146610. [CrossRef] [PubMed]
5. Mane, R.; Chouhan, T.; Guan, C. BCI for stroke rehabilitation: Motor and beyond. *J. Neural Eng.* **2020**, *17*, 041001. [CrossRef] [PubMed]
6. Mansour, S.; Ang, K.K.; Nair, K.P.S.; Phua, K.S.; Arvaneh, M. Efficacy of Brain–Computer Interface and the Impact of Its Design Characteristics on Poststroke Upper-limb Rehabilitation: A Systematic Review and Meta-analysis of Randomized Controlled Trials. *Clin. EEG Neurosci.* **2022**, *53*, 79–90. [CrossRef] [PubMed]
7. Ma, P.; Dong, C.; Lin, R.; Ma, S.; Liu, H.; Lei, D.; Chen, X. Effect of Local Network Characteristics on the Performance of the SSVEP Brain-Computer Interface. *IRBM* **2023**, *44*, 100781. [CrossRef]
8. Saeidi, M.; Karwowski, W.; Farahani, F.V.; Fiok, K.; Taiar, R.; Hancock, P.A.; Al-Juaid, A. Neural Decoding of EEG Signals with Machine Learning: A Systematic Review. *Brain Sci.* **2021**, *11*, 1525. [CrossRef]
9. La Rocca, M.; Laporta, A.; Clemente, L.; Ammendola, E.; Delussi, M.D.; Ricci, K.; Tancredi, G.; Stramaglia, S.; de Tommaso, M. Galcanezumab treatment changes visual related EEG connectivity patterns in migraine patients. *Cephalalgia* **2023**, *43*, 03331024231189751. [CrossRef]
10. Saha, S.; Mamun, K.A.; Ahmed, K.; Mostafa, R.; Naik, G.R.; Darvishi, S.; Khandoker, A.H.; Baumert, M. Progress in Brain Computer Interface: Challenges and Opportunities. *Front. Syst. Neurosci.* **2021**, *15*, 578875. [CrossRef]
11. Pandarinath, C.; Nuyujukian, P.; Blabe, C.H.; Sorice, B.L.; Saab, J.; Willett, F.R.; Hochberg, L.R.; Shenoy, K.V.; Henderson, J.M. High performance communication by people with paralysis using an intracortical brain-computer interface. *eLife* **2017**, *6*, e18554. [CrossRef]
12. Kaiju, T.; Doi, K.; Yokota, M.; Watanabe, K.; Inoue, M.; Ando, H.; Takahashi, K.; Yoshida, F.; Hirata, M.; Suzuki, T. High Spatiotemporal Resolution ECoG Recording of Somatosensory Evoked Potentials with Flexible Micro-Electrode Arrays. *Front. Neural Circuits* **2017**, *11*, 20. [CrossRef] [PubMed]
13. Wolpaw, J.R. Brain–computer interfaces. In *Handbook of Clinical Neurology*; Elsevier: Amsterdam, The Netherlands, 2013; Volume 110, pp. 67–74. [CrossRef]
14. Zhao, Z.-P.; Nie, C.; Jiang, C.-T.; Cao, S.-H.; Tian, K.-X.; Yu, S.; Gu, J.-W. Modulating Brain Activity with Invasive Brain–Computer Interface: A Narrative Review. *Brain Sci.* **2023**, *13*, 134. [CrossRef] [PubMed]
15. Wolpaw, J.R.; Birbaumer, N.; McFarland, D.J.; Pfurtscheller, G.; Vaughan, T.M. Brain–computer interfaces for communication and control. *Clin. Neurophysiol.* **2002**, *113*, 767–791. [CrossRef] [PubMed]
16. Sciaraffa, N.; Di Flumeri, G.; Germano, D.; Giorgi, A.; Di Florio, A.; Borghini, G.; Vozzi, A.; Ronca, V.; Babiloni, F.; Aricò, P. Evaluation of a New Lightweight EEG Technology for Translational Applications of Passive Brain-Computer Interfaces. *Front. Hum. Neurosci.* **2022**, *16*, 901387. [CrossRef]
17. Roman-Gonzalez, A. EEG Signal Processing for BCI Applications. In *Human—Computer Systems Interaction: Backgrounds and Applications 2*; Hippe, Z.S., Kulikowski, J.L., Mroczek, T., Eds.; Advances in Intelligent and Soft Computing; Springer: Berlin/Heidelberg, Germany, 2012; Volume 98, pp. 571–591. [CrossRef]

18. Sörnmo, L.; Laguna, P. *Bioelectrical Signal Processing in Cardiac and Neurological Applications*; Elsevier Academic Press: Amsterdam, The Netherlands, 2005.
19. De Tommaso, M.; Pecoraro, C.; Sardaro, M.; Serpino, C.; Lancioni, G.; Livrea, P. Influence of aesthetic perception on visual event-related potentials. *Conscious. Cogn.* **2008**, *17*, 933–945. [CrossRef] [PubMed]
20. De Tommaso, M.; La Rocca, M.; Quitadamo, S.G.; Ricci, K.; Tancredi, G.; Clemente, L.; Gentile, E.; Ammendola, E.; Delussi, M. Central effects of galcanezumab in migraine: A pilot study on Steady State Visual Evoked Potentials and occipital hemodynamic response in migraine patients. *J. Headache Pain* **2022**, *23*, 52. [CrossRef]
21. Ayaz, H.; Izzetoglu, M.; Bunce, S.; Heiman-Patterson, T.; Onaral, B. Detecting cognitive activity related hemodynamic signal for brain computer interface using functional near infrared spectroscopy. In Proceedings of the 2007 3rd International IEEE/EMBS Conference on Neural Engineering, Kohala Coast, HI, USA, 2–5 May 2007; IEEE: Piscataway, NJ, USA, 2007; pp. 342–345. [CrossRef]
22. Wilcox, T.; Biondi, M. fNIRS in the developmental sciences. *WIRES Cogn. Sci.* **2015**, *6*, 263–283. [CrossRef]
23. La Rocca, M.; Clemente, L.; Gentile, E.; Ricci, K.; Delussi, M.; De Tommaso, M. Effect of Single Session of Anodal M1 Transcranial Direct Current Stimulation—TDCS—On Cortical Hemodynamic Activity: A Pilot Study in Fibromyalgia. *Brain Sci.* **2022**, *12*, 1569. [CrossRef]
24. Coyle, S.; Ward, T.; Markham, C.; McDarby, G. On the suitability of near-infrared (NIR) systems for next-generation brain–computer interfaces. *Physiol. Meas.* **2004**, *25*, 815–822. [CrossRef]
25. Di Dio, C.; Ardizzi, M.; Massaro, D.; Di Cesare, G.; Gilli, G.; Marchetti, A.; Gallese, V. Human, Nature, Dynamism: The Effects of Content and Movement Perception on Brain Activations during the Aesthetic Judgment of Representational Paintings. *Front. Hum. Neurosci.* **2016**, *9*, 705. [CrossRef] [PubMed]
26. Thakral, P.P.; Moo, L.R.; Slotnick, S.D. A neural mechanism for aesthetic experience. *NeuroReport* **2012**, *23*, 310–313. [CrossRef] [PubMed]
27. Battaglia, F.; Lisanby, S.H.; Freedberg, D. Corticomotor Excitability during Observation and Imagination of a Work of Art. *Front. Hum. Neurosci.* **2011**, *5*, 79. [CrossRef] [PubMed]
28. Proverbio, A.M.; Riva, F.; Zani, A. Observation of Static Pictures of Dynamic Actions Enhances the Activity of Movement-Related Brain Areas. *PLoS ONE* **2009**, *4*, e5389. [CrossRef] [PubMed]
29. Uchitel, J.; Vidal-Rosas, E.E.; Cooper, R.J.; Zhao, H. Wearable, Integrated EEG–fNIRS Technologies: A Review. *Sensors* **2021**, *21*, 6106. [CrossRef] [PubMed]
30. Li, R.; Li, S.; Roh, J.; Wang, C.; Zhang, Y. Multimodal Neuroimaging Using Concurrent EEG/fNIRS for Poststroke Recovery Assessment: An Exploratory Study. *Neurorehabilit. Neural Repair* **2020**, *34*, 1099–1110. [CrossRef] [PubMed]
31. Chopra, A.; Cavalieri, T.A.; Libon, D.J. Dementia Screening Tools for the Primary Care Physician. *Clin. Geriatr.* **2007**, *15*, 38–45.
32. Machado, A.; Cai, Z.; Pellegrino, G.; Marcotte, O.; Vincent, T.; Lina, J.-M.; Kobayashi, E.; Grova, C. Optimal positioning of optodes on the scalp for personalized functional near-infrared spectroscopy investigations. *J. Neurosci. Methods* **2018**, *309*, 91–108. [CrossRef]
33. Delpy, D.T.; Cope, M.; Zee, P.V.D.; Arridge, S.; Wray, S.; Wyatt, J. Estimation of optical pathlength through tissue from direct time of flight measurement. *Phys. Med. Biol.* **1988**, *33*, 1433–1442. [CrossRef]
34. Maris, E.; Oostenveld, R. Nonparametric statistical testing of EEG- and MEG-data. *J. Neurosci. Methods* **2007**, *164*, 177–190. [CrossRef]
35. Liu, Z.; Shore, J.; Wang, M.; Yuan, F.; Buss, A.; Zhao, X. A systematic review on hybrid EEG/fNIRS in brain-computer interface. *Biomed. Signal Process. Control* **2021**, *68*, 102595. [CrossRef]
36. Sadjadi, S.M.; Ebrahimzadeh, E.; Shams, M.; Seraji, M.; Soltanian-Zadeh, H. Localization of Epileptic Foci Based on Simultaneous EEG–fMRI Data. *Front. Neurol.* **2021**, *12*, 645594. [CrossRef] [PubMed]
37. Formaggio, E.; Tonellato, M.; Antonini, A.; Castiglia, L.; Gallo, L.; Manganotti, P.; Masiero, S.; Del Felice, A. Oscillatory EEG-TMS Reactivity in Parkinson Disease. *J. Clin. Neurophysiol.* **2023**, *40*, 263–268. [CrossRef] [PubMed]
38. Holtzer, R.; Rakitin, B.C.; Steffener, J.; Flynn, J.; Kumar, A.; Stern, Y. Age effects on load-dependent brain activations in working memory for novel material. *Brain Res.* **2009**, *1249*, 148–161. [CrossRef] [PubMed]
39. Doi, T.; Makizako, H.; Shimada, H.; Park, H.; Tsutsumimoto, K.; Uemura, K.; Suzuki, T. Brain activation during dual-task walking and executive function among older adults with mild cognitive impairment: A fNIRS study. *Aging Clin. Exp. Res.* **2013**, *25*, 539–544. [CrossRef] [PubMed]
40. Chaparro, G.; Balto, J.M.; Sandroff, B.M.; Holtzer, R.; Izzetoglu, M.; Motl, R.W.; Hernandez, M.E. Frontal brain activation changes due to dual-tasking under partial body weight support conditions in older adults with multiple sclerosis. *J. Neuroeng. Rehabil.* **2017**, *14*, 65. [CrossRef] [PubMed]
41. Mahoney, J.R.; Holtzer, R.; Izzetoglu, M.; Zemon, V.; Verghese, J.; Allali, G. The role of prefrontal cortex during postural control in Parkinsonian syndromes a functional near-infrared spectroscopy study. *Brain Res.* **2016**, *1633*, 126–138. [CrossRef] [PubMed]
42. Chatterjee, A.; Vartanian, O. Neuroaesthetics. *Trends Cogn. Sci.* **2014**, *18*, 370–375. [CrossRef]
43. Niu, H.; Li, X.; Chen, Y.; Ma, C.; Zhang, J.; Zhang, Z. Reduced Frontal Activation during a Working Memory Task in Mild Cognitive Impairment: A Non-Invasive Near-Infrared Spectroscopy Study. *CNS Neurosci. Ther.* **2013**, *19*, 125–131. [CrossRef]
44. Yeung, M.K.; Sze, S.L.; Woo, J.; Kwok, T.; Shum, D.H.; Yu, R.; Chan, A.S. Reduced Frontal Activations at High Working Memory Load in Mild Cognitive Impairment: Near-Infrared Spectroscopy. *Dement. Geriatr. Cogn. Disord.* **2016**, *42*, 278–296. [CrossRef]

45. Papaliagkas, V.T.; Kimiskidis, V.K.; Tsolaki, M.N.; Anogianakis, G. Cognitive event-related potentials: Longitudinal changes in mild cognitive impairment. *Clin. Neurophysiol.* **2011**, *122*, 1322–1326. [CrossRef] [PubMed]

46. Babiloni, C.; Lizio, R.; Del Percio, C.; Marzano, N.; Soricelli, A.; Salvatore, E.; Ferri, R.; Cosentino, F.I.; Tedeschi, G.; Montella, P.; et al. Cortical Sources of Resting State EEG Rhythms are Sensitive to the Progression of Early Stage Alzheimer's Disease. *J. Alzheimer's Dis.* **2013**, *34*, 1015–1035. [CrossRef] [PubMed]

47. Polich, J. Updating P300: An integrative theory of P3a and P3b. *Clin. Neurophysiol.* **2007**, *118*, 2128–2148. [CrossRef] [PubMed]

48. Medvidovic, S.; Titlic, M.; MarasSimunic, M. P300 Evoked Potential in Patients with Mild Cognitive Impairment. *Acta Inform. Medica* **2013**, *21*, 89–92. [CrossRef] [PubMed]

49. Papadaniil, C.D.; Kosmidou, V.E.; Tsolaki, A.; Tsolaki, M.; Kompatsiaris, I.; Hadjileontiadis, L.J. Cognitive MMN and P300 in mild cognitive impairment and Alzheimer's disease: A high density EEG-3D vector field tomography approach. *Brain Res.* **2016**, *1648*, 425–433. [CrossRef] [PubMed]

50. Howe, A.S.; Bani-Fatemi, A.; De Luca, V. The clinical utility of the auditory P300 latency subcomponent event-related potential in preclinical diagnosis of patients with mild cognitive impairment and Alzheimer's disease. *Brain Cogn.* **2014**, *86*, 64–74. [CrossRef]

51. Fix, S.T.; Arruda, J.E.; Andrasik, F.; Beach, J.; Groom, K. Using visual evoked potentials for the early detection of amnestic mild cognitive impairment: A pilot investigation. *Int. J. Geriatr. Psychiatry* **2015**, *30*, 72–79. [CrossRef] [PubMed]

52. Bennys, K.; Rondouin, G.; Benattar, E.; Gabelle, A.; Touchon, J. Can Event-Related Potential Predict the Progression of Mild Cognitive Impairment? *J. Clin. Neurophysiol.* **2011**, *28*, 625–632. [CrossRef]

53. Gozke, E.; Tomrukcu, S.; Erdal, N. Visual Event-Related Potentials in Patients with Mild Cognitive Impairment. *Int. J. Gerontol.* **2016**, *10*, 190–192. [CrossRef]

54. Leder, H.; Belke, B.; Oeberst, A.; Augustin, D. A model of aesthetic appreciation and aesthetic judgments. *Br. J. Psychol.* **2004**, *95*, 489–508. [CrossRef]

55. Clemente, L.; Gasparre, D.; Alfeo, F.; Battista, F.; Abbatantuono, C.; Curci, A.; Lanciano, T.; Taurisano, P. Theory of Mind and Executive Functions in Individuals with Mild Cognitive Impairment or Healthy Aging. *Brain Sci.* **2023**, *13*, 1356. [CrossRef] [PubMed]

56. Höfel, L.; Jacobsen, T. Electrophysiological indices of processing aesthetics: Spontaneous or intentional processes? *Int. J. Psychophysiol.* **2007**, *65*, 20–31. [CrossRef] [PubMed]

57. Righi, S.; Orlando, V.; Marzi, T. Attractiveness and affordance shape tools neural coding: Insight from ERPs. *Int. J. Psychophysiol.* **2014**, *91*, 240–253. [CrossRef] [PubMed]

 *sensors*

Article

Improving Aerosol Characterization Using an Optical Particle Counter Coupled with a Quartz Crystal Microbalance with an Integrated Microheater

Emiliano Zampetti *, Maria Aurora Mancuso, Alessandro Capocecera, Paolo Papa and Antonella Macagnano

Institute of Atmospheric Pollution Research—National Research Council (IIA-CNR), Research Area of Rome 1, Strada Provinciale 35d, 9-00010 Montelibretti, Italy; alessandro.capocecera@cnr.it (A.C.); antonella.macagnano@cnr.it (A.M.)

* Correspondence: emiliano.zampetti@cnr.it

Abstract: Aerosols, as well as suspended particulate matter, impact atmospheric pollution, the climate, and human health, directly or indirectly. Particle size, chemical composition, and other aerosol characteristics are determinant factors for atmospheric pollution dynamics and more. In the last decade, low-cost devices have been widely used in instrumentation to measure aerosols. However, they present some issues, such as the problem of discriminating whether the aerosol is composed of liquid particles or solid. This issue could lead to errors in the estimation of mass concentration in monitoring environments where there is fog. In this study, we investigate the use of an optical particle counter (OPC) coupled to a quartz crystal microbalance with an integrated microheater (H-QCM) to enhance measurement performances. The H-QCM was used not only to measure the collected mass on its surface but also, by using the integrated microheater, it was able to heat the collected mass by performing heating cycles. In particular, we tested the developed system with aerosolized saline solutions of sodium chloride (NaCl), with three decreasing concentrations of salt and three electronic cigarette solutions (e-liquid), with different concentrations of propylene glycol and glycerin mixtures. The results showed that the OPC coherently counted the salt dilution effects, and the H-QCM output confirmed the presence of liquid and solid particles in the aerosols. In the case of e-liquid aerosols, the OPC counted the particles, and the HQCM output highlighted that in the aerosol, there were no solid particles but a liquid phase only. These findings contribute to the refinement of aerosol measurement methodologies by low-cost sensors, fostering a more comprehensive understanding.

Keywords: aerosol; OPC; QCM; sensors; particulate matter

Citation: Zampetti, E.; Mancuso, M.A.; Capocecera, A.; Papa, P.; Macagnano, A. Improving Aerosol Characterization Using an Optical Particle Counter Coupled with a Quartz Crystal Microbalance with an Integrated Microheater. *Sensors* **2024**, *24*, 2500. https://doi.org/10.3390/s24082500

Academic Editor: Stephen Holler

Received: 12 March 2024
Revised: 4 April 2024
Accepted: 11 April 2024
Published: 13 April 2024

1. Introduction

An aerosol is a colloidal suspension of fine solid or liquid particles in a gas and has been investigated in several research fields. In medicine, it is mainly studied as a method for drug delivery [1]. In material and chemical science, it is studied as a method to produce or functionalize materials [2]. Aerosols, as well as particulate matter or suspended particulate matter (PMx), impact atmospheric pollution, the climate, and human health, either directly or indirectly. For these reasons, the number of publications on atmospheric aerosols has dramatically increased in the last decade [3]. Aerosols play an important role in environmental pollution, and the measurements of the particles' sizes or their chemical composition are crucial for understanding atmospheric pollution dynamics [4–6]. Knowledge of aerosols is continually deepening thanks to scientific research, but concepts such as deliquescence transition, efflorescence, coagulation, nucleation, or hygroscopic growth of salt are not fully understood when the size is submicrometric [7].

In atmospheric pollution monitoring (or PM), aerosols are among the most important pollutants, and several devices are available to perform automatic measurements, such as scanning mobility particle sizer (SMPS), quartz crystal microbalance cascade, beta gauge, and other vibrating devices [8–10]. While these devices provide accurate and reproducible responses, they often occupy significant volumes, and their management costs are not sustainable for use in large-scale mapping applications. In fact, most of these mentioned measurement systems require a sampling system upstream of the instrument to treat the air sample (e.g., dryers, thermal conditioning sampler, filters, etc.), which cannot be easily miniaturized and have a non-negligible impact on costs and energy consumption [11,12]. For these reasons, the research community is addressing the study and development of low-cost sensors or sensor systems [13]. These devices, with their small size, low cost, and low power consumption, are suitable for application scenarios where a large number of sampling points are needed [14–16]. Optical particle counters (OPCs) and quartz crystal microbalances (QCMs) are low-cost devices widely employed to measure and study aerosols. OPCs count particles with an equivalent diameter in certain ranges using the laser scattering method [17]. QCMs measure the thickness or mass of collected substance on its electrode surface by ultrasonic vibration of the piezoelectric crystal [18–20]. These devices are widely used in biosensing, medical, space, and pollutant monitoring applications [21–24].

As described by Görner et al., a well-calibrated OPC is suitable for aerosol mass concentration monitoring in the workplace [25]. Additionally, in a paper by Hand, Jenny L., and Sonia M. Kreidenweis, OPC data were used to implement a new data analysis method to retrieve the refractive index and effective density from aerosol size distribution data [26].

However, measurements carried out only by OPC can be influenced by the different density or composition of the aerosol, particularly when data about mass concentration are desirable. In fact, to calculate the mass concentration of aerosols, the OPC uses a fixed particle mass density value that can vary depending on the aerosol composition [27,28]. QCM and OPC are powerful tools that can provide valuable information about the properties and behaviour of particles in a wide range of applications. Using them together can provide even more comprehensive insights. As described in the research paper by Kyeong-Rak Lee et al. [29], an OPC and QCM can be used together in certain applications to provide complementary information about the properties and behaviour of particles.

In this work, we propose combining a low-cost OPC with a modified QCM and integrating a microheater and a microresistance temperature detector on its surface (H-QCM) as reported in our previous paper [30]. The developed sensor system produces both particle counting and the possibility of obtaining information regarding the presence of liquid or solid phases in the analyzed aerosol sample. In particular, we tested the developed system with aerosolized saline solutions of sodium chloride (NaCl), with three decreasing concentrations of salt and three electronic cigarette solutions (e-liquid), with different concentrations of propylene glycol and glycerin mixtures. The results highlighted the possibility of using the QCM combined with OPC to determine the presence or absence of a solid phase in the tested aerosol.

This improvement could be applied in real cases where the measurement of aerosols (or PM) may not be entirely reliable using only low-cost OPCs, without any kind of sample processing system, due to the aerosol composition or interference agents such as fog or oil vapours.

2. Methods and Materials

2.1. Solid and Liquid Aerosols

An aerosol is a colloidal suspension of fine solid or liquid particles in a gas [6,31]. An example of a solid aerosol is aerosol salt. Specifically, NaCl aerosol refers to a type of aerosol containing sodium chloride particles generated via various methods such as salt sprays, nebulizers, or humidifiers. These particles vary in size, ranging from fractions of a sub-micrometre to a few millimetres, and they may be uniformly or unevenly distributed throughout the gas.

The salty particles of NaCl easily absorb moisture from the air, making them susceptible to deliquescence and efflorescence. These phenomena occur when the salty particles, now in a wet form, either dissolve into a liquid solution (deliquescence) or leave behind salt crystals on a solid surface as the water evaporates (efflorescence). The incidence of these dynamics depends on factors like humidity and temperature [32].

Instead, an example of a liquid aerosol is the aerosolization of a mixture of propylene glycol (PG) and vegetable glycerin (VG), commonly used as e-liquid for electronic cigarettes. This mixture (when vaporized) produces an aerosol that users inhale. Properties of this aerosol depend on the ratio of PG to VG in the e-liquid. PG produces small liquid particles and less dense aerosol than VG, while VG produces bigger particles and a denser aerosol [33,34].

The OPC is able to detect particles using light scattering and faces a fundamental limitation in differentiating the physical state of the particles. While it provides valuable information about particle size in aerosols, it cannot distinguish between solid and liquid particles solely based on light scattering [25,29].

To address this limitation, especially for the study of liquid and saline aerosols, a combination of the OPC and the H-QCM [30,35] could be very interesting.

The salt particles may initially be found in an aerosol phase, and as soon as they come into contact with the QCM surface, they may undergo phase transition and form a solid deposit.

According to Sauerbrey's Equation (1), the deposition of salt particles on the QCM surface can lead to an increase in the mass of the QCM, which can be measured as a change in the resonant frequency of the crystal as follows:

$$\Delta f = -\frac{2\,fo^2}{A\,\sqrt{v_q\,\rho_q}}\Delta m = -\frac{C_f}{A}\Delta m \tag{1}$$

Δf refers to the frequency shift due to the changing mass Δm; C_f represents a constant, where f_0 is the fundamental resonating frequency; v_q is the velocity of propagation of the transverse wave in the plane of the quartz; ρ_q is the density of the quartz; and A is the effective area of the electrode [36].

The deposition of aerosol salt on an H-QCM can be influenced by various factors, including the concentration and size distribution of the salt particles in the aerosol, the relative humidity of the air, and the temperature of the H-QCM surface [35]. In fact, considering these parameters, phenomena like efflorescence and deliquescence could potentially occur.

Propylene glycol (PG) and vegetable glycerin (VG) can be mixed to produce a liquid aerosol that is detectable on a quartz crystal microbalance (QCM). In this case, detecting liquid particulate matter introduces additional considerations. Liquids may exhibit behaviours such as wetting or adherence, affecting how they distribute and adhere to the QCM surface. This can result in a more nuanced change in mass compared to solid particles. Additionally, it is important to note that under certain conditions, the liquid aerosol could evaporate from the QCM surface.

While the core principles of mass detection can be applied to both solid and liquid particulate matter on a QCM [37], their specific dynamics and interactions can differ. Understanding these subtleties is crucial for interpreting QCM data accurately in studies involving both solid and liquid aerosols to obtain analytical results.

2.2. Reagents and Samples Preparation

The chemicals used to prepare the samples, which included two different sets of solutions, were vegetable glycerol (VG, CAS No. 56-81-5), propylene glycol (PG, CAS No. 57-55-6), and sodium chloride (NaCl, CAS No. 7647-14-5) purchased from Merck (Darmstadt, Germany). A stock solution of NaCl (0.15 M, approximately 0.90%) at a physiological concentration (NaClphy) was prepared by weighing 2.2104 g of NaCl in 250 mL of distilled water (H_2Odist) as the solvent. Two additional solutions were prepared by diluting the stock solution by half (NaCl 1:2, 0.075 M) and tenfold (NaCl 1:10, 0.015 M), respectively. After aerosolization, this set of solutions (NaClphy, NaCl1:2, NaCl1:10) simulated saline aerosols, where fine particles of NaCl were suspended in the air at different concentrations.

In addition, another set of three solutions was prepared using different ratios of propylene glycol (PG) and glycerol (VG): liq80:20, liq50:50, and liq20:80 of PG and VG, respectively. This time, liquid aerosols were generated, where fine particles of PG:VG were suspended in the air. To summarize, six samples were prepared, three of which are related to solid aerosol (NaClphy, NaCl1:2, and NaCl1:10) and three of which are related to liquid aerosol (liq80:20, liq50:50, and liq20:80). In order to observe the differences between liquid droplets and solid particulate in the samples, a solution of NaCl and one of e-liquid were nebulized over an optically polished quartz slice. Details of the nebulization system will be described in the following paragraph. Using an optical microscope (Leica DM2700 M, By Leica, Milan, Italy) with a $100\times/0.85$ magnification objective and a 22 mm field of view, liquid droplets of PG:VG were observed (Figure 1a), while solid particles of NaCl were observed (Figure 1b) after aerosolization on a quartz crystal slice. In particular, the photo on the right was captured after a heating treatment of the nebulized sample quartz. In fact, without heating, even the sample with NaCl appeared in the form of liquid droplets.

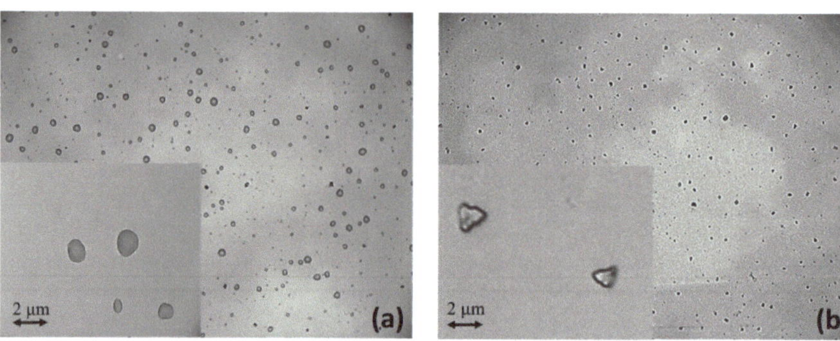

Figure 1. Microscope images of liquid droplets of PG:VG (**a**) and solid particles of NaCl deposited over quartz crystal slice after aerosolization (**b**).

2.3. Measurement Setup

For these experiments, an OPC (OPC-N2 by Alphasense Ltd., Great Notley, Braintree Essex CM77 7AA, Rayne, UK) and an H-QCM were used together to provide complementary information about the properties of aerosols (e.g., liquid or solid phases). Figure 2 reports the scheme of the measurement setup used to perform all tests presented (Figure S1 in the Supplementary Materials shows a photograph of the setup).

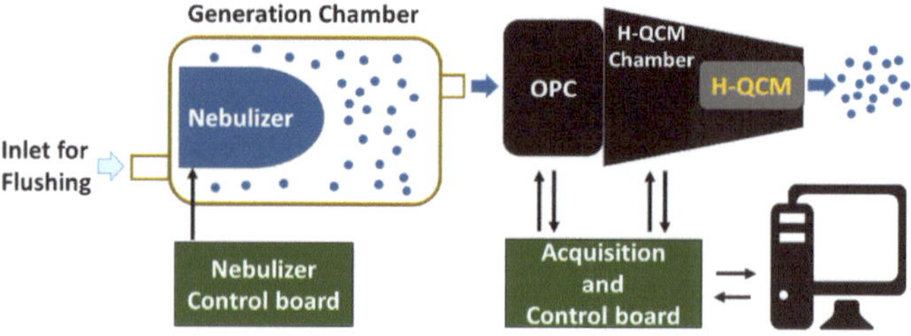

Figure 2. Measurement setup schematic block. The nebulizer was based on ultrasonic vibrating mesh, and the OPC outlet was connected to H-QCM by a suitable adapter (H-QCM chamber).

Samples were aerosolized using a piezo vibrating mesh nebulizer (1100 holes, 6 um of diameter, and 110 kHz of vibrating frequency) [38] placed inside a chamber (generation chamber) and controlled by a nebulizer control board. The OPC withdrew an aerosol sample (for a duration of 1 min) from the generation chamber using its integrated fan. Following this, the sampled aerosol reached the H-QCM connected to the output of the OPC by a suitable mechanical adapter (H-QCM chamber). The OPC data and the resonant frequency of H-QCM were acquired by the acquisition and control board. Finally, a PC unit was used to store all the data. The frequency shift $\Delta F = f(t) - f0$ was used to signify output data of H-QCM where $f0$ was the resonant frequency obtained without any mass deposition. Figure S2 in the Supplementary Materials shows a detailed view of the developed sensor system.

As previously described, the H-QCM, based on AT-cut quartz crystal, was used to measure the mass of the aerosol sampled by OPC. At the same time, the integrated heater of the H-QCM heated the collected mass on the surface of the sensor in order to discern the aerosol characteristics. The microheater was a double omega-shaped thin film (one on the top and one on the bottom of the crystal) that was connected to a temperature controller that measured the temperature and regulated the power supplied. The temperature of the heater was calculated by the controller using a calibration curve concerning the heater electrical resistance and temperature (measured by several micro-thermocouples during the calibration activities) [39,40]. During the heating, the crystal frequency changed following the behaviour of AT-cut crystal, and then, after the heating, the frequency returned to the previous value [41]. The graph reported in Figure S3 of the Supplementary Materials illustrates the frequency shift observed at various temperatures provided by the integrated heater without deposited mass. The steps of temperature were set by the acquisition and control board to perform the calibration. A maximum value of $\overline{\Delta f} = 2909$ Hz was reached for $\overline{T} = 180\ °C$ when 1.1 W of power was provided to the heater in the presence of an air flux produced by the OPC fan (dynamic behaviour of the frequency during the heating step is reported in the Supplementary Materials in Figure S4).

3. Results and Discussions

The following results were obtained during the test following the nebulization of saline aerosol (solid aerosol), and PG/VG aerosol (liquid aerosol) will be presented. In particular, in Sections 3.1 and 3.3, we reported the data obtained from the OPC, and in Sections 3.2 and 3.4, the results concerning the H-QCM.

3.1. Measurements of the Saline Aerosol with OPC

When an aerosol containing salt particles is introduced to an OPC, the salt particles may be detected and counted as individual particles. The detection of salt particles depends on several factors including the size and concentration of the particles and the refractive index. Salt particles typically have a high refractive index compared to other aerosol particles, which can make them more easily detectable by an OPC [42,43].

The Count Mean Diameter (CMD) was derived from measurements taken with the OPC for the three solutions; it can be obtained by calculating a weighted average based on the number of particles in each size range. Then, CMD provides an estimation of the average size of saline aerosol particles, allowing us to gain a better understanding of the particle size distribution [44].

The counts related to the three NaCl solutions were plotted (Figure 3). Specifically, on the y-axis, normalized counts are plotted, and on the x-axis, the bins corresponding to different diameters are represented (Figure S5 of the Supplementary Materials reports an example of the counts as a function of the aerodynamic diameter (d) of NaCl 1:10 and distilled water, H_2Odist).

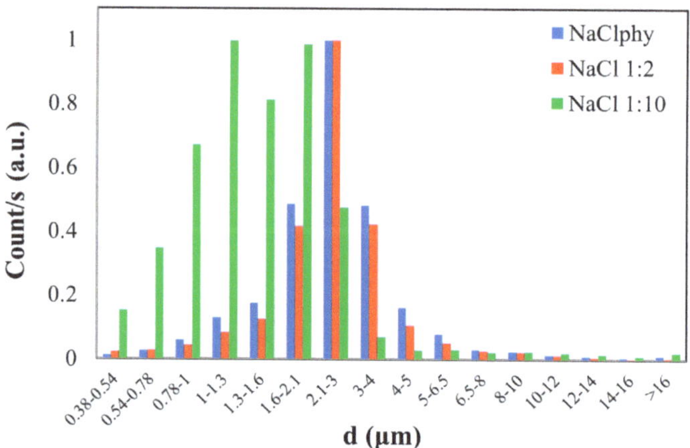

Figure 3. Normalized graph of counts as a function of the aerodynamic diameter (d) of NaCl particles with different concentrations (NaCl$_{phy}$, NaCl$_{1:2}$, and NaCl$_{1:10}$).

The blue histogram represents the size distribution of particles in the NaCl physiological solution where the calculated CMD is 2.90 ± 0.45 μm for this sample. The red histogram refers to the dimensional distribution of the NaCl$_{1:2}$ solution, which has half the concentration compared to the physiological one. In this case, a CMD of 2.83 ± 0.44 μm was obtained. The graph of the latest solution diluted with a 1:10 ratio to the physiological one (NaCl$_{1:10}$) is represented by the green histogram. The calculated CMD is 1.75 ± 0.39 μm, and this value slightly deviates from the first two, considering experimental error as well. This result may be related to the dilution of this latest solution compared to the initial one.

Both NaCl$_{phy}$ and NaCl$_{1:2}$ exhibit a CMD value that is comparable within experimental error, indicating a comparable size distribution. The particles in the more concentrated solution may initially have larger sizes due to the higher solute concentration. Consequently, during nebulization, the OPC may detect both larger and smaller particles, resulting in a broader size distribution. Conversely, in the case of the diluted solution, the particles may be initially smaller. As the droplets fragment during nebulization, predominantly smaller particles may be generated. The OPC is likely to primarily detect smaller particles, leading to a size distribution that is more concentrated around smaller diameters.

3.2. Measurements of the Saline Aerosol with H-QCM

The H-QCM offers a complementary approach, measuring frequency variations during the collection of aerosols to discriminate between solid and liquid phases. The collection of saline aerosol particles on its electrode induces a frequency shift correlated to the added mass, as shown in Equation (1). The H-QCM provides a real-time and sensitive method for monitoring the dynamics of saline aerosol particles. Furthermore, the heating process facilitated by the integrated heater enables the evaporation of the solvent (H$_2$Odist) from the prepared solutions. This allows for the correlation of the frequency shift exclusively to the solid mass deposited on the quartz. This approach allows for the discrimination between the presence of solid and liquid particulates on the surface.

Figure 4 shows the frequency shift chronogram obtained for the NaCl$_{phy}$ sample. Before turning on the heater, the observed Δf relates to an aerosol deposition, a mix of solid and liquid phases. In this state, the signal does not settle, unlike after heating. This could be due to the dynamic processes on the H-QCM surface, where both solid NaCl and liquid H$_2$O are present. In fact, when salt arrives in the form of an aerosol, it will not be in its solid crystalline state but will be surrounded by a certain concentration of water [45]. This condition can result in surface dynamics that delay the stabilization of the signal. For instance, phenomena such as deliquescence and aggregation, where salt particles may cluster together or adsorb water on the surface, can occur. This can result in the formation of more complex structures or the creation of a thicker and more homogeneous layer. Furthermore, variations in surface tension may still occur, affecting the water's ability to wet the surface or influencing other behaviours related to surface properties [46].

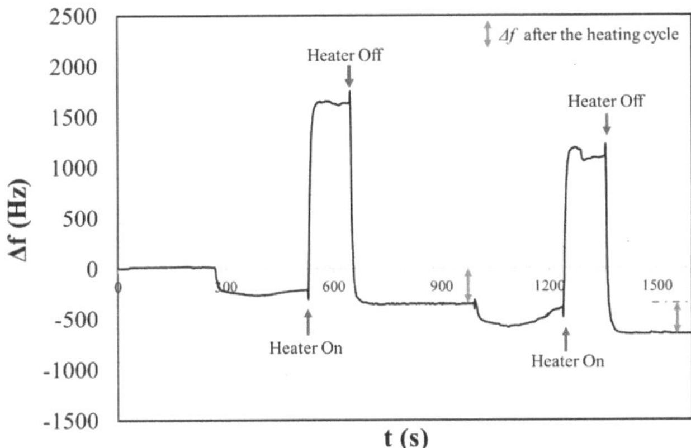

Figure 4. The chronogram of NaCl physiological aerosol shows two consecutive depositions before and after heating on a H-QCM.

After three minutes, the surface heater was turned on, resulting in a rapid positive variation in Δf and was turned off after two minutes (rapid decreasing and stabilizing at a frequency of Δf = 354 Hz). The reached temperature resulted in the evaporation of water, and NaCl was present as a solid crystalline form. During this process, the compound may undergo changes in its crystal structure. The phenomenon of efflorescence is often associated with the evaporation of water containing dissolved salts, which leads to the crystallization of salts on the surface of the material. The frequency shift remains stable after the heating cycle. In fact, when heated, the bound water is released in the form of vapour, leaving behind anhydrous sodium chloride, which is devoid of water. A second measurement was performed consecutively, and once again, it was possible to observe the first and second rapid variation in the frequency following the activation and deactivation of the heater, stabilizing at a frequency of Δf = 301 Hz.

It is possible to note that before heating cycles in both measurements, the frequency shift is smaller compared to that observed after heating. This could be correlated with the sizes of the salt crystals that initially contain hydration water. The dimensions of some crystals might be larger than 2 microns and therefore may not be detected by the 10 MHz QCM, as discussed in our previous article [47]. Conversely, after heating, a larger frequency shift suggests the evaporation of H_2O from the salt particle, resulting in smaller dimensions that can be detected by the QCM. This is consistent with the results obtained from the OPC, where the CMD for NaClphy was measured to be 2.90 ± 0.45 μm.

Figures S6 and S7 of the Supplementary Materials present examples of chronograms of the $NaCl_{1:2}$ and $NaCl_{1:10}$ saline solutions. The behaviour of the $NaCl_{1:2}$ solution is entirely analogous to that of the physiological solution, including the dimensional effect, consistent with the values obtained from the calculation of the CMD (2.90 μm and 2.83 μm, respectively). For $NaCl_{1:10}$ aerosols, the majority of the crystals are presumably below 2 μm, as detected by measurements with the OPC, resulting in a CMD of 1.75 μm. For this reason, the decrease in Δf is related to the evaporation of water and, consequently, to the loss of mass from the quartz crystal surface (Equation (1)). The integration of the OPC and H-QCM enhances our understanding of the particulate matter an aerosol is composed of. The OPC, which relies on light scattering, may encounter challenges when dealing with salt particles, as discussed in the preceding paragraphs regarding humidity and aggregation phenomena. Meanwhile, the H-QCM, with its heated surface, provides real-time sensitivity and the ability to discriminate between solid and liquid phases. The analysis of NaCl solutions emphasizes OPC's capability to detect diverse particle sizes, complemented by the H-QCM's observations of frequency changes during nebulization and particle dynamics.

3.3. Measurements of the Liquid Aerosol with OPC

When a liquid aerosol passes through the OPC, the droplets interact with the laser beam, operating similarly to when dealing with saline aerosols.

This implies that measurements obtained using an OPC do not discriminate between aerosol phases. To obtain data that closely represents real conditions, a correction for the density of the different particulate matter detected is necessary. Indeed, one of the main limitations of the OPC is that particles of different substances may have different densities. This means that even if two particles have the same optical diameter, they could have different volumes or aerodynamic masses due to their density [48]. Also, in this case, the integration of an OPC with a H-QCM could enhance the understanding of the aerosol's characteristics.

In Figure 5, the green histogram shows the size distribution of the PG/VG liquid aerosol in an 80:20 ratio, with a found CMD of 0.90 ± 0.17 μm. The red histogram represents the dimensional distribution of the 50:50 PG/VG, resulting in a CMD of 1.01 ± 0.15 μm. In the last case, the blue histogram illustrates the normalized count distribution of the PG/VG 20:80, yielding a CMD of 1.13 ± 0.15 μm. The dimensional distribution is nearly similar for all three samples with different ratios of PG/VG, considering the obtained CMD value and the calculated errors.

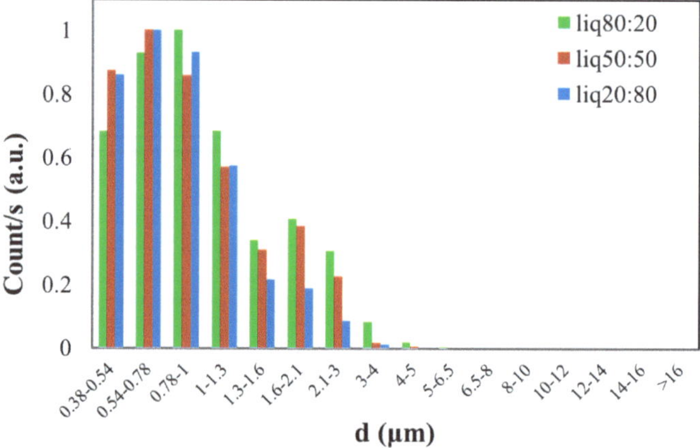

Figure 5. Normalized graph of counts as a function of the aerodynamic diameter (d) of e-liquid droplets with different PG/VG ratios (liq80:20, liq50:50, liq20:80).

3.4. Measurements of the Liquid Aerosol with QCM

The liquid aerosol induces a frequency shift of H-QCM related to the deposition of mass, although the liquid aerosol is expected to undergo natural evaporation (at room temperature) from the crystal surface over a specific time frame. However, by utilizing the integrated heater, it becomes possible to accelerate the evaporation of liquid with high evaporation temperature. This allows us to discern whether the phase of an aerosol detected by the H-QCM is solid or liquid, a distinction that is not achievable with a standalone OPC (Section 3.3).

In Figure 6, we present the resulting chronogram of two distinct depositions before and after heating. The aerosol generated within the nebulization chamber was aspirated for one minute by the pump of the OPC. Subsequently, the aerosol flow was directed onto the surface of the QCM. After three minutes of stabilizing the frequency shift, the heater was activated, reaching a temperature of about 170 °C in two minutes. This was followed by a waiting period of five minutes to ensure signal stabilization. The experiment reveals that before activating the heater, a frequency shift occurs that might be mistakenly associated with the presence of solid particulate ($\Delta f \neq 0$). However, during the heating process, the liquid aerosol droplets evaporate from the surface, restoring the initial resonance frequency. We obtained the same behaviour for the liq80:20 and liq20:80 samples. Figures S8 and S9 in Supplementary Materials show examples of chronograms obtained for the other two PG/VG liquid solutions. Table 1 summarizes the results of experiments of saline and liquid aerosols. In particular, the table reports average values and deviations obtained by several repetitions of measurements.

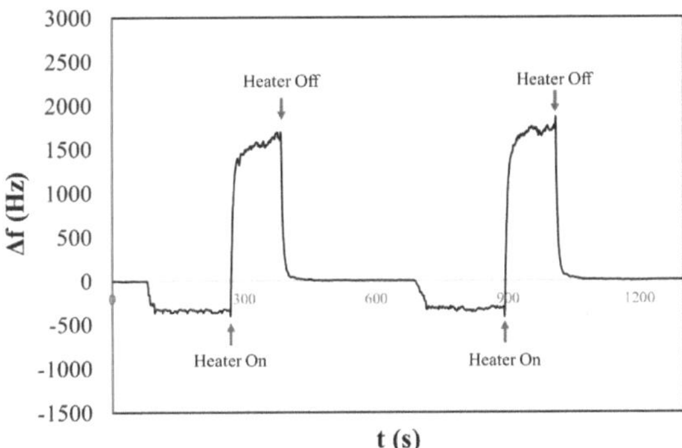

Figure 6. The chronogram of PG/VG 50:50 shows two consecutive depositions before and after heating on a H-QCM.

Table 1. Summary of the average frequency shifts for saline and liquid aerosols.

	Before Heating $\overline{\Delta f}$ (Hz)	After Heating $\overline{\Delta f}$ (Hz)	CMD (µm)
NaCl phy	219 ± 105	358 ± 40	2.90 ± 0.45
NaCl 1:2	112 ± 125	194 ± 21	2.83 ± 0.44
NaCl 1:10	381 ± 258	96 ± 30	1.75 ± 0.39
liq 80:20	161 ± 5	/	0.90 ± 0.17
liq 50:50	337 ± 13	/	1.01 ± 0.15
liq 20:80	441 ± 29	/	1.13 ± 0.15

Regarding the saline aerosol, it is evident that before heating, the average value of frequency shift ($\overline{\Delta f}$) does not follow the trend of salt concentration in the solutions. In fact, the $\overline{\Delta f}$ should be higher for the more concentrated saline solution and lower for the one diluted by a factor of ten. Additionally, the average frequency value shows a significant standard deviation, reflecting the possibility of different surface dynamics processes at the interface (e.g., the crystal electrode surface and sample). These dynamics could contribute to lower reproducibility of the deposition when both solid and liquid phases are present.

After heating cycles, the $\overline{\Delta f}$ values not only align with the saline concentration trend in different solutions (NaCl phy > NaCl 1:2 > NaCl 1:10) but also exhibit a lower standard deviation compared to the previous measurements. This indicates improved reproducibility of measurements after heating cycles.

In the case of liquid aerosols, the observed $\overline{\Delta f}$ values might be mistakenly attributed to the measurements of solid particles. However, after heating, restoration of the initial frequency indicates the absence of residual mass on the surface of the H-QCM, associated with the evaporation of the liquid. This information obtained by using H-QCM could be used as important feedback for OPC measurements. In simpler terms, if, after heating cycles, the frequency shift persists, the OPC data are correct and are not affected by the presence of a liquid phase of aerosol. In future work, we will analyze the performances of the H-QCM (e.g., sensitivity, the limit of detection, reproducibility, etc.) to assess the possibility of correcting the OPC output data when expressed in terms of mass concentration ($\mu g/m^3$).

The proposed OPC+H-QCM device could be employed in the field of PM measurement, particularly in environments with marine aerosol or fog, which limits the use of simple OPCs without the use of an air sample treatment system before performing the count. Moreover, such systems require minimal space and energy for their operation.

In the chemical processing industry, exhaust fumes from process reactor stacks simultaneously contain aerosols with different phases. The OPC+H-QCM device could be useful for analyzing these fumes, enabling a thermogravimetric analysis (TGA) alongside total particulate matter dimensional counting.

Although this study presents preliminary results, the proposed tool could be useful for analyzing the fumes produced by e-cigarettes. It could help in identifying solid phases inhaled by users and correlating the quantity and sizes of the particles to some potential pathologies.

4. Conclusions

This study investigated the responses of an OPC coupled with an H-QCM to saline and liquid aerosols. Regarding the saline aerosol, the OPC demonstrated that the CMD values for NaClphy and NaCl 1:2 were comparable within experimental error, suggesting similar size distributions, while NaCl 1:10 exhibited a slightly different trend, potentially attributed to dilution effects. The H-QCM, focusing on the saline aerosol, elucidated the dynamics during and after heating. In particular before heating, the frequency shifts suggested the coexistence of solid and liquid phases, influenced by surface phenomena such as deliquescence and aggregation. After heating, the separation of solid and liquid phases became evident, resulting in improved reproducibility in frequency shifts and allowing for a clearer interpretation of the deposition process. In the case of liquid aerosols (PG/VG mixtures), the OPC provided consistent CMD values across different ratios, emphasizing the need for density corrections when OPC is used alone. Conversely, the H-QCM demonstrated its ability to discriminate between solid and liquid phases. Overall, this study highlights the complementary strengths of OPC and H-QCM in aerosol analysis. While the OPC is utilized for size distribution characterization, the H-QCM provides real-time insights into phase discrimination aerosols. These findings contribute to the refinement of aerosol measurement using a low-cost sensor, fostering a more comprehensive understanding.

Supplementary Materials: The following supporting information can be downloaded at https://www.mdpi.com/article/10.3390/s24082500/s1. Figure S1. Measurement setup with the nebulization chamber and the coupled OPC+H-QCM system. Figure S2. Photograph of the developed sensor system (a). The H-QCM was fixed by a support in the H-QCM measurement chamber (b). Figure S3. H-QCM frequency shifts versus temperature (12 increasing steps) without deposited mass. The frequency and temperature errors were 1 Hz and 0.1 °C, respectively. Figure S4. The H-QCM exhibited frequency variations as temperature increased, beginning from 22 °C. At the maximum temperature reached, approximately 181 °C, the frequency shift was about 3000 Hz. Figure S5. Counts as a function of the aerodynamic diameter (d) of NaCl 1:10 and distilled water (H_2Odist). Figure S6. Chronogram of frequency shift due to two consecutive NaCl1:2 aerosol measurements. In particular, the graph highlighted the behaviour before and after heating cycles. Figure S7. Chronogram of frequency shift due to two consecutive NaCl1:10 aerosol measurements. In particular, the graph highlighted the behaviour before and after heating cycles. Figure S8. Chronogram of frequency shift due to two consecutive liq80:20 aerosol measurements. In particular, the graph highlighted the behaviour before and after heating cycles. Figure S9. Chronogram of frequency shift due to two consecutive liq20:80 aerosol measurements. In particular, the graph highlighted the behavior before and after heating cycles.

Author Contributions: Conceptualization, E.Z. and M.A.M.; Investigation, E.Z. and M.A.M.; Data curation, A.C. and P.P.; Resources, E.Z.; Supervision, E.Z. and A.M. All authors have read and agreed to the published version of the manuscript.

Funding: This research received no external funding.

Institutional Review Board Statement: Not applicable.

Informed Consent Statement: Not applicable.

Data Availability Statement: All data that support the findings of this study are available after the reasonable request to the corresponding author.

Acknowledgments: The authors thank Alessandro Capocecera for his technical contribution to the development of both the software interface and Arduino scripts.

Conflicts of Interest: The authors declare no conflicts of interest.

References

1. Dolovich, M.B.; Dhand, R. Aerosol drug delivery: Developments in device design and clinical use. *Lancet* **2011**, *377*, 1032–1045. [CrossRef] [PubMed]
2. Gurav, A.; Kodas, T.; Pluym, T.; Xiong, Y. Aerosol processing of materials. *Aerosol Sci. Technol.* **1993**, *19*, 411–452. [CrossRef]
3. Pöschl, U. Atmospheric aerosols: Composition, transformation, climate and health effects. *Angew. Chem. Int. Ed. Engl.* **2005**, *44*, 7520–7540. [CrossRef] [PubMed]
4. Myhre, G.; Myhre, C.E.L.; Samset, B.H.; Storelvmo, T. Aerosols and their Relation to Global Climate and Climate Sensitivity. *Nat. Educ. Knowl.* **2013**, *4*, 7.
5. Spurny, K.R. Methods of Aerosol Measurement before the 1960s. *Aerosol Sci. Technol.* **1998**, *29*, 329–349. [CrossRef]
6. Tripathi, S.N.; Tare, V.; Chinnam, N.; Srivastava, A.K.; Dey, S.; Agarwal, A.; Lal, S. Measurements of atmospheric parameters during Indian Space Research Organization Geosphere Biosphere Programme Land Campaign II at a typical location in the Ganga basin: 1. Physical and optical properties. *J. Geophys. Res.* **2006**, *111*, D23209. [CrossRef]
7. Cheng, Y.; Su, H.; Koop, T.; Mikhailov, E.; Pöschl, U. Size dependence of phase transitions in aerosol nanoparticles. *Nat. Commun.* **2015**, *6*, 5923. [CrossRef] [PubMed]
8. McMurry, P.H. A review of atmospheric aerosol measurements. *Atmos. Environ.* **2000**, *34*, 1959–1999. [CrossRef]
9. Amaral, S.S.; de Carvalho Costa, J.A., Jr.; Pinheiro, M.A.M.C. An overview of particulate matter measurement instruments. *Atmosphere* **2015**, *6*, 1327–1345. [CrossRef]
10. Kangasluoma, J.; Cai, R.; Jiang, J.; Deng, C.; Stolzenburg, D.; Ahonen, L.R.; Lehtipalo, K. Overview of measurements and current instrumentation for 1–10 nm aerosol particle number size distributions. *J. Aerosol Sci.* **2020**, *148*, 105584. [CrossRef]
11. Vincent, J.H. *Aerosol Sampling: Science, Standards, Instrumentation and Applications*; John Wiley & Sons: Hoboken, NJ, USA, 2007.
12. Wiedensohler, A.; Birmili, W.; Putaud, J.P.; Ogren, J. Recommendations for aerosol sampling. In *Aerosol Science: Technology and Applications*; Wiley: Hoboken, NJ, USA, 2013; pp. 45–59. [CrossRef]
13. Sousan, S.; Koehler, K.; Thomas, G.; Park, J.H.; Hillman, M.; Halterman, A.; Peters, T.M. Inter-comparison of low-cost sensors for measuring the mass concentration of occupational aerosols. *Aerosol Sci. Technol.* **2016**, *50*, 462–473. [CrossRef] [PubMed]
14. Giordano, M.R.; Malings, C.; Pandis, S.N.; Presto, A.A.; McNeill, V.F.; Westervelt, D.M.; Subramanian, R. From low-cost sensors to high-quality data: A summary of challenges and best practices for effectively calibrating low-cost particulate matter mass sensors. *J. Aerosol Sci.* **2021**, *158*, 105833. [CrossRef]
15. Alfano, B.; Barretta, L.; Del Giudice, A.; De Vito, S.; Di Francia, G.; Esposito, E.; Polichetti, T. A review of low-cost particulate matter sensors from the developers' perspectives. *Sensors* **2020**, *20*, 6819. [CrossRef] [PubMed]
16. Chao, C.Y.; Zhang, H.; Hammer, M.; Zhan, Y.; Kenney, D.; Martin, R.V.; Biswas, P. Integrating fixed monitoring systems with low-cost sensors to create high-resolution air quality maps for the Northern China Plain Region. *ACS Earth Space Chem.* **2021**, *5*, 3022–3035. [CrossRef]
17. Liu Benjamin, Y.H.; Berglund, R.N.; Agarwal, J.K. Experimental studies of optical particle counters. *Atmos. Environ. 1967* **1974**, *8*, 717–732.
18. Steinem, C.; Janshoff, A. *Piezoelectric Sensors*; Springer Science & Business Media: Berlin/Heidelberg, Germany, 2007; Volume 5.
19. Crilley, L.R.; Shaw, M.; Pound, R.; Kramer, L.J.; Price, R.; Young, S.; Lewis, A.C.; Pope, F.D. Evaluation of a low-cost optical particle counter (Alphasense OPC-N2) for ambient air monitoring, Atmos. *Meas. Tech.* **2018**, *11*, 709–720. [CrossRef]
20. Chen, M.; Romay, F.J.; Li, L.; Naqwi, A.; Marple, V.A. A novel quartz crystal cascade impactor for real-time aerosol mass distribution measurement. *Aerosol Sci. Technol.* **2016**, *50*, 971–983. [CrossRef]
21. Vashist, S.K.; Vashist, P. Recent Advances in Quartz Crystal Microbalance-Based Sensors. *J. Sens.* **2011**, *2011*, 571405. [CrossRef]
22. Afzal, A.; Mujahid, A.; Schirhagl, R.; Bajwa, S.Z.; Latif, U.; Feroz, S. Gravimetric Viral Diagnostics: QCM Based Biosensors for Early Detection of Viruses. *Chemosensors* **2017**, *5*, 7. [CrossRef]
23. Scaccabarozzi, D.; Saggin, B.; Tarabini, M.; Palomba, E.; Longobardo, A.; Zampetti, E. Thermo-mechanical design and testing of a microbalance for space applications. *Adv. Space Res.* **2014**, *54*, 2386–2397. [CrossRef]
24. Zampetti, E.; Papa, P.; Bearzotti, A.; Macagnano, A. Pocket Mercury-Vapour Detection System Employing a Preconcentrator Based on Au-TiO$_2$ Nanomaterials. *Sensors* **2021**, *21*, 8255. [CrossRef] [PubMed]

25. Görner, P.; Simon, X.; Bémer, D.; Lidén, G. Workplace aerosol mass concentration measurement using optical particle counters. *J. Environ. Monit.* **2012**, *14*, 420–428. [CrossRef] [PubMed]
26. Hand, J.L.; Sonia, M.K. A new method for retrieving particle refractive index and effective density from aerosol size distribution data. *Aerosol Sci. Technol.* **2002**, *36*, 1012–1026. [CrossRef]
27. McMeeking, G.R.; Kreidenweis, S.M.; Carrico, C.M.; Lee, T.; Collett, J.L., Jr.; Malm, W.C. Observations of smoke-influenced aerosol during the Yosemite Aerosol Characterization Study: Size distributions and chemical composition. *J. Geophys. Res. Atmos.* **2005**, 110. [CrossRef]
28. Crilley, L.R.; Singh, A.; Kramer, L.J.; Shaw, M.D.; Alam, M.S.; Apte, J.S.; Pope, F.D. Effect of aerosol composition on the performance of low-cost optical particle counter correction factors. *Atmos. Meas. Tech.* **2020**, *13*, 1181–1193. [CrossRef]
29. Lee, K.-R.; Kim, Y.-J. Portable multilateral measurement system employing Optical Particle Counter and one-stage Quartz Crystal Microbalance to measure PM10. *Sens. Actuators A Phys.* **2022**, *333*, 113272. [CrossRef]
30. Zampetti, E.; Macagnano, A.; Papa, P.; Bearzotti, A.; Petracchini, F.; Paciucci, L.; Pirrone, N. Exploitation of an integrated microheater on QCM sensor in particulate matter measurements. *Sens. Actuators A Phys.* **2017**, *264*, 205–211. [CrossRef]
31. Moosmüller, H.; Chakrabarty, R.K.; Arnott, W.P. Aerosol light absorption and its measurement: A review. *J. Quant. Spectrosc. Radiat. Transf.* **2009**, *110*, 844–878. [CrossRef]
32. Li, X.; Gupta, D.; Eom, H.J.; Kim, H.; Ro, C.U. Deliquescence and efflorescence behavior of individual NaCl and KCl mixture aerosol particles. *Atmos. Environ.* **2014**, *82*, 36–43. [CrossRef]
33. Li, L.; Lee, E.S.; Nguyen, C.; Zhu, Y. Effects of propylene glycol, vegetable glycerin, and nicotine on emissions and dynamics of electronic cigarette aerosols. *Aerosol Sci. Technol.* **2020**, *54*, 1270–1281. [CrossRef]
34. Jordt, S.E.; Jabba, S.; Ghoreshi, K.; Smith, G.J.; Morris, J.B. Propylene Glycol and Glycerin in E-Cigarettes Elicit Respiratory Irritation Responses and Modulate Human Sensory Irritant Receptor Function. *Am. J. Respir. Crit. Care Med.* **2019**, *199*, A4169.
35. Jang, I.R.; Jung, S.I.; Lee, G.; Park, I.; Kim, S.B.; Kim, H.J. Quartz crystal microbalance with thermally-controlled surface adhesion for an efficient fine dust collection and sensing. *J. Hazard. Mater.* **2022**, *424*, 127560. [CrossRef]
36. Sauerbrey, G. Verwendung von Schwingquarzen zur Wigung Dünner Schichten und zur Mikrowigung. *Z. Phys.* **1959**, *155*, 206–222. [CrossRef]
37. Kanazawa, K.K.; Gordon, J.G. The oscillation frequency of a quartz resonator in contact with liquid. *Anal. Chim. Acta* **1985**, *175*, 99–105. [CrossRef]
38. Moon, S.-H.; Chang, K.H.; Park, H.M.; Park, B.J.; Yoo, S.K.; Nam, K.C. Effects of Driving Frequency and Voltage on the Performances of Vibrating Mesh Nebulizers. *Appl. Sci.* **2021**, *11*, 1296. [CrossRef]
39. Scaccabarozzi, D.; Saggin, B.; Magni, M.; Corti, M.G.; Zampetti, E.; Palomba, E.; Longobardo, A.; Dirri, F. Calibration in cryogenic conditions of deposited thin-film thermometers on quartz crystal microbalances. *Sens. Actuators A Phys.* **2021**, *330*, 112878. [CrossRef]
40. Magni, M.; Scaccabarozzi, D.; Palomba, E.; Zampetti, E.; Saggin, B. Characterization of Thermal Gradient Effects on a Quartz Crystal Microbalance. *Sensors* **2022**, *22*, 7256. [CrossRef]
41. Brice, J.C. Crystals for quartz resonators. *Rev. Mod. Phys.* **1985**, *57*, 105. [CrossRef]
42. Niedermeier, D.; Wex, H.; Voigtländer, J.; Stratmann, F.; Brüggemann, E.; Kiselev, A.; Heintzenberg, J. LACIS-measurements and parameterization of sea-salt particle hygroscopic growth and activation. *Atmos. Chem. Phys.* **2008**, *8*, 579–590. [CrossRef]
43. Tang, I.N. Chemical and size effects of hygroscopic aerosols on light scattering coefficient. *J. Geophys. Res.* **1996**, *101*, 19245–19250. [CrossRef]
44. Sarangi, B.; Aggarwal, S.G.; Sinha, D.; Gupta, P.K. Aerosol effective density measurement using scanning mobility particle sizer and quartz crystal microbalance with the estimation of involved uncertainty. *Atmos. Meas. Tech.* **2016**, *9*, 859–875. [CrossRef]
45. Zhang, C.; Fen, G.; Sui, S. Study on behaviour of QCM sensor in loading variation. *Sens. Actuators B Chem.* **1997**, *40*, 111–115. [CrossRef]
46. Chao, H.J.; Huang, W.C.; Chen, C.L.; Chou, C.C.K.; Hung, H.M. Water Adsorption vs Phase Transition of Aerosols Monitored by a Quartz Crystal Microbalance. *ACS Omega* **2020**, *5*, 31858–31866. [CrossRef] [PubMed]
47. Zampetti, E.; Mancuso, M.A.; Dirri, F.; Palomba, E.; Papa, P.; Capocecera, A.; Bearzotti, A.; Macagnano, A.; Scaccabarozzi, D. Effects of Oscillation Amplitude Variations on QCM Response to Microspheres of Different Sizes. *Sensors* **2023**, *23*, 5682. [CrossRef]
48. Kovilakam, M.; Deshler, T. On the accuracy of stratospheric aerosol extinction derived from in situ size distribution measurements and surface area density derived from remote SAGE II and HALOE extinction measurements. *J. Geophys. Res. Atmos.* **2015**, *120*, 8426–8447. [CrossRef]

Article

Self-Diagnostic and Self-Compensation Methods for Resistive Displacement Sensors Tailored for In-Field Implementation [†]

Federico Mazzoli [1,*], Davide Alghisi [2] and Vittorio Ferrari [1]

[1] Department of Information Engineering, University of Brescia, Via Branze 38, 25123 Brescia, Italy; vittorio.ferrari@unibs.it
[2] Gefran SpA, Via Cave 11, 25050 Provaglio d'Iseo, Italy; davide.alghisi@gefran.com
[*] Correspondence: f.mazzoli002@unibs.it; Tel.: +39-3331272335
[†] The paper is an extended and updated version of the contribution presented by the same authors at Eurosensors XXXV, Lecce, Italy, 10–13 September 2023, entitled "Self-Diagnostic Method for Resistive Displacement Sensors".

Abstract: This paper presents a suitably general model for resistive displacement sensors where the model parameters depend on the current sensor conditions, thereby capturing wearout and failure, and proposes a novel fault detection method that can be seamlessly applied during sensor operation, providing self-diagnostic capabilities. On the basis of the estimation of model parameters, an innovative self-compensation method is derived to increase the accuracy of sensors subject to progressive wearout. The proposed model and methods have been validated by both numerical simulations and experimental tests on two real resistive displacement sensors, placed in undamaged and faulty conditions, respectively. The fault detection method has shown an accuracy of 97.2%. The position estimation error is $< \pm 0.2\%$ of the full-scale span for the undamaged sensor, while the self-compensation method successfully reduces the position estimation error from $\pm 15\%$ to approximately $\pm 2\%$ of the full-scale span for the faulty sensor.

Keywords: smart sensor; self-validating sensor; resistive displacement sensors

Citation: Mazzoli, F.; Alghisi, D.; Ferrari, V. Self-Diagnostic and Self-Compensation Methods for Resistive Displacement Sensors Tailored for In-Field Implementation. *Sensors* **2024**, *24*, 2594. https://doi.org/10.3390/s24082594

Academic Editor: Andrea Cataldo

Received: 4 March 2024
Revised: 16 April 2024
Accepted: 16 April 2024
Published: 18 April 2024

1. Introduction

Displacement sensors are widely used in different applications and fields, such as industrial [1,2], medical [3,4] and automotive [5,6]. Within the industrial field, the role of displacement sensors in control applications is gaining importance since the degree of industrial automation is continuously growing. In this context, there is a constant interest in displacement sensors with increasing intelligence, robust measurement methods, and improved performances [7]. In particular, sensor reliability in producing accurate measurement data is crucial. Augmenting sensor diagnostic capabilities plays an increasingly important role, as undetected failures may negatively affect the industrial process into which the sensor is inserted, resulting in efficiency loss, downtimes or even threatening life safety [8].

Several solutions can be considered to prevent sensor failure and enhance sensor diagnostic coverage. A first approach is the adoption of a Preventive Maintenance (PM) policy, which implies sensor calibration and maintenance activities run on a regular time basis to avoid failures before they occur [9]. Determining in advance when a sensor will enter the wearout phase is challenging, as predictions typically rely on a theoretical failure rate rather than on the current sensor conditions. While PM can reduce unexpected process downtime, it can lead to costly and unnecessary procedures and repairs.

A second possible approach involves employing Fault Detection and Identification (FDI) theory [10–13]. FDI techniques recognize system faults in process components by monitoring system inputs and outputs in order to generate residuals highlighting discrepancies. Thresholds, determined through experimental tests, aid in fault symptom identification. Analytical redundancy techniques [14–16] detect and identify sensor failures from the exploitation of analytical models built on a specific process, which includes actuators and plants as well as sensors. However, FDI techniques require a deep knowledge of the entire process and a significant effort in model development. Furthermore, the model must be updated as the process changes. The economic effort conveyed in the model design is not easily amortized since the model is related to a specific process. Moreover, technical staff must be available for maintaining and updating the model as the process evolves over time.

A third possible approach to detect and identify sensor faults is through the use of Machine Learning (ML) techniques [17–19]. The ML models are trained on sensor data to identify patterns or features indicative of faults. The performance of ML techniques are dependent on the quality and quantity of the available training data. In fact, the training data could be limited, biased, or not representative of all possible faults, which can limit the detectable faults. Moreover, the ML approach mainly focuses on detecting the sensor faults through certain time and/or frequency domain signal characteristic features. In particular, faults can be identified by changes in signal behavior, e.g., bias, noise, spikes, rather than by identifying underlying sensor-specific causes for the fault [20].

An alternative approach could be the adoption of smart sensors inherently capable of providing self-diagnostic information related to the alteration of their conditions due to incipient wearout or failure. The diagnostic information provided might be exploited by suitable Predictive Maintenance (PdM) techniques to optimize sensor maintenance scheduling and minimize plant downtime [21]. As opposed to PM approaches, self-diagnostic sensors could avoid both unnecessary maintenance costs and plant downtime due to undetected sensor failures. Moreover, in contrast to many FDI techniques, self-diagnostic sensors place the focus within the sensor itself to make it a more reliable component, leaving out the additional system parts of a given industrial process. Therefore, in principle, a generic process can be freely modified without requiring any change in the diagnostic techniques embedded in the self-diagnostic sensor. As such, the research on displacement sensors equipped with self-diagnostic functionalities is a topic of relevant importance, which, however, has been explored to a somewhat limited extent. In particular, resistive displacement sensors would benefit from solutions for embedding self-diagnostic capabilities. However, to the authors' best knowledge, the current state of the art does not yet adequately cover the development of a resistive displacement sensor model for self-diagnostic and self-compensation purposes.

In this context, this work innovatively proposes a simple yet effective model for a generic resistive linear displacement sensor where the model parameters directly depend on the current sensor conditions, thereby capturing wearout, damaging, and failure occurrence. Furthermore, a method for the continuous estimation of the model parameters is presented. The method originally offers the advantage that it can be applied while the sensor is kept in operation online, i.e., without functionally disconnecting the sensor from the monitored process. The estimated parameters are exploited to detect failures by the proposed fault detection technique, which aims at detecting faults by identifying the underlying sensor-specific cause. The continuously updated parameters are used by the self-compensation method to compensate for the position estimation error arising over time due to progressive wearout and cumulative damage. This paper considerably extends the preliminary work previously presented [22], deepening the study and analysis of both the model for a resistive linear displacement sensor and related self-diagnostic and self-compensation methods.

A schematic comparison between the most common methods to detect and identify sensor faults and the approach proposed in this work is reported in Table 1.

The paper is organized as follows. A generic resistive linear displacement sensor model along with the proposed parameter estimation method, method for compensating position estimation errors, and fault detection method are described in Section 2. In Section 3 the numerical simulations carried out and the experimental setup arranged to test the proposed model and methods are described and the results are reported. The obtained experimental results are discussed in Section 4. The conclusions are drawn in Section 5.

Table 1. Schematic comparison of the most common methods to detect and identify sensor faults and the approach proposed in this work.

Approach	Description	Properties
Preventive Maintenance	Regular maintenance based on theoretical failure rate to avoid failures before they occur.	It is costly and prone to cause unnecessary repairs.
Analytical redundancy techniques	Detection and identification of sensor failures from models built on a specific process.	It requires deep knowledge of the entire process. The model must be updated as the process changes.
Machine Learning	Detection and identification of sensor faults through models trained on sensor data.	Faults are identified by changes in certain signal characteristic features rather than by underlying sensor-specific causes.
Method proposed in this work	Detection and identification of sensor failures from models built on the sensor itself, independently of the processes.	It can be cost effective and the effort for model development can be easily amortized. It offers augmented sensor diagnostic coverage.

2. Materials and Methods

2.1. Resistive Displacement Sensor Model

Resistive displacement sensors measure the linear or angular displacement of an object by sensing a resistance change related to the object position variation. With specific reference to linear displacement sensors along the direction x, as shown in Figure 1a, the sensing element is essentially composed of a resistive track, a sliding cursor, which is constrained to move along the longitudinal axis of the sensor, and a conductive track. The resistive and conductive tracks are electrically connected to each other through the sliding cursor, which is mechanically linked to the moving object under measurement.

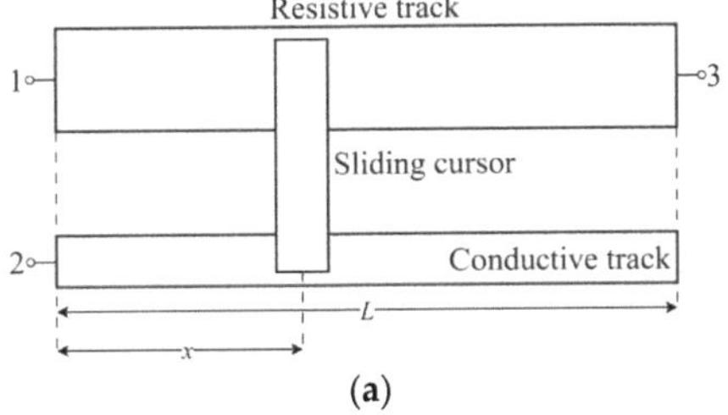

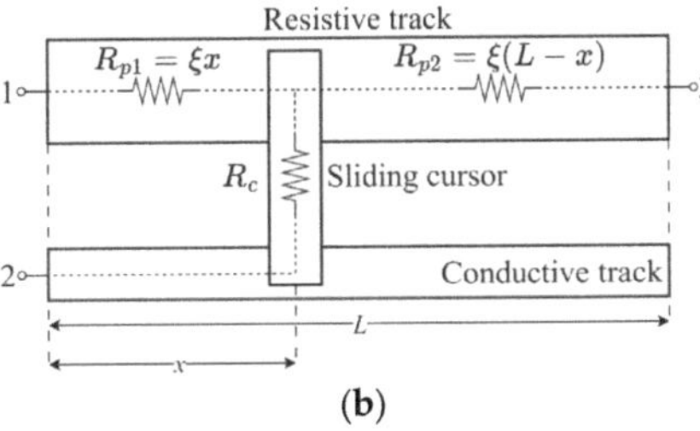

Figure 1. (**a**) Schematization of a generic resistive linear displacement sensor with (**b**) the corresponding resistive circuit modeling.

The most common causes of failure for linear sensors include the following:

- Wear: sensor components like the sliding cursor and the resistive/conductive track can wear out or become contaminated with dust, dirt, or other debris, causing the sensor to malfunction or eventually fail.
- Mechanical damage: shock, vibrations or impacts can damage or even break the sliding cursor or the resistive/conductive track.
- Environmental factors: the sensor can be affected by temperature changes or gradients, pressure variations impairing sensor sealing, moisture infiltration, or other causes producing changes in resistance, which leads to measurement errors up to complete failure.
- Aging: the electrical properties of materials forming the resistive/conductive track and the cursor might change over time, causing sensor resistance drift or output instability, leading to inaccurate measurements up to an unacceptable level.

Considering the above-mentioned failure modes, a tailored sensor model is presented below to provide information about the actual sensor conditions and its possible alterations over time. As shown in Figure 1b, the sensing element is modeled by a resistive circuit composed of three resistors. The resistor R_c represents the contact resistance between the sliding cursor and the resistive track. The resistors R_{p1} and R_{p2} represent the resistive track resistances originated from its partitioning caused by the sliding cursor according to its position. Resistances R_{p1} and R_{p2} as a function of the sliding cursor position x are given by the following:

$$R_{p1} = (\rho/S)x = \xi x \tag{1}$$

$$R_{p2} = (\rho/S)(L - x) = \xi(L - x) \tag{2}$$

where ρ, S, L, and $\xi = \rho/S$ are the electrical resistivity, section area, length, and resistance per unit length of the resistive track, respectively. The conductive track is assumed to have zero resistance.

The resistances measured between the sensor terminal pairs (1, 2), (2, 3), and (1, 3) can be expressed as follows:

$$R_{12} = R_{p1} + R_c = \xi x + R_c \tag{3}$$

$$R_{23} = R_c + R_{p2} = R_c + \xi(L - x) \tag{4}$$

$$R_{13} = R_{p1} + R_{p2} = \xi x + \xi(L - x) = \xi L \tag{5}$$

Equations (3) and (4) can be combined obtaining the following:

$$\Sigma = R_{12} + R_{23} = \xi x + R_c + R_c + \xi(L - x) = 2R_c + \xi L \tag{6}$$

where the term Σ is independent from the sliding cursor position x at which the resistances R_{12} and R_{23} are measured.

Therefore, the sensor model is expressed by combining Equations (5) and (6) in matrix form as follows:

$$Ap = R \rightarrow \begin{bmatrix} 2 & L \\ 0 & L \end{bmatrix} \begin{bmatrix} R_c \\ \xi \end{bmatrix} = \begin{bmatrix} \Sigma \\ R_{13} \end{bmatrix}, \tag{7a}$$

$$p = A^{-1}R \rightarrow \begin{bmatrix} R_c \\ \xi \end{bmatrix} = \begin{bmatrix} \frac{1}{2}\Sigma - \frac{1}{2}R_{13} \\ \frac{1}{L}R_{13} \end{bmatrix} = \begin{bmatrix} \frac{1}{2}(R_{12} + R_{23}) - \frac{1}{2}R_{13} \\ \frac{1}{L}R_{13} \end{bmatrix} \tag{7b}$$

where the vector R is formed by the resistance terms Σ and R_{13}, the vector p includes the sensor parameters R_c and ξ, and the square matrix A only depends on the resistive track length L. Notably, the linear model of Equation (7a), expressed in its inverted form in Equation (7b), is independent from the measurand displacement x. As such, it advantageously allows to estimate p by only knowing the resistive track length L and the resistances R_{12}, R_{23}, and R_{13} measured at any unknown cursor position x.

The parameters R_c and ξ offer the advantage to directly relate the measurable sensor electrical characteristics with the elements that are most susceptible to wear and faults, i.e., the sliding cursor and resistive track [23]. Although more parameters could be taken into consideration to obtain a more detailed sensor model, R_c and ξ already allow the identification of the most common causes of wearout and failure for resistive displacement sensors, at the same time avoiding excessive model complexity. In addition, the estimation of R_c and ξ can be advantageously performed without the knowledge of the cursor position x, allowing their continuous estimation during sensor normal operation.

The parameters p are used to detect alterations in sensor conditions up to sensor faults. For example, the wearout caused by the contact force between the sliding cursor and the resistive track can be considered. Repeated cycles of slides and/or an excessive contact force leads to a reduction in resistive track thickness by abrasion, and in turn a reduction in section S, with possible overheating and local oxidation and wear [24]. Compared to the initial conditions of a brand-new sensor, the reduction in S implies an increase in ξ. The resistive element wear debris deposited between the sliding cursor and the resistive element increases the contact resistance R_c because the contact area decreases, and the debris acts as an insulator. In addition, creep and fatigue in the sliding cursor brush cause the contact force to decrease, again increasing R_c [23,25]. With respect to possible mechanical damage, the breakout of the sliding cursor opens the electrical contact between the resistive and conductive tracks, leading R_{12} and R_{23} to infinity, as well as R_c. Other possible mechanical damages could be cracks in the resistive track, which typically increase the resistance R_{13} and thus ξ. Additionally, the resistive track ages, leading to a change in resistivity ρ, which reflects in a change of ξ. Meanwhile, the slider cursor ages by oxidizing, causing R_c to increase.

The model assumes operation at constant temperature. Such an assumption is not a substantial limitation in practice, as in the majority of industrial applications, the sensor, after a thermal transient at the machine start-up, works at a reasonably constant temperature thereof. For this reason, though track resistivity may in general depend on temperature, the proposed model can be conveniently restricted to thermal steady-state operation without suffering unacceptable inaccuracies.

Sensor condition alterations and failures are detected by monitoring the evolution of R_c and ξ with respect to their initial values, i.e., those of a brand-new undamaged sensor. Initial values of R_c and ξ are design parameters subject to the typical variability of the manufacturing process. In the first analysis, the detection of fault conditions can be accomplished by comparing each parameter with a respective nominal value taken as the alarm threshold. Due to product and process variability, the definition of *a priori* nominal alarm threshold could be ineffective for diagnostic monitoring of sensor conditions. For this reason, it is proposed to identify specific alarm thresholds for each individual sensor.

2.2. Parameter Estimation

To determine the initial parameters and monitor their evolution during the sensor operation, the proposed estimation method derives the parameters of the vector p online, i.e., without functionally disconnecting the sensor from the monitored process, and without knowing the cursor position x.

Online estimation is a substantial advantage of the proposed method and is mandatory for its effective practical application since both the initial parameters and their variations during operation in the field must be determined.

As illustrated in the flow chart of Figure 2, the estimation method can be divided into two phases. In the first phase, denoted as the measuring phase, the resistance set represented by the three-component vector $S_R = [R_{12}, R_{23}, R_{13}]$ measured at an unknown cursor position x is used to evaluate the resistance vector $R = [\Sigma, R_{13}]$. In the second phase, denoted as the estimation phase, the previously evaluated R is used to compute p as reported in Equation (7b).

In the proposed method, each resistance set S_R, measured at an unknown position x, is used to derive a correspondent estimation of p. The estimated parameters reflect the sensor local conditions at the unknown position x where S_R is measured. This provides indirect indications of possible nonhomogeneities.

As an alternative, p can be computed using the average of N resistance sets S_R measured in N different cursor positions. Provided that the N positions are sufficiently spread, this results in spatial averaging along the resistive track with a correspondent reduction the residual dependence of p on the cursor position x.

In this work, the parameters p are estimated starting from each single resistance set S_R, i.e., $N = 1$. This offers the advantage of maintaining a biunivocal correspondence between p and the position x without introducing spatial averaging.

Since the parameters p are estimated starting from the direct measurement of the resistances R_{12}, R_{23}, and R_{13}, the lower the measurement uncertainty of the resistance measurement, the better the parameter estimation will be. The uncertainties $u_c(R_c)$ and $u_c(\xi)$ resulting from the resistance uncertainties $u(R_{12})$, $u(R_{12})$, and $u(R_{13})$ can be considered in defining a fault detection method that is robust with respect to measured data. From Equation (7b), the composite uncertainties $u_c(R_c)$ and $u_c(\xi)$ are given by the following:

$$u_c(R_c) = \sqrt{\frac{1}{4}u^2(R_{12}) + \frac{1}{4}u^2(R_{23}) + \frac{1}{4}u^2(R_{13})} \tag{8}$$

$$u_c(\xi) = \sqrt{\frac{1}{L^2}u^2(R_{13})} \tag{9}$$

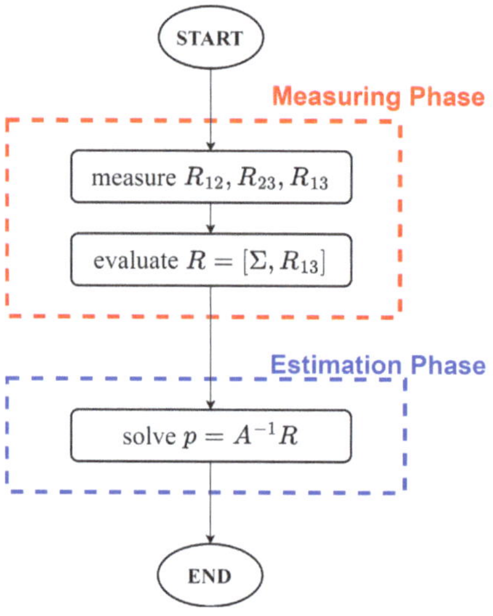

Figure 2. Parameter estimation method flow chart.

2.3. Cursor Position Estimation and Compensation

There are several methods to obtain the cursor position of a resistive displacement sensor. Considering the schematization of a generic resistive displacement sensor of Figure 1a, one common method involves exciting the sensor with a known constant voltage at the terminal pair (1, 3) while reading at high-input impedance the voltage divider output at the terminal pair (2, 3), which varies as a function of the cursor position x. This method does not account for the sensor conditions represented by parameters p. Actually, the reading is to first order independent from p as long as ξ is uniform along the sensor stroke L. As no information on the sensor conditions is extracted, sensor faults cannot be anticipated nor predicted.

On the other hand, the resistances R_{12} and R_{23} depend on both the cursor position x at which they are measured and the parameters p. By manipulating Equations (3) and (4), the actual cursor position x can be expressed as follows:

$$x = (R_{12} - R_{23})/2\xi + L/2 \tag{10}$$

An estimated cursor position x' can be derived from Equation (10) by employing the current parameter estimates $p_c = [R_{cc}, \xi_c]$. The closer the match between p_c and p, the lower the difference will be between x' and x. The position estimation error e_{pos} due to the difference between the actual unknown ξ and its current estimation ξ_c is as follows:

$$e_{pos} = x' - x = \frac{R_{12} - R_{23}}{2}\left(\frac{\xi - \xi_c}{\xi_c \xi}\right) \tag{11}$$

where R_{12} and R_{23} are the resistances measured in the current sensor conditions. Since the parameters vary as the sensor conditions change, it is crucial that the current parameters p_c are continuously updated since x' depends on ξ_c.

If instead of p_c the nominal parameters $p_0 = [R_{c0}, \xi_0]$ relative to an undamaged sensor are used as estimates of p during the sensor life, the estimated position $x_0\prime$ is subject to an estimation error e_{pos0} given by the following:

$$e_{pos0} = x_0' - x = \left(x_0' - x'\right) + \left(x' - x\right) = \frac{R_{12} - R_{23}}{2}\left(\frac{\xi_c - \xi_0}{\xi_c \xi_0}\right) + e_{pos} = \delta + e_{pos} \quad (12)$$

The position estimation error e_{pos0} can be seen as the sum of e_{pos} given by Equation (11) and an additive error δ related to the difference between the current parameter ξ_c and ξ_0 employed for estimating position.

The additive error δ is zero when $\xi_c = \xi_0$, i.e., when ξ does not vary from the undamaged conditions. It also shows that the error is higher the farther the cursor is from $L/2$. As the position error e_{pos} does not depend on R_c as long as R_c is finite and Equation (11) remains valid, the proposed estimation method is unaffected by R_c.

Starting from Equation (12), if the estimate of ξ is constantly updated, the position estimation error e_{pos0} can be reduced to e_{pos} by nulling the additive position estimation error δ. In this perspective, the method proposed to reduce the position estimation error can be seen as a self-compensation method.

2.4. Fault Detection

The parameters R_c and ξ are known to play a crucial role in defining the conditions of the resistive displacement sensor as they are directly affected by sensor wearout and faults [23]. Therefore, the proposed fault detection method focuses on assessing progressive sensor wearout or sudden failure by comparing the variation in R_c and ξ with reference values by means of alarm thresholds. Innovatively, this approach is built on a model that allows determination of the sensor conditions during its operation, without disconnecting it from the host system and without requiring knowledge of the cursor position x.

As illustrated in the flow chart of Figure 3, faults are detected on the basis of parameter variations $\Delta p = [\Delta R_c, \Delta \xi] = |p_c - p_0|$, where $p_c = [R_{cc}, \xi_c]$ and $p_0 = [R_{c0}, \xi_0]$ represent the sensor parameters in the current and initial undamaged conditions, respectively. The nominal parameters p_0 are calculated as the spatial average of the initial parameters R_c and ξ estimated along the entire sensor stroke. The parameters p_0 could be determined during factory calibration or directly in the field, during the first installation. It is expected that R_c and ξ variations are mostly related to sliding cursor and resistive track faults, respectively.

Fault detection is based on threshold surpassing. Specifically, if both ΔR_c and $\Delta \xi$ are lower than their respective alarm thresholds T_1 and T_2, then no fault is detected. Differently, the sensor is faulty, and the fault could affect the resistive track, the sliding cursor, or both depending on which threshold is surpassed.

The thresholds T_1 and T_2 can be set empirically on the basis of ensemble standard deviations $\sigma(R_{c0})$ and $\sigma(\xi_0)$ derived from a suitably large population representative of undamaged sensors, or from each individual sensor at the installation time in the field. Assuming that model parameters for an undamaged sensor are random variables with normal distribution, it is reasonable to suppose that changes that exceed the range of $\pm 3\sigma(R_{c0})$ and $\pm 3\sigma(\xi_0)$ are ascribable to an occurred alteration in the sensor conditions due to wearout or fault.

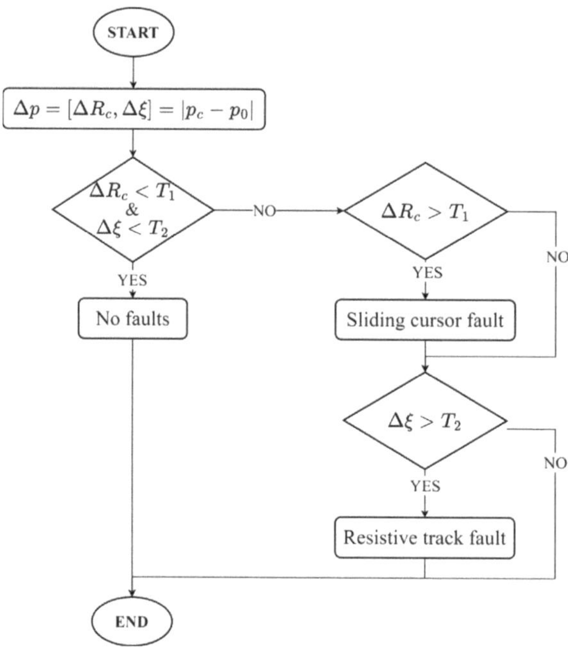

Figure 3. Fault detection method flow chart.

3. Results

The sensor model, along with the parameter estimation, the self-compensation, and the fault detection methods have been simulated and experimentally tested.

The simulations have estimated the parameters from a set of resistance values corrupted with pseudorandom numerical noise, while in the experimental tests the parameters have been derived from measurements on a Device Under Test (DUT), i.e., a resistive displacement sensor, placed in two different conditions, namely undamaged and faulty.

The term undamaged here refers to a sensor that has never been used and has been stored in compliance with the requirements described in the sensor datasheet. Instead, the term faulty here indicates a sensor subject to a resistive track and sliding cursor wearout, where the resistive track wear along x is virtually uniform. Wear has been considered among the failures listed in Section 2.1 as it represents the main and most common failure mode.

The simulations have been aimed at evaluating the parameter estimation variability with respect to the measurement accuracy of resistances R_{12}, R_{23}, and R_{13}. The experimental tests have been aimed at assessing the fitting accuracy of the sensor model defined as the difference between the measured resistances R_{12}, R_{23}, and R_{13} and those synthetized from the sensor model. Furthermore, the estimated parameters have been employed to evaluate the self-compensation method in reducing the position estimation error and the effectiveness of alarm thresholds T_1 and T_2 in detecting wear failure.

3.1. Simulation Results

Simulations have been performed in MATLAB 2022b to repeatedly estimate the model parameters R_c and ξ from a set of resistances R_{12}, R_{23}, and R_{13} corrupted with numerical noise. The parameter mean values μ and standard deviations σ have been calculated from the estimates across the repetitions taken at different positions.

The resistances R_{12}, R_{23}, and R_{13} forming the resistance set S_R have been synthetized from their respective analytical expressions reported in Equations (3)–(5), fixing the ideal parameters $\bar{p} = \left[\overline{R_c}, \overline{\xi} \right] = [100\ \Omega;\ 50{,}000\ \Omega/\text{m}]$, the sensor stroke $L = 100$ mm and sweeping the displacement from $x = 0$ to $x = L$ with a step of 1 mm. The values chosen for $\overline{p}$ are representative of the actual sensor that has been experimentally tested as detailed in Section 3.2. In particular, $\overline{R_c} = 100\ \Omega$ includes non-idealities like the conductive track resistance and the resistance of the connecting terminals. A normally distributed pseudorandom numerical noise with a standard deviation $\sigma_n = 1\ \Omega$ has been added to corrupt R_{12}, R_{23}, and R_{13}.

For each considered cursor position x, the parameters have been estimated from the corresponding resistance set S_R. Then, the mean values μ and standard deviations σ with respect to x have been calculated from the estimates. The obtained results are reported in Table 2 as the difference between $\overline{p}$ and the estimated parameter mean values μ, along with their standard deviations σ. Figures 4 and 5 report the parameter estimation errors as a function of the position along the sensor stroke.

Table 2. Parameter estimation errors and standard deviations σ obtained by ranging x from 0 to L with a 1 mm step ($M = 100$ parameter estimations), assuming as a reference the ideal parameters $\overline{p} = \left[\overline{R_c}, \overline{\xi}\right] = [100\ \Omega;\ 50{,}000\ \Omega/\text{m}]$. The uncertainties of the parameter mean values μ and standard deviations σ are derived from the parameter composite uncertainties $u_c(R_c)$ and $u_c(\xi)$.

R_c Estimation Error $\overline{R_c} - \mu(R_c)\ [\Omega]$	R_c Standard Deviation $\sigma(R_c)$	ξ Estimation Error $\overline{\xi} - \mu(\xi)\ [\Omega/\text{m}]$	ξ Standard Deviation $\sigma(\xi)$
0.1 ± 0.1	0.8 ± 0.1	-1 ± 1	11.5 ± 0.7

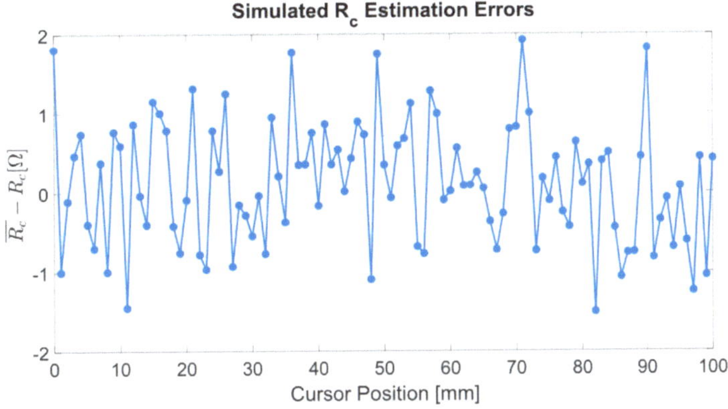

Figure 4. Estimation error of the parameter R_c versus the position x along the sensor stroke.

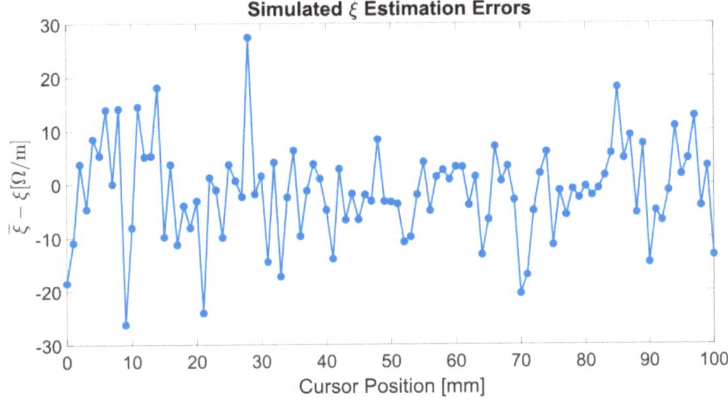

Figure 5. Estimation error of the parameter ξ versus the position x along the sensor stroke.

The simulation evaluates the variability of estimated parameters with respect to the measurement accuracy of resistances R_{12}, R_{23}, and R_{13}. The parameter estimation is adequate for the application scope because the percentage deviations between the ideal parameters $\bar{p}$ and the mean values of the estimated parameters are less than $\pm 0.1\%$. Furthermore, the parameter standard deviations $\sigma(R_c)$ and $\sigma(\xi)$ agree with the corresponding composite uncertainties $u_c(R_c) = \sqrt{3}/2\ \Omega$ and $u_c(\xi) = 10\ \Omega/\text{m}$, obtained from Equations (8) and (9) by setting $u(R_{12}) = u(R_{23}) = u(R_{13}) = \sigma_n = 1\ \Omega$.

3.2. Experimental Setup

The DUT is a resistive linear displacement sensor (Gefran PK), with a displacement full-scale span (FSS) $L = 100$ mm, a nominal resistive track resistance of 5 kΩ and an independent linearity error of $\pm 0.05\%$ FSS.

Sensor wearout has been intentionally provoked by sliding the cursor back and forth throughout the whole sensor stroke for 10^6 cycles. Figure 6 shows the stress-test machine adopted to induce wear, while Figure 7 shows the wearout effect produced on the resistive track.

Figure 6. Stress-test machine employed to induce wearout in the resistive track and sliding cursor.

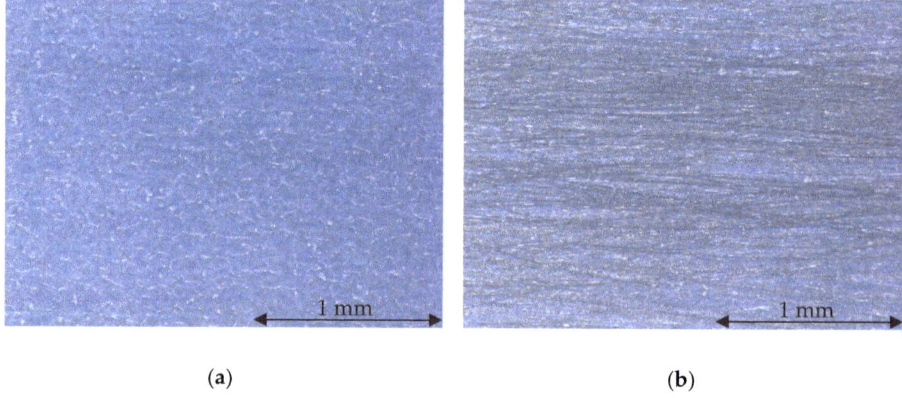

(a)   (b)

Figure 7. Images obtained with a digital microscopy system (Leica DMS300) of the undamaged (**a**) and wearout (**b**) resistive track.

The assembled experimental setup comprises a linear positioning stage, a digital multimeter, and a personal computer, as shown in Figures 8 and 9.

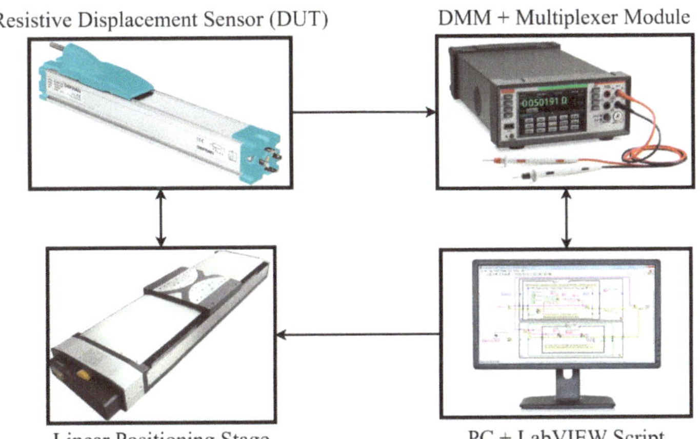

Figure 8. Block diagram of the adopted experimental setup.

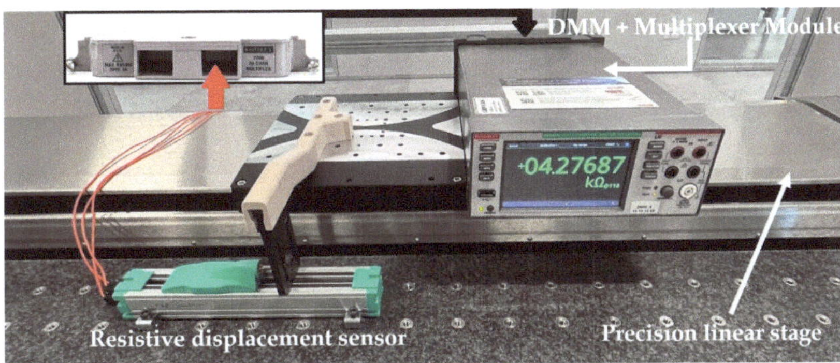

Figure 9. Photo of the assembled experimental setup.

The sensor parameters are estimated from the resistance sets S_R measured in M different cursor positions along the sensor stroke. Resistances have been measured by means of a 6.5-digit digital multimeter (Keithley DAQ6510, Cleveland, OH, USA) equipped with a multiplexer module (Keithley 7700), providing a specified resistance measurement accuracy $u(R) = 1\ \Omega$ in the 10 kΩ range of interest.

The sensor cursor displacement has been set by a precision linear positioning stage (Physik Instrumente LS-270, Karlsruhe, Germany) with a calibrated position accuracy in the order of ± 1 μm. A LabVIEW script has been developed to implement the parameter estimation method and handle the experimental setup. The script is designed to move the linear stage to M different positions and trigger the digital multimeter to measure the components of S_R at each position.

3.3. Experimental Results

Experimental tests have been carried out to estimate the parameters of an undamaged (U) sensor and a faulty (F) sensor, denoted as $p_\mathbf{u} = [R_{cu}, \xi_u]$ and $p_\mathbf{f} = [R_{cf}, \xi_f]$, respectively. For the undamaged and faulty sensors, the respective resistance sets S_{Ru} and S_{Rf} have been repeatedly measured M times for the position x ranging from 0 to L with a 1 mm step resulting in $M = 100$, as schematized in Figure 10. From each set S_{Ru} and S_{Rf}, the parameters $p_\mathbf{u}$ and $p_\mathbf{f}$ have been estimated according to Equation (7b).

Table 3 reports the mean values $\mathbf{P_u} = [\mu(R_{cu}), \mu(\xi_u)]$ and $\mathbf{P_f} = [\mu(R_{cf}), \mu(\xi_f)]$, and standard deviations $\sigma_\mathbf{u} = [\sigma(R_{cu}), \sigma(\xi_u)]$ and $\sigma_\mathbf{f} = [\sigma(R_{cf}), \sigma(\xi_f)]$ for $p_\mathbf{u}$ and $p_\mathbf{f}$ obtained over the M repetitions at varying x. From $\mathbf{P_u}$ and $\mathbf{P_f}$ inserted into Equations (3)–(5), the terms

$\mathbf{R_{mu}} = [R_{12mu}, R_{23mu}, R_{13mu}]$ and $\mathbf{R_{mf}} = [R_{12mf}, R_{23mf}, R_{13mf}]$ have been obtained, where the subscript m indicates outcomes from the model. Hence the model residuals have been computed for the undamaged and faulty sensor as $\mathbf{r_u} = [r_{12u}, r_{23u}, r_{13u}] = \mathbf{S_{Ru}} - \mathbf{R_{mu}}$ and $\mathbf{r_f} = [r_{12f}, r_{23f}, r_{13f}] = \mathbf{S_{Rf}} - \mathbf{R_{mf}}$, respectively. The residuals $\mathbf{r_u}$ and $\mathbf{r_f}$ have been evaluated for all the M repetitions at which $\mathbf{S_{Ru}}$ and $\mathbf{S_{Rf}}$ have been measured. The root mean square errors (RMSE) of the residuals are also reported in Table 3. Figure 11 shows the estimated parameters $\mathbf{p_u}$ and $\mathbf{p_f}$ while Figure 12a,b report the residuals $\mathbf{r_u}$ and $\mathbf{r_f}$ as a function of repetition index ranging from 1 to $M = 100$.

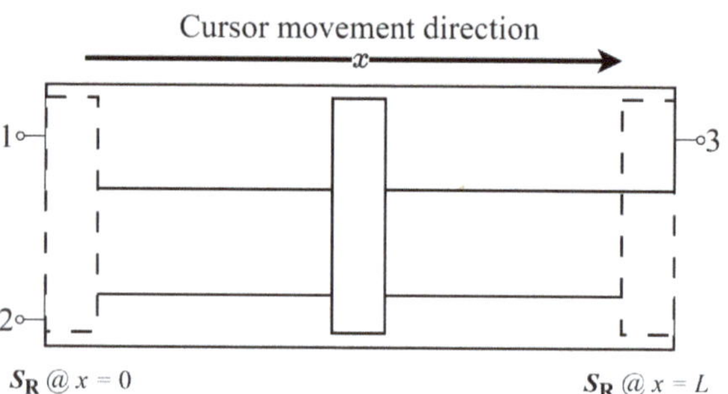

Figure 10. Schematization of $\mathbf{S_{Ru}}$ and $\mathbf{S_{Rf}}$ measurement processes. The resistance sets $\mathbf{S_{Ru}}$ and $\mathbf{S_{Rf}}$ are measured from the undamaged and faulty sensor over M repetitions taken for x varying from 0 to L with a 1 mm step.

Table 3. Parameter mean values μ and standard deviations σ, along with the RMSE of the residuals r_{12}, r_{23} and r_{13} obtained in the experimental test. Parameter estimations are performed from the resistance sets $\mathbf{S_R}$ measured $M = 100$ times for x ranging from 0 to L with a 1 mm step. Since the resistance measurement uncertainty is $u(R) = 1\ \Omega$, the parameter composite uncertainties are $u_c(R_c) = 0.9\ \Omega$ and $u_c(\xi) = 10\ \Omega/m$. The uncertainties of the parameter mean values μ, standard deviations σ, and RMSE of the residuals r_{12}, r_{23} and r_{13} are derived from $u_c(R_c)$, $u_c(\xi)$, and $u(R)$.

Sensor Condition	R_c Mean Value $\mu(R_c)$ [Ω]	R_c Standard Deviation $\sigma(R_c)$ [Ω]	ξ Mean Value $\mu(\xi)$ [Ω/m]	ξ Standard Deviation $\sigma(\xi)$ [Ω/m]	r_{12} Root Mean Square Error RMSE(r_{12}) [Ω]	r_{23} Root Mean Square Error RMSE(r_{23}) [Ω]	r_{13} Root Mean Square Error RMSE(r_{13}) [Ω]
Undamaged (U)	123.0 ±0.1	4.5 ± 0.1	49,058 ± 1	1.3 ± 0.7	10.9 ± 0.1	10.9 ± 0.1	0.1 ± 0.1
Faulty (F)	161.6 ± 0.1	21.9 ± 0.1	62,142 ± 1	9.9 ± 0.7	88.6 ± 0.1	83.1 ± 0.1	1.1 ± 0.1

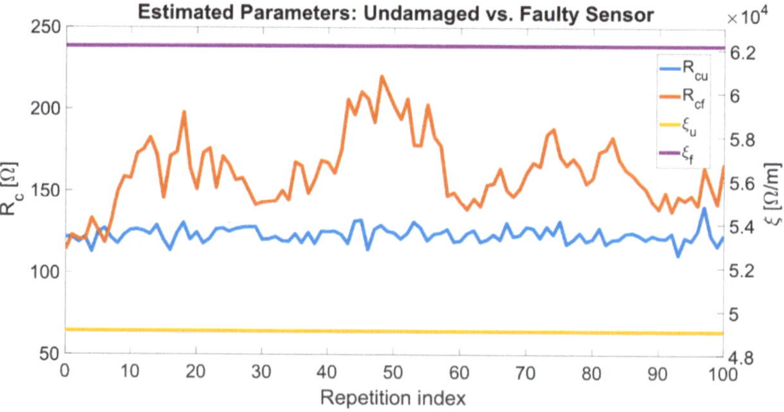

Figure 11. Parameters $\mathbf{p_u}$ and $\mathbf{p_f}$ estimated, respectively, from the measured resistance sets $\mathbf{S_{Ru}}$ and $\mathbf{S_{Rf}}$ as a function of the repetition index ranging from 1 to $M = 100$.

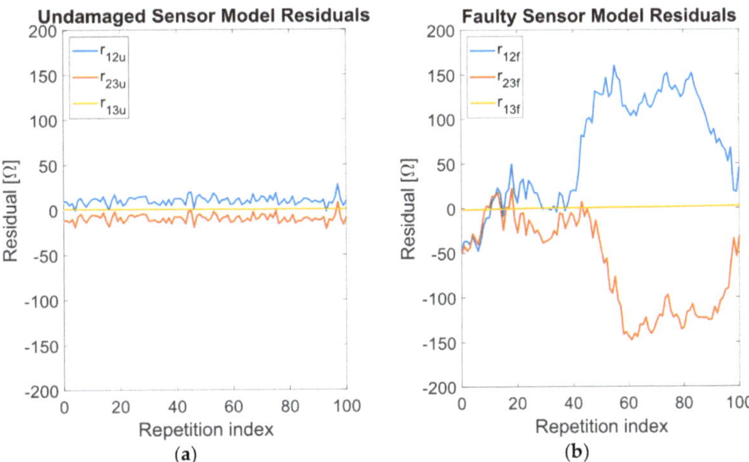

Figure 12. Residuals $\mathbf{r_u} = S_{\mathbf{Ru}} - R_{\mathbf{mu}}$ (a) and $\mathbf{r_f} = S_{\mathbf{Rf}} - R_{\mathbf{mf}}$ (b) resulting from the measured $S_{\mathbf{Ru}}$ and $S_{\mathbf{Rf}}$ as a function of the repetition index ranging from 1 to $M = 100$.

The position estimation error e_{pos} is evaluated for both the U and F sensors, for all the cursor positions at which $S_{\mathbf{Ru}}$ and $S_{\mathbf{Rf}}$ are measured. The true, i.e., actual, cursor position x is assumed to be given by the linear positioning stage taken as the reference. The position estimations x' are derived from Equation (10) by employing $\mu(\xi_{\mathrm{u}})$ and $\mu(\xi_{\mathrm{f}})$ for the U and F sensor, respectively. Figure 13 shows the position estimation error e_{pos} for both the U and F sensors as a function of x. Figure 14 shows e_{pos} obtained for the faulty sensor both with and without the application of the self-compensation method described in Section 2.3, where the uncompensated cursor positions are intentionally estimated using $\mu(\xi_{\mathrm{u}})$ in order to simulate a mismatch between the nominal, i.e., initial, and the current sensor conditions.

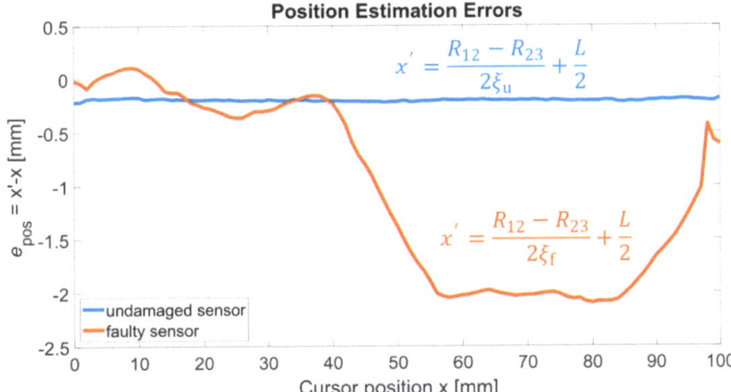

Figure 13. Position estimation error e_{pos} for to the undamaged (U) and faulty (F) sensors versus the actual cursor position x.

Lastly, the fault detection method discriminates the fault and no-fault conditions on the basis of the parameter variations $\Delta p_{\mathbf{u}} = |p_{\mathbf{u}} - \mathbf{P_u}|$ and $\Delta p_{\mathbf{f}} = |p_{\mathbf{f}} - \mathbf{P_u}|$. The fault detection thresholds T_1 and T_2 have been set based on the $p_{\mathbf{u}}$ standard deviations $\sigma_{\mathbf{u}}$, resulting in $T_1 = 3\sigma(R_{\mathrm{cu}}) = 13.5\ \Omega$ and $T_2 = 3\sigma(\xi_{\mathrm{u}}) = 3.9\ \Omega/\mathrm{m}$.

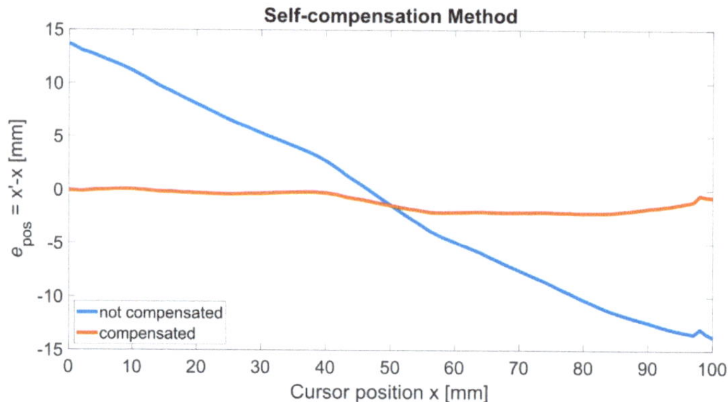

Figure 14. Position estimation error e_{pos} for to the faulty (F) sensor, with and without the self-compensation method applied, versus the actual cursor position x. The uncompensated cursor positions have been deliberately estimated using $\mu(\xi_u)$ to simulate a discrepancy between the nominal and current sensor conditions.

4. Discussion

Considering the simulation results described in Section 3.1, the normally distributed pseudorandom numerical noise added to R_{12}, R_{23}, and R_{13} and the resulting parameter variability are reflected in the parameter composite uncertainties $u_c(\xi)$ and $u_c(R_c)$ expressed in Equations (8) and (9). Assuming the resistance uncertainties equal to the noise standard deviation, i.e., $u(R_{12}) = u(R_{23}) = u(R_{13}) = \sigma_n = 1\ \Omega$, the parameter composite uncertainties result $u_c(R_c) = \sqrt{3}/2\ \Omega$ and $u_c(\xi) = 10\ \Omega/\text{m.}$, which agree with the simulation results $\sigma(R_c) = 0.8\ \Omega$ and $\sigma(\xi) = 11\ \Omega/\text{m.}$

The standard deviations of the parameters R_c and ξ can be analyzed to compare the variability of the parameters obtained in the simulations with those from experimental results. Although the noise standard deviation σ_n in simulations is taken equal to the resistance measurement uncertainty $u(R)$, i.e., $u(R) = \sigma_n = 1\ \Omega$, the experimental standard deviations of R_c, $\sigma(R_{cu}) = 4.5\ \Omega$ and $\sigma(R_{cf}) = 21.9\ \Omega$, are larger than $\sigma(R_c) = 0.8\ \Omega$ for simulations.

This is consistent with the fact that $\sigma(R_{cu})$ and $\sigma(R_{cf})$ cannot be imputed exclusively to $u(R)$, in fact they can be ascribed to the nonideality of R_{12} and R_{23} trends as a function of x due to the presence of faults, described by the residuals r_{12} and r_{23}. Similarly, the residual r_{13} explains the difference between the experimental standard deviations $\sigma(\xi_u) = 1.3\ \Omega$ and $\sigma(\xi_f) = 9.9\ \Omega$ with the simulated one $\sigma(\xi) = 11.5\ \Omega$.

Considering the experimental results described in Section 3.3, Figure 11 shows a visible dependence of R_{cf} on the repetition index and, in turn, versus position x. Such dependence shows nonuniform variations along the resistive track length L, with regions where the contact resistance between the sliding cursor and the resistive track is higher. The microscope analysis confirms that the regions where the contact resistance is higher match with the regions where the resistive track is mostly worn. Therefore, the proposed method theoretically allows detection of local faults through R_c trend analysis.

The sensor model goodness of fit is evaluated on the basis of the residuals r_u and r_f. As shown in Figure 12a,b, for the undamaged sensor the RMSEs of r_u are $\sim 0.2\%$ of the resistive track nominal resistance, which is adequately low for fault detection and estimation of position x'. For the faulty sensor, the RMSEs of r_f increase compared to the undamaged case due to the trends of R_{12} and R_{23} as a function of x caused by nonhomogeneous resistive track wear along its length.

As shown in Figure 13, the position estimation error e_{pos} for the undamaged sensor is approximately constant throughout the sensor stroke, with a mean value of 0.2 mm. For the faulty sensor, e_{pos} is an order of magnitude higher spanning a range of roughly

2 mm. As reported in Figure 14, the self-compensation method successfully reduces e_{pos} to a span of about 2 mm in the faulty case versus the value of 30 mm obtained intentionally using $\mu(\xi_u)$ to represent the case when the parameter estimates are not updated to the current values.

Given the chosen thresholds $T_1 = 3\sigma(R_{cu})$ and $T_2 = 3\sigma(\xi_u)$, the results obtained in applying the fault detection method are shown in the confusion matrices of Figure 15.

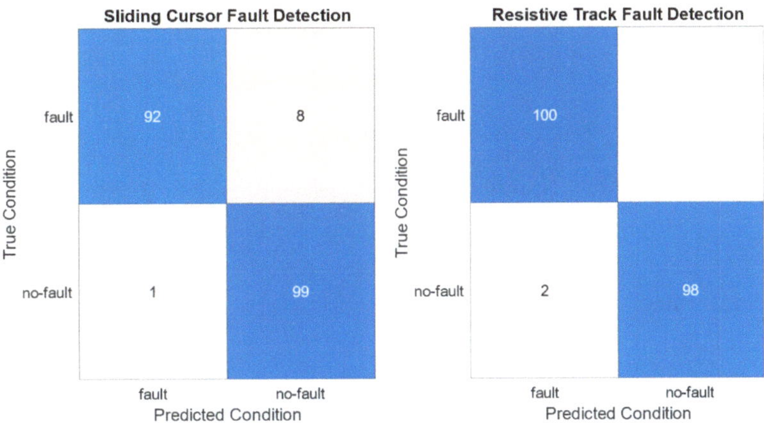

Figure 15. Confusion matrices for the sliding cursor and resistive track fault detections.

The fault detection method has been evaluated using accuracy, precision, and recall performance metrics. Accuracy measures the fault detection correctness, precision evaluates the number of actual faults identified, while recall assesses the effectiveness in detecting faults among all instances of faults. The performance metrics obtained for the resistive track and sliding cursor fault detections are reported in Table 4.

Table 4. Accuracy, precision and recall performance metrics for the sliding cursor and resistive track fault detections. Accuracy = (TP + TN)/(TP + TN + FP + FN), precision = TP/(TP + FP), and recall = TP/(TP + FN), where TP, TN, FP, and FN are the true positive, true negative, false positive, and false negative fault predictions, respectively.

Type of Fault Detected	Accuracy	Precision	Recall
Sliding Cursor	95.5%	98.9%	92.0
Resistive Track	99.0%	98.0%	100%

The false negative sliding cursor fault detections are due to the variability of R_c as a function of x caused by nonhomogeneous wear of the resistive track. A possible solution could be acting on threshold values. For example, lowering T_1 to equal $\sigma(R_{cu})$ results in the sliding cursor fault detection recall increasing to 96.0%. However, threshold lowering makes the fault detection method more prone to false positives, leading the precision to drop to 78.6%.

Moreover, the uncertainty of the standard deviations of the parameters in the thresholds T_1 and T_2 has an impact on the accuracy and precision metrics for resistive track fault detection. In particular, accuracy and precision vary within [97.5%; 100%] and [95.2%; 100%], respectively.

5. Conclusions

This work introduces a suitably general model for resistive displacement sensors and proposes thereof an innovative fault detection method that can be applied with the sensor in operation providing self-diagnostic capabilities. Based on the output of the self-diagnostic step, a self-compensation method with automatic updating is in turn derived to increase the accuracy in sensors subject to progressive wearout. The proposed model and methods have been validated by both numerical simulations and experimental tests on real devices.

The position is estimated by exploiting the sensor model and the initial value of ξ. The self-compensation method reduces the position estimation error by considering the difference between the current and initial values of ξ, which causes a position estimation error. The self-compensation method successfully reduces the position estimation error e_{pos} in the faulty case, intentionally obtained using $\mu(\xi_u)$, from a span of 30 mm to a span of 2 mm. The proposed fault detection method evaluates the progressive sensor wearout or sudden failure by comparing the variation in the current parameters R_c and ξ with respect to their initial values against the alarm thresholds T_1 and T_2, respectively. The fault detection method, for $T_1 = 3\sigma(R_{cu})$ and $T_2 = 3\sigma(\xi_u)$, has shown a resistive track and sliding cursor fault detection accuracy of 99.0% and 95.5%, respectively. Lowering T_1 from $3\sigma(R_{cu})$ to $\sigma(R_{cu})$ has increased the sliding cursor fault detection recall to 96.0%, involving the precision to drop to 78.6% as the threshold lowering makes the fault detection method more prone to false positives.

The future work will involve the implementation of the proposed methods and their testing in a dynamic industrial process.

6. Patents

There is a patent application resulting from the work reported in this manuscript [26].

Author Contributions: Conceptualization, D.A. and V.F.; investigation, F.M.; methodology, F.M., D.A. and V.F.; supervision, D.A. and V.F.; visualization, F.M.; writing—original draft, F.M.; writing—review and editing, D.A. and V.F. All authors have read and agreed to the published version of the manuscript.

Funding: This research received no external funding.

Institutional Review Board Statement: Not applicable.

Informed Consent Statement: Not applicable.

Data Availability Statement: Data presented in this article are available on request from the corresponding author.

Conflicts of Interest: Author Davide Alghisi was employed by the company Gefran SpA. The remaining authors declare that the research was conducted in the absence of any commercial or financial relationships that could be construed as a potential conflict of interest.

References

1. Valenzuela, R.; Vazquez, M.; Hernando, A. A Position Sensor Based on Magnetoimpedance. *J. Appl. Phys.* **1996**, *79*, 6549–6551. [CrossRef]
2. Aiestaran, P.; Dominguez, V.; Arrue, J.; Zubia, J. A Fluorescent Linear Optical Fiber Position Sensor. *Opt. Mater.* **2009**, *31*, 1101–1104. [CrossRef]
3. Nagaoka, T.; Uchiyama, A. Development of a Small Wireless Position Sensor for Medical Capsule Devices. In Proceedings of the 26th Annual International Conference of the IEEE Engineering in Medicine and Biology Society, San Francisco, CA, USA, 1–5 September 2004; Volume 1, pp. 2137–2140.
4. Bebek, O.; Cavusoglu, C.M. Whisker-like Position Sensor for Measuring Physiological Motion. *IEEE/ASME Trans. Mechatron.* **2008**, *13*, 538–547. [CrossRef]
5. Datlinger, C.; Hirz, M. Benchmark of Rotor Position Sensor Technologies for Application in Automotive Electric Drive Trains. *Electronics* **2020**, *9*, 1063. [CrossRef]
6. Kirchhoff, M.R.; Boese, C.; Güttler, J.; Feldmann, M.; Büttgenbach, S. Innovative High-Precision Position Sensor Systems for Robotic and Automotive Applications. *Procedia Chem.* **2009**, *1*, 501–504. [CrossRef]
7. Reininger, T.; Welker, F.; von Zeppelin, M. Sensors in Position Control Applications for Industrial Automation. *Sens. Actuators A Phys.* **2006**, *129*, 270–274. [CrossRef]
8. Zhu, M.; Li, J.; Wang, W.; Chen, D. Self-Detection and Self-Diagnosis Methods for Sensors in Intelligent Integrated Sensing System. *IEEE Sens. J.* **2021**, *21*, 19247–19254. [CrossRef]

9. Basri, E.I.; Abdul Razak, I.H.; Ab-Samat, H.; Kamaruddin, S. Preventive Maintenance (PM) Planning: A Review. *J. Qual. Maint. Eng.* **2017**, *23*, 114–143. [CrossRef]
10. Betta, G.; Pietrosanto, A. Instrument Fault Detection and Isolation: State of the Art and New Research Trends. *IEEE Trans. Instrum. Meas.* **2000**, *49*, 100–107. [CrossRef]
11. Isermann, R. Model-Based Fault-Detection and Diagnosis—Status and Applications. *Annu. Rev. Control* **2005**, *29*, 71–85. [CrossRef]
12. Samy, I.; Postlethwaite, I.; Gu, D.-W. Survey and Application of Sensor Fault Detection and Isolation Schemes. *Control Eng. Pract.* **2011**, *19*, 658–674. [CrossRef]
13. Thirumarimurugan, M.; Bagyalakshmi, N.; Paarkavi, P. Comparison of Fault Detection and Isolation Methods: A Review. In Proceedings of the 10th International Conference on Intelligent Systems and Control, ISCO 2016, Coimbatore, India, 7–8 January 2016; Institute of Electrical and Electronics Engineers Inc.: Piscataway, NJ, USA, 2016; pp. 1–6.
14. Lou, X.; Willsky, A.S.; Verghese, G.C. Optimally Robust Redundancy Relations for Failure Detection in Uncertain Systems. *Automatica* **1986**, *22*, 333–344. [CrossRef]
15. Clark, R.N.; Fosth, D.C.; Walton, V.M. Detecting Instrument Malfunctions in Control Systems. *IEEE Trans. Aerosp. Electron. Syst.* 1975; AES-11, 465–473. [CrossRef]
16. Willsky, A.S.; Deyst, J.J.; Crawford, B.S. Adaptive Filtering and Self-Test Methods for Failure Detection and Compensation. In Proceedings of the Joint Automatic Control Conference, Austin, TX, USA, 18–21 June 1974; pp. 637–645.
17. Mohapatra, D.; Subudhi, B.; Daniel, R. Real-Time Sensor Fault Detection in Tokamak Using Different Machine Learning Algorithms. *Fusion Eng. Des.* **2020**, *151*, 111401. [CrossRef]
18. Warriach, E.U.; Tei, K. Fault Detection in Wireless Sensor Networks: A Machine Learning Approach. In Proceedings of the 16th IEEE International Conference on Computational Science and Engineering, CSE 2013, Sydney, NSW, Australia, 3–5 December 2013; pp. 758–765.
19. Jan, S.U.; Lee, Y.D.; Koo, I.S. A Distributed Sensor-Fault Detection and Diagnosis Framework Using Machine Learning. *Inf. Sci.* **2021**, *547*, 777–796. [CrossRef]
20. Henry, M.P.; Clarke, D.W. The Self-Validating Sensor: Rationale, Definitions and Examples. *Control Eng. Pract.* **1993**, *1*, 585–610. [CrossRef]
21. Çinar, Z.M.; Nuhu, A.A.; Zeeshan, Q.; Korhan, O.; Asmael, M.; Safaei, B. Machine Learning in Predictive Maintenance towards Sustainable Smart Manufacturing in Industry 4.0. *Sustainability* **2020**, *12*, 8211. [CrossRef]
22. Mazzoli, F.; Alghisi, D.; Ferrari, V. Self-Diagnostic Method for Resistive Displacement Sensors. *Proceedings* **2024**, *97*, 164. [CrossRef]
23. Saitoh, M.; Osada, K. Long Life Potentiometric Position Sensor—Its Material and Application. *SAE Trans.* **1991**, *100*, 251–262.
24. Hattangadi, A.A. Potentiometer Failures. In *Failure Prevention of Plant and Machinery*; McGraw-Hill Education: New York, NY, USA, 2004; ISBN 9780070483095.
25. Todd, C.D. *The Potentiometer Handbook: Users' Guide to Cost-Effective Applications*; McGraw-Hill: New York, NY, USA, 1975; ISBN 978-0070066908.
26. Mazzoli, F.; Alghisi, D.; Ferrari, V. Position Sensor Configured to Monitor the Condition and/or the Wear Level of That Position Sensor During Its Use.

 *sensors*

Article

Co$_3$O$_4$-Based Materials as Potential Catalysts for Methane Detection in Catalytic Gas Sensors [†]

Olena Yurchenko [1,*], Patrick Diehle [2], Frank Altmann [2], Katrin Schmitt [1,3] and Jürgen Wöllenstein [1,3]

[1]   Fraunhofer Institute for Physical Measurement Techniques (IPM), 79110 Freiburg, Germany; katrin.schmitt@ipm.fraunhofer.de (K.S.); juergen.woellenstein@ipm.fraunhofer.de (J.W.)

[2]   Fraunhofer Institute for Microstructure of Materials and Systems (IMWS), 06120 Halle, Germany; patrick.diehle@imws.fraunhofer.de (P.D.); frank.altmann@imws.fraunhofer.de (F.A.)

[3]   Department of Microsystems Engineering (IMTEK), University of Freiburg, 79110 Freiburg, Germany

*   Correspondence: olena.yurchenko@ipm.fraunhofer.de

†   Presented at the Eurosensors XXXIV, Lecce, Italy, 10–13 September 2023.

Abstract: The present work deals with the development of Co$_3$O$_4$-based catalysts for potential application in catalytic gas sensors for methane (CH$_4$) detection. Among the transition-metal oxide catalysts, Co$_3$O$_4$ exhibits the highest activity in catalytic combustion. Doping Co$_3$O$_4$ with another metal can further improve its catalytic performance. Despite their promising properties, Co$_3$O$_4$ materials have rarely been tested for use in catalytic gas sensors. In our study, the influence of catalyst morphology and Ni doping on the catalytic activity and thermal stability of Co$_3$O$_4$-based catalysts was analyzed by differential calorimetry by measuring the thermal response to 1% CH$_4$. The morphology of two Co$_3$O$_4$ catalysts and two Ni$_x$Co$_{3-x}$O$_4$ with a Ni:Co molar ratio of 1:2 and 1:5 was studied using scanning transmission electron microscopy and energy dispersive X-ray analysis. The catalysts were synthesized by (co)precipitation with KOH solution. The investigations showed that Ni doping can improve the catalytic activity of Co$_3$O$_4$ catalysts. The thermal response of Ni-doped catalysts was increased by more than 20% at 400 °C and 450 °C compared to one of the studied Co$_3$O$_4$ oxides. However, the thermal response of the other Co$_3$O$_4$ was even higher than that of Ni$_x$Co$_{3-x}$O$_4$ catalysts (8% at 400 °C). Furthermore, the modification of Co$_3$O$_4$ with Ni simultaneously brings stability problems at higher operating temperatures ($\geq$400 °C) due to the observed inhomogeneous Ni distribution in the structure of Ni$_x$Co$_{3-x}$O$_4$. In particular, the Ni$_x$Co$_{3-x}$O$_4$ with high Ni content (Ni:Co ratio 1:2) showed apparent NiO separation and thus a strong decrease in thermal response of 8% after 24 h of heat treatment at 400 °C. The reaction of the Co$_3$O$_4$ catalysts remained quite stable. Therefore, controlling the structure and morphology of Co$_3$O$_4$ achieved more promising results, demonstrating its applicability as a catalyst for gas sensing.

Keywords: cobalt oxide; catalyst; catalytic sensors; morphology

Citation: Yurchenko, O.; Diehle, P.; Altmann, F.; Schmitt, K.; Wöllenstein, J. Co$_3$O$_4$-Based Materials as Potential Catalysts for Methane Detection in Catalytic Gas Sensors. *Sensors* **2024**, *24*, 2599. https://doi.org/10.3390/s24082599

Academic Editor: Boris Tartakovsky

Received: 29 February 2024
Revised: 11 April 2024
Accepted: 13 April 2024
Published: 18 April 2024

1. Introduction

Catalytic gas sensors, known as pellistors, are commercially available sensors used to detect flammable gases and to safely monitor gas concentrations below the lower explosive limit (LEL) [1–4]. The main area of application for pellistors is the identification of explosion risk, which arises if the concentration of flammable gases exceeds the lower explosive limit [2]. Especially light alkanes such as methane, propane, and butane are of particular interest for safety monitoring. Methane is an essential gas for various technologies, and as a main component of natural gas, it is still widely used as energy source in industrial applications and homes [5–7]. Propane and butane are the components of liquefied petroleum gas (LPG) commonly used in heating systems, cooking appliances, and vehicles [8]. Gas leaks in the home are extremely dangerous for people and pose a high risk of property damage [9].

Pellistors were first described by Alan Baker in the early 1960s [10]. A pellistor consists of two sensor elements: the active sensor coated with a catalyst layer and the passive sensor element without any catalyst. The passive sensor only responds to the changes in environmental conditions, such as humidity and temperature, while the active sensor interacts with target gases. Because of these interactions, the target gases are oxidized, and heat evolves, which is detected as thermal response of the pellistor. To activate catalytic oxidation of target gases, the sensor is operated at a specified temperature [11,12].

The oxidation of methane requires high operation temperatures ($\geq 450\,^\circ$C) due to high activation energy caused by the high stability of its C–H bonds [13,14]. However, high operation temperatures accelerate the aging and degradation of the catalysts. Propane and butane, as well as other flammable gases, are more reactive gases that are oxidized at much lower operation temperatures than methane and are therefore easier to detect [1]. Therefore, the focus in development of new catalysts for catalytic gas sensors is on suitable materials for detecting methane. Aluminum oxide-based catalysts containing a large amount of catalytically highly active Pd and Pt metals are used in commercially available pellistors to ensure the proper detection of gases, primarily of inert methane. The high metal content is used to reduce operating temperature and ensure the required sensor life. This is mainly due to the strong tendency of Pd-based catalysts to deactivate during operation. In view of the scarcity of precious metals, these catalysts should be substituted by metal-oxide catalysts containing none or only a low loading of noble metals. Metal oxides from transition metals (Co, Mn, Fe, Cr, Ni, Cu, etc.) with perovskite or spinel structure as well as transition metal-substituted hexa-aluminates are considered as alternative candidates to noble metal-based catalysts for oxidation of light hydrocarbons, mainly methane, at lean conditions [7,15–19]. Low concentrations applied at lean conditions in catalysis prevail also in gas sensing for safety monitoring, thus the target concentration range (the LEL range) for methane detection is between 0.2% and 4%. Transition metal oxides offer an advantage over catalytically inert alumina due to their intrinsic catalytic activity. Their catalytic activity is attributed to the multiple oxidation states (e.g., Co^{3+}/Co^{2+}, Fe^{2+}/Fe^{2+}, Cu^{2+}/Cu^{+}, and $Mn^{4+}/Mn^{3+}/Mn^{2+}$), lattice defects (i.e., oxygen vacancies), high oxygen storage capacity, and oxygen transfer capability [19]. The multiple oxidation states in the crystal structures are considered as active sites for adsorption and activation of reactant molecules and oxygen [19]. High oxygen storage capacity and oxygen transfer capability help to restore the surface-adsorbed oxygen species consumed through the catalytic oxidation by release of lattice oxygen species and formation of oxygen vacancies.

Co_3O_4-based materials are the most promising catalysts for oxidation of methane among all transition metal oxides. Co_3O_4 possesses a spinel-type structure with two oxidation states (Co^{2+}/Co^{3+}). The easy reduction of Co^{3+} to Co^{2+} in Co_3O_4 facilitates the formation of oxygen vacancies as surface defects at low temperatures, promoting oxygen mobility. High oxygen mobility and highly active surface oxygen species of Co_3O_4, whose formation is supported by oxygen vacancies, contribute to the activation and breaking of the C–H bond at low temperature [14,15,17,20–22]. In addition, the dissociation of the C–H bond is the rate-controlling step during oxidation of saturated hydrocarbons. The specific electronic structure of Co_3O_4 with partially filled d-orbitals of Co^{3+} and Co^{2+} (d^6 and d^7, respectively) lead to the reduction of the activation energy for methane dissociation via direct interaction of C–H orbitals with d-type orbitals of Co cations [23].

Moreover, the incorporation of hetero atoms such as Ce, Ni, Cr, etc., into the spinel structure of Co_3O_4 can improve further its catalytic ability. Heteroatoms induce lattice distortion, which could increase the amount of active surface oxygen species and generate more structural defects (e.g., oxygen vacancies) [24]. On the other hand, doping of Co_3O_4 with another metallic element can improve its textural properties [19]. Especially, cobalt–nickel mixed oxides, i.e., $Ni_xCo_{3-x}O_4$, effectively reduce oxygen vacancy formation energies compared to pure Co_3O_4, thus increasing catalytic activity [21,22,25,26]. However, researchers have disagreed on which nickel to cobalt molar ratio (Ni:Co) exhibits the best catalytic performance in methane oxidation. $NiCo_2O_4$ with the high Ni:Co ratio of 1:2

was found by many authors to be the most active catalyst compared to other $Ni_xCo_{3-x}O_4$ or pure oxides [15,16,21]. Yet, other authors have reported reduced catalyst stability at temperatures > 400 °C because of sintering and partial decomposition to NiO accompanied with performance degradation when the Ni:Co ratio is high. Reducing the Ni:Co ratio can slow down thermal deactivation of $Ni_xCo_{3-x}O_4$ catalysts, improving their long-term stability at higher operation temperatures (450 °C) [27,28].

However, it was demonstrated by several studies on pure Co_3O_4 catalysts that morphology and textural properties also play a decisive role in catalytic activity due to variations in the porosity, crystal structure, formation of surface defects, and surface-active species [14,16,28–31]. In general, the surface atomic configurations and surface defects of catalyst particles can change the adsorption and desorption properties (e.g., surface oxygen bond strength) influencing the catalytic activity of the surfaces [23]. Furthermore, morphology can strongly influence the stability of catalysts. Lyu at al. reported that three-dimensionally ordered, mesoporous Co_3O_4 is not appropriate for application in pellistors due to the high instability of its structure under pellistor operation conditions. Especially, the mesoporous Co_3O_4 catalyst functionalized with Au-Pd nanoparticles exhibited strong deactivation because of strong metal oxide sintering [32]. Thus, the suitable structure and morphology of the metal oxide had a crucial impact on performance of the catalytic sensors. Despite the high interest in metal oxides as catalysts for catalytic combustion, metal oxide materials have hardly been tested for use in catalytic gas sensors [32,33]. Although the catalytic oxidation of target gases underlies the sensor's response, the results of standard catalyst studies using reactors cannot be easily transferred to catalytic gas sensors because the catalysts are prepared for examination according to the requirements for the catalytic bed reactors, e.g., by dilution with other components, pelletization, etc. [18]. Moreover, not all available catalyst morphologies are suitable for application in sensors. For example, hierarchical morphologies and micrometer-sized particles are not appropriate for sensors. In addition, the thermal conductivity of the catalyst layer largely determines the thermal response of the catalytic gas sensor. Therefore, the catalyst morphology is extremely important for achieving optimal thermal conductivity, and potential catalysts for catalytic gas sensors should be examined under appropriate conditions.

In the present work, we investigated the impact of the doping of Co_3O_4 with Ni on the catalyst's morphology and catalytic response towards lean methane (1 vol%) along with its short-term stability under 400 °C and 450 °C. Two pure Co_3O_4 and two Ni-modified cobalt oxide catalysts, i.e., $Ni_xCo_{3-x}O_4$, with two different Ni to Co ratios (1:2 and 1:5), were synthetized by co-precipitation techniques commonly used for the synthesis of pure and heterogeneous metal oxide catalysts due to its simple preparation procedure and easy scaling-up in industrial production [21,26,30]. The given Ni to Co ratios were chosen to examine the effect of Ni quantity on the thermal response and stability of Co_3O_4 catalysts because, according to literature, the catalysts with a 1:2 ratio showed mostly the highest activity, and those with a 1:5 ratio had improved stability [16,28]. In the case of Co_3O_4, N_2 or air was used during synthesis to obtain particles of different morphology according to [16], which reported on the atmosphere having a key role in controlling the particle morphology of $Ni_xCo_{3-x}O_4$ catalysts.

The thermal response of the catalysts was measured in the temperature range pertaining to pellistor applications by means of differential scanning calorimetry (DSC). In DSC, the thermal signal related to the heat evolved by methane oxidation is measured. Details concerning the DSC method and the procedure adapted for catalysts examinations can be found in our recent publication [34]. Previous investigations on one of the examined catalysts ($NiCo_2O_4$) confirmed that DSC measurements can be used to check the catalytic activity of potential catalysts for pellistors [35].

2. Materials and Methods

Pure (mCo_3O_4 and sCo_3O_4) and Ni-doped Co_3O_4 ($NiCo_2O_4$ with Ni:Co ratio of 1:2; $Ni_{0.5}Co_{2.5}O_4$ with Ni:Co ratio of 1:5) catalysts were synthesized by the same precipitating procedure as described in [21]. mCo_3O_4 was synthesized in N_2 atmosphere, as reported in the original publication, while sCo_3O_4 was prepared in air. In the case of $NiCo_2O_4$ and $Ni_{0.5}Co_{2.5}O_4$, N_2 atmosphere was applied. Furthermore, a part of the Co precursor was substituted by a Ni precursor ($Ni(NO_3)_2$ 6 H_2O) in order to obtain the specific Ni:Co ratio (1:2 or 1:5). The composition of the final $NiCo_2O_4$ catalyst was examined by energy dispersive X-ray analysis. The mean Ni:Co ratio in $NiCo_2O_4$ corresponded to the ratio used in the synthesis (1:2).

The mCo_3O_4 sample was synthesized by the following precipitating procedure: $Co(NO_3)_2$ $6H_2O$ (17.46 g; Carl Roth, Karlsruhe, Germany; >98%) was dissolved in 100 mL deionized water at 23 °C. Then, a KOH solution (1 mol L^{-1}, 300 mL) was added under the bubbling of nitrogen gas and continuous, strong stirring. A precipitate was collected and washed three times with hot, deionized water (60 °C), followed by drying at 130 °C for 24 h. The obtained solid was ground to powder in a mortar and further calcined at 350 °C in air for 24 h, forming the synthesized catalyst, which was denoted as mCo_3O_4.

To examine the effect of calcination temperature on the catalytic activity, a part of sCo_3O_4 was investigated directly after drying at 130 °C for 24 h (denoted as $nc\_sCo_3O_4$), and a part was additionally calcined at 400 °C for 24 h (denoted as $nc\_sCo_3O_4$ 24 h 400 °C).

A scanning transmission electron microscope (STEM) equipped with an in-lens secondary electron detector (SE) and an energy dispersive X-ray (EDX) detector (Hitachi HF 5000, Hitachi, Tokyo, Japan) was used to visualize the morphology and to analyze the composition of the catalysts. The SE detector gives access to surface information and visualizes the surface topography of the sample. In the following, the images are described as SE-STEM since the images were acquired in STEM mode, although the electrons are not transmitting the sample.

Crystalline structure and phases of catalysts were checked by means of X-ray diffraction (XRD) analyses (Empyrean, Malvern Panalytical Ltd., Malvern, UK) using Cu Kα radiation.

Differential scanning calorimetry (DSC) (STA 409 CD-QMS 403/5 SKIMMER, Netzsch, Selb, Germany) was used to examine the thermal response of the catalysts to 1% methane in dry air. The investigations using DSC can only be carried out method specifically in a dry gas atmosphere. DSC is, in its principle, very similar to pellistor measurements. In DSC, the temperature difference between the empty reference pan and the sample pan filled with a powder is detected at increasing temperature (dynamic conditions) as an electrical signal normalized to the sample weight (μV mg^{-1}).

To investigate catalytic activity, the measurement procedure applied for standard measurements was adapted [34]. In our experiments, the same procedure was used as for testing the gas sensors. The temperature difference between sample and reference pan was measured at isothermal conditions, when first, dry compressed air was introduced into the system to record a baseline and then 1% methane (1 vol% CH_4 in synthetic air, Air Liquide, Düsseldorf, Germany). At each temperature level, the baseline recorded in air was set to zero [34]. Figure S1 (Supplementary Materials) shows an exemplary DSC signal obtained for sCo_3O_4 catalyst upon exposure to 1% methane (orange bars) at a predefined temperature program (red line). The height of the signal was considered for catalyst characterization. The dynamic behavior, such as response or recovery time, was not analyzed since it depends on the instrumentation or sensor used and does not provide any valuable information in this case.

The measured DSC voltage corresponds to the heat produced during catalytic oxidation. The experiments were performed on 8 g catalyst using an aluminum pan at a constant gas flow rate of 100 mL min^{-1}. To investigate the thermal stability, samples were held in compressed dry air containing 400 ppm CO_2 at 400 °C or 450 °C for 24 h and 12 h, respectively. Air was chosen instead of synthetic air to achieve conditions more realistic for sensor applications. The catalytic activity was measured before and after thermal treatment of samples using the same temperature profile.

3. Results and Discussion

3.1. Structure and Morphology of Investigated Catalysts

The X-ray diffraction patterns of undoped and Ni-doped Co_3O_4 oxides are shown in Figure 1. The catalysts are fine crystalline with small crystallites or many defects in crystallites, which is visible from the line broadening of XRD diffraction peaks. The main peaks can be assigned to the cubic Co_3O_4 spinel structure (No. 04-025-8553). No peaks related to $Co(OH)_2$ or partially oxidized $CoO(OH)$ were observed in the XRD patterns of Co_3O_4 (Figure 1a), indicating the complete transformation of $Co(OH)_2$ to Co_3O_4. However, XRD patterns of all catalysts contain some peaks, e.g., at $2\Theta = 33.3°$, $40.3°$, $53.2°$, and $58.4°$, which cannot be assigned to any crystalline phase. No mentions of these additional peaks were found in the literature, although the crystallinity of catalysts in the literature is significantly higher. The patterns of mCo_3O_4 and sCo_3O_4 catalysts reveal minor variations in the crystal structure apart from the peak at $2\Theta = 19.0°$ assigned to the (111) plane, which is absent in mCo_3O_4. DRIFT and DFT calculations revealed that the (111)-plane of Co_3O_4 is responsible for the facile activation of the C–H bond and its high activity [18]. X-ray diffraction of Ni-doped Co_3O_4 oxides (Figure 1b) showed that both $Ni_xCo_{3-x}O_4$ catalysts have the same spinel structure as Co_3O_4. That coincides with the results from Tao et al. [21], who reported that $Ni_xCo_{3-x}O_4$ catalysts synthesized by co-precipitation with KOH exhibit the same diffraction pattern as Co_3O_4. This reveals the integration of Ni cations into the crystallographic lattice of spinel Co_3O_4. However, in contrast to the literature, some peaks assigned to NiO ((200) at $2\Theta = 43.3°$ and (220) at $2\Theta = 62.9°$) were found in the $NiCo_2O_4$ catalyst. Additionally, the lattice planes (442), (533), and (622) of Co_3O_4 are not well formed in $NiCo_2O_4$.

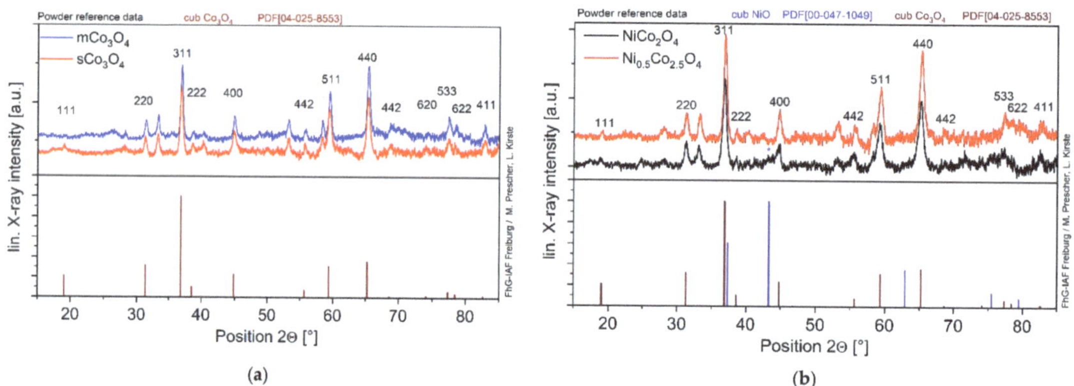

Figure 1. XRD patterns of (**a**) undoped mCo_3O_4 and sCo_3O_4 oxides as well as (**b**) Ni-doped oxides $NiCo_2O_4$ and $Ni_{0.5}Co_{2.5}O_4$; the reference data are listed below.

The dried, non-calcined_sCo_3O_4 and calcined sCo_3O_4 oxides (Figure 2) exhibit similar XRD patterns revealing that the conversion of $Co(OH)_2$ into Co_3O_4 was mostly completed after catalyst drying at 130 °C. After calcination at 350 °C, only one additional peak appears at $2\Theta = 40.1°$. Thus, most alterations happening during the calcination step concern the catalyst surface, such as surface defects and oxygen species.

In Figure 3, the morphologies of Co_3O_4 and Ni-doped $Ni_xCo_{3-x}O_4$ catalysts are shown. mCo_3O_4 and sCo_3O_4 demonstrate similar morphology with small differences. mCo_3O_4 (Figure 3a) reveals rather angular particles of irregular size and shape, while for sCo_3O_4 (Figure 3b), particles with sheet morphology are present in addition to small nanoparticles. Moreover, mCo_3O_4 particles contain a high number of holes inside (Figure 4a,b), significantly more than sCo_3O_4 particles (Figure 4c,d), as the comparison of the corresponding HAADF-STEM images clearly shows. Thus, the morphological investigations indicate that mCo$_3$O$_4$ and sCo$_3$O$_4$ materials have different particle morphology and structure, so differences in their interactions with reactants and heat transport are expected.

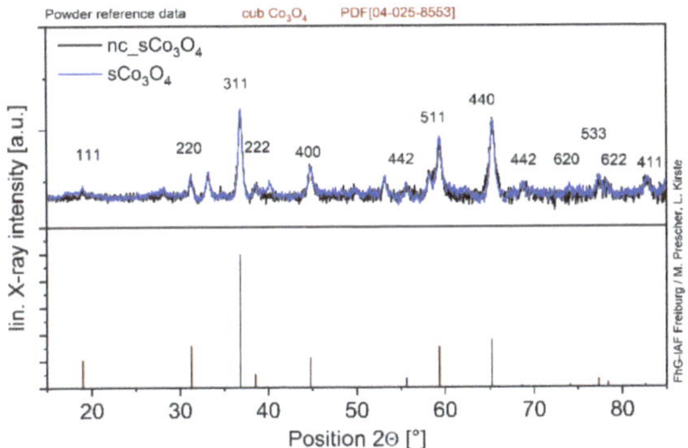

Figure 2. XRD patterns of the dried and calcined sCo$_3$O$_4$ oxides in comparison; the reference data are listed below.

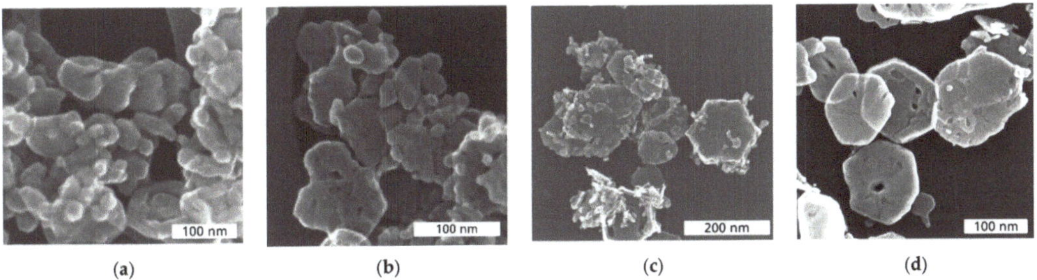

Figure 3. SE-STEM images of (**a**) mCo$_3$O$_4$; (**b**) sCo$_3$O$_4$; (**c**) NiCo$_2$O$_4$; (**d**) Ni$_{0.5}$Co$_{2.5}$O$_4$ metal-oxides.

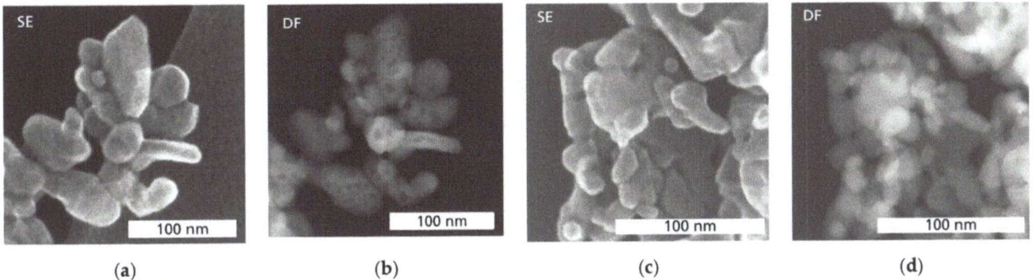

Figure 4. SE-STEM (**a,c**) and HAADF-STEM (**b,d**) images in comparison with mCo$_3$O$_4$ (**a,b**) and sCo$_3$O$_4$ (**c,d**).

Both Ni$_x$Co$_{3-x}$O$_4$ catalysts exhibit a hexagonal sheet morphology (Figure 3c,d), which differs from the morphology of pure Co$_3$O$_4$ catalysts (Figure 3a,b). Such a morphology is beneficial for the catalysis due to high macroporosity and better gas transport inside the catalyst layer. However, the sheet's quality of NiCo$_2$O$_4$ catalyst is very low. Moreover, NiCo$_2$O$_4$ (Figure 3c) reveals in addition to the defective hexagonal sheets an accumulation of small particles and debris. A high number of diverse defects in the structure of NiCo$_2$O$_4$ is well visible in HAADF-STEM images (compare Figure 5a,b). In contrast to that, well-formed nanoplatelets of hexagonal shape with a porous surface and holes inside can be observed for Ni$_{0.5}$Co$_{2.5}$O$_4$ (Figures 3d and 5c,d).

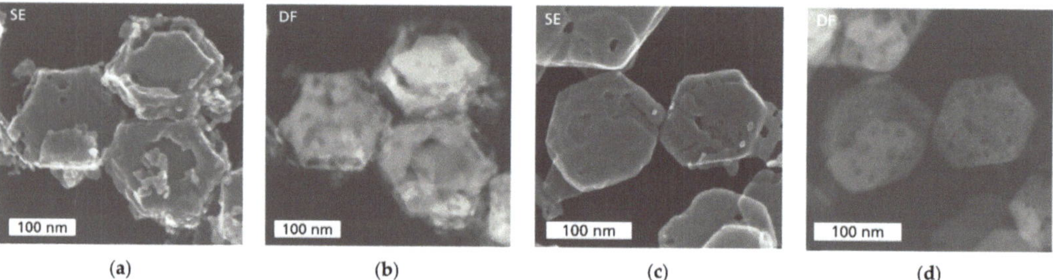

Figure 5. SE-STEM (**a,c**) and HAADF-STEM (**b,d**) images in comparison with NiCo$_2$O$_4$ (**a,b**) and Ni$_{0.5}$Co$_{2.5}$O$_4$ (**c,d**).

The EDX analysis of mCo$_3$O$_4$ (Figure 6a–c) and sCo$_3$O$_4$ (Figure 6d–f) catalysts shows that the elemental distributions of Co and O coincides, and hence, no segregation was observed. Moreover, no further elements were detected, excluding impurities.

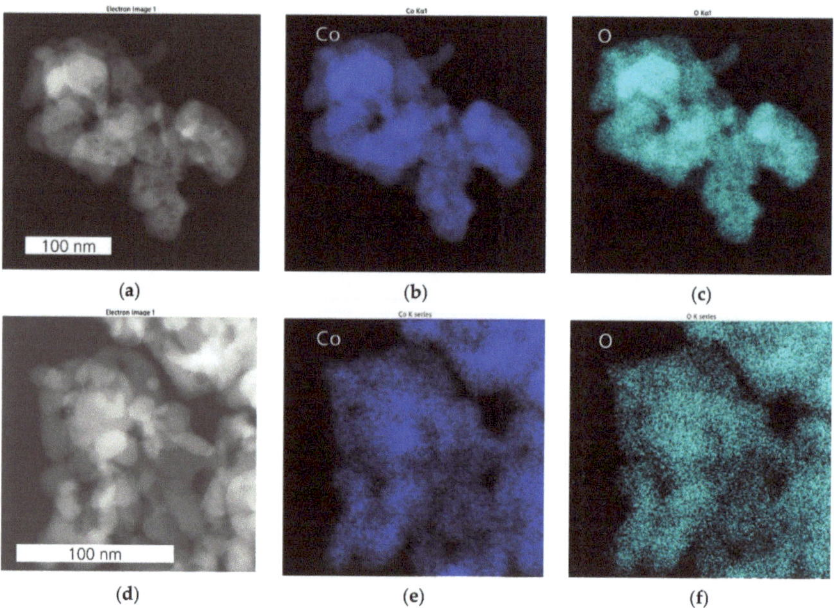

Figure 6. EDX analysis of Co$_3$O$_4$ catalysts: HAADF-STEM images of mCo$_3$O$_4$ (**a**) and sCo$_3$O$_4$ (**d**); the corresponding elemental distribution maps of Co (**b,e**) and O (**c,f**) in respective metal-oxides.

EDX elemental mapping of NiCo$_2$O$_4$ catalyst (Figure 7) demonstrates that Co and O are distributed evenly throughout the nanoplates, whereas Ni is predominantly concentrated at the edge of the hexagonal plates as well as in the nanoparticles and debris. Less nickel is found at the center of the hexagonal plates, revealing strongly pronounced phase separation of Ni as NiO nanoparticles. The phase segregation of NiO and a high number of debris indicate that an excess of Ni disrupts the Ni integration into the Co$_3$O$_4$ lattice. The investigations reveal that NiCo$_2$O$_4$ contains a mixture of two metal oxides: the mixed oxide Ni$_x$Co$_{3-x}$O$_4$ with variable Ni:Co content and the pure NiO.

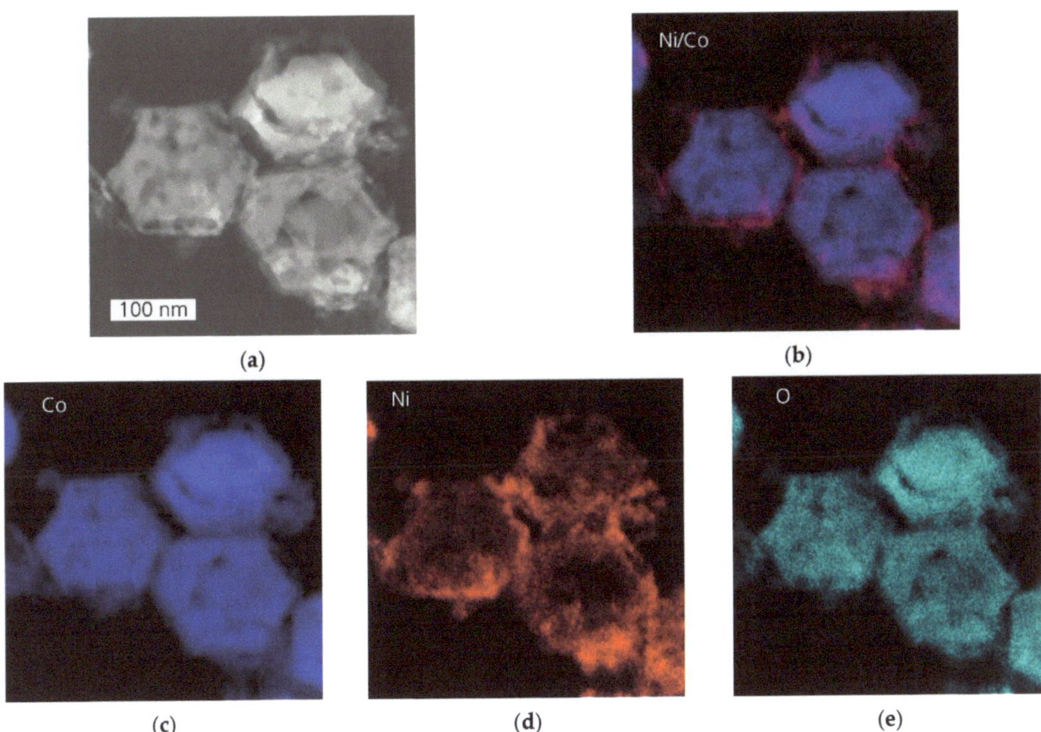

Figure 7. EDX analysis of NiCo$_2$O$_4$ catalyst: (**a**) HAADF-STEM image; (**b**) overlay of the elemental distribution maps of Ni and Co; the elemental distribution maps of Co (**c**), Ni (**d**), and O (**e**).

EDX elemental mapping of Ni$_{0.5}$Co$_{2.5}$O$_4$ oxide indicates that the distribution of Co, O, and Ni is mainly uniform inside hexagonal plates (Figure 8). Thus, Ni is well integrated into the structure of Co$_3$O$_4$ oxide. Consequently, the application of a lower Ni:Co ratio results in a doped spinel structure, i.e., Ni$_x$Co$_{3-x}$O$_4$, with highly uniform morphology and composition. However, for some particles, a slight inhomogeneous distribution of nickel in the hexagonal plates is still observed.

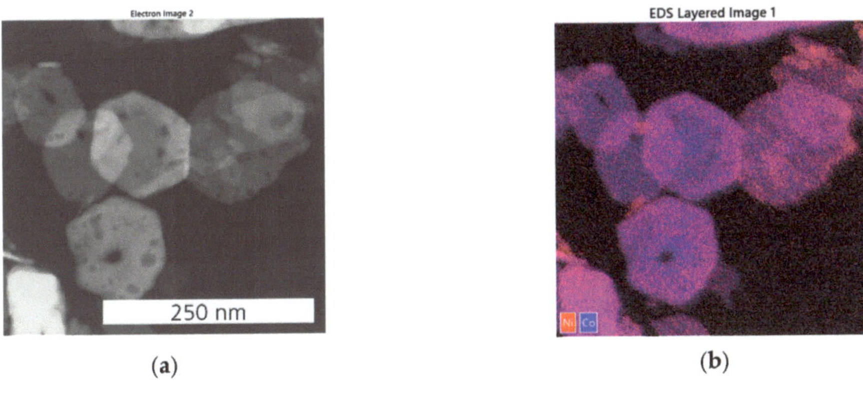

Figure 8. *Cont.*

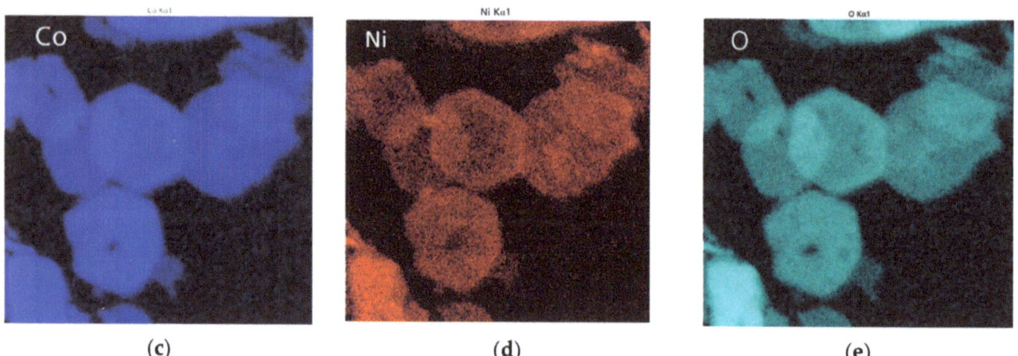

(c)　　　　　　　　　　(d)　　　　　　　　　　(e)

Figure 8. EDX analysis of $Ni_{0.5}Co_{2.5}O_4$ catalyst: (**a**) HAADF-STEM image; (**b**) overlay of the elemental distribution maps of Ni and Co; the elemental distribution of Co (**c**), Ni (**d**), and O (**e**).

3.2. Characterization of Catalytic Response

Figure 9 shows the results of DSC measurements on the four catalysts obtained in temperature range between 250 °C and 450 °C with 1 vol% methane. As expected, the DSC voltage increases with higher temperature. At 250 °C and 300 °C, the DSC voltages were comparably low for all catalysts (0.25–0.32 μV mg^{-1} at 300 °C), caused by the low activity of metal oxide catalysts at such low temperatures. Only when the temperature increases above 300 °C does the difference in the catalyst activity become substantial. The Ni-modified catalysts, namely $NiCo_2O_4$ and $Ni_{0.5}Co_{2.5}O_4$, exhibit clearly higher DSC signals than pure mCo_3O_4, particularly at 450 °C (3.3 μV mg^{-1} vs. 2.7 μV mg^{-1}), which is in agreement with the literature reports about the higher activity of $Ni_xCo_{3-x}O_4$ catalysts compared to pure Co_3O_4 [16,27]. The differences in activity between $NiCo_2O_4$ and $Ni_{0.5}Co_{2.5}O_4$ are, however, not significant. The higher Ni ratio in $NiCo_2O_4$ does not cause the expected increase in catalytic activity. This can be explained by the segregation of NiO, which exhibits a much lower activity than $Ni_xCo_{3-x}O_4$ and Co_3O_4 [16,28].

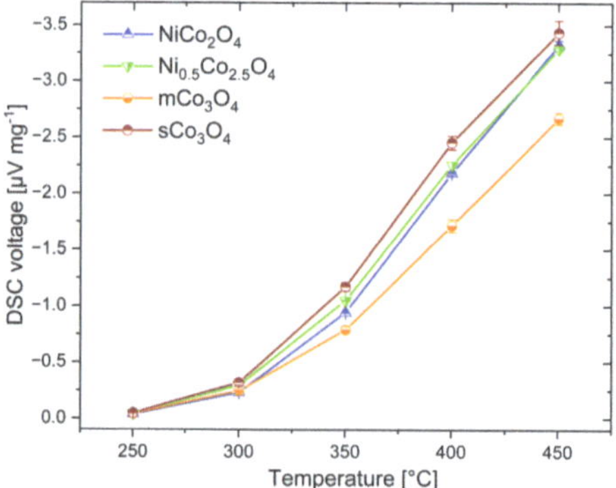

Figure 9. Temperature-dependent DSC response of the four investigated catalysts to 1% CH_4, with error bars giving the standard deviation from three measurements.

Interestingly, sCo_3O_4, precipitated in air, seems to slightly outperform the activity of the Ni-modified oxides $NiCo_2O_4$ and $Ni_{0.5}Co_{2.5}O_4$ at all temperatures. The considerable differences in response for mCo_3O_4 and sCo_3O_4 must be caused by their different morphology and structure, as seen in Figures 3a,b and 4 [29,36]. Obviously, the high structural defect density of mCo_3O_4 (Figure 2b) does not improve the catalytic activity of Co_3O_4, whereas the sheet morphology of sCo_3O_4 seems to have a positive impact on its activity. It is notable that sCo_3O_4 exhibits a higher variance in activity compared to mCo_3O_4, which correlates with the variations in particle morphology observed for sCo_3O_4.

Figure 10 shows the activity of the two Co_3O_4 catalysts before and after the treatment in pressurized air at 400 and 450 °C for 24 and 12 h, respectively. mCo_3O_4 and sCo_3O_4 show either no or a slight decrease or even an increase in the DSC signal after the treatment, revealing their stability under test conditions. The increase in the thermal response after heat treatment can be explained by an improved contact between individual particles in the layer, causing better thermal conductivity and thus better transfer of reaction heat to thermopiles of the detector.

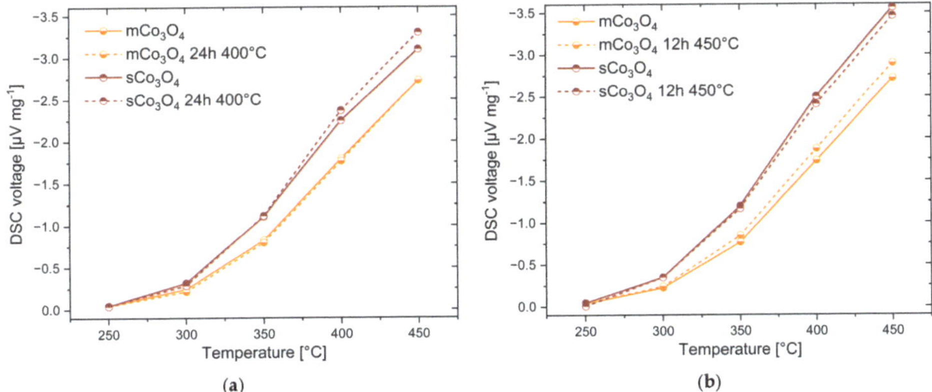

Figure 10. Results of the stability investigations: temperature-dependent DSC response of two Co_3O_4 catalysts to 1% methane measured before and after treatment at 400 °C (**a**) and at 450 °C (**b**).

Regarding XRD and morphology investigations, it is likely that the significantly improved thermal response of sCo_3O_4 compared to mCo_3O_4 is caused by low defect density and sheet morphology. It is also possible that the (111) lattice plane present in sCo_3O_4 contributes to its improved catalytic activity compared to mCo_3O_4.

In contrast to Co_3O_4, $NiCo_2O_4$ and $Ni_{0.5}Co_{2.5}O_4$ (Figure 11) show a reduced DSC signal as a result of the heat treatment. Particularly, the $NiCo_2O_4$ catalyst experiences a considerable decrease in catalytic activity already after a short treatment at both temperatures. The most pronounced decrease from 3.4 $\mu V\ mg^{-1}$ to 3.0 $\mu V\ mg^{-1}$ was observed for $NiCo_2O_4$ at 450 °C after treatment at 450 °C (Figure 11b). After thermal treatment at 400 °C (Figure 11a), the DSC signal decrease is lower, although it still remains high (difference of 0.3 $\mu V\ mg^{-1}$ at 400 °C). $Ni_{0.5}Co_{2.5}O_4$ demonstrated higher thermal stability than $NiCo_2O_4$. After thermal treatment at 400 °C (Figure 11a), the DSC signal difference for $Ni_{0.5}Co_{2.5}O_4$ before and after treatment is twice as low (0.15 $\mu V\ mg^{-1}$ at 400 °C). The highest decrease in the DSC signal from 3.3 $\mu V\ mg^{-1}$ to 3.1 $\mu V\ mg^{-1}$ was also obtained at 450 °C after treatment at 450 °C (Figure 11b). The high susceptibility of $Ni_xCo_{3-x}O_4$ materials to high temperatures agrees with earlier reports [27,28].

The fast deterioration of the initial activity in $NiCo_2O_4$ correlates with a strongly pronounced NiO segregation and its defective structure (Figure 7). $Ni_{0.5}Co_{2.5}O_4$ contained fewer structural defects and no NiO segregation (Figure 8) but showed a slightly inhomogeneous Ni distribution inside hexagonal plates. The latter one can affect the catalyst's stability at higher operation temperatures. This is because higher operation temperatures usually promote further phase segregation, causing catalyst destabilization and decreasing catalyst activity. Otherwise, the thermal stability of 2D materials may be limited compared to nanoparticles [37]. It is conceivable that the stability of $Ni_xCo_{3-x}O_4$ catalysts could be further improved when Ni ions are completely incorporated into the structure of Co_3O_4. This could be achieved by using stabilizing additives or by applying alternative synthesis methods.

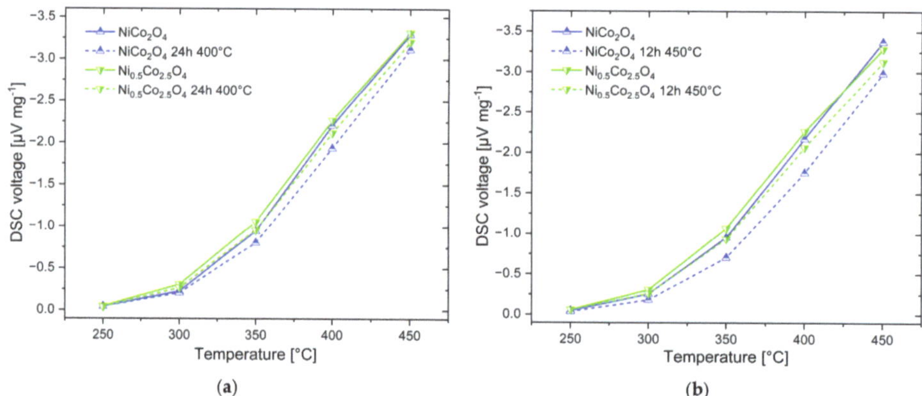

Figure 11. Results of the stability investigations: temperature-dependent DSC response of two $Ni_xCo_{3-x}O_4$ catalysts to 1% methane measured before and after treatment at 400 °C (**a**) and at 450 °C (**b**).

In addition to the stability experiments performed at 400 °C and 450 °C, the impact of the calcination on the catalyst activity was examined. The most active sCo_3O_4 was chosen for the examination. All investigated catalysts were dried at 130 °C and subsequently calcined at 350 °C. To see the effect of calcination, sCo_3O_4 was additionally examined directly after drying without a calcination step (denoted as $nc\_sCo_3O_4$). For further comparison, non-calcined sCo_3O_4 was then calcined at 400 °C for 24 h (denoted as $nc\_sCo_3O_4$ 24 h 400 °C). Figure 12 shows the temperature-dependent DSC signals of original sCo_3O_4, $nc\_sCo_3O_4$, and $nc\_sCo_3O_4$ for 24 h at 400 °C.

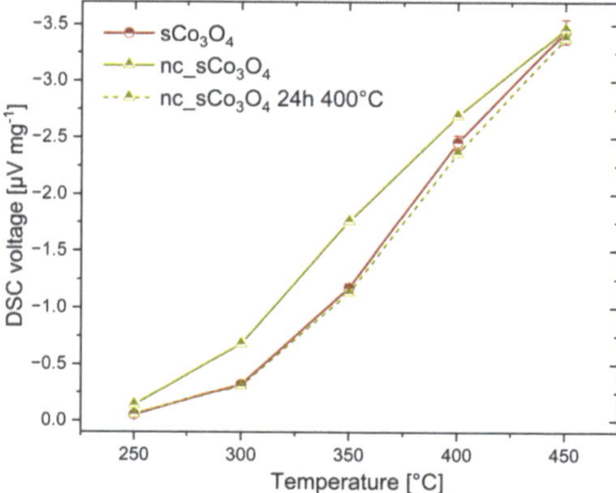

Figure 12. Effect of calcination temperature on the temperature-dependent DSC signal to 1% methane, demonstrated for sCo_3O_4.

$nc\_sCo_3O_4$ revealed a considerably higher signal than sCo_3O_4, especially in the low-temperature range between 250 °C and 350 °C. The DSC signal at 350 °C increased from 1.8 µV mg^{-1} for $nc\_sCo_3O_4$ to 1.2 µV mg^{-1} for sCo_3O_4. The $nc\_sCo_3O_4$ sample calcined at 400 °C showed comparable activity to the original sCo_3O_4. Such an effect of calcination temperature on the activity of sCo_3O_4 demonstrates that a Co_3O_4-based catalyst features temperature-sensitive catalytic functionalities. Regarding the XRD investigations (Figure 2), it is obvious that higher temperatures cause a lower amount of active surface oxygen present on the surface of Co_3O_4 catalysts. Reduced surface oxygen as active species in the C–H bond dissociation has a negative effect on the catalytic activity of Co_3O_4 at low operation temperatures. Otherwise, catalyst calcination at temperatures higher than later operating temperatures is required to stabilize the structure of Co_3O_4 catalysts despite the associated reduced activity [38]. Nevertheless, the optimal operation temperature for Co_3O_4-based catalysts without precise metals is in the low-temperature range, below $\leq 350\,^{\circ}C$.

The next step in the development of the metal-oxide catalysts for catalytic gas sensors is the examination of metal-decorated Co_3O_4 catalysts. Metal oxides offer several advantages over aluminum oxide as support material, such as stronger interactions with metal nanoparticles. Thus, the advantages and disadvantages of such kind of catalysts should be carefully examined.

4. Conclusions

In conclusion, the investigations of the thermal response of Ni-doped and undoped Co_3O_4 metal oxides synthetized by co-precipitation towards 1% methane have shown that Ni doping improves the catalytic response of Co_3O_4. However, control of the structure and morphology of Co_3O_4 has a comparable impact on the activity and thermal response compared to doping, as was shown using sCo_3O_4 catalyst with optimized morphology. Moreover, the doping of the spinel structure of Co_3O_4 with nickel requires a balanced Ni:Co ratio and highly controllable synthesis conditions to reduce or eliminate the appearance of phase segregation, which is mainly responsible for the low thermal stability of $Ni_xCo_{3-x}O_4$ catalysts. Because of its higher thermal stability, pure Co_3O_4 is preferable as a catalyst and for further modification with precise metals, which are applied to increase catalyst activity and sensor response at lower operation temperatures. Besides the detection of methane, Co_3O_4-based materials are of special interest for the detection of hydrogen since hydrogen can usually be detected at noticeably lower temperatures than hydrocarbons due to its significantly higher reactivity.

Supplementary Materials: The following supporting information can be downloaded at: https://www.mdpi.com/article/10.3390/s24082599/s1, Figure S1: DSC signal of sCo_3O_4 catalyst measured in the temperature range between 250 and 450 $^{\circ}C$.

Author Contributions: Conceptualization, O.Y.; methodology, O.Y. and P.D.; investigation, O.Y. and P.D.; writing—original draft preparation, O.Y.; writing—review and editing, O.Y., P.D., K.S., F.A. and J.W.; funding acquisition, J.W. and F.A. All authors have read and agreed to the published version of the manuscript.

Funding: This research was funded by the Fraunhofer Gesellschaft within the FluMEMS project and the Federal Ministry of Education and Research within European funding program Eurostars in project MEscal E! 113779 as well as within the project HySABi (03EN5009C).

Institutional Review Board Statement: Not applicable.

Informed Consent Statement: Not applicable.

Data Availability Statement: Data are contained within the article and Supplementary Materials.

Conflicts of Interest: The authors declare no conflicts of interest.

References

1. Bársony, I.; Ádám, M.; Fürjes, P.; Lucklum, R.; Hirschfelder, M.; Kulinyi, S.; Dücső, C. Efficient catalytic combustion in integrated micropellistors. *Meas. Sci. Technol.* **2009**, *20*, 124009. [CrossRef]
2. Szulczyński, B.; Gębicki, J. Currently Commercially Available Chemical Sensors Employed for Detection of Volatile Organic Compounds in Outdoor and Indoor Air. *Environments* **2017**, *4*, 21. [CrossRef]
3. Roslyakov, I.; Kolesnik, I.; Evdokimov, P.; Skryabina, O.; Garshev, A.; Mironov, S.; Stolyarov, V.; Baranchikov, A.; Napolskii, K. Microhotplate catalytic sensors based on porous anodic alumina: Operando study of methane response hysteresis. *Sens. Actuators B* **2021**, *330*, 129307. [CrossRef]
4. Samotaev, N.; Pisliakov, A.; Gorshkova, A.; Dzhumaev, P.; Barsony, I.; Ducso, C.; Biro, F. Al_2O_3 nanostructured gas sensitive material for silicon based low power thermocatalytic sensor. *Mater. Today Proc.* **2020**, *30*, 443–447. [CrossRef]
5. Lorenzo-Bayona, J.L.; León, D.; Amez, I.; Castells, B.; Medic, L. Experimental Comparison of Functionality between the Main Types of Methane Measurement Sensors in Mines. *Energies* **2023**, *16*, 2207. [CrossRef]

6. Choya, A.; Rivas, B.d.; González-Velasco, J.R.; Gutiérrez-Ortiz, J.I.; López-Fonseca, R. Oxidation of residual methane from VNG vehicles over Co_3O_4-based catalysts: Comparison among bulk, Al_2O_3-supported and Ce-doped catalysts. *Appl. Catal. B Environ.* **2018**, *237*, 844–854. [CrossRef]

7. He, L.; Fan, Y.; Bellettre, J.; Yue, J.; Luo, L. A review on catalytic methane combustion at low temperatures: Catalysts, mechanisms, reaction conditions and reactor designs. *Renew. Sustain. Energy Rev.* **2020**, *119*, 109589. [CrossRef]

8. Ma, L.; Geng, Y.; Chen, X.; Yan, N.; Li, J.; Schwank, J.W. Reaction mechanism of propane oxidation over Co_3O_4 nanorods as rivals of platinum catalysts. *Chem. Eng. J.* **2020**, *402*, 125911. [CrossRef]

9. Wang, B.; Zhao, X.; Zhang, Y.; Wang, Z. Research on High Performance Methane Gas Concentration Sensor Based on Pyramid Beam Splitter Matrix. *Sensors* **2024**, *24*, 602. [CrossRef] [PubMed]

10. Baker, A. Apparatus for Detecting Combustible Gases Having an Electrically Conductive Member Enveloped in a Refractory Material. U.S. Patent 3,092,799, 4 June 1963.

11. White, R. The Pellistor Is Dead? Long Live the Pellistor! Available online: www.envirotech-online.com/article/environmental-laboratory/7/sgx-sensortech/the-pellistor-is-dead-nbsplong-live-the-pellistor/1699 (accessed on 28 February 2024).

12. Doncaster, A. A Discussion on Pellistor Gas Sensor Responses. Annual Buyers Guide. 2009. Available online: https://www.chromatographytoday.com/article/gas-detection/8/clairair/a-discussion-on-pellistor-gas-sensor-responses/467 (accessed on 18 March 2021).

13. Florea, O.G.; Stănoiu, A.; Gheorghe, M.; Cobianu, C.; Neațu, F.; Trandafir, M.M.; Neațu, Ș.; Florea, M.; Simion, C.E. Methane Combustion Using Pd Deposited on CeO_x-MnO_x/La-Al_2O_3 Pellistors. *Materials* **2020**, *13*, 4888. [CrossRef] [PubMed]

14. Jiang, D.; Khivantsev, K.; Wang, Y. Low-Temperature Methane Oxidation for Efficient Emission Control in Natural Gas Vehicles: Pd and Beyond. *ACS Catal.* **2020**, *10*, 14304–14314. [CrossRef]

15. Lim, T.H.; Cho, S.J.; Yang, H.S.; Engelhard, M.; Kim, D.H. Effect of Co/Ni ratios in cobalt nickel mixed oxide catalysts on methane combustion. *Appl. Catal. A Gen.* **2015**, *505*, 62–69. [CrossRef]

16. Dai, Y.; Kumar, V.P.; Zhu, C.; Wang, H.; Smith, K.J.; Wolf, M.O.; MacLachlan, M.J. Bowtie-Shaped $NiCo_2O_4$ Catalysts for Low-Temperature Methane Combustion. *Adv. Funct. Mater.* **2019**, *29*, 1807519. [CrossRef]

17. Zheng, Y.; Yu, Y.; Zhou, H.; Huang, W.; Pu, Z. Combustion of lean methane over Co_3O_4 catalysts prepared with different cobalt precursors. *RSC Adv.* **2020**, *10*, 4490–4498. [CrossRef] [PubMed]

18. Yao, J.; Shi, H.; Sun, D.; Lu, H.; Hou, B.; Jia, L.; Xiao, Y.; Li, D. Facet-Dependent Activity of Co_3O_4 Catalyst for C3H8 Combustion. *ChemCatChem* **2019**, *11*, 5570–5579. [CrossRef]

19. Lin, H.; Liu, Y.; Deng, J.; Jing, L.; Dai, H. Methane Combustion over the Porous Oxides and Supported Noble Metal Catalysts. *Catalysts* **2023**, *13*, 427. [CrossRef]

20. Cargnello, M.; Delgado Jaén, J.J.; Hernández Garrido, J.C.; Bakhmutsky, K.; Montini, T.; Calvino Gámez, J.J.; Gorte, R.J.; Fornasiero, P. Exceptional activity for methane combustion over modular $Pd@CeO_2$ subunits on functionalized Al_2O_3. *Science* **2012**, *337*, 713–717. [CrossRef] [PubMed]

21. Tao, F.F.; Shan, J.; Nguyen, L.; Wang, Z.; Zhang, S.; Zhang, L.; Wu, Z.; Huang, W.; Zeng, S.; Hu, P. Understanding complete oxidation of methane on spinel oxides at a molecular level. *Nat. Commun.* **2015**, *6*, 65. [CrossRef] [PubMed]

22. Bao, L.; Chang, L.; Yao, L.; Meng, W.; Yu, Q.; Zhang, X.; Liu, X.; Wang, X.; Chen, W.; Li, X. Acid etching induced defective Co_3O_4 as an efficient catalyst for methane combustion reaction. *New J. Chem.* **2021**, *45*, 3546–3551. [CrossRef]

23. Liotta, L.F.; Wu, H.; Pantaleo, G.; Venezia, A.M. Co_3O_4 nanocrystals and Co_3O_4–MO_x binary oxides for CO, CH_4 and VOC oxidation at low temperatures: A review. *Catal. Sci. Technol.* **2013**, *3*, 3085–3102. [CrossRef]

24. Ercolino, G.; Stelmachowski, P.; Grzybek, G.; Kotarba, A.; Specchia, S. Optimization of Pd catalysts supported on Co_3O_4 for low-temperature lean combustion of residual methane. *Appl. Catal. B Environ.* **2017**, *206*, 712–725. [CrossRef]

25. Zhang, Z.; Hu, X.; Zhang, Y.; Sun, L.; Tian, H.; Yang, X. Ultrafine PdO_x nanoparticles on spinel oxides by galvanic displacement for catalytic combustion of methane. *Catal. Sci. Technol.* **2019**, *9*, 6404–6414. [CrossRef]

26. Shao, C.; Li, W.; Lin, Q.; Huang, Q.; Pi, D. Low Temperature Complete Combustion of Lean Methane over Cobalt-Nickel Mixed-Oxide Catalysts. *Energy Technol.* **2017**, *5*, 604–610. [CrossRef]

27. Ren, Z.; Wu, Z.; Song, W.; Xiao, W.; Guo, Y.; Ding, J.; Suib, S.L.; Gao, P.-X. Low temperature propane oxidation over Co_3O_4 based nano-array catalysts: Ni dopant effect, reaction mechanism and structural stability. *Appl. Catal. B Environ.* **2016**, *180*, 150–160. [CrossRef]

28. Chen, J.; Zou, X.; Rui, Z.; Ji, H. Deactivation Mechanism, Countermeasures, and Enhanced CH_4 Oxidation Performance of Nickel/Cobalt Oxides. *Energy Technol.* **2020**, *8*, 1900641. [CrossRef]

29. Chen, Z.; Wang, S.; Liu, W.; Gao, X.; Gao, D.; Wang, M.; Wang, S. Morphology-dependent performance of Co_3O_4 via facile and controllable synthesis for methane combustion. *Appl. Catal. A Gen.* **2016**, *525*, 94–102. [CrossRef]

30. Pu, Z.; Zhou, H.; Zheng, Y.; Huang, W.; Li, X. Enhanced methane combustion over Co_3O_4 catalysts prepared by a facile precipitation method: Effect of aging time. *Appl. Surf. Sci.* **2017**, *410*, 14–21. [CrossRef]

31. Yurchenko, O.; Pernau, H.-F.; Engel, L.; Bierer, B.; Jägle, M.; Wöllenstein, J. Impact of particle size and morphology of cobalt oxide on the thermal response to methane examined by thermal analysis. *J. Sens. Sens. Syst.* **2021**, *10*, 37–42. [CrossRef]

32. Lyu, X.; Yurchenko, O.; Diehle, P.; Altmann, F.; Wöllenstein, J.; Schmitt, K. Accelerated Deactivation of Mesoporous Co_3O_4-Supported Au–Pd Catalyst through Gas Sensor Operation. *Chemosensors* **2023**, *11*, 271. [CrossRef]

33. Simion, C.E.; Florea, O.G.; Florea, M.; Neaţu, F.; Neaţu, Ş.; Trandafir, M.M.; Stănoiu, A. CeO_2:Mn_3O_4 Catalytic Micro-Converters Tuned for CH_4 Detection Based on Catalytic Combustion under Real Operating Conditions. *Materials* **2020**, *13*, 2196. [CrossRef]

34. Yurchenko, O.; Pernau, H.-F.; Engel, L.; Wöllenstein, J. Differential thermal analysis techniques as a tool for preliminary examination of catalyst for combustion. *Sci. Rep.* **2023**, *13*, 9792. [CrossRef] [PubMed]

35. Bierer, B.; Grgić, D.; Yurchenko, O.; Engel, L.; Pernau, H.-F.; Jägle, M.; Reindl, L.; Wöllenstein, J. Low-power sensor node for the detection of methane and propane. *J. Sens. Sens. Syst.* **2021**, *10*, 185–191. [CrossRef]

36. Hu, L.; Peng, Q.; Li, Y. Selective Synthesis of Co_3O_4 Nanocrystal with Different Shape and Crystal Plane Effect on Catalytic Property for Methane Combustion. *J. Am. Chem. Soc.* **2008**, *130*, 16136–16137. [CrossRef] [PubMed]

37. Datye, A.K.; Votsmeier, M. Opportunities and challenges in the development of advanced materials for emission control catalysts. *Nat. Mater.* **2020**, *20*, 1049–1059. [CrossRef] [PubMed]

38. Sanchis, R.; García, A.; Ivars-Barceló, F.; Taylor, S.H.; García, T.; Dejoz, A.; Vázquez, M.I.; Solsona, B. Highly Active Co_3O_4-Based Catalysts for Total Oxidation of Light C_1-C_3 Alkanes Prepared by a Simple Soft Chemistry Method: Effect of the Heat-Treatment Temperature and Mixture of Alkanes. *Materials* **2021**, *14*, 7120. [CrossRef] [PubMed]

Article

On the Influence of Humidity on a Thermal Conductivity Sensor for the Detection of Hydrogen

Sophie Emperhoff [1,2,*], Matthias Eberl [2], Tim Dwertmann [2] and Jürgen Wöllenstein [1,3]

1   Department of Microsystems Engineering (IMTEK), Albert-Ludwigs-Universität Freiburg, 79110 Freiburg im Breisgau, Germany; juergen.woellenstein@imtek.uni-freiburg.de
2   Infineon Technologies AG, 81549 Neubiberg, Germany; tim.dwertmann@gmail.com (T.D.)
3   Fraunhofer Institute for Physical Measurement Techniques (IPM), 79110 Freiburg, Germany
*   Correspondence: sophie.emperhoff@infineon.com

Abstract: Thermal conductivity sensors face an omnipresent cross-influence through varying humidity levels in real-life applications. We present the results of investigations on the influence of humidity on a hydrogen thermal conductivity sensor and approaches for predicting the behavior of thermal conductivity towards humidity. A literature search and comparison of different mixing equations for binary gas mixtures were carried out. The theoretical results were compared with experimental results from three different thermal conductivity sensors with mixtures of water vapor in nitrogen. The mixing equations show a large discrepancy between each other. Some of the models predict a continuously decreasing thermal conductivity and some predict an increasing thermal conductivity for increasing levels of humidity. Our measurements indicate an increase in thermal conductivity followed by a decrease after reaching a peak value. It is shown that the measured behavior is reproducible with different sensors. Depending on the sensor, this corresponds to an error up to 2 vol.% in the measured hydrogen value. The measured behavior is consistent with only one of the three models. Compared to this model, our own sensor shows a maximum deviation of 1.4%. Mixing equations for gas mixtures must be chosen carefully, taking into consideration whether mixing partners include polar or non-polar molecules. Some simplified mixing equations cannot be used to calculate the thermal conductivity of water vapor in air or nitrogen.

Keywords: thermal conductivity; hydrogen; water vapor; humidity; cross-sensitivity; MEMS

Citation: Emperhoff, S.; Eberl, M.; Dwertmann, T.; Wöllenstein, J. On the Influence of Humidity on a Thermal Conductivity Sensor for the Detection of Hydrogen. *Sensors* **2024**, *24*, 2697. https://doi.org/10.3390/s24092697

Academic Editors: Bruno Ando, Luca Francioso and Pietro Siciliano

Received: 13 March 2024
Revised: 16 April 2024
Accepted: 19 April 2024
Published: 24 April 2024

1. Introduction

With increasing efforts being made to develop alternative drive systems, solutions such as electric and fuel cell vehicles are shifting into focus. Hydrogen as a fuel is seen as an important component of decarbonization in areas where electrification is not practical or possible [1]. Germany, for example, even provides a national hydrogen strategy to support achieving climate neutrality [2]. This development naturally leads to an increased interest in sensing solutions for hydrogen. The presence of hydrogen in concentrations between 4 and 77 vol.% can lead to explosive gas mixtures. Hydrogen leakage sensors can be used to avoid endangering passengers or bystanders in the event of leakages in a fuel cell vehicle. Battery monitoring can be another application for this type of sensor. In the event of a thermal runaway of a lithium-ion battery, various gases, including hydrogen, are released from the battery. In a study with 51 batteries, the concentration of hydrogen among the gases released was 22.27% on average [3]. That means hydrogen sensors can also be used to detect a thermal runaway event.

We can find many measurement principles for hydrogen sensors. Thermal conductivity is especially suited to hydrogen detection since the thermal conductivity of hydrogen is seven times higher than air at room temperature. Electrochemical or catalytic sensors may show higher sensitivities and a better selectivity towards hydrogen but also show drawbacks like a short lifetime and vulnerability to poisoning, respectively [4]. Applications like leakage detection of hydrogen in a vehicle require a certain lifetime and stability of the sensor which can be offered by a physical based measurement principle such as thermal conductivity. In safety relevant applications, the impact of cross-influences such as temperature, pressure and humidity also gain importance since they directly impact the accuracy of the sensor. In this study, we focus on the impact of humidity on thermal conductivity sensors. Humidity here corresponds to another gas in the gas mixture and thus contributes to the overall thermal conductivity. Thermal conductivity is a non-selective measurement principle as every material has its own thermal conductivity value. In the context of hydrogen measurements, this can lead to a distortion of the sensor signal if the proportion of humidity changes. Even if the hydrogen concentration remains constant, the sensor signal can either increase or falsely decrease due to an increased proportion of another gas, in this case moisture. This leads to problems, particularly in safety-relevant applications, as either false-positive or false-negative results can be obtained. We have investigated the influence of humidity theoretically and experimentally to determine the potential error of hydrogen sensor readings due to humidity.

This theoretical study provided numerous approaches for calculating the thermal conductivity of a gas mixture. With models for binary gas mixtures and models that have been introduced specifically for gas mixtures of water vapor and air, a theoretical comparison was carried out and then compared to experimental results. The available models use different input parameters or fitting parameters. We could identify a large discrepancy between those approaches. Most importantly, the expectation contradicted the measured behavior. The thermal conductivity of water vapor is lower than the thermal conductivity of dry air. Using a simple mixing equation based on the volumetric proportions of a gas mixture, this would lead to a decrease in the thermal conductivity of the overall mixture. It seems that many models cannot sufficiently describe the thermal conductivity of water vapor and air mixtures. This mixture of a polar and a non-polar gas leads to the more complex behavior of thermal conductivity due to the collision behavior of the molecules and their degrees of freedom.

In this study, we compared common mixing equations with measurements of our own thermal conductivity sensor for measuring hydrogen and compared them with other commercially available thermal conductivity sensors. We investigated the influence of humidity on different thermal conductivity sensors and evaluated whether common mixing equations can be used to describe and predict the behavior of those sensors.

Therefore, we conducted gas measurements between 25 and 85 °C with humidity variations between 0% and 95% relative humidity at atmospheric pressure. The results of all sensors were compared to the available mixing models.

2. Materials and Methods

2.1. Theoretical Background

Thermal conductivity is a material property that describes the ability of a material to conduct heat. Generally, the thermal conductivity of gases is very low compared to liquid or solid materials. Thermal conductivity values are often only available for pure gases, and data points for mixtures are limited. Since the target application requires measuring different concentrations of hydrogen in air, the focus is on the resulting thermal conductivity of a gas mixture. Measured values at a specific concentration, temperature, and pressure are hard to find. Therefore, methods with which the thermal conductivity of gas mixtures can be calculated are being sought.

In this study, two different gas mixtures were in focus, the target mixture of the sensor being an air–hydrogen mixture in order to investigate the cross-influence of a water vapor and air mixture. Both mixtures are considered binary gas mixtures in which air represents one component of the mixture (constant composition). In the literature, we can find different equations and approaches to estimating the thermal conductivity of binary gas mixtures. These equations are of varying complexity. We only concentrate on gas mixing equations for binary gas mixtures.

Udoetok presents a simple approach using the molar ratios of the gases and their corresponding thermal conductivity. The resulting thermal conductivity of the gas mixture is directly dependent on the molar ratios of the components in the gas mixture. Udoetok derives this from a model in which he considers thermal resistances and their parallel and serial connection [5].

$$k_{mix} = 0.5 \times \frac{k_A k_B}{x_A k_B + x_B k_A} + 0.5 \times (x_A k_A + x_B k_B) \tag{1}$$

where k_{mix} is the resulting thermal conductivity of components A and B with their thermal conductivities k_A and k_B, respectively. x_A and x_B are the molar fraction of each component in the mixture. This approach has already been investigated for binary gas mixtures of rare gases in a study by Mathur, Tondon and Saxena [6]. Zhukov and Pätz consider this approach for the calculation of a mixture of hydrogen and oxygen [7].

Mason and Saxena provide a general gas mixing equation which according to them is derived from kinetic theory by well-defined approximations [8]. Additional to the molar fractions and the thermal conductivities of the pure components, the molecular weights and viscosities at the corresponding temperature are needed for calculating a resulting thermal conductivity.

$$k_{mix} = \frac{k_A}{1 + \left(G_{AB}\frac{x_B}{x_A}\right)} + \frac{k_B}{1 + \left(G_{BA}\frac{x_A}{x_B}\right)} \tag{2}$$

The parameters G_{AB} and G_{BA} are calculated using the viscosities μ_A and μ_B and molar masses M_A and M_B for the gases A and B, respectively.

$$G_{AB} = \frac{\gamma}{2\sqrt{2}} \times \left(1 + \frac{M_A}{M_B}\right)^{-0.5} \times \left[1 + \left(\frac{\mu_A M_B}{\mu_B M_A}\right)^{0.5} \times \left(\frac{M_A}{M_B}\right)^{0.25}\right]^2 \tag{3}$$

$$G_{BA} = \frac{\gamma}{2\sqrt{2}} \times \left(1 + \frac{M_B}{M_A}\right)^{-0.5} \times \left[1 + \left(\frac{\mu_B M_A}{\mu_A M_B}\right)^{0.5} \times \left(\frac{M_B}{M_A}\right)^{0.25}\right]^2 \tag{4}$$

This equation can be used for polyatomic gases in binary gas mixtures, but polar gases are excluded in the study of Mason and Saxena. The factor γ is an empirical factor and defined as 1.065 for non-polar gas mixtures [8].

Tsilingiris proposed a gas mixing equation for gas mixtures containing air and water vapor [9]. The base for this equation is the equation proposed by Mason and Saxena as described above. Tsilingiris adapted the empirical factor and conducted subsequent substitutions. He also added empirical equations to calculate the thermal conductivities and viscosities for water vapor and air for different temperatures.

$$k_{mix} = \frac{(1 - x_v) \times k_{air}}{(1 - x_v) + x_v \times \phi_{av}} + \frac{x_v \times k_v}{x_v + (1 - x_v) \times \phi_{va}} \tag{5}$$

With x_v being the molar fraction of water vapor in air and the thermal conductivities k_{air} and k_v of air and water vapor, respectively. The paramterers ϕ_{av} and ϕ_{va} can be obtained as follows:

$$\phi_{av} = \frac{\sqrt{2}}{4} \times \left(1 + \frac{M_{air}}{M_v}\right)^{-0.5} \times \left[1 + \left(\frac{\mu_{air}}{\mu_v}\right)^{0.5} \times \left(\frac{M_v}{M_{air}}\right)^{0.25}\right]^2 \tag{6}$$

$$\phi_{va} = \frac{\sqrt{2}}{4} \times \left(1 + \frac{M_v}{M_{air}}\right)^{-0.5} \times \left[1 + \left(\frac{\mu_v}{\mu_{air}}\right)^{0.5} \times \left(\frac{M_{air}}{M_v}\right)^{0.25}\right]^2 \tag{7}$$

Melling et al. also proposed a model for calculating the thermal conductivity of a gas mixture containing air and water vapor [10]. The model includes a set of empirical equations to calculate the viscosities and thermal conductivities of air and water vapor at a temperature range from 100 °C and 200 °C. The gas mixing equation was taken from Mason and Saxena, but Melling et al. proposed a different empirical factor for this specific gas mixture containing a polar and a non-polar gas. Thermal conductivity is calculated using Equations (2)–(4), where the gases *A* and *B* represent air and water vapor, respectively. They state that a γ of 0.8 best fits the available measurement data.

Some of the presented models are based on the same original equation but use different values for the parameter γ. Melling et al. state that the most suitable value for this parameter is 0.8 for mixtures of polar and non-polar gases. Tsilingiris uses factor 1 for this polar and non-polar gas mixture of water vapor and air. Tondon and Saxena generally suggest a value of 0.85 for polar and non-polar gas mixtures [11]. Mason and Saxena use 1.065 in their standard equation for binary gas mixtures containing only non-polar gases.

Tondon and Saxena compared different approaches for calculating the thermal conductivity of polar and non-polar gas mixtures [11]. The conclusion of this study was that all four methods lead to an adequate calculation of thermal conductivity values with an error of around 2%. One method, which they call the approximate method, is the approach by Mason and Saxena from Equation (2), which is also used by Melling et al. Another approach among those investigated is the method by Lindsay and Bromley [12]. As they could not identify a clearly superior approach, our comparison does not include any further methods. The presented equations can be used for binary gas mixtures. Calculations of multicomponent gas mixtures have been performed by Muckenfuss and Curtiss and with a modification by Mason and Saxena [13,14].

Figure 1 presents a graphical comparison of the different mixing equations for a binary gas mixture containing hydrogen in nitrogen (a) and water vapor in air (b). Nitrogen is used as a substitute for air since air consists of 78 vol.% of nitrogen. Both mixing equations for hydrogen show a linear behavior for the selected range from 0 to 2 vol.% of hydrogen. The larger the hydrogen concentration, the larger the difference between both calculations. In the case of the water vapor and air mixture, it is important to notice that all three equations show a very different outcome. Since the thermal conductivity of water vapor is lower than that of air, the equation by Udoetok leads to a decrease in the thermal conductivity with an increasing molar fraction of water vapor. Tsilingiris also predicts a decrease in the thermal conductivity for higher concentrations of water vapor but at a different slope than Udoetok. A very different behavior is shown in the equation of Melling et al., where the thermal conductivity of the water vapor–air mixture first increases, reaches a maximum, and then decreases again.

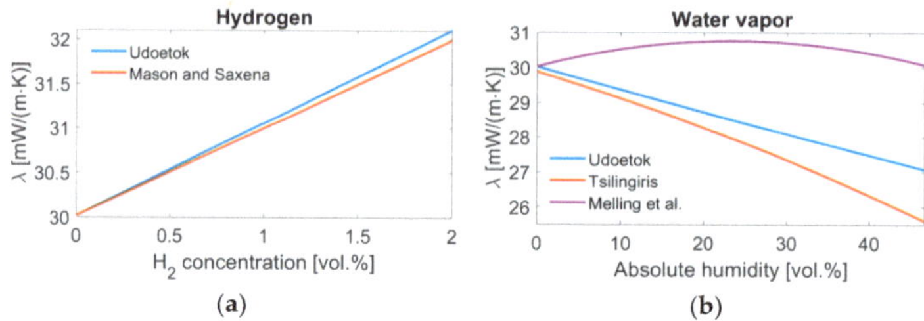

Figure 1. Comparison of the resulting thermal conductivity at 80 °C of gas mixtures containing (**a**) hydrogen in nitrogen as calculated by Equation (1) from Udoetok and Equations (2)–(4) from Mason and Saxena and (**b**) water vapor in air as calculated by Equation (1) from Udoetok, Equations (5)–(7) from Tsilingiris and Equations (2)–(4) using the empirical factor from Melling et al. [5,8–10].

2.2. Thermal Conductivity Sensors

The behavior of our MEMS-based thermal conductivity sensor towards humidity was investigated. The MEMS structure has already been introduced by Emperhoff et al. [15].

The sensor structure will hereafter be referred to as the IFX sensor. It is based on a three-layer structure wherein a structured silicon wafer is sandwiched between two structured glass wafers. This stacking leads to two cavities with two silicon heaters each, which act as heater and resistor. One cavity is hermetically sealed through the two glass wafers, whereas the other cavity remains open with a gas inlet in the bottom wafer (Figure 2a). That means two resistors are used as a reference in a constant atmosphere and the other two resistors are exposed to gas changes. The resistors are connected as a Wheatstone bridge to allow a stable and low-noise measurement of resistance changes (Figure 2b). The resistors are both heaters and sensor elements. When a gas enters the open cavity, the thermal conductivity of the gas changes. This leads to a temperature change at the heater, as it influences how well the heat is dissipated. This causes a change in resistance, which can be measured via the Wheatstone bridge.

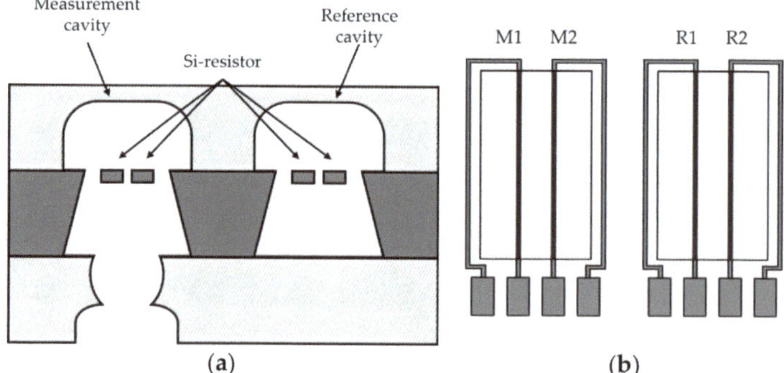

Figure 2. (**a**) Schematic cross-section of the MEMS structure used, as presented in [15]. (**b**) Schematic top view of the sensor as presented in [15]. The four resistors are connected as a Wheatstone bridge where two of them (M1, M2) are in a measurement cavity and the other two (R1, R2) are in a closed reference cavity.

Additionally, two commercially available sensors with which to compare the IFX sensor were selected. These are the CO_2 sensor STC31 by Sensirion (Sensirion AG, Staefa, Switzerland) and the H_2 sensor PGS1000 by Posifa Technologies (Posifa Technologies, San Jose, CA, USA) [16,17]. All sensors are based on thermal conductivity as their measurement principle. Figure 3 shows a schematic overview of the two reference sensors. Figure 3a shows the structure of the Sensirion STC31 CO_2 sensor. It is assumed that a heater structure in the middle is surrounded by two temperature-sensitive resistors at different distances from the heater. The PGS1000 hydrogen sensor consists of a heat source and a thermopile on a membrane which are above a gas cavity/heat sink (Figure 3b).

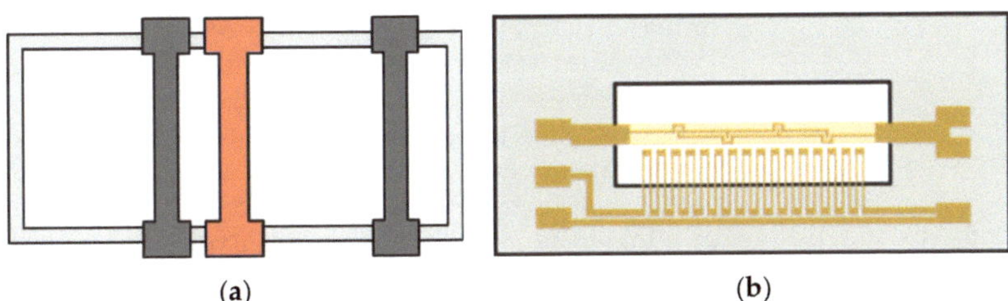

(a)   (b)

Figure 3. Schematic layout of reference thermal conductivity sensors with (**a**) the Sensirion STC31 on the left and (**b**) the Posifa Technologies PGS1000 on the right [16,17].

2.3. Gas Measuring Station

All measurements were conducted using the gas measurement setup shown in Figure 4. The HovaCAL® (IAS GmbH, Oberursel, Germany) is a calibration gas generator and can produce dry and defined humidified gas mixtures. The gas mixing unit is optimized for hydrogen gas mixtures and provides highly accurate hydrogen–nitrogen mixtures. The water dosage is administered via precise syringes and fed to the gas flow through an evaporator.

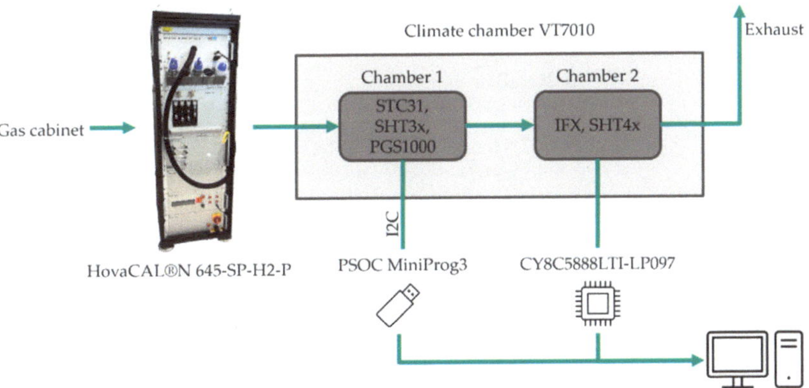

Figure 4. Schematic overview of the utilized measurement setup consisting of a gas-mixing unit (HovaCAL®) and a climate chamber containing two separate measurement chambers for the sensors to be tested.

The preheated and humidified gas mixture is fed into the climate chamber via a heated line to avoid condensation. Two separate measurement chambers containing the devices under test are located within the climate chamber. Each chamber contains a temperature and humidity reference sensor to measure the ambient conditions in close proximity to the sensors. The Sensirion STC31 and the PGS1000 are placed in the first chamber. A humidity and temperature reference sensor are included in the STC31 module. Both sensors can be read out via an I2C bus. The communication is conducted with a PSOC MiniProg3 (Infineon Technologies AG, Neubiberg, Germany). The second chamber includes the IFX sensor and a SHT4x relative humidity sensor (Sensirion AG, Staefa, Switzerland). The IFX sensor is powered and read out via the Cypress microcontroller CY8C5888LTI-LP097 (Infineon Technologies AG, Neubiberg, Germany).

2.4. Measurement Methods

In this study, the behavior of thermal conductivity sensors towards humidity was investigated. To evaluate and compare the different thermal conductivity sensors, a calibration measurement with hydrogen in nitrogen was carried out first as all sensors deliver different sensor outputs. The STC31 is a CO_2 sensor, and the output is given as a CO_2 concentration in ppm. The PGS1000 is a hydrogen sensor which provides the hydrogen concentration as a percentage. The IFX thermal conductivity sensor provides a raw voltage signal.

The measurements were conducted in two steps. First, a calibration measurement was conducted using hydrogen in nitrogen with different concentrations. A humidity measurement with increasing water vapor concentrations in nitrogen followed the calibration measurement.

The measurements were carried out with a total flow rate of 1500 mL/min and repeated at different temperatures between 25 and 85 °C. The calibration measurement uses nitrogen as carrier gas and adds hydrogen to receive the concentrations 0%, 1%, and 2% of hydrogen. In the following measurements, the signal output of all sensors is converted into a hydrogen concentration using the measured sensitivity to hydrogen for each sensor at each temperature. Directly following the hydrogen measurement, the humidity measurement was conducted. The water vapor concentration was increased stepwise from 0 to 90% relative humidity.

The change in the original sensor output of each sensor is transformed into a hydrogen concentration on the one hand and a change in the thermal conductivity value on the other hand. Using the sensor sensitivity of the calibration measurement with hydrogen, all sensor signals are converted into a hydrogen equivalent. The expected thermal conductivity of a hydrogen–nitrogen mixture from 0 to 2 vol.% is calculated using the mixing equation by Mason and Saxena with a γ of 1.065, which, according to Zhukov and Pätz, is most suitable for hydrogen mixtures [7]. The change in the sensor signal of the hydrogen equivalent is then transformed into a corresponding change in thermal conductivity.

3. Results

Humidity Response

Two out of three of the presented models predicted a decrease in thermal conductivity for increasing concentrations of water vapor. This means that there were large deviations between the results of those equations. We investigated how our thermal conductivity sensor behaved in actual measurement conditions. To verify the results, we not only tested the IFX sensor but also tested sensors from other producers, using hydrogen–nitrogen mixtures and water vapor–nitrogen mixtures. In the theory section, we saw that increasing the share of hydrogen in nitrogen leads to an increase in the thermal conductivity of the gas mixture. Within our measuring range, this behavior can be regarded as linear.

This means that according to two out of the three presented models, a signal change in the other direction is expected for humid gas mixtures. To simplify the presentation of the measurement results, the signals are displayed as hydrogen equivalents.

Initially, the IFX thermal conductivity sensor sample was measured alone to investigate the impact of humidity on the IFX sensor signal. Figure 5 shows the result of a characterization at four different temperatures, with increasing humidity levels between 0 and 90% relative humidity. The sensor output is displayed as a hydrogen equivalent at the respective temperature. The offset due to temperature changes is compensated. The first thing to note is that the sensor responded with a positive hydrogen equivalent for increasing levels of humidity. This indicates an increase in the thermal conductivity (contrary to what some mixing equations predicted). Furthermore, as temperature increased, so did the non-linearity of the sensor signal in response to the relative humidity. At some point, this even led to a turning point at which the signal decreased again and dropped below a 0% hydrogen equivalent. The higher the temperature, the larger the maximum hydrogen equivalent caused by the water vapor.

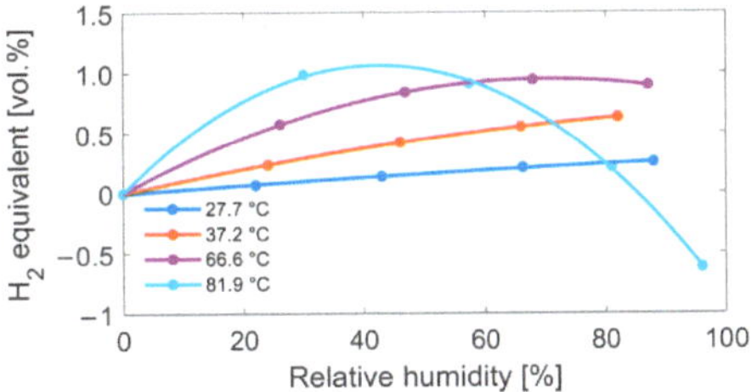

Figure 5. Humidity measurement of the IFX sensor at four different temperatures with increasing humidity levels between 0 and 90% relative humidity.

Figure 6 shows the measurement data at 81.9 °C compared to the introduced models. While comparing the models from the theory section with our measurement data, we can see that only the model by Melling et al. showed an increase in the thermal conductivity with increasing relative humidity [10]. The mixing equation of Udoetok and Tsilingiris resulted in a decrease in the thermal conductivity of water vapor and air mixtures, as was intuitively expected, since the thermal conductivity of water vapor is lower than that of air [5,9].

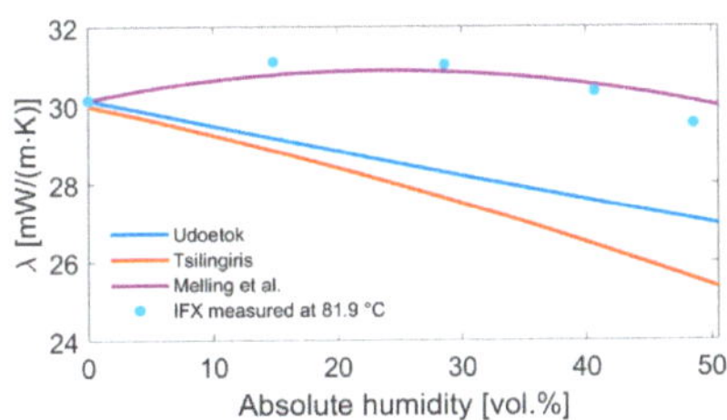

Figure 6. IFX sensor measurement compared to gas mixing equations from Udoetok, Tsilingiris and Melling et al. for the thermal conductivity of water vapor in air at 81.9 °C [5,9,10].

To ensure that this result was not just a misleading measurement or fault caused by the sensor system, the humidity measurement was repeated, and two external sensors based on thermal conductivity were added to the measurement setup as described in Figure 4. Those sensors are the STC31 CO_2 sensor by Sensirion and the PGS1000 H_2 sensor by Posifa Technologies. The measurements were performed as described in Section 2.

Figure 7 shows the humidity measurements at four different temperatures. Mean values for every measurement point are displayed for each sensor. The results are given as hydrogen equivalents over the measured relative humidity. A polynomial curve of second order is fitted through the measurement points. The first thing to notice is that the humidity response parallels the hydrogen response for all three sensors. Second, the non-linearity increases at larger temperatures. Third, all three sensors show the same general behavior but with different magnitudes. The largest humidity response is shown by the STC31, which reaches up to 2 vol.% of hydrogen. The IFX and PGS1000 are quite comparable, especially at lower humidity levels, and show a maximum error between 1.1 and 1.2 vol.% of hydrogen.

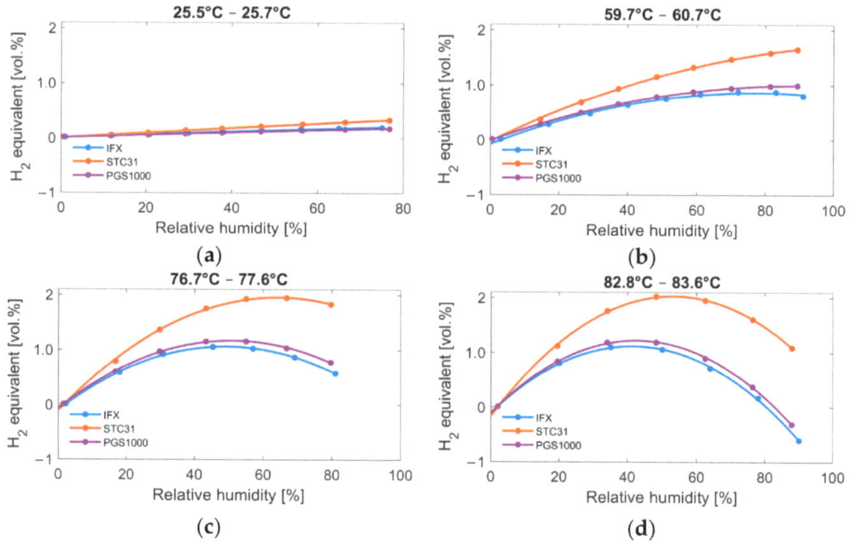

Figure 7. Measurement with three different thermal conductivity sensors (IFX, STC31 and PGS1000) at four different temperatures (**a**) 25.5–25.7 °C, (**b**) 59.7–60.7 °C, (**c**) 76.7–77.6 °C and (**d**) 82.8–83.6 °C, with varying humidity from 0 to 95% relative humidity.

Figure 8 displays the change in thermal conductivity as calculated with the equation from Melling et al. [10] Four different temperatures are displayed, depending on (a) the relative humidity between 0% and 100% (temperature offset compensated). The general behavior is quite comparable to what we see in the measurement results, with the two main characteristics of an increasing thermal conductivity and an increasing non-linearity at higher temperatures. Relative humidity results in a different absolute humidity at different temperatures; Figure 8b displays the thermal conductivity of a water vapor in air mixture depending on the absolute humidity in vol.%. The change in the thermal conductivity depending on the absolute humidity almost shows a parabola. That means that the change in thermal conductivity can be regarded as mostly dependent on the absolute humidity. The relative humidity is converted to an absolute humidity using following approach:

$$AH = RH \times \frac{P_s(t)}{P_b} \tag{8}$$

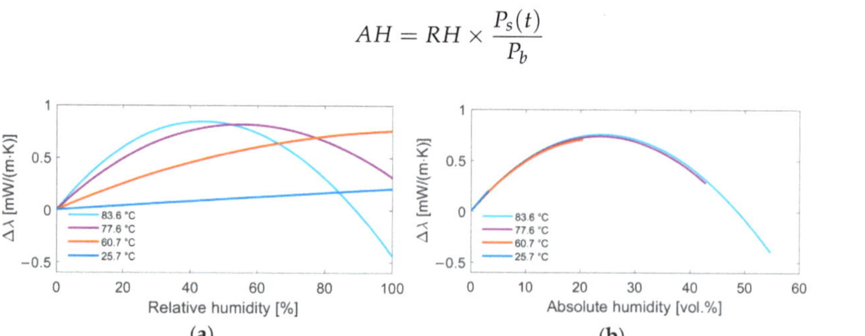

Figure 8. Result of the mixing equation by Melling et al. displaying the change in thermal conductivity of a water vapor–air mixture depending on (**a**) relative humidity and (**b**) absolute humidity in vol.% at four different temperatures between 25 and 85 °C [10].

The absolute humidity AH in vol.% is calculated using the relative humidity RH in %, the saturation vapor pressure P_s in mbar at the reference temperature t, and the reference pressure P_b in mbar. The saturation vapor pressure is calculated with the Goff–Gratch equation.

Similar to the display of the model in Figure 8, Figure 9 shows the measurement results depending on the absolute humidity to evaluate whether the sensor output is in fact also mostly dependent on the absolute humidity share in air. We can see that the output seems to match well in low humidity ranges. However, at higher humidities, the results at the same absolute humidity but at different temperatures start to deviate. This applies to all three sensors, albeit to different degrees.

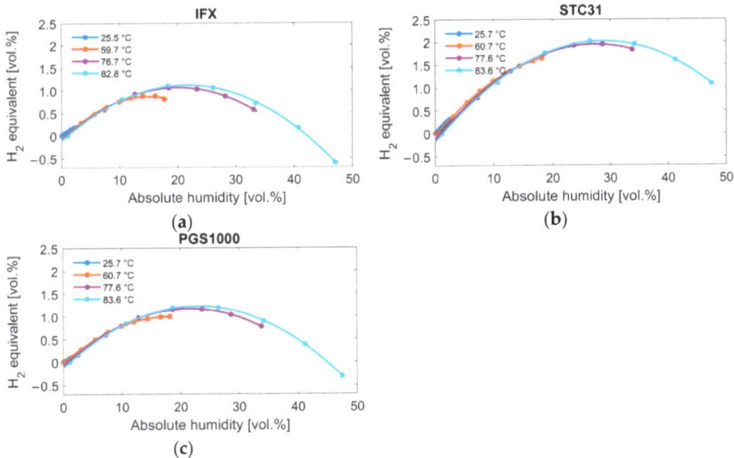

Figure 9. Result of the humidity measurement of the (**a**) IFX, (**b**) STC31, and (**c**) PGS1000 thermal conductivity sensors displayed as a hydrogen equivalent and depending on the absolute humidity in vol.%, as calculated using Equation (8).

The same measurement data are transformed into a thermal conductivity value for a direct comparison to the model by Melling et al. [10]. The equation by Mason and Saxena is used to calculate a corresponding thermal conductivity change for a certain hydrogen concentration. This can be used to convert the hydrogen equivalent of the humidity measurements to an expected change in thermal conductivity. This delta is added to the thermal conductivity of 100% of the carrier gas at the respective temperature. Figure 10 shows the result of this display for all three sensors. The increase in the humidity level in the gas mixture seems to lead to an increase in thermal conductivity until the curve reaches its maximum and the thermal conductivity decreases again. Generally, the model shows the same behavior. A certain degree of uncertainty remains, particularly with the position of the maximum value and the magnitude, which is different for each sensor.

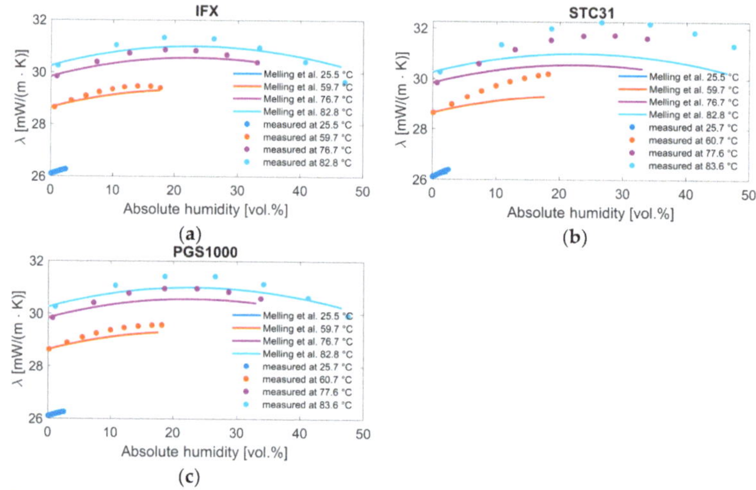

Figure 10. Thermal conductivity of water vapor and air mixtures at different temperatures as calculated with the equation of Melling et al. [10] compared to the measurement results of the (**a**) IFX, (**b**) STC31, and (**c**) PGS1000 thermal conductivity sensors. The results are displayed as a corresponding thermal conductivity value.

To verify our measurement data, earlier measurement results by Gruess and Schmick and Vargaftik [18,19], which were taken from Tsilingiris [9], were compared to measured data from the IFX sensor as well as the presented mixing equations at a temperature of 80 °C (Figure 11). The IFX measurement was slightly above 80 °C with a temperature of 82.9 °C. The experimental data by Gruess and Schmick and Vargaftik are well in line with our own results.

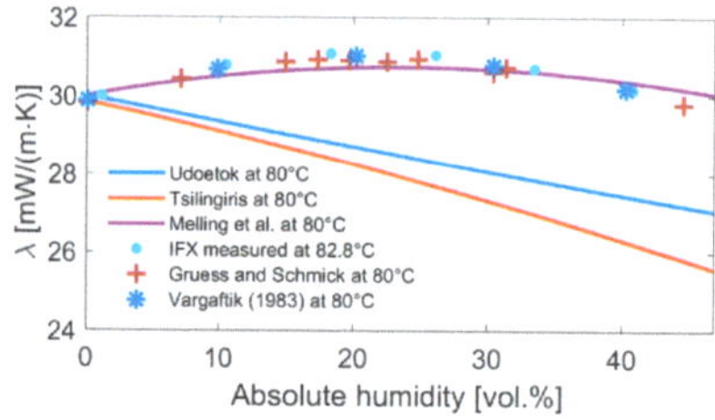

Figure 11. IFX measurements at 82.8 °C and different mixing equations for the thermal conductivity of water vapor in air at 80 °C compared to experimental data from Gruess and Schmick (1928) and Vargaftik (1983) taken from Tsilingiris [5,9,10,18,19].

4. Discussion

The three models presented showed different thermal conductivities for water vapor in air mixtures. Two of them are specifically for humid air mixtures. The third model is a simple mixing equation based on molar ratios. Intuitively, water vapor should decrease the thermal conductivity of the mixture since the thermal conductivity of pure water vapor is lower than the thermal conductivity of dry air. As a polar molecule with hydrogen bonds, water often behaves differently from other materials. During our measurements, we observed an increase in thermal conductivity with increasing levels of humidity, similar to what the mixing equation of Melling et al. indicated. At higher humidity levels, the thermal conductivity decreased again. Tsilingiris, on the other hand, predicted a decrease in thermal conductivity. The best alignment of experimental data with the models was achieved with the mixing equation from Melling et al. In general, experimental data for gas mixtures of water vapor and air are scarce for the intended temperature and pressure range. Experimental data were mostly cited from Gruess and Schmick and were taken from Tsilingiris to compare them with our own results. Tsilingiris also provides experimental data from Vargaftik. A deviation between experimental data and calculated values was acknowledged, but the model functioned sufficiently well. The smallest deviation between experimental and modeled data was found with the approach of Melling et al. [10]. This applies both to our own measurement results and to earlier data from Gruess and Schmick and Vargaftik which were taken from Tsilingiris [9,18,19]. When choosing a mixing equation for the thermal conductivity of gas mixtures, it is important to be aware of the boundary conditions. Mixtures of polar and non-polar gases cannot be predicted using simple equations such as the one from Udoetok [5]. We were able to show that the thermal conductivity of humid air first increases compared to dry air, although many publications still state that it decreases continuously.

Two commercially available sensors were investigated and compared to experimental results from our own sensor. All the investigated sensors are based on the measurement of thermal conductivity. Thermal conductivity sensors consist of a heat source and a temperature-sensing element which indicates changes in thermal conductivity through measuring temperature changes. Those two components can be separate, as in the STC31 and the PGS1000, or combined, as in the IFX sensor. If hydrogen or humidity are present in the vicinity of the sensing elements, the change in thermal conductivity causes a change in the temperature of the sensor element. In the case of the IFX sensor, this resulted in a change in resistance, which could be measured through the Wheatstone bridge. All three sensors qualitatively showed the same behavior towards humidity. The IFX and PGS1000 were also similar in the magnitude of the effect, which corresponded to 1.1 to 1.2 vol.% in hydrogen measurement values. The STC31 showed the largest effect towards humidity, with an equivalent of 2 vol.% in the hydrogen measurement value at maximum. The difference in the signal level can partly be explained by different heater temperatures. When the heater is activated, this creates a microclimate around the heater. This causes a temperature gradient between heater temperature and ambient temperature. Thermal conductivity is a temperature-dependent material property. The temperature gradient can be different for different gases. This can lead to an altered ratio of the thermal conductivities of hydrogen and water vapor. We assumed that the STC31 had the highest heater temperature compared to the other two sensors.

The behavior of all three sensors towards humidity fit best to the model proposed by Melling et al. [10]. The IFX, STC31, and PGS1000 showed a maximum error of 1.4%, 4.5%, 1.7%, respectively. There are several possible reasons for these deviations. The mixing models represent only thermal conductivity. In our measurements, we must also consider different types of temperature transport, which are always a combination of convection, conductivity, and heat radiation. Forced convection should be at a minimal level because of the sensors' geometry and housing. Free convection, on the other hand, is present inside the measurement cavities. Furthermore, we must differentiate between thermal conductivity and thermal diffusivity. Thermal conductivity still contains the specific heat capacity and density of the gas. Mixing models cannot contain the whole complexity of those factors, especially since the extent of each factor varies with the sensor type and geometry.

5. Conclusions

In this paper, we investigated the impact of humidity on thermal conductivity sensors. This included a search of the literature for different mixing equations for calculating the thermal conductivity of binary gas mixtures and comparative measurements with different thermal conductivity sensors. Out study of the literature did not provide a clear result regarding the behavior of the thermal conductivity value for mixtures with varying humidity. Different approaches provide different solutions. Some models predict a continuous decrease and others an increase in thermal conductivity with increasing humidity, which is followed by a decrease at very high absolute humidities. However, the measurement results of three different sensors show a behavior comparable to the model of Melling et al. that indicates an increase followed by a decrease after thermal conductivity reaches a maximum. In our measurements, the sensor signals caused by humidity reached up to 2 vol.% in the hydrogen measurement value at maximum. The behavior of all three sensors was comparable, even if they displayed different degrees of change. Deviations between the experimental data of the three sensors and the model by Melling et al. [10] range between 1.4% and 4.5%. Earlier measurements by Gruess and Schmick or Vargaftik [18,19] are in line with these results. The behavior of the thermal conductivity depending on humidity is also shown in a study by Kimura, who even uses this behavior to measure the absolute humidity in air [20].

The results were compared with the identified mixing equations to evaluate whether these equations can be used to predict the behavior of thermal conductivity sensors and to show the actual behavior towards humidity for real-life applications. The thermal conductivity of the gas mixture containing water vapor and air or nitrogen first increased and then decreased again after it reached its maximum value. Simplified mixing models cannot be used to describe this behavior. Mixing models must be adjusted to fit the present gases, as in the approach by Melling et al., where the empirical factor of the equation by Mason and Saxena was adapted to a gas mixture containing water vapor and air [8,10].

More actual thermal conductivity measurements would be helpful for a more accurate adaption of the mixing equation to water vapor–air mixtures as the available experimental data are still limited in lower temperature and pressure ranges, as investigated in this study. Especially for measurements with the investigated sensors, we must also consider the microclimate of the heater structures, as the actual temperature around the heater does not correspond to the measured temperature and leads to a gradient surrounding the heater. Furthermore, measurements with the available sensors are not only influenced by thermal conductivity but could also include other types of temperature transport such as convection and heat radiation or sensor-specific characteristics and dynamic behaviors. The proposed gas mixing models cannot cover the whole complexity of those factors.

The results herein will help the implementation of thermal conductivity sensors in real-life applications. Thermal conductivity is mostly dependent on absolute humidity if it is temperature-compensated. It can be used as a simple way of compensating for humidity. We have shown that the observed behavior of our thermal conductivity sensor towards humidity is a general phenomenon that can be reproduced with other sensors and previous measurement data from other authors, even if some models still suggest a different behavior. Further investigations are planned to find adequate ways to improve models for sensor applications and to implement methods of compensating for humidity.

Author Contributions: Conceptualization, S.E. and M.E.; methodology, S.E. and M.E.; validation, S.E.; investigation, S.E. and T.D.; writing—original draft preparation, S.E.; writing—review and editing, S.E., M.E. and J.W.; visualization, S.E.; supervision, M.E. and J.W. All authors have read and agreed to the published version of the manuscript.

Funding: This research was funded by the German Federal Ministry of Economic Affairs and Climate Action (BMWK), grant number 03EN5009A.

Institutional Review Board Statement: Not applicable.

Informed Consent Statement: Not applicable.

Data Availability Statement: Restrictions apply to the availability of these data. The datasets presented in this article are not readily available because of company restrictions. Requests to access the datasets should be directed to the corresponding author.

Acknowledgments: We acknowledge support by the Open Access Publication Fund of the University of Freiburg.

Conflicts of Interest: The authors S.E., M.E. and T.D. were employed by Infineon Technologies AG. The remaining authors declare that the research was conducted in the absence of any commercial or financial relationships that could be construed as a potential conflict of interest.

References

1. Robinson, C.; Davies, A.; Wainwright, J. *Global Hydrogen Balance: Outlook to 2050—Inflections and Green Rules*; IHS Markit: London, UK, 2022.
2. *National Hydrogen Strategy Update*; Federal Ministry for Economic Affairs and Climate Action: Berlin, Germany, 2023.
3. Shahid, S. A Review of Thermal Runaway Prevention and Mitigation Strategies for Lithium-Ion Batteries. *Energy Convers. Manag.* **2022**, *16*, 100310. [CrossRef]
4. Hübert, T.; Boon-Brett, L.; Black, G.; Banach, U. Hydrogen Sensors—A Review. *Sens. Actuators B Chem.* **2011**, *157*, 329–352. [CrossRef]
5. Udoetok, E.S. Thermal conductivity of binary mixtures of gases. *Front. Heat Mass Transf.* **2013**, *4*, 1–5. [CrossRef]
6. Mathur, S.; Tondon, P.K.; Saxena, S.C. Thermal Conductivity of Binary, Ternary and Quaternary Mixtures of Rare Gases. *Mol. Phys.* **1967**, *12*, 569–579. [CrossRef]
7. Zhukov, V.P.; Pätz, M. On Thermal Conductivity of Gas Mixtures Containing Hydrogen. *Heat Mass Transf.* **2017**, *53*, 2219–2222. [CrossRef]
8. Mason, E.A.; Saxena, S.C. Approximate Formula for the Thermal Conductivity of Gas Mixtures. *Phys. Fluids* **1958**, *1*, 361. [CrossRef]
9. Tsilingiris, P.T. Thermophysical and Transport Properties of Humid Air at Temperature Range between 0 and 100 °C. *Energy Convers. Manag.* **2008**, *49*, 1098–1110. [CrossRef]
10. Melling, A.; Noppenberger, S.; Still, M.; Venzke, H. Interpolation Correlations for Fluid Properties of Humid Air in the Temperature Range 100 °C to 200 °C. *J. Phys. Chem. Ref. Data* **1997**, *26*, 1111–1123. [CrossRef]
11. Tondon, P.K.; Saxena, S.C. Calculation of Thermal Conductivity of Polar-Nonpolar Gas Mixtures. *Appl. Sci. Res.* **1968**, *19*, 163–170. [CrossRef]
12. Lindsay, A.L.; Bromley, L.A. Thermal Conductivity of Gas Mixtures. *Ind. Eng. Chem.* **1950**, *42*, 1508–1511. [CrossRef]
13. Muckenfuss, C.; Curtiss, C.F. Thermal Conductivity of Multicomponent Gas Mixtures. *J. Chem. Phys.* **1958**, *29*, 1273–1277. [CrossRef]
14. Mason, E.A.; Saxena, S.C. Thermal Conductivity of Multicomponent Gas Mixtures. II. *J. Chem. Phys.* **1959**, *31*, 511–514. [CrossRef]
15. Emperhoff, S.; Eberl, M.; Barraza, J.P.; Brandl, F.; Wöllenstein, J. Differential Thermal Conductivity Hydrogen Sensor. In Proceedings of the Sensor and Measurement Science International 2023, Nürnberg, Germany, 8–11 May 2023; pp. 65–66.
16. Datasheet STC31—CO_2 Sensor Based on Thermal Conductivity 2020. Available online: https://sensirion.com/media/documents/7B1D0EA7/61652CD0/Sensirion_Thermal_Conductivity_Datasheet_STC31_D1_1.pdf (accessed on 16 December 2023).
17. PGS1000 Series—MEMS Thermal Conductivity Hydrogen Sensor 2022. Available online: https://posifatech.com/wp-content/uploads/2020/11/Datasheet_PGS1000_MEMS_TC_H2_RevA_C0.5.pdf (accessed on 16 December 2023).
18. Gruess, H.; Schmick, H. *Wissenschaftliche Veröffentlichungen aus dem Siemens-Konzern*; Springer: Berlin, Germany, 1928; Volume 7.
19. Vargaftik, N. *Handbook of Physical Properties of Liquids and Gases—Pure Substances and Mixtures*; Hemispere: Washington, DC, USA, 1983.
20. Kimura, M. Absolute-Humidity Sensing Independent of the Ambient Temperature. *Sens. Actuators Phys.* **1996**, *55*, 7–11. [CrossRef]

Article

Measuring Surface Electromyography with Textile Electrodes in a Smart Leg Sleeve †

Federica Amitrano [1,‡], Armando Coccia [1,*,‡], Gaetano Pagano [2], Arcangelo Biancardi [1], Giuseppe Tombolini [3], Vito Marsico [4] and Giovanni D'Addio [1]

[1]   Bioengineering Unit, Telese Terme Institute, Istituti Clinici Scientifici Maugeri IRCCS, 82037 Telese Terme, Italy; federica.amitrano@icsmaugeri.it (F.A.); gianni.daddio@icsmaugeri.it (G.D.)
[2]   Bioengineering Unit, Bari Institute, Istituti Clinici Scientifici Maugeri IRCCS, 70124 Bari, Italy; gaetano.pagano@icsmaugeri.it
[3]   Tombolini Officine Ortopediche, 74121 Taranto, Italy
[4]   Orthopaedics Unit, Bari Institute, Istituti Clinici Scientifici Maugeri IRCCS, 70124 Bari, Italy; vito.marsico@icsmaugeri.it
*   Correspondence: armando.coccia@icsmaugeri.it
†   This paper is an extended version of our paper published in Proceedings of the XXXV EUROSENSORS Conference, Lecce, Italy, 10–13 September 2023.
‡   These authors contributed equally to this work.

Abstract: This paper presents the design, development, and validation of a novel e-textile leg sleeve for non-invasive Surface Electromyography (sEMG) monitoring. This wearable device incorporates e-textile sensors for sEMG signal acquisition from the lower limb muscles, specifically the anterior tibialis and lateral gastrocnemius. Validation was conducted by performing a comparative study with eleven healthy volunteers to evaluate the performance of the e-textile sleeve in acquiring sEMG signals compared to traditional Ag/AgCl electrodes. The results demonstrated strong agreement between the e-textile and conventional methods in measuring descriptive metrics of the signals, including area, power, mean, and root mean square. The paired data t-test did not reveal any statistically significant differences, and the Bland–Altman analysis indicated negligible bias between the measures recorded using the two methods. In addition, this study evaluated the wearability and comfort of the e-textile sleeve using the Comfort Rating Scale (CRS). Overall, the scores confirmed that the proposed device is highly wearable and comfortable, highlighting its suitability for everyday use in patient care.

Keywords: e-textile; textile-based electrode; surface electromyography; EMG; wearable sensors; comfort rating scale; comfort assessment

Citation: Amitrano, F.; Coccia, A.; Pagano, G.; Biancardi, A.; Tombolini, G.; Marsico, V.; D'Addio, G. Measuring Surface Electromyography with Textile Electrodes in a Smart Leg Sleeve. *Sensors* **2024**, *24*, 2763. https://doi.org/10.3390/s24092763

Academic Editors: Bruno Ando, Luca Francioso and Pietro Siciliano

Received: 15 March 2024
Revised: 23 April 2024
Accepted: 23 April 2024
Published: 26 April 2024

1. Introduction

Technological advancements have transformed the healthcare sector, with wearable technologies emerging as a key component of patient monitoring and health management [1,2]. These wearable devices, characterised by their constant connectivity, comfort, and discreet integration into daily life, are rapidly becoming indispensable tools in medical diagnostics and therapy [3,4]. The evolution from simple step counting to advanced monitoring of vital signs such as blood pressure and potential arrhythmias underscores the growing ability of wearable technologies to seamlessly integrate into our daily lives while improving access to healthcare and disease prevention [5]. The integration of wearable technologies extends from sports medicine, as demonstrated by Skazalski et al. [1], who used wearable devices to monitor the jumping load in elite volleyball players, to extreme conditions, where Chen et al. [2] developed methods to detect heat stroke. Similarly, Dooley et al. [6] compared and validated key consumer devices for measuring exercise intensity, highlighting the role of wearable devices in fitness and health tracking.

The use of wearable devices in healthcare, such as smartwatches, electronic bracelets, and sensor-embedded garments, represents a move towards patient-centred care. These devices enable real-time monitoring of crucial health metrics such as heart rate, blood pressure, and body temperature, facilitating immediate medical intervention when necessary [3,7].

Outstanding research has highlighted the potential of wearable technologies in the medical field, particularly their application in continuous health monitoring and disease prevention [8]. Wearable devices offer a unique combination of convenience, real-time data analysis, and personalisation of medical treatments, thereby revolutionising patient care and disease management.

However, a major challenge in the field of wearable technology is the integration of sensors with everyday clothing to improve usability and comfort. This challenge has led to the development of Electronic Textiles (e-textiles) or Smart Textiles which seamlessly integrate electronic components into textile materials and use innovative materials such as graphene to demonstrate the effectiveness of e-textiles in continuous health monitoring [9–11]. Despite their potential, challenges related to signal distortion and reduced stability with repeated washing remain critical issues to be addressed, highlighting the need for further advances in the field [12]. E-textiles represent a groundbreaking advancement, as they react and adapt to environmental stimuli through the integration of smart materials into their structure. This innovation presents several advantages over traditional electronic devices, including direct contact with the skin, flexibility and adaptability to the human body's contours, and cost-effectiveness thanks to reusability and washability that is comparable regular clothing [8].

Recent studies, such as the systematic review of e-textiles in biomedical applications by Cesarelli et al., have highlighted the role of wearable biosensors and fabric-based devices in health monitoring and disease management, offering new paradigms in patient care [13]. Their comprehensive review illustrates the various applications and potentialities of e-textiles in various medical settings, reinforcing the importance of these technologies in improving patient comfort and autonomy in health management. However, the same review identified a common limitation regarding the small study populations used to test these novel devices. Typically, devices are tested on a single volunteer. This is likely correlated with the other relevant issue, which concerns the early stage of development of the technologies presented in the literature thus far.

sEMG is of great importance due to its wide range of applications and benefits. Among other interesting applications, during biomechanical analysis of movements and practical rehabilitation it can provide a simple and objective quantitative assessment of muscle function with high information content [14–16]. sEMG has emerged as a suitable application of e-textile technology considering the need to detect the muscle action potentials with electrodes directly on the skin [17]. However, the main challenge remains developing electrodes and devices that are accurate and reliable while also being comfortable and easy for patients to use. Recently, attention has shifted towards the use of flexible materials and conductive fabrics, which promise to overcome many of the limitations of traditional electrodes. Recent studies have explored the use of non-invasive flexible electrodes for sEMG acquisition, exploiting designs inspired by biological structures to improve adhesion and reduce interference during signal acquisition [18]. For example, the creation of adhesive microstructures on the surface of electrodes, inspired by natural mechanisms such as tree roots or marine organisms, has been shown to significantly improve both the adhesion and extension capability of electrodes. Another significant area of research concerns the use of conductive fabrics such as Conductive Composite Silicone Material (CCSM), which have demonstrated superior performance compared to screen-printed materials in terms of durability and resistance to deformation during use [19]. Conductive plating on flexible fabrics offers an interesting alternative, with a conductive coating on each individual fibre of the fabric to improve strength and flexibility.

In addition to these innovations in materials and fabrication techniques, it is crucial to consider the integration of these advances with wearable devices that can be easily adopted by users. This approach requires a focus on functionality and reliability as well as on ergonomics and user comfort, aspects that are crucial to ensure the wide adoption and sustained use of these devices.

In this context, several approaches to creating prototype e-textile devices have been identified in the literature. For instance, Ohiri et al. [19] proposed a modular suit designed to extensively measure muscle activity, with sensors on key muscle groups such as the torso, arms, legs, and back. Similarly, Goncu-Berk et al. [20] developed prototype t-shirts with varying sleeve lengths made of stretch polyester knit fabric and e-textile electrodes sewn with conductive thread. Alizadeh-Meghrazi et al. [21] proposed a sleeve covering the entire forearm that integrates knitted textile electrodes using conductive silicone rubber-based filaments. Ozturk and Yapici [22,23] analysed the performance of wearable graphene-based electrodes in monitoring the muscular activity the upper and lower limbs, showing acceptable Signal to Noise Ratio (SNR) values. Similar results regarding graphene-based electrodes were obtained by Awan et al. [24] in a cohort of eight healthy controls. Other studies have explored different structures and materials for textile electrodes, as shown by Katherine Le et al. [25], who analysed the effect of different types of conductive pastes and textile electrode structures on biopotential monitoring performance. Furthermore, Milad Alizadeh-Meghrazi et al. [26] examined the importance of skin–electrode impedance and embroidery technique in the effectiveness of sEMG textile electrodes. A different approach was followed by Choudry et al. [27], who used flexible conductive threads stitched on fabric to design textile-based piezoresistive sensors embedded inside a garment to measure muscle activity based on the small pressure changes exerted by muscles. These different approaches highlight the continuing innovation in the design and application of e-textiles, as further demonstrated by the previously mentioned systematic literature review [13] highlighting the expanding scope and capabilities of e-textiles in the biomedical field.

Furthermore, a topic of debate in scientific discussions concerns the size of the electrodes. Kim et al. [28] reported that electrodes with diameters of 20 and 30 mm outperformed those with smaller dimensions in terms of SNR and baseline noise. This finding is consistent with other studies that have shown lower electrode–skin impedance and better sEMG signal quality for electrodes with a larger surface area [29–32]. Despite these advantages, it is important to consider that increasing the size and thereby reducing the inter-electrode distance can lead to an increase in sEMG cross-talk and production complexity. The device presented in this work embeds small-diameter (10 mm) circular electrodes with increased skin contact pressure to overcome this issue. In fact, the studies by Kim et al. [28] and Taji et al. [33] suggest that increasing the clothing pressure on skin can lead to better performance with smaller electrodes. An increased contact area can be achieved by adjusting the tightness of the clothing using arm or leg sleeves [29] or by inserting pads or foams of various thicknesses between the electrodes and the substrate fabrics [34,35]. We considered both the solutions to improve the pressure on the electrodes; the proposed device is an adjustable sleeve made of elastic fabric, and the electrodes are located on foams fixed on the substrate fabric. On the other hand, excessive increases in the pressure of the device on the body district can cause discomfort and pain in the wearer, as reported by An et al. [29]. Therefore, in this study we assessed the wearability and comfort levels of the device by means of the Comfort Rating Scale.

The focus of this manuscript is on the development and validation of a novel wearable device, specifically, a textile leg sleeve with e-textile embedded sensors for sEMG. The proposed textile sleeve aims to provide a non-invasive, comfortable, and efficient tool to capture sEMG signals from specific muscles of the lower limb during gait in order to contribute to improved patient care and monitoring [17,36]. The sEMG sleeve is developed for integration into an ankle–foot orthotic device for patients with ankle dorsiflexion deficits. These supports have positive effects on walking [37]; however, it is of interest to assess the function of the muscles involved (i.e., the anterior tibialis and lateral gastrocnemius), which are the principal muscles responsible for foot dorsiflexion. Analysis of their activity, particularly during walking, can help to to improve biomechanical modeling of walking with the aim of diagnosing and monitoring clinical conditions [38]. This manuscript provides a comprehensive analysis, beginning with the design requirements and specifications of the e-textile band, followed by the fabrication process of the e-textile sleeve. Insight into the validation of the device using statistical methods on a study population of eleven healthy volunteers is presented, and the results of the study are reported and discussed with reference to the present scientific literature. This research aims to demonstrate the feasibility and effectiveness of wearable devices based on e-textiles in medical diagnostics and patient monitoring, which could pave the way for future innovations in healthcare technology.

2. Materials and Methods

2.1. Textile Sleeve for sEMG

The wearable device for acquiring sEMG signals is represented by an adjustable sleeve with removable e-textile electrodes, designed to create an adaptable device for various leg sizes that can be washed and reused. The electrodes for detecting sEMG signals are connected at specific points on the elastic sleeve, which is worn below the knee to capture signals from the anterior tibial and lateral gastrocnemius muscles.

The device is made of elastic fabric with a rectangular structure. The short ends are closed on themselves with an adjustable hook closure. This allows for sleeve adjustment and perfect adherence to various calf sizes. The initial structure is a rectangle measuring 300 mm in length and 240 mm in width, which is then folded and sewn onto itself to achieve a rectangle measuring 300 mm in length and 120 mm in width. Hook closures are then sewn onto the lateral edges of this rectangle. Regarding the outer side of the sleeve, six clips with a diameter of 13 mm are sewn for connection to the acquisition system. These clips are of the same size as those on standard Ag/AgCl electrodes used for biosignal collection, ensuring compatibility with sEMG acquisition systems.

For the inner side, which adheres to the calf, six clips with a diameter of 15 mm are sewn, onto which removable electrodes are applied. The electrode placement follows the Surface Electromyography for the Non-Invasive Assessment of Muscles (SENIAM) guidelines for sEMG detection of the anterior tibial and lateral gastrocnemius muscles [39]. The clips on the inner and outer sides of the sleeve are electrically connected to establish a connection between the electrode and the acquisition system.

Four removable electrodes with a diameter of 10 mm are used for acquiring the electromyography (EMG) signal from the two muscles of interest, along with two reference removable electrodes with a larger diameter (15 mm). The textile electrodes are made from Silver Fiber Knitted Fabric (Suzhou Yu Gao Radiation Protection Technology Co., Ltd., Suzhou, China), a conductive knitted fabric with a resistance of less than 1 Ω per foot in any direction across the fabric, and are wrapped in a soft non-conductive thickness to improve adhesion to the skin, then sewn onto a clip (Figure 1). This design allows for the electrodes to be removed when the acquisition is complete, allowing the sleeve to be washed.

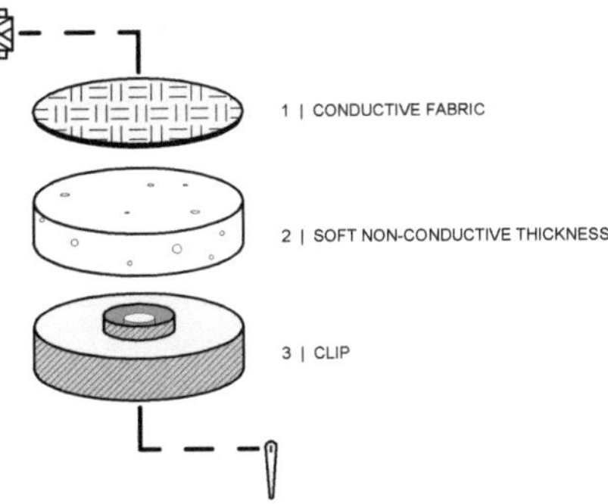

Figure 1. The textile electrode consists of three elements: a textile fabric placed on a non-conductive pad and then sewn to a metal clip.

2.2. Experimental Setup

The study involved eleven healthy volunteers (eight women and three men, age 25.7 ± 1.7, height 168.3 ± 5.9 cm, weight 62.2 ± 5.9 kg) and evaluated the performance of the e-textile sleeve for sEMG in comparison to Ag/AgCl electrodes. The evaluation was performed by comparing the performance of the two types of electrodes in detecting sEMG signal characteristics during Maximal Voluntary Isometric Contraction (MVIC) of the lower limb muscles considered in the analysis. MVICs are commonly used, as they allow for comparison of muscle activity levels across muscles, tasks, and individuals while limiting the inhomogeneity due to intrinsic and extrinsic factors [40].

Each volunteer performed two measurement sessions, one with the textile sleeve and one with the standard electrodes, in random order to avoid ordering effects on the results. The session protocol involved the subject performing five MVICs, each lasting 5 s, with a rest of 30 s between contractions [40]. Specifically, EMG signals from the two target muscles, namely, the anterior tibialis and lateral gastrocnemius, were recorded separately. The acquisition was conducted with the assistance of a timer. When the timer was started, the subject remained in the resting position for 30 s. After the rest period, a start signal was given to the subject, who performed the task and the maximum isometric contraction for 5 s. Subsequently, the subject returned to the resting position and the protocol was repeated in the same manner for five cycles. For acquisition of the EMG signal from the anterior tibialis (Figure 2), the subject was seated and supported the leg, with the ankle joint in dorsiflexion and the foot in eversion without extension of the big toe (sensor locations: tibialis anterior; see the SENIAM guidelines). The MVIC of the lateral gastrocnemius (Figure 3) was obtained with the subject assuming an upright unipodal position, with the knee fully extended (0°) and the ankle in maximum plantar flexion [40]. Participants repeated the protocol with both limbs. Three out eleven participants had minor impairments of the left leg which affected the correct performance of the tasks. The total number of trials analysed was 19.

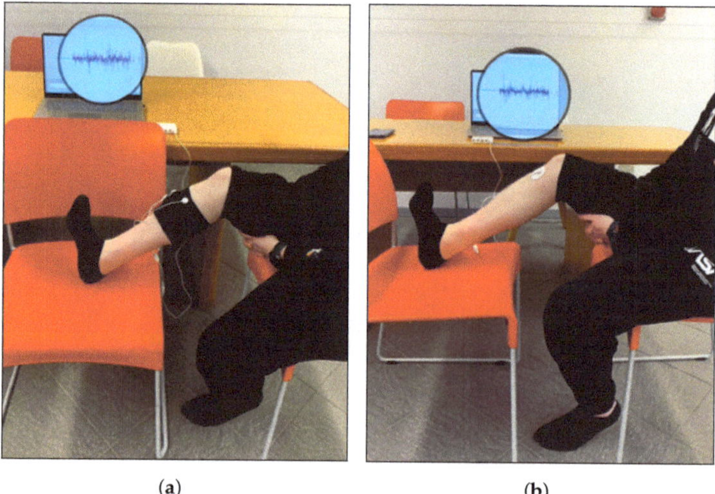

(a)   (b)

Figure 2. Experimental setup for acquisition of anterior tibialis EMG signal by (**a**) band with e-textile electrodes and (**b**) conventional pre-gelled electrodes.

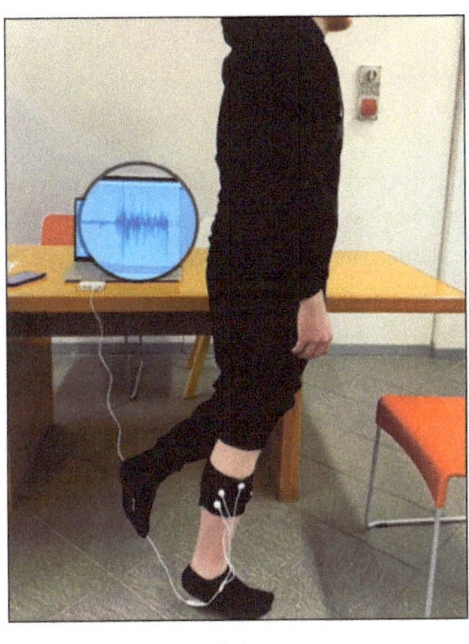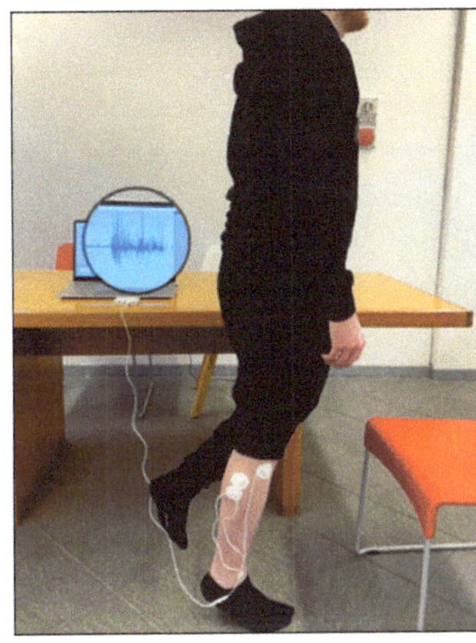

(a)   (b)

Figure 3. Experimental setup for acquisition of lateral gastrocnemius EMG signal by (**a**) band with e-textile electrodes and (**b**) conventional pre-gelled electrodes.

2.3. EMG Acquisition and Feature Extraction

EMG data were collected using the assembled EMG Sensor BITalino (r)evolution BLE system (PLUX Wireless Biosignals S.A., Lisbon, Portugal) designed for real-time physiological data recording together with the OpenSignals (r)evolution software (public build 2022-05-16). The system was used both in connection to the novel wearable sleeve with e-textile electrodes and with standard Ag/AgCl electrodes for surface EMG in order to fix the acquisition protocol and focus on the differences between standard and textile electrodes. Figure 4 schematically illustrates the connection setup employed during the acquisition with the textile sleeve. The plug-in textile electrodes are attached to the internal side of the unit, while the conductive clips on the outer side are connected to the BITalino unit through a standard electrode cable. EMG signals are wirelessly transmitted to a personal computer, where they are stored and processed. Two muscles of particular interest were studied, namely, the anterior tibialis and lateral gastrocnemius. These muscles play a crucial role in human locomotion, being the major muscles responsible for foot dorsiflexion [41–43]. The muscles on the right side were monitored, as the participants were all right-handed. Electrodes were placed on these muscles following SENIAM recommendations [39]. EMG signals were processed in MATLAB R2023a (Mathworks Inc., Natick, MA, USA) to extract quantitative metrics estimating muscle fatigue exerted during the tasks. The parameters describing the EMG signals considered in this analysis are as follows: area of the signal, power of the signal, mean value, and Root Mean Square (RMS). These quantitative metrics were extracted from the rectified signal during the 5 s contraction windows and then averaged over a trial.

2.4. Statistical Analysis

In this study, a statistical analysis was conducted to compare the performances of the e-textile electrodes with the reference pre-gelled Ag/AgCl electrodes for sEMG signal acquisition. To assess whether a statistically significant difference existed between the signals acquired with the two methods, either the paired *t*-test or its non-parametric form, the Wilcoxon Mann–Whitney test, was applied according to the results of the Shapiro–Wilk test for normality of the data.

In addition, the Bland–Altman method was used to assess the agreement between the two measurement techniques. This is the most popular method for measuring agreement between two measurement systems [44]. It plots the differences between the two sets of measurements against their averages, allowing for identification of the bias (the mean difference) and the Limits of Agreement (LoA), calculated as the bias ±1.96 times the standard deviation of the differences [45,46]. If the differences between methods do not have a normal and/or symmetric distribution, then the LoAs are considered to be between the 2.5% and 97.5% percentiles. The Bland–Altman plot additionally helps to visualize any proportional or constant systematic error and to identify patterns or anomalies in the data. Significant statistical errors are said to be present if the LoA of bias does not contain any zero values. Bland and Altman suggested that the agreement between the methods being tested should be accepted if this interval contains a zero value [45]. These statistical tools provided a comprehensive framework for comparing the performance of the two types of electrodes used in this study, facilitating an understanding of their relative effectiveness in EMG signal acquisition. Statistical analyses were performed using R, version 4.0.3 (R Foundation, Vienna, Austria).

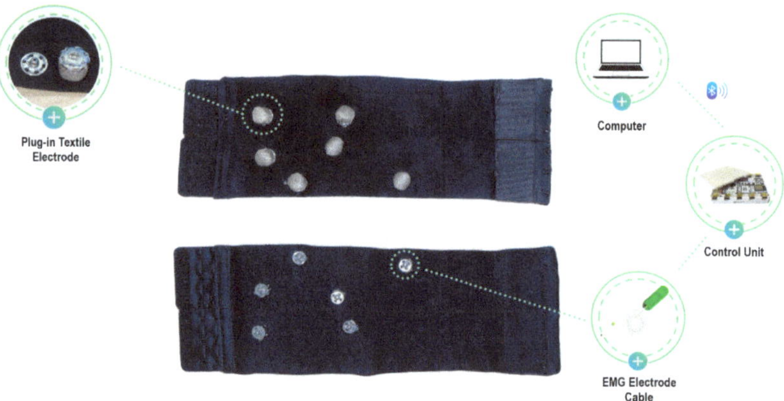

Figure 4. Smart textile sleeve: the upper part shows the internal view of the textile electrodes, while the lower part shows the external view connected to the acquisition system.

3. Results

Analysis of the agreement between the two measurement methods was addressed by performing the statistical tests described above. Table 1 shows all of the metrics describing the data. Descriptive statistics, i.e., the mean values and standard deviation, are provided for each parameter, each muscle, and each measurement system. The results of the statistical tests, i.e., the normality test and *t*-test for paired data, are indicated in terms of *p*-value. The level of significance was set to 0.05, with *p*-values higher than the threshold considered to be not significant (ns in the table). Paired *t*-tests were run in parametric or non-parametric form after obtaining the results of the normality test. The hypothesis of no difference between the systems was tested, with *p*-values lower than the statistical threshold suggesting the rejection of agreement between the systems and *p*-values higher than the level of significance meaning that the differences are not statistically significant and are the result of random measurement errors.

The same table shows the descriptive numerical values derived from the Bland–Altman analysis. The bias represents the average of the differences between the measures calculated by the systems, and is provided with the values of the LoA. In the graphs in Figures 5 and 6, the bias is shown by solid red lines; the dashed red lines represent the corresponding confidence intervals. The LoA values reported in the table are shown in the graphical representations as dashed black lines. They are estimated as the 2.5 and 97.5 percentiles of the differences, as the differences do not have a symmetrical Gaussian distribution.

The results indicate general agreement for all of the analysed muscle parameters. Statistical tests for paired data did not detect a significant difference between the measurements extracted from the sEMG signals of the lower limb muscles during walking with the two measurement methods. The Bland–Altman plot shows substantial agreement between the systems, as both the confidence interval and the LoA of the bias contain zeros for all analysed parameters. The recorded differences in the extracted metrics from the signals acquired from the tibialis anterior muscle are lower than those from the lateral gastrocnemius. The recorded biases are always less than 7% of the mean parameter value for the tibialis anterior muscle and less than 12% for the gastrocnemius lateralis muscle. However, the amplitude ranges are narrower for the latter muscle than for the former. All of the biases have negative values, indicating that the signals recorded with the textile electrodes present higher magnitude and that the extracted parameters are slightly overestimated compared to those recorded with standard electrodes.

Table 1. Descriptive statistics of the datasets and results of the statistical tests and analyses.

Area (mV ms)	Anterior Tibialis		Lateral Gastrocnemius	
	Textile Sleeve	Ag/AgCl Electrodes	Textile Sleeve	Ag/AgCl Electrodes
Mean± SD	198 ± 90	195 ± 76	205 ± 100	190 ± 76
Paired *t*-test		0.314 ns		0.186 ns
Bias		-3.27		-14.9
Lower LoA		-87.5		-92.8
Upper LoA		138		69.2

Power (mV ms)	Anterior Tibialis		Lateral Gastrocnemius	
	Textile Sleeve	Ag/AgCl Electrodes	Textile Sleeve	Ag/AgCl Electrodes
Mean $\pm$ SD	$1.88 \cdot 10^4 \pm 1.48 \cdot 10^4$	$1.76 \cdot 10^4 \pm 1.26 \cdot 10^4$	$2.01 \cdot 10^4 \pm 1.95 \cdot 10^4$	$1.68 \cdot 10^4 \pm 1.28 \cdot 10^4$
Paired *t*-test		0.184 ns		0.134 ns
Bias		$-1.17 \cdot 10^3$		$-1.92 \cdot 10^3$
Lower LoA		$-1.80 \cdot 10^3$		$-1.37 \cdot 10^4$
Upper LoA		$2.45 \cdot 10^3$		$9.00 \cdot 10^3$

Mean (mV)	Anterior Tibialis		Lateral Gastrocnemius	
	Textile Sleeve	Ag/AgCl Electrodes	Textile Sleeve	Ag/AgCl Electrodes
Mean $\pm$ SD	39.7 ± 18.1	39.1 ± 15.3	41.1 ± 20.1	38.1 ± 15.2
Paired *t*-test		0.314 ns		0.185 ns
Bias		-0.649		-3.00
Lower LoA		-17.5		-18.5
Upper LoA		27.6		13.9

RMS (mV)	Anterior Tibialis		Lateral Gastrocnemius	
	Textile Sleeve	Ag/AgCl Electrodes	Textile Sleeve	Ag/AgCl Electrodes
Mean $\pm$ SD	55.5 ± 25.3	54.8 ± 21.5	57 ± 28	52.8 ± 22.3
Paired *t*-test		0.334 ns		0.179 ns
Bias		-0.727		-4.27
Lower LoA		-24.4		-23.9
Upper LoA		40.7		20.3

Tibialis

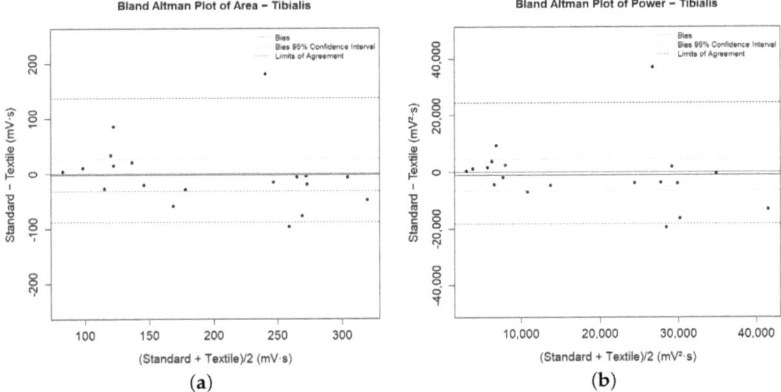

Figure 5. *Cont.*

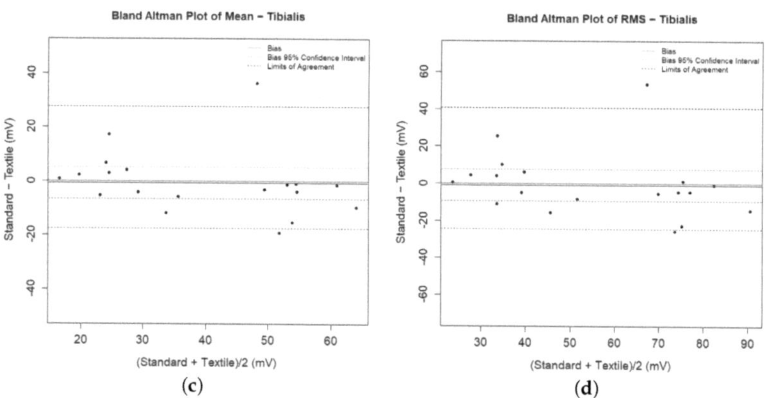

Figure 5. Bland-Altman Plots for parameters related to the Tibialis muscle: (**a**) Area; (**b**) Power; (**c**) Mean and (**d**) RMS parameters.

Gatrocnemius

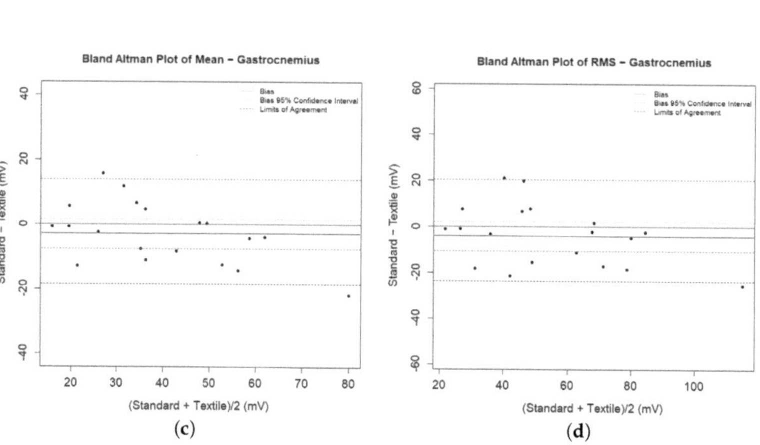

Figure 6. Bland–Altman plots for parameters related to the gastrocnemius muscle: (**a**) area; (**b**) power; (**c**) mean; (**d**) RMS parameters.

Comfort Assessment

In addition to validating technical performance, we conducted a wearability and comfort assessment to evaluate the system's acceptance by end users and identify areas for design improvement.

Evaluating a device's wearability is a multidimensional analysis, as wearable devices can affect the wearer in various ways. When considering the effects of wearing something, it is important to take comfort into account. The level of comfort can be influenced by various factors, including the size and weight of the device, its impact on movement, and any discomfort it may cause.

Knight and Baber (2005) proposed that comfort should be measured across multiple dimensions, including psychological responses such as embarrassment or anxiety in addition to physical factors. To achieve this, they developed the CRS [47].

The CRS offer a convenient tool for evaluating the comfort of wearable devices. The CRS aims to provide a comprehensive assessment of the wearer's comfort status by measuring comfort across six dimensions, as described in Table 2. To rate perceptions of comfort, the scorer indicates their level of agreement with the statements in the 'description' column of Table 2 on a scale from 0 (low) to 20 (high). The scores are interpreted based on the five Wearability Levels (WLs) proposed by Knight et al. (2006) [48], which are obtained by dividing the scales into equal parts. The meaning of each level is shown in Table 3. According to Knight and Baber (2005), the range used in their study was considered large enough to elicit a variety of responses that could be analysed in detail [47]. The participants in our study were invited to complete the CRS and provide their judgments. All subjects were guided by the same interviewer using standardised instructions. Figure 7 shows the scores assigned for each field by the subjects involved in our study.

Table 2. Description of CRS fields and results.

Title	Description	Mean	Std.
Emotion	I am worried about how I look when I wear this device. I feel tense or on edge because I am wearing the device.	3.91	3.08
Attachment	I can feel the device on my body. I can feel the device moving.	4.09	2.70
Harm	The device is causing me some harm. The device is painful to wear.	2.00	2.05
Perceived change	Wearing the device makes me feel physically different. I feel strange wearing the device.	3.27	2.41
Movement	The device affects the way I move. The device inhibits or restricts my movement.	4.09	2.74
Anxiety	I do not feel secure wearing the device.	0.364	0.674

Table 3. Wearability levels for interpretation of CRS scores.

Wearability Level	CRS Score	Outcome
WL1	0–4	System is wearable
WL2	5–8	System is wearable, but changes may be necessary, further investigation is needed
WL3	9–12	System is wearable, but changes are advised, uncomfortable
WL4	13–16	System is not wearable, fatiguing, very uncomfortable
WL5	17–20	System is not wearable, extremely stressful, and potentially harmful

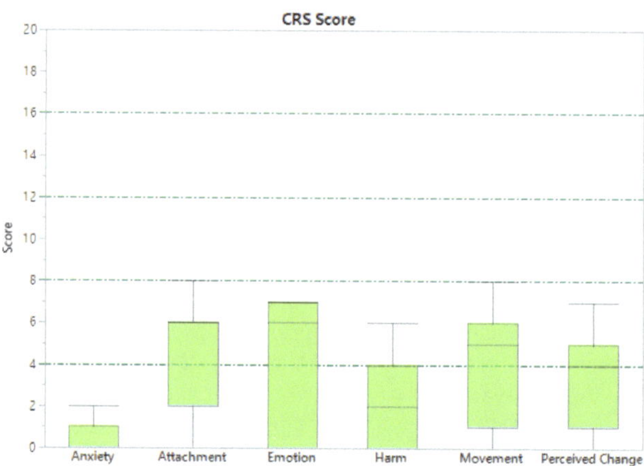

Figure 7. Distribution of CRS scores provided by users.

The evaluation was conducted on a small sample of eleven individuals; therefore, these results should be considered preliminary. Table 2 reports the mean value and standard deviation of the scores. All of the fields were in the range of WL1, indicating that the system is wearable and changes are not required. The highest values are for attachment and movement (4.09); however this could be attributed to the wired electronic device for acquisition rather than the textile sleeve, which the users may have perceived as an embedded part of the device. Lower values were registered for the subjective perception of emotion and perceived change (3.91 and and 3.27, respectively). Subjects reported no discomfort or pain related to the use of the device (harm score 2.00). The lowest value was for anxiety (0.364), with all of the subjects indicating they felt safe using the device. The overall results confirm the high wearability and comfort of the proposed device, despite its being a prototype with low production costs. In order to better identify the device's wearability level and ways to improve it, future analysis will aim to assess its comfort more extensively by testing it on a wider cohort of subjects.

4. Discussion

The current study introduces an e-textile sleeve for sEMG measurement that is designed to fit different calf sizes and to be both washable and reusable. The device integrates circular electrodes in a silver knitted fabric with a diameter of 10 mm.

For the purpose of performance analysis in measuring sEMG, this study compared the performance of the leg sleeve with pre-gelled Ag/AgCl electrodes with eleven healthy volunteers participating in MVIC testing. This methodology is in line with approaches used in recent studies exploring the effectiveness of textile electrodes in sEMG monitoring [25,26].

The results indicate good agreement between the two types of electrodes in measuring characteristic metrics of sEMG signals from the lower leg muscles. Agreement was confirmed by means of paired-data statistical tests and quantitative and qualitative analyses of Bland–Atman plots. All of the sEMG parameters demonstrate correspondence between the two measurement methods we tested. There were no statistically significant differences between the groups of measurements obtained with the two methods, and the average biases were negligible compared to the typical values of the metrics.

As discussed in the Section 1 of this work, the feasibility of sEMG measurements using dry textile electrodes has been analysed in several works that have focused mainly on the type of material used and their effect on the quality of the recorded signal. The results were generally comforting, and indicate that the use of textiles for smart clothes in biosignal detection is a solution that may have an important future.

Generally, the production process of these sensors, usually described in works focused on materials, is very complex and requires collaboration from various industries that are not accustomed to collaborating. This represents a significant barrier to large-scale production [49]. Therefore, in this study we aimed to test a commercial, readily available, and affordable fabric. The electrodes were manually created in order to demonstrate the feasibility of sEMG measurements with minimal resources and a simple production process. It seems clear from the scientific literature that the limitations are non-technical and that the acquired signal quality is satisfactory. However, a common limitation is that most studies have focused on developing new technologies and only tested the resulting devices on a single healthy subject. Only a few studies have considered larger study populations to validate the device and move towards broader diffusion of the technology. With this in mind, we expanded our analysis to eleven subjects and prioritised evaluating the wearability and comfort of the device for users. When implementing new technologies, it is important to consider the technical performance, impact on users, and applicability in real-world contexts. Our evaluation conducted through the CRS yielded positive results, despite the prototype nature and low production costs of the device. These aspects are significant, particularly considering that smart clothes for biosignal detection may replace standard methods in out-of-lab measurements where comfort and ease of use are necessary. The results of our wearability and comfort analysis indicate that the users did not experience pain from the device and suggested no significant modifications. To the best of our knowledge, no other reference studies have quantitatively investigated these aspects on wearable prototype devices for recording sEMG. Regardless, the collected answers fall within the ranges indicated by the questionnaire developers as indicating the two highest levels of wearability.

Although larger than the samples usually analysed in similar studies, the sample size in this study was still relatively small and the subjects fell within a narrow age range, which may not represent wider variations in the general population. For future research, it would be worth exploring the use of these textile sleeves in a larger and more diverse sample. As a further limitation, this study focused on four features extracted from the analysis of sEMG signals in time domain, which is a useful approach for characterising the intensity and duration of muscle activations. In future research, further interest may concern the analysis of other parameters, such as frequency content, signal-to-noise ratio, or measures of muscular fatigue. In addition, it would be interesting to evaluate the effectiveness of these electrodes in dynamic applications and under exercise conditions, which would be a suitable method for analysing muscle activation during walking or dynamic tasks. In particular, a more comfortable solution than classical methods involving adhesive electrodes and heavy instrumentation could favour remote monitoring applications, which, together with telemedicine, are growing in popularity today.

Author Contributions: Conceptualization, G.D., A.C. and F.A.; data curation, A.B., F.A. and G.P.; methodology, A.C., F.A. and V.M.; software, F.A., A.C. and A.B.; formal analysis, F.A. and A.C.; investigation, A.C., F.A. and V.M.; resources, G.T., G.P. and V.M.; writing—original draft preparation, A.C. and F.A.; writing—review and editing, G.D.; visualization, A.C., F.A., G.P. and V.M.; supervision, G.D.; project administration, G.D. and G.T.; funding acquisition, G.D.A., G.P. and G.T. All authors have read and agreed to the published version of the manuscript.

Funding: This research was partially funded by APTIS—the Advanced Personalized Three-dimensional printed Sensorized orthosis project, granted by the Italian Ministry of Economic Development.

Institutional Review Board Statement: The study was conducted in accordance with the Declaration of Helsinki as part of the APTIS—Advanced Personalized Three-dimensional printed Senorized orthosis project, approved by Italian Minister of Economic Development (F/190055/02/X44 D.M. n.2510 of 25 June 2020).

Informed Consent Statement: Informed consent was obtained from all subjects involved in the study.

Data Availability Statement: Data available on request due to privacy and legal restrictions.

Conflicts of Interest: The authors declare no conflicts of interest. The funding agency had no role in the design of the study; in the collection, analyses, or interpretation of data; in the writing of the manuscript; or in the decision to publish the results'.

Abbreviations

The following abbreviations are used in this manuscript:

BMI	Body Mass Index
CCSM	Conductive Composite Silicone Material
CRS	Comfort Rating Scale
EMG	Electromyography
LoA	Limits of Agreement
MVIC	Maximal Voluntary Isometric Contraction
RMS	Root Mean Square
sEMG	Surface Electromyography
SENIAM	Surface Electromyography for Non-Invasive Assessment of Muscles
WL	Wearability Level
SNR	Signal-to-Noise Ratio

References

1. Skazalski, C.; Whiteley, R.; Hansen, C.; Bahr, R. A valid and reliable method to measure jump-specific training and competition load in elite volleyball players. *Scand. J. Med. Sci. Sports* **2018**, *28*, 1578–1585. [CrossRef] [PubMed]
2. Chen, S.T.; Lin, S.S.; Lan, C.W.; Hsu, H.Y. Design and Development of a Wearable Device for Heat Stroke Detection. *Sensors* **2017**, *18*, 17. [CrossRef] [PubMed]
3. Engin, M.; Demirel, A.; Engin, E.Z.; Fedakar, M. Recent developments and trends in biomedical sensors. *Measurement* **2005**, *37*, 173–188. [CrossRef]
4. Amitrano, F.; Coccia, A.; Donisi, L.; Pagano, G.; Cesarelli, G.; D'Addio, G. Gait Analysis using Wearable E-Textile Sock: An Experimental Study of Test-Retest Reliability. In Proceedings of the 2021 IEEE International Symposium on Medical Measurements and Applications (MeMeA), Lausanne, Switzerland, 23–25 June 2021; pp. 1–6. [CrossRef]
5. Smith, A.A.; Li, R.; Tse, Z.T.H. Reshaping healthcare with wearable biosensors. *Sci. Rep.* **2023**, *13*, 4998. [CrossRef] [PubMed]
6. Dooley, E.E.; Golaszewski, N.M.; Bartholomew, J.B. Estimating Accuracy at Exercise Intensities: A Comparative Study of Self-Monitoring Heart Rate and Physical Activity Wearable Devices. *JMIR mHealth uHealth* **2017**, *5*, e34. [CrossRef] [PubMed]
7. Coccia, A.; Amitrano, F.; Donisi, L.; Cesarelli, G.; Pagano, G.; Cesarelli, M.; D'Addio, G. Design and validation of an e-textile-based wearable system for remote health monitoring. *Acta Imeko* **2021**, *10*, 220. [CrossRef]
8. DeRossi, D.; Lymberis, A. Guest Editorial New Generation of Smart Wearable Health Systems and Applications. *IEEE Trans. Inf. Technol. Biomed.* **2005**, *9*, 293–294. [CrossRef] [PubMed]
9. Karim, N.; Afroj, S.; Tan, S.; He, P.; Fernando, A.; Carr, C.; Novoselov, K.S. Scalable Production of Graphene-Based Wearable E-Textiles. *ACS Nano* **2017**, *11*, 12266–12275. [CrossRef] [PubMed]
10. Chen, G.; Zhao, X.; Andalib, S.; Xu, J.; Zhou, Y.; Tat, T.; Lin, K.; Chen, J. Discovering giant magnetoelasticity in soft matter for electronic textiles. *Matter* **2021**, *4*, 3725–3740. [CrossRef] [PubMed]
11. Fang, Y.; Chen, G.; Bick, M.; Chen, J. Smart textiles for personalized thermoregulation. *Chem. Soc. Rev.* **2021**, *50*, 9357–9374. [CrossRef] [PubMed]
12. Iqbal, S.M.A.; Mahgoub, I.; Du, E.; Leavitt, M.A.; Asghar, W. Advances in healthcare wearable devices. *NPJ Flex. Electron.* **2021**, *5*, 9. [CrossRef]
13. Cesarelli, G.; Donisi, L.; Coccia, A.; Amitrano, F.; D'Addio, G.; Ricciardi, C. The E-Textile for Biomedical Applications: A Systematic Review of Literature. *Diagnostics* **2021**, *11*, 2263. [CrossRef] [PubMed]
14. Kotov-Smolenskiy, A.M.; Khizhnikova, A.E.; Klochkov, A.S.; Suponeva, N.A.; Piradov, M.A. Surface EMG: Applicability in the Motion Analysis and Opportunities for Practical Rehabilitation. *Hum. Physiol.* **2021**, *47*, 237–247. [CrossRef]
15. Donisi, L.; Capodaglio, E.; Pagano, G.; Amitrano, F.; Cesarelli, M.; Panigazzi, M.; D'Addio, G. Feasibility of Tree-based Machine Learning algorithms fed with surface electromyographic features to discriminate risk classes according to NIOSH. In Proceedings of the 2022 IEEE International Symposium on Medical Measurements and Applications (MeMeA), Messina, Italy, 22–24 June 2022; pp. 1–6. [CrossRef]
16. Coccia, A.; Capodaglio, E.M.; Amitrano, F.; Gabba, V.; Panigazzi, M.; Pagano, G.; D'Addio, G. Biomechanical Effects of Using a Passive Exoskeleton for the Upper Limb in Industrial Manufacturing Activities: A Pilot Study. *Sensors* **2024**, *24*, 1445. [CrossRef] [PubMed]
17. Merletti, R.; Aventaggiato, M.; Botter, A.; Holobar, A.; Marateb, H.; Vieira, T.M. Advances in Surface EMG: Recent Progress in Detection and Processing Techniques. *Crit. Rev. Biomed. Eng.* **2010**, *38*, 305–345. [CrossRef]

18. Cheng, L.; Li, J.; Guo, A.; Zhang, J. Recent advances in flexible noninvasive electrodes for surface electromyography acquisition. *NPJ Flex. Electron.* **2023**, *7*, 39. [CrossRef]
19. Ohiri, K.A.; Pyles, C.O.; Hamilton, L.H.; Baker, M.M.; McGuire, M.T.; Nguyen, E.Q.; Osborn, L.E.; Rossick, K.M.; McDowell, E.G.; Strohsnitter, L.M.; et al. E-textile based modular sEMG suit for large area level of effort analysis. *Sci. Rep.* **2022**, *12*, 9650. [CrossRef] [PubMed]
20. Goncu-Berk, G.; Tuna, B.G. The Effect of Sleeve Pattern and Fit on E-Textile Electromyography (EMG) Electrode Performance in Smart Clothing Design. *Sensors* **2021**, *21*, 5621. [CrossRef] [PubMed]
21. Alizadeh-Meghrazi, M.; Sidhu, G.; Jain, S.; Stone, M.; Eskandarian, L.; Toossi, A.; Popovic, M.R. A Mass-Producible Washable Smart Garment with Embedded Textile EMG Electrodes for Control of Myoelectric Prostheses: A Pilot Study. *Sensors* **2022**, *22*, 666. [CrossRef]
22. Ozturk, O.; Yapici, M.K. Muscular Activity Monitoring and Surface Electromyography (sEMG) with Graphene Textiles. In Proceedings of the 2019 IEEE SENSORS, Montreal, QC, Canada, 27–30 October 2019; pp. 1–4. [CrossRef]
23. Ozturk, O.; Yapici, M.K. Surface Electromyography With Wearable Graphene Textiles. *IEEE Sens. J.* **2021**, *21*, 14397–14406. [CrossRef]
24. Awan, F.; He, Y.; Le, L.; Tran, L.L.; Han, H.D.; Nguyen, L.P. ElectroMyography Acquisition System Using Graphene-based e-Textiles. In Proceedings of the 2019 International Symposium on Electrical and Electronics Engineering (ISEE), Ho Chi Minh, Vietnam, 10–12 October 2019; pp. 59–62. [CrossRef]
25. Le, K.; Soltanian, S.; Narayana, H.; Servati, A.; Servati, P.; Ko, F. Roll-to-roll fabrication of silver/silver chloride coated yarns for dry electrodes and applications in biosignal monitoring. *Sci. Rep.* **2023**, *13*, 21182. [CrossRef] [PubMed]
26. Alizadeh-Meghrazi, M.; Ying, B.; Schlums, A.; Lam, E.; Eskandarian, L.; Abbas, F.; Sidhu, G.; Mahnam, A.; Moineau, B.; Popovic, M.R. Evaluation of dry textile electrodes for long-term electrocardiographic monitoring. *BioMed. Eng. Online* **2021**, *20*, 68. [CrossRef] [PubMed]
27. Choudhry, N.A.; Rasheed, A.; Ahmad, S.; Arnold, L.; Wang, L. Design, Development and Characterization of Textile Stitch-Based Piezoresistive Sensors for Wearable Monitoring. *IEEE Sens. J.* **2020**, *20*, 10485–10494. [CrossRef]
28. Kim, S.; Lee, S.; Jeong, W. EMG Measurement with Textile-Based Electrodes in Different Electrode Sizes and Clothing Pressures for Smart Clothing Design Optimization. *Polymers* **2020**, *12*, 2406. [CrossRef] [PubMed]
29. An, X.; Tangsirinaruenart, O.; Stylios, G.K. Investigating the performance of dry textile electrodes for wearable end-uses. *J. Text. Inst.* **2019**, *110*, 151–158. [CrossRef]
30. Marozas, V.; Petrenas, A.; Daukantas, S.; Lukosevicius, A. A comparison of conductive textile-based and silver/silver chloride gel electrodes in exercise electrocardiogram recordings. *J. Electrocardiol.* **2011**, *44*, 189–194. [CrossRef]
31. Li, G.; Wang, S.; Duan, Y.Y. Towards gel-free electrodes: A systematic study of electrode-skin impedance. *Sens. Actuators B Chem.* **2017**, *241*, 1244–1255. [CrossRef]
32. Puurtinen, M.M.; Komulainen, S.M.; Kauppinen, P.K.; Malmivuo, J.A.V.; Hyttinen, J.A.K. Measurement of noise and impedance of dry and wet textile electrodes, and textile electrodes with hydrogel. In Proceedings of the 2006 International Conference of the IEEE Engineering in Medicine and Biology Society, New York, NY, USA, 30 August–3 September 2006; pp. 6012–6015. [CrossRef]
33. Taji, B.; Shirmohammadi, S.; Groza, V. Measuring skin-electrode impedance variation of conductive textile electrodes under pressure. In Proceedings of the 2014 IEEE International Instrumentation and Measurement Technology Conference (I2MTC), 12–15 May 2014; pp. 1083–1088, ISSN 1091-5281. [CrossRef]
34. Pani, D.; Achilli, A.; Spanu, A.; Bonfiglio, A.; Gazzoni, M.; Botter, A. Validation of Polymer-Based Screen-Printed Textile Electrodes for Surface EMG Detection. *IEEE Trans. Neural Syst. Rehabil. Eng.* **2019**, *27*, 1370–1377. [CrossRef] [PubMed]
35. Cömert, A.; Honkala, M.; Hyttinen, J. Effect of pressure and padding on motion artifact of textile electrodes. *BioMed. Eng. Online* **2013**, *12*, 26. [CrossRef] [PubMed]
36. Farago, E.; MacIsaac, D.; Suk, M.; Chan, A.D.C. A Review of Techniques for Surface Electromyography Signal Quality Analysis. *IEEE Rev. Biomed. Eng.* **2023**, *16*, 472–486. [CrossRef] [PubMed]
37. Amitrano, F.; Coccia, A.; Pagano, G.; Biancardi, A.; Tombolini, G.; D'Addio, G. Effects of Ankle-Foot Orthosis on Balance of Foot Drop Patients. In *Studies in Health Technology and Informatics*; Hägglund, M., Blusi, M., Bonacina, S., Nilsson, L., Cort Madsen, I., Pelayo, S., Moen, A., Benis, A., Lindsköld, L., Gallos, P., Eds.; IOS Press: Amsterdam, The Netherlands, 2023. [CrossRef]
38. Coccia, A.; Amitrano, F.; Pagano, G.; Dileo, L.; Marsico, V.; Tombolini, G.; D'Addio, G. Biomechanical modelling for quantitative assessment of gait kinematics in drop foot patients with ankle foot orthosis. In Proceedings of the 2022 IEEE International Symposium on Medical Measurements and Applications (MeMeA), Messina, Italy, 22–24 June 2022; pp. 1–6. [CrossRef]
39. Hermens, H.J.; Freriks, B.; Merletti, R.; Stegeman, D.; Joleen, B.; Günter, R.; Disselhorst-Klug, C.; Hägg, G. European recommendations for surface electromyography. *Roessingh Res. Dev.* **1999**, *8*, 13–54.
40. Schwartz, C.; Wang, F.C.; Forthomme, B.; Denoël, V.; Brüls, O.; Croisier, J.L. Normalizing gastrocnemius muscle EMG signal: An optimal set of maximum voluntary isometric contraction tests for young adults considering reproducibility. *Gait Posture* **2020**, *82*, 196–202. [CrossRef] [PubMed]
41. Kimata, K.; Otsuka, S.; Yokota, H.; Shan, X.; Hatayama, N.; Naito, M. Relationship between attachment site of tibialis anterior muscle and shape of tibia: Anatomical study of cadavers. *J. Foot Ankle Res.* **2022**, *15*, 54. [CrossRef] [PubMed]
42. Yong, J.R.; Dembia, C.L.; Silder, A.; Jackson, R.W.; Fredericson, M.; Delp, S.L. Foot strike pattern during running alters muscle-tendon dynamics of the gastrocnemius and the soleus. *Sci. Rep.* **2020**, *10*, 5872. [CrossRef]

43. Ferris, R.M.; Hawkins, D.A. Gastrocnemius and Soleus Muscle Contributions to Ankle Plantar Flexion Torque as a Function of Ankle and Knee Angle. *Sports Inj. Med.* **2020**, *4*, 63.
44. Zaki, R.; Bulgiba, A.; Ismail, R.; Ismail, N.A. Statistical Methods Used to Test for Agreement of Medical Instruments Measuring Continuous Variables in Method Comparison Studies: A Systematic Review. *PLoS ONE* **2012**, *7*, e37908. [CrossRef] [PubMed]
45. Altman, D.G.; Bland, J.M. Measurement in Medicine: The Analysis of Method Comparison Studies. *Statistician* **1983**, *32*, 307. [CrossRef]
46. Bland, J.M.; Altman, D.G. Measuring agreement in method comparison studies. *Stat. Methods Med. Res.* **1999**, *8*, 135–160. [CrossRef] [PubMed]
47. Knight, J.F.; Baber, C. A Tool to Assess the Comfort of Wearable Computers. *Hum. Factors J. Hum. Factors Ergon. Soc.* **2005**, *47*, 77–91. [CrossRef] [PubMed]
48. Knight, J.F.; Deen-Williams, D.; Arvanitis, T.N.; Baber, C.; Sotiriou, S.; Anastopoulou, S.; Gargalakos, M. Assessing the Wearability of Wearable Computers. In Proceedings of the 2006 10th IEEE International Symposium on Wearable Computers, Montreux, Switzerland, 11–14 October 2006; pp. 75–82, ISSN 2376-8541. [CrossRef]
49. Dunne, L. Smart Clothing in Practice: Key Design Barriers to Commercialization. *Fash. Pract.* **2010**, *2*, 41–65. [CrossRef]

sensors

Article

Concept Drift Mitigation in Low-Cost Air Quality Monitoring Networks

Gerardo D'Elia [1,2,*], Matteo Ferro [3], Paolo Sommella [2], Sergio Ferlito [1], Saverio De Vito [1] and Girolamo Di Francia [1]

[1] TERIN-SSI-EDS Laboratory, ENEA CR-Portici, P. le E. Fermi 1, 80055 Portici, Italy; sergio.ferlito@enea.it (S.F.); saverio.devito@enea.it (S.D.V.); girolamo.difrancia@enea.it (G.D.F.)

[2] Department of Industrial Engineering (DIIn), University of Salerno, Via Giovanni Paolo II, 132, 84084 Fisciano, Italy; psommella@unisa.it

[3] Hippocratica Imaging S.r.l., Via Giulio Pastore, 32, 84131 Salerno, Italy; mferro@hippocratica-imaging.it

* Correspondence: gerardo.delia@enea.it; Tel.: +39-081-7723-616

Abstract: Future air quality monitoring networks will integrate fleets of low-cost gas and particulate matter sensors that are calibrated using machine learning techniques. Unfortunately, it is well known that concept drift is one of the primary causes of data quality loss in machine learning application operational scenarios. The present study focuses on addressing the calibration model update of low-cost NO_2 sensors once they are triggered by a concept drift detector. It also defines which data are the most appropriate to use in the model updating process to gain compliance with the relative expanded uncertainty (REU) limits established by the European Directive. As the examined methodologies, the general/global and the importance weighting calibration models were applied for concept drift effects mitigation. Overall, for all the devices under test, the experimental results show the inadequacy of both models when performed independently. On the other hand, the results from the application of both models through a stacking ensemble strategy were able to extend the temporal validity of the used calibration model by three weeks at least for all the sensor devices under test. Thus, the usefulness of the whole information content gathered throughout the original co-location process was maximized.

Keywords: air quality network; concept drift; general calibration; global calibration; importance weighting; relative expanded uncertainty; calibration model update

Citation: D'Elia, G.; Ferro, M.; Sommella, P.; Ferlito, S.; De Vito, S.; Di Francia, G. Concept Drift Mitigation in Low-Cost Air Quality Monitoring Networks. *Sensors* **2024**, *24*, 2786. https://doi.org/10.3390/s24092786

Academic Editor: Erwin Peiner

Received: 5 March 2024
Revised: 16 April 2024
Accepted: 25 April 2024
Published: 27 April 2024

1. Introduction

Scientific and technical communities have long recognized the potential of exploiting low-cost sensors in modern air quality monitoring networks [1]. The main derived advantage is the possibility of deploying a significant number of sensing nodes at reasonable costs while densifying the sparse regulatory monitoring network. This in turn allows for long-sought high-density assessments of air pollution phenomena known to exhibit inherent high spatio-temporal variance [2]. The final outcomes include enhanced awareness, focused, efficient, and shared remediation policies, improving urban planning and reducing environmental and health inequalities. However, the large-scale application of such a technology is still hindered by inherent data quality issues. Correcting sensors' sensitivity to non-target gases and environmental variables is mandatory to achieve the needed accuracy [3]. Moreover, the deterioration in the quality of the measurements occurring during long-term operational deployment decreases the data quality perception for these devices. Often, before any real-life deployment, a costly co-location time at certified reference stations is needed to obtain sufficient data to derive sensor calibration functions. The calibration obtained in the controlled environment failed to provide the requested robustness. Unfortunately, sensor fabrication variance forces this process to be repeated for each and any sensor by determining overall high costs and logistic bottlenecks. Moreover, every ab initio-derived calibration function will fail in the long term due to changes in sensor characteristics like poisoning or ageing [4].

Machine learning (ML) algorithms are recognized as a very convenient methodology to derive calibration functions [5]. Unfortunately, due to the operational scenario in which the sensor is forced to operate, the well-known independent and identically distributed samples hypothesis (*i.i.d. assumption*) is certainly violated. Indeed, the calibrated devices will have to operate in different conditions from those learned during the short-term co-location phase. Factors like seasonal environmental variations, anthropogenic forces, micrometeorology-related fluctuations in pollutant concentration levels, sensor aging, and so on will determine significant changes in the ongoing operative conditions [6]. As such, the data distribution model is a paramount issue [7]. The change in the type (and its location and spread parameters) of distribution will negatively impact the accuracy of the model output [8]. When this condition occurs, it defines the presence of a *concept drift* [9,10]. Concept drift is in fact a condition in which the statistical distribution of the target and input variables of a machine learning model has changed or continuously changed over time. The analysis of concept drift occurrence has given rise to the need for machine learning algorithms that are capable to learn in so-called "non-stationary environments" [6].

How to efficiently update the calibration model, after a concept drift condition has been recognized, is still an open question. Remote calibration and hierarchical network management are two of the most often proposed approaches.

With the term *Remote Calibration*, researchers indicate continuous or repeated recalibration schemes relying on a reference data stream from remote stations exploiting particular conditions and hypotheses. After a seminal paper by Tsujita et al. in 2005, different teams involved in air quality monitoring using low-cost sensor technologies have recently suggested that data coming from remote regulatory grade stations should be exploited, in low-spatial-variance conditions, to correct the effects of a drifting calibration in low-cost air quality networks [11]. It is actually reasonable to suppose a uniform spatial distribution of the pollutant concentration in a restricted geographical area at certain times when local emissions are negligible [12,13]. Based on this hypothesis and, therefore, thanks to data from a nearby regulatory station, in the case of uniformly low pollutant concentrations (e.g., during early morning hours), it could be possible to correct the baseline response of targeted gas sensors [14]. However, the generalization of such encouraging method findings has to be carefully evaluated since the spatial and temporal variance of the recorded phenomena may be significantly different depending on the geographic context negatively affecting the performance of remote calibration approaches [4].

Hierarchical networks, including "golden" reference stations (proxies), on top have also been proposed as a solution to the problem of the continuous calibration of low-cost air quality monitoring systems (LCAQMSs). Usually, regulatory instrumentations are on top in hierarchy, while some well-calibrated low-cost nodes are also entered into intermediate positions. These latter nodes which are limited in number can be regularly recalibrated, offering a quasi-gold reference data stream which can be used to continuously recalibrate other low-cost nodes. This can be achieved, for example, by regularly collocating these nodes with the latter by relocating the proxy nodes for short calibration-oriented periods. Alternatively, by carefully designing node deployment, proxy measurements can be used to remotely adjust the calibration of the low-cost nodes. One notable example is the Breathe London project, in which quasi-golden node data streams are exploited remotely [15]. Another way to practically implement this is by matching the mean and standard deviations between the measures of the proxy itself and low-cost sensor data over a fixed time window [16]. Based on this methodology, the Aeroqual Inc. company (Auckland, New Zealand) has developed a virtual calibration service for its own products (www.aeroqual.com (accessed on 23 April 2024)).

At present, continuous calibration strategies exploiting remote data are feasible and cost-effective procedures for continuous recalibration, keeping the required accuracy level for long-term deployments. Hourly averaged pollution data are, in fact, frequently published by regulatory institutions for the convenience of citizens and other users. This means that such data are reachable for everyone through API REST services. One possible implementation to be considered is MOMA (the acronym for MOment MAtching), which is the commercial name of the virtual calibration service distributed by the Aeroqual Inc. company. This service updates the calibration at fixed time intervals at an early development stage [17]. To apply a similar workflow, we took such a service into consideration and explored the possibility of triggering calibration by updating using a concept drift detector. This methodology will allow for the performance of recalibration only when specifically needed, saving precious computational resources and avoiding unwanted overfitting errors.

In order to select the suitable strategies to mitigate the effects of the concept drift, the following three main steps have been carried out: (i) understanding which data should be used to update the model; (ii) evaluation and validation of the general calibration model and importance weighting calibration model in the presence of a concept drift; (iii) attempting to improve the quality of pollution estimations data with the stacking ensemble technique in air quality network scenarios.

2. Materials and Methods

2.1. Experimental Campaign

The LCAQMS employed is based on an electrochemical sensors array using Alphasense (Braintree, UK) A4 class sensor units, respectively, targeted to carbon monoxide (CO-A4), nitrogen dioxide (NO2-A43F), and ozone (O3-A431). Relative humidity (RH) and temperature (T) sensors complete the sensing array. The electrochemical gas sensors array is mounted on an analog front-end provided by Alphasense that outputs the signals related to the concentration of monitored air gases. The working electrode (WE) and the auxiliary electrode (AE) signals of each type of sensor are acquired and converted by an ST Microelectronics (Geneva, Switzerland) Nucleo LK432KC board. The node measurements are packeted in a proper JSON and sent via low-energy Bluetooth to a Raspberry Pi (Galles, UK) 3B+ Model. A Wi-Fi router shares the WAN connectivity services with the Raspberry Pi, allowing the data to reach the MONGO DB database server for storage.

Four nodes, named AQ6, AQ8, AQ11, and AQ12, have been deployed against a mobile regulatory air quality monitoring station for a two-months-long winter co-location campaign completed in the city of Portici (Naples, Italy) from 2 January to 2 March 2020. During the campaign, the AQ8 sensor broke, and therefore it will not be considered.

This research study is focused only on nitrogen dioxide NO_2, which is considered one of the most dangerous pollutants, meaning that the value of the NO_2 concentration is our target variable Y. In fact, the experimental data stored on the server were averaged on an hourly basis and used to calibrate the sensor nodes with an ad hoc multilinear regression model, characterized by six input features X: relative humidity (RH), temperature (Temp), AE and WE signals of NO_2 sensor, AE and WE signals of O_3 sensor [18].

2.2. Concept Drift Detection Trigger

Let (X, Y) be couples of input-target samples and $P_t(X, Y)$ be its joint probability distribution at the instant time t; a concept drift results if $P_t(X, Y) \neq P_{t+1}(X, Y)$. This formula can be separated by means of Bayes' law in the following relation:

$$P_t(X)P_t(Y|X) \neq P_{t+1}(X)P_{t+1}(Y|X) \tag{1}$$

A spatial or temporal change in the characteristics of the features X, but also a change in the relationship linking the inputs to the target variable, is the cause of the decline in the performance of the model used. Therefore, the greater the difference, the greater the error that has been committed. It becomes imperative to be able to identify when a similar event is occurring. Statistical tools are one of the most widely used and well-established methods for accomplishing these tasks. A comprehensive guide to the concept drift detection algorithms is available in [10].

An immediate and simple implementation of a concept drift detection block can be implemented (both on the node hardware and on the back-end server-side software) using the two-sample Kolmogorov–Smirnov test (TSKS test), which is able to evaluate whether two samples come from the same distribution without making any assumption regarding the type of distribution. In the case of low-cost air quality sensors calibrated using ML techniques, the TSKS test can be used to determine whether both the distributions of the input features X and the distributions of the output variable Y during the training phase remain the same also throughout the test phase. Of course, concept drift has occurred, and a recalibration request is triggered if the distributions turn out to be different. Since X-values are always available during the LCAQMS operational scenario, it is possible to detect concept drift even without a reference instrument by comparing the X-distributions obtained during co-location with those obtained during the operational phase (test phase). The same Is true for the target variable. The value in co-location is compared with the value predicted by the model in the test phase. Furthermore, if it is possible to access reference station data from a remote calibration perspective, the same approach is obvious for the output variable. According to the heuristic rule described in [19], a recalibration request is only triggered when the maximum difference between the cumulative density functions of the training and test distributions of the temperature and/or NO_2 concentration exceeds the threshold of 0.3.

By applying this approach to the experimental data collected, divided into eight batches (T1 to T8), each one week long, the presence of two concept drifts has been detected. The first is just before the end of the T4 slot time, while the second drift event happens in T8, but it is not considered in such analysis. A graphic representation of the concept drift existing in the dataset under analysis is shown in Figure 1. The green dots represent the trend of the "concept" that characterizes the target variable and an input variable (temperature, in this case) during the co-location period and, therefore, during the training process. The red dots represent the trend of the same variables after the concept drift detection; therefore, the "new concept" characterizes the test set after only one month.

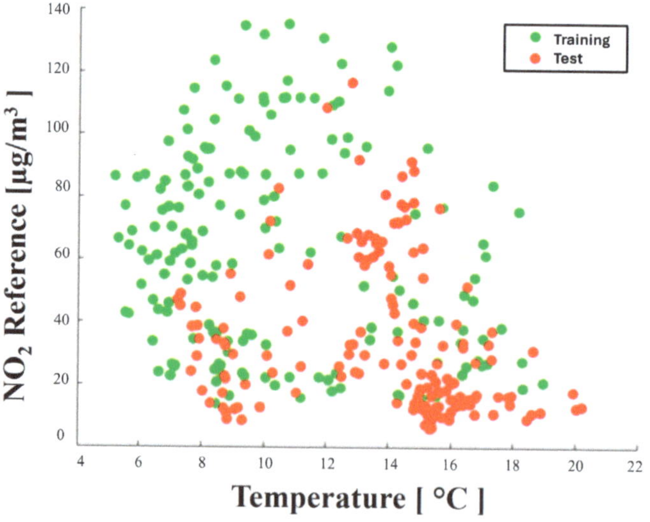

Figure 1. Evidence of concept drift highlighted on the co-location samples of the target and input variables.

2.3. Calibration Model Update Data Selection

To tackle the problem of concept drift handling, data were analyzed to find those containing the useful information to update the calibration model.

Let t_0 be the time at which the concept drift detector block implemented in the air quality monitoring network sends an alarm signal to the network administrator, meaning that the current calibration model is obsolete and needs to be updated. Now the first question is to understand which data should be used for the update. Three possible reference datasets can be used within this scope: the dataset preceding the concept drift alert, called "Last" and identified with the T3 time slot (Figure 2); the dataset following the concept drift alert, called "Next" (T5 in Figure 2); or part of both called "Mixed" that contains the time instant t_0 and is reported as the T4 time slot (Figure 2) [20].

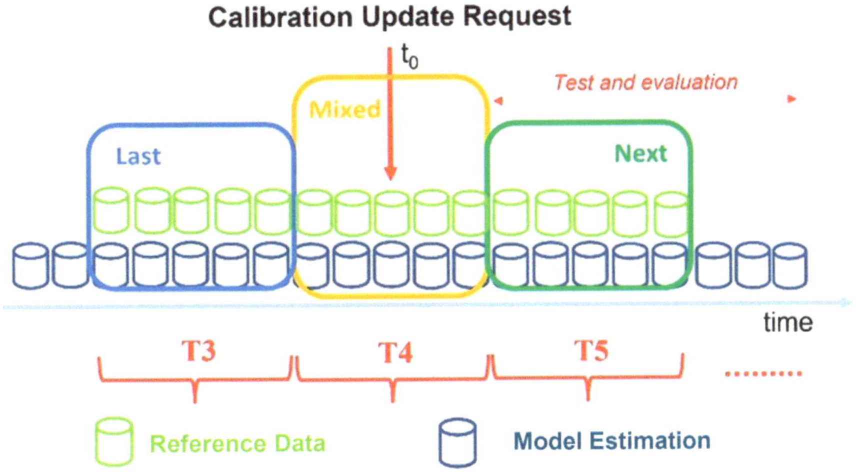

Figure 2. Data selection for calibration update when a recalibration request arrives from the concept drift detector.

Our attention will focus on trying to restore the compliant data quality. The following tests are fulfilled by updating the model using on-hand reference data (in the proper time windows) and then evaluating the performance of the new model in T5–T6–T7.

2.4. Calibration Update without Reference Data

Bringing back the data quality at the regulatory level after the detection of a concept drift event remains the final goal of our study. This means the mitigation of the effects related to the concept drift. To make the most out of the information content of the co-location data, two new calibration models beyond the ad hoc one were extracted from it. The following techniques were explored: the general (also named global or, less often, universal) calibration model, and the importance weighting calibration model.

2.4.1. General Calibration Model

This methodology has been introduced in recent years as an attempt to reduce the calibration costs [21,22]. It consists of identifying and applying a general calibration model to all the nodes involved in the network, thus avoiding the need for additional ad hoc calibrations. As known, the raw output of most electrochemical gas sensors consists of the voltage (mV) measured at the working and auxiliary electrodes that is representative of the measured gas concentration. Let us consider n co-located sensors, and then the median between these n values is calculated. The same procedure is applied for temperature and humidity values. The set of medians of all the single quantities (input variables) involved in model creation accounts for the training set which the global model is trained on. Such a procedure is depicted in Figure 3.

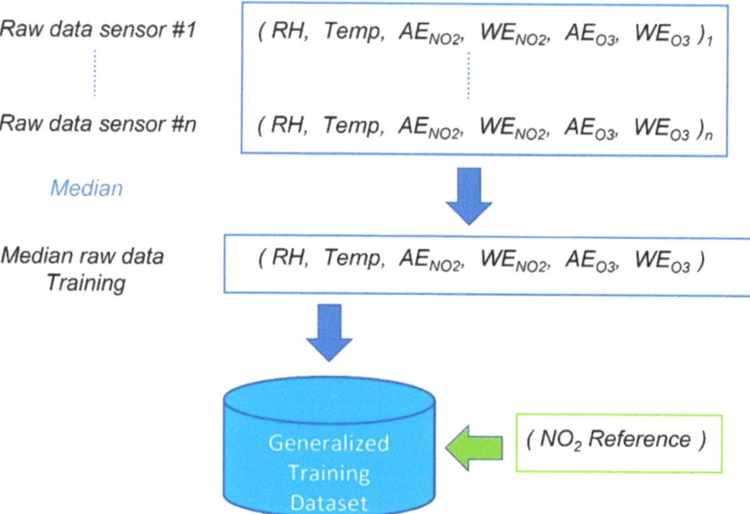

Figure 3. Procedure for creating the training set of the general calibration model.

Such an approach can incorporate the inherent variability of each sensor into a single model, as previously investigated to contrast the effects of concept drift on metal oxide (MOX) gas sensors [23]. The same approach was applied to the NO_2 electrochemical gas sensors in our case study.

2.4.2. Importance Weighting Calibration Model

The idea behind this procedure is to "weigh" the test set samples in order to "match" the distribution used during the training phase. Once the weights are obtained, these will be applied in the training process, obtaining a new calibration model [24].

The importance of a sample (i.e., the "weight") is calculated as the ratio between the probability density functions (*pdf*) of the test and training set (Figure 4). If such a ratio $w(x)$ is equal to 1, this means that the sample has the same degree of importance in both the test and the training set, while if $w(x) > 1$, the considered sample is more important in describing the test set rather than the training set.

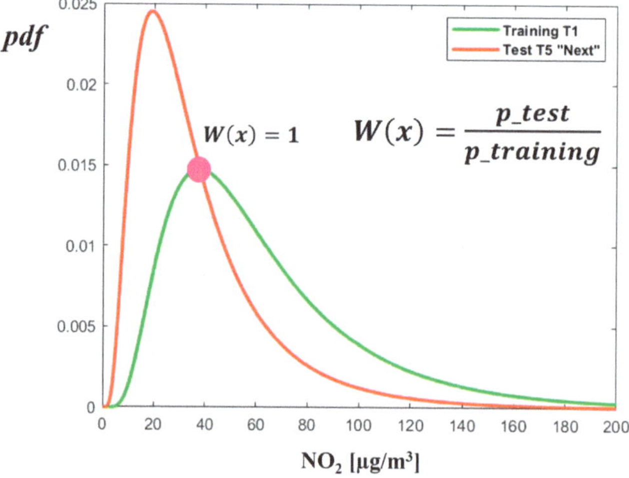

Figure 4. How the weights are used in the fitting process in MATLB fitlm (*x*, *y*, 'Weight', w) function.

In the present analysis, the application of the importance of the weighting calibration model was limited to the target variable only, and the weights were obtained using the ad hoc model prediction of the time slot T4 (the last useful time slot before the effects of the concept drift would become disruptive).

2.4.3. Stacking Ensemble Calibration Model

Recently, the stacking ensemble technique has been successfully applied in domain adaptation under concept drift due to its ability to reduce the deviation and variance in neural networks, thus producing robust predictions. Such a strategy has also already been applied in air quality sensor networks [25,26].

Stacking ensemble consists of combining the outputs of several models produced by different algorithms (generally called *base learners* or also *weak models*) in order to increase the total accuracy and increase generalization. The estimations of the basic learners are combined in a second level, called meta-learner, which, in the case of regression problems, can be a simple linear regression. In our case study, instead of implementing additional algorithms such as base learners, we exploited the estimations of nitrogen dioxide resulting from the general calibration model and the importance weighting calibration model. Figure 5 describes the proposed architectural scheme in detail. The reference data (labels) for the meta-learner training phase will be requested to the air quality network in the T4 time slot. Such an approach is known as remote calibration, as described in the Introduction, and in our case study, its usage avoids new co-locations.

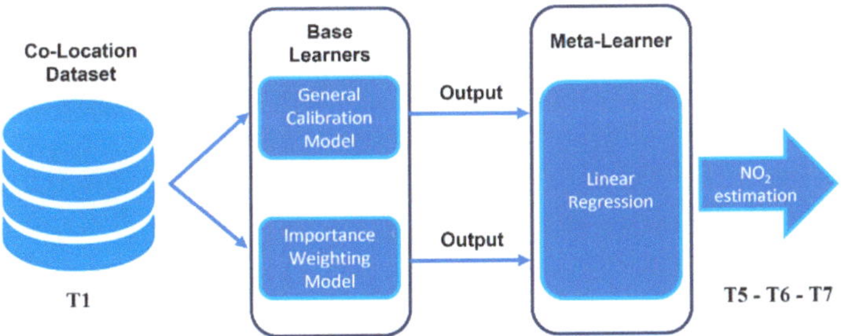

Figure 5. The proposed stacking ensemble architecture.

2.4.4. Performance Evaluation

According to the European Directive [27], if a NO_2 low-cost sensor is used in an air quality network for indicative measurements, it should present a relative expanded uncertainty (REU) below 25%. In this analysis, we labeled this positive operational scenario with "PASS", while "FAIL" identified REU > 25%. The REU and the REU plots, reported in the Supplementary Materials, were computed with an appropriate MATLAB script on the basis of the formulas and the methodology described in [19,28]. In our analysis, another main parameter that was taken into account was the pollutant concentration value, at which the REU graph intercepts the data quality objective (DQO) threshold line at 25%. Obviously, the following choice between different models would be oriented towards the model that shows the lowest possible REU, which is translated into the model's better capability to mitigate the effects of the concept drift.

3. Results and Discussion

To improve the readability of this section, the reader can refer directly to the following tables, which summarize the outcomes of the REU plots, which are detailed for completeness in the Supplementary Material.

3.1. Results with Reference Data

It would be obvious to come to the conclusion that the "Next" approach is to be preferred over the others because it is after the alert that the new operating conditions fully manifest themselves, independently, from the metrics used for the assessment.

However, in order to learn the new operating domain, it is necessary to wait for the necessary time to acquire a certain number of samples to train the new model. This means continuing, in the meantime, to invalidate the data released by the node (which would be a PASS level of REU up to T4 and FAIL from T5 onward). An important drawback is represented by the delayed time needed to obtain reference data before having a running updated model. It could be possible to wait for T5 reference data and then embed the new concept into the new updated model. However, such a strategy, even if allowing for good performances in T6 and T7, would lead to a loss of continuity in the data quality in T5. The goal, instead, is to achieve a continuity in data quality by wishing the REU in T5–T6–T7 to be equal to PASS. Therefore, it would be helpful if the "Last" and/or "Mixed" approaches presented acceptable REU values.

Tables 1 and 2 summarize the results obtained from the REU plots in the "Mixed" and "Last" scenarios and emphasize their ability to mitigate the concept drift effects and lead to more accurate and precise NO_2 estimations. Both tables show, for each time slot and for each device, the assigned REU labels (PASS, FAIL) as well as the value corresponding to the intersection of the pollutant concentration value with the data quality objective (DQO) threshold line at 25%, as disposed by the European Directive [27]. The missing values in the tables are related to the REU plot, failing to reach the minimum threshold of 25%.

Table 1. Summary of REU results for the "Mixed" scenario.

	T5		T6		T7	
	REU	**Value** $[\mu g/m^3]$	**REU**	**Value** $[\mu g/m^3]$	**REU**	**Value** $[\mu g/m^3]$
AQ6	FAIL	-	PASS	65	PASS	70
AQ11	PASS	60	PASS	80	PASS	80
AQ12	PASS	55	PASS	55	PASS	51

Table 2. Summary of REU results for the "Last" scenario.

	T5		T6		T7	
	REU	**Value** $[\mu g/m^3]$	**REU**	**Value** $[\mu g/m^3]$	**REU**	**Value** $[\mu g/m^3]$
AQ6	FAIL	-	PASS	80	FAIL	-
AQ11	PASS	55	FAIL	-	FAIL	-

Observing the REU plots from T5 to T7 for the AQ12 device in the "Mixed" approach, it can be seen that the REU drops below 25% from 55 $\mu g/m^3$ onwards in T5 and T6 (Figures S1.1 and S1.2), while it drops from 51 $\mu g/m^3$ in T7 (Figure S1.3). Surely such a situation is to be preferred since it allows for the continuity of data quality, extending the hoped PASS REU value for 3 more weeks. Regarding the AQ11 device, the "Mixed" approach worked similarly to AQ12, but the inherent variability of the device negatively affected the model performance, meaning it failed to repair the effects of the concept drift. In fact, it can be seen that the intersections with the DQO threshold line at 25% are higher, at 60 $\mu g/m^3$ in T5 (Figure S1.4) and 80 $\mu g/m^3$ in T6 and T7 (Figures S1.5 and S1.6), respectively. On the other hand, in the "Last" approach, AQ11 exhibits good performances only in T5 (Figure S1.7), while it does not reach the desired REU levels in T6 and T7 (Figures S1.8 and S1.9). This means that the intrinsic variability of AQ11 and the operating conditions in T3 used for the fitting of the new model are too different from those present after the concept drift. The AQ6 device in the "Mixed" approach presents a FAIL in T5, while in T6 and T7, a PASS is reached at 65 and 70 $\mu g/m^3$, respectively (Figures S1.10–S1.12). The "Last" scenario for AQ6 does not work, except for in T6, but at a higher pollution level of 80 $\mu g/m^3$ (Figures S1.13–S1.15). The outcomes for the AQ12 device in the "Last" scenario are missing due to data transmission issues.

From the analysis carried out, it can be concluded that all in all, if there is a possibility of drawing on the data of a regulatory reference station, it is possible to update the calibration model after the trigger signal using the "*Mixed*" approach, i.e., selecting part of the reference data before the concept drift alert and remaining data after it.

3.2. Results without Reference Data: General Calibration Model

Proceeding similarly to the previous case, it is essential to evaluate the REU to understand if the general calibration model is able to positively mitigate the effects of the concept drift. Table 3 details the obtained results, while the REU plots in the Supplementary Materials are also shown in comparison with the ad hoc calibration model.

Table 3. Summary of REU results for the general calibration model.

	T5		T6		7	
	REU	**Value [$\mu g/m^3$]**	**REU**	**Value [$\mu g/m^3$]**	**REU**	**Value [$\mu g/m^3$]**
AQ6	PASS	45	PASS	63	PASS	80
AQ11	FAIL	-	FAIL	-	FAIL	-
AQ12	FAIL	-	FAIL	-	FAIL	-

Applying the general calibration model to the AQ6 node, the REU plot drops below 25% at 45 $\mu g/m^3$, suggesting that the global calibration model is efficient in mitigating the concept drift consequences in the T5 time slot (Figure S2.1). This value shifts at higher concentration levels in T6 and T7, specifically at 63 $\mu g/m^3$ and 80 $\mu g/m^3$, respectively (Figures S2.2 and S2.3). Although the REU does not fall within a considerable range of values below 25% in T6 and T7, it must be observed that such a situation presents better results than the ad hoc model. Therefore, a first applicative scenario would be to use the global model for AQ6 from T5 onwards, instead of continuing with the ad hoc one. For the AQ12 device, the general calibration model matches the ad hoc model performance (Figures S2.4–S2.6). The same result has been obtained for particulate matter sensors in terms of the mean absolute error (MAE) in [29]. The AQ11 intrinsic node variability makes this instrument too different from the others; therefore, a general model built in this way is unable to work properly for our purposes. Indeed, its performance is too far also from the performance of the ad hoc model; thus, we chose not to report the REU plots. Future research projects could investigate this aspect and evaluate whether it is better to build more global models among similar devices by means of clustering [30].

3.3. Results without Reference Data: Importance Weighting Calibration Model

As can be seen in Table 4, re-weighing the target variable in T4 produces a considerable improvement in the performance of the AQ11 device. In fact, looking at the REU plot of AQ11 in T5, the intersection with the 25% value was observed at 40 $\mu g/m^3$ (Figure S3.4), while in T6 and T7, it did not quite reach the DQOs, although the importance weighting calibration model performs slightly better than the ad hoc one (Figures S3.5 and S3.6). As far as the AQ6 and AQ12 devices, the REU plots equaled those of the ad hoc model in some time slots, while they were worse in others (Figures S3.1–S3.3 and S3.7–S3.9).

Table 4. Summary of REU results for the importance weighting calibration model.

	T5		T6		T7	
	REU	**Value** $[\mu g/m^3]$	**REU**	**Value** $[\mu g/m^3]$	**REU**	**Value** $[\mu g/m^3]$
AQ6	FAIL	-	FAIL	-	FAIL	-
AQ11	PASS	40	FAIL	-	FAIL	-
AQ12	FAIL	-	FAIL	-	FAIL	-

Summarizing the application of the importance weighting calibration model brings the AQ11 device in T5 back to the allowed REU values.

3.4. Stacking Ensemble Calibration Model Results

Table 5 shows the summary of the REU values for the stacking ensemble calibration model. The inference to be drawn comparing the outcomes of Table 5 with those of Table 1 is that the application of the stacking ensemble technique lowers the value at which the REU reaches 25%, making the mitigation of the concept drift effects more robust in all the analyzed cases. In principle, the use of better-performing base learner algorithms could contribute to further error reductions. The proposed solution, although not yet optimal, constitutes a good compromise, since it guarantees the continuity of the data's quality even after the detection of concept drifts. Furthermore, operating within an air quality network, the implementation of the proposed stacking ensemble approach requires only the reference data in T4, meaning that any other re-co-location could be avoided in a remote calibration approach.

Table 5. Summary of REU results for the stacking ensemble calibration model.

	T5		T6		T7	
	REU	**Value** $[\mu g/m^3]$	**REU**	**Value** $[\mu g/m^3]$	**REU**	**Value** $[\mu g/m^3]$
AQ6	PASS	44	PASS	56	PASS	65
AQ11	PASS	44	PASS	70	PASS	76
AQ12	PASS	51	PASS	54	PASS	45

4. Conclusions

In this study, we attempted to achieve the data quality objectives in the measurements of environmental pollutants through machine-learning-calibrated low-cost gas sensors. Through our results, we hope to accelerate the spread of this technology in smart city applications. The core of our work is a concept drift adaptation in the context of air quality monitoring networks but by exploring the possibility of using the whole useful information content of the data collected in the course of the co-location.

Another noteworthy aspect relates to the choice of calibration models used. With the general calibration model, we wanted to incorporate the intrinsic variability of each sensor into a single calibration model. On the other hand, with the importance weighting calibration model, we tried to reduce the prediction error by matching the distributions between training and test using accurate weights. Unfortunately, the general calibration model and the importance weighting calibration model show promising results only for some devices and only in the one time slot in this specific dataset. On the contrary, their use in a stacking ensemble architecture succeeds in mitigating the concept drift effects by allowing an extension of the calibration robustness both for all devices and in all time slots after concept drift. For example, by turning off the ad hoc model and turning on the stacking ensemble model, the node is able to output reliable data.

Even though the proposed methodology offers encouraging results, future research is encouraged to test other models in the first layer of the stacking ensemble to increase performance but also to test the assessment and validation of the methodology using wider datasets with a greater number of devices.

Finally, an important aspect to remark is the feasibility of the implementation on the backend side of the platform of an air quality network of the proposed approach. This might deliver a continuous calibration service. Recently, indeed the issue of evaluating and monitoring machine learning models from validation to production has come to the fore. Therefore, numerous start-ups have implemented their own platform and began offering services in many application areas [31]. The low-cost air quality monitoring network could be one of such applications.

Supplementary Materials: The following supporting information can be downloaded at: https://www.mdpi.com/article/10.3390/s24092786/s1, Figures S1.1–S1.15: REU plots of reference data selection for calibration update; Figures S2.1–S2.6: REU plots of general calibration model; Figures S3.1–S3.9: REU plots of importance weighting calibration model; Figures S4.1–S4.9: REU plots of stacking ensemble calibration model.

Author Contributions: Conceptualization, G.D. and P.S.; investigation G.D., M.F. and P.S.; methodology, S.F. and S.D.V.; data curation, G.D., S.F. and M.F.; validation, G.D., S.F. and P.S.; supervision, G.D.F.; funding acquisition, P.S. and S.D.V. All authors have read and agreed to the published version of the manuscript.

Funding: This research received partial funding by the European Commission through POR Campania FESR Research and Innovation programs, ARMONIA Project, CUP B57H22003350007 and also from the European Union's Horizon 2020 Research and Innovation program, VIDIS Project, under grant agreement No 952433.

Institutional Review Board Statement: Not applicable.

Informed Consent Statement: Not applicable.

Data Availability Statement: Supporting data are currently available on request.

Acknowledgments: The authors wish to thank Paolo D'Auria of ARPA Campania for supporting the co-location campaign.

Conflicts of Interest: The authors declare no conflicts of interest.

References

1. Morawska, L.; Thai, P.K.; Liu, X.; Asumadu-Sakyi, A.; Ayoko, G.; Bartonova, A.; Bedini, A.; Chai, F.; Christensen, B.; Dunbabin, M.; et al. Applications of low-cost sensing technologies for air quality monitoring and exposure assessment: How far have they gone? *Environ. Int.* **2018**, *116*, 286–299. [CrossRef] [PubMed]
2. Schneider, P.; Castell, N.; Vogt, M.; Dauge, F.R.; Lahoz, W.A.; Bartonova, A. Mapping urban air quality in near real-time using observations from low-cost sensors and model information. *Environ. Int.* **2017**, *106*, 234–247. [CrossRef] [PubMed]
3. Karagulian, F.; Barbiere, M.; Kotsev, A.; Spinelle, L.; Gerboles, M.; Lagler, F.; Redon, N.; Crunaire, S.; Borowiak, A. Review of the performance of low-cost sensors for air quality monitoring. *Atmosphere* **2019**, *10*, 506. [CrossRef]
4. van Zoest, V.; Osei, F.B.; Stein, A.; Hoek, G. Calibration of low-cost NO_2 sensors in an urban air quality network. *Atmos. Environ.* **2019**, *210*, 66–75. [CrossRef]
5. Topalović, D.B.; Davidović, M.D.; Jovanović, M.; Bartonova, A.; Ristovski, Z.; Jovašević-Stojanović, M. In search of an optimal in-field calibration method of low-cost gas sensors for ambient air pollutants: Comparison of linear, multilinear and artificial neural network approaches. *Atmos. Environ.* **2019**, *213*, 640–658. [CrossRef]
6. Ditzler, G.; Roveri, M.; Alippi, C.; Polikar, R. Learning in Nonstationary Environments: A Survey. *IEEE Comput. Intell. Mag.* **2015**, *10*, 12–25. [CrossRef]
7. De Vito, S.; Esposito, E.; Castell, N.; Schneider, P.; Bartonova, A. On the robustness of field calibration for smart air quality monitors. *Sens. Actuators B Chem.* **2020**, *310*, 127869. [CrossRef]
8. Masey, N.; Gillespie, J.; Ezani, E.; Lin, C.; Wu, H.; Ferguson, N.S.; Hamilton, S.; Heal, M.R.; Beverland, I.J. Temporal changes in field calibration relationships for Aeroqual S500 O_3 and NO_2 sensor-based monitors. *Sens. Actuators B Chem.* **2018**, *273*, 1800–1806. [CrossRef]
9. Quinonero-Candela, J.; Sugiyama, M.; Schwaighofer, A.; Lawrence, N.D. *Dataset Shift in Machine Learning*; Mit Press: Cambridge, MA, USA, 2008; ISBN 9780262545877.
10. Lu, J.; Liu, A.; Dong, F.; Gu, F.; Gama, J.; Zhang, G. Learning under concept drift: A review. *IEEE Trans. Knowl. Data Eng.* **2018**, *31*, 463–473. [CrossRef]

11. Tsujita, W.; Yoshino, A.; Ishida, H.; Moriizumi, T. Gas sensor network for air-pollution monitoring. *Sens. Actuators B Chem.* **2005**, *110*, 304–311. [CrossRef]

12. Popoola, O.A.M.; Carruthers, D.; Lad, C.; Bright, V.B.; Mead, M.I.; Stettler, M.E.J.; Saffell, J.R.; Jones, R.L. Use of networks of low-cost air quality sensors to quantify air quality in urban settings. *Atmos. Environ.* **2018**, *194*, 58–70. [CrossRef]

13. Heimann, I.; Bright, V.B.; McLeod, M.W.; Mead, M.I.; Popoola, O.A.M.; Stewart, G.B.; Jones, R.L. Source attribution of air pollution by spatial scale separation using high spatial density networks of low-cost air quality sensors. *Atmos. Environ.* **2015**, *113*, 10–19. [CrossRef]

14. Barcelo-Ordinas, J.M.; Ferrer-Cid, P.; Garcia-Vidal, J.; Ripoll, A.; Viana, M. Distributed Multi-Scale Calibration of Low-Cost Ozone Sensors in Wireless Sensor Networks. *Sensors* **2019**, *19*, 2503. [CrossRef] [PubMed]

15. BL-Pilot-Final-Technical-Report. Available online: https://globalcleanair.org/wp-content/blogs.dir/95/files/2021/05/BL-Pilot-Final-Technical-Report.pdf (accessed on 23 April 2024).

16. Miskell, G.; Salmond, J.A.; Williams, D.E. Solution to the Problem of Calibration of Low-Cost Air Quality Measurement Sensors in Networks. *ACS Sens.* **2018**, *3*, 832–843. [CrossRef] [PubMed]

17. Weissert, L.F.; Henshaw, G.S.; Williams, D.E.; Feenstra, B.; Lam, R.; Collier-Oxandale, A.; Papapostolou, V.; Polidori, A. Performance evaluation of MOMA (MOment MAtching)—A remote network calibration technique for PM2.5 and PM10 sensors. *Atmos. Meas. Tech.* **2023**, *16*, 4709–4722. [CrossRef]

18. Esposito, E.; D'Elia, G.; Ferlito, S.; De Vito, S.; Di Francia, G. Optimal Field Calibration of Multiple IoT Low Cost Air Quality Monitors: Setup and Results. In Proceedings of the Computational Science and Its Applications—ICCSA 2020, Cagliari, Italy, 1–4 July 2020; Lecture Notes in Computer Science; Springer: Berlin/Heidelberg, Germany; Volume 12253. [CrossRef]

19. D'Elia, G.; Ferro, M.; Sommella, P.; De Vito, S.; Ferlito, S.; D'Auria, P.; Di Francia, G. Influence of Concept Drift on Metrological Performance of Low-Cost NO_2 Sensors. *IEEE Trans. Instrum. Meas.* **2022**, *71*, 1–11. [CrossRef]

20. Baier, L.; Reimold, J.; Kühl, N. Handling Concept Drift for Predictions in Business Process Mining. In Proceedings of the IEEE 22nd Conference on Business Informatics (CBI), Antwerp, Belgium, 22–24 June 2020; pp. 76–83. [CrossRef]

21. Malings, C.; Tanzer, R.; Hauryliuk, A.; Kumar, S.P.; Zimmerman, N.; Kara, L.B.; Presto, A.A.; Subramanian, R. Development of a general calibration model and long-term performance evaluation of low-cost sensors for air pollutant gas monitoring. *Atmos. Meas. Tech.* **2019**, *12*, 903–920. [CrossRef]

22. De Vito, S.; D'Elia, G.; Di Francia, G. Global Calibration Models Match Ad-Hoc Calibrations Field Performances in Low-Cost Particulate Matter Sensors. In Proceedings of the 2022 IEEE International Symposium on Olfaction and Electronic Nose (ISOEN), Aveiro, Portugal, 29 May–1 June 2022; pp. 1–4.

23. Fonollosa, J.; Fernández, L.; Gutiérrez-Gálvez, A.; Huerta, R.; Marco, S. Calibration transfer and drift counteraction in chemical sensor arrays using Direct Standardization. *Sens. Actuators B Chem.* **2016**, *236*, 1044–1053. [CrossRef]

24. Sugiyama, M.; Krauledat, M.; MÃžller, K.R. Covariate shift adaptation by importance weighted cross validation. *J. Mach. Learn. Res.* **2007**, *8*, 985–1005.

25. Yuan, L.; Li, H.; Xia, B.; Gao, C.; Liu, M.; Yuan, W.; You, X. Recent Advances in Concept Drift Adaptation Methods for Deep Learning. In Proceedings of the 31st International Joint Conference on Artificial Intelligence, Vienna, Austria, 23–29 July 2022; International Joint Conferences on Artificial Intelligence Organization: Vienna, Austria, 2022; pp. 5654–5661.

26. Bagkis, E.; Kassandros, T.; Karatzas, K. Learning Calibration Functions on the Fly: Hybrid Batch Online Stacking Ensembles for the Calibration of Low-Cost Air Quality Sensor Networks in the Presence of Concept Drift. *Atmosphere* **2022**, *13*, 416. [CrossRef]

27. European Directive 2008/50/EC. Available online: http://eur-lex.europa.eu/legal-content/EN/TXT/PDF/?uri=CELEX:32008L0050&from=en (accessed on 18 January 2024).

28. European Commission. *Guide to the Demonstration of Equivalence of Ambient Air Monitoring Methods*; Report by an EC Working, Group on Guidance; European Commission: Brussels, Belgium, 2010.

29. De Vito, S.; D'Elia, G.; Ferlito, S.; Di Francia, G.; Davidović, M.D.; Kleut, D.; Stojanović, D.; Jovaševic-Stojanović, M. A Global Multi-Unit Calibration as a Method for Large Scale IoT Particulate Matter Monitoring Systems Deployments. *IEEE Trans. Instrum. Meas.* **2024**, *73*, 1–16. [CrossRef]

30. Smith, K.R.; Edwards, P.M.; Ivatt, P.D.; Lee, J.D.; Squires, F.; Dai, C.; Peltier, R.E.; Evans, M.J.; Sun, Y.; Lewis, A.C. An Improved Low-Power Measurement of Ambient NO_2 and O_3 Combining Electrochemical Sensor Clusters and Machine Learning. *Atmos. Meas. Tech.* **2019**, *12*, 1325–1336. [CrossRef]

31. Symeonidis, G.; Nerantzis, E.; Kazakis, A.; Papakostas, G.A. MLOps—Definitions, Tools and Challenges. In Proceedings of the 12th IEEE Annual Computing and Communication Workshop and Conference (CCWC 2022), Las Vegas, NV, USA, 26–29 January 2022; pp. 453–460.

Article

Applications of Chipless RFID Humidity Sensors to Smart Packaging Solutions

Viviana Mulloni [1,*], Giada Marchi [1], Andrea Gaiardo [1], Matteo Valt [1], Massimo Donelli [2] and Leandro Lorenzelli [1]

[1] Center for Sensors and Devices, Fondazione Bruno Kessler, 38123 Trento, Italy; gmarchi@fbk.eu (G.M.); gaiardo@fbk.eu (A.G.); mvalt@fbk.eu (M.V.); lorenzel@fbk.eu (L.L.)

[2] Department of Civil Environmental and Mechanical Engineering, University of Trento, 38123 Trento, Italy; massimo.donelli@unitn.it

* Correspondence: mulloni@fbk.eu

Abstract: Packaging solutions have recently evolved to become smart and intelligent thanks to technologies such as RFID tracking and communication systems, but the integration of sensing functionality in these systems is still under active development. In this paper, chipless RFID humidity sensors suitable for smart packaging are proposed together with a novel strategy to tune their performances and their operating range. The sensors are flexible, fast, low-cost and easy to fabricate and can be read wirelessly. The sensitivity and the humidity range where they can be used are adjustable by changing one of the sensor's structural parameters. Moreover, these sensors are proposed as double parameter sensors, using both the frequency shift and the intensity variation of the resonance peak for the measure of the relative humidity. The results show that the sensitivity can vary remarkably among the sensors proposed, together with the operative range. The sensor suitability in two specific smart packaging applications is discussed. In the first case, a threshold sensor in the low-humidity range for package integrity verification is analyzed, and in the second case, a more complex measurement of humidity in non-hermetic packages is investigated. The discussion shows that the sensor configuration can easily be adapted to the different application needs.

Keywords: chipless RFID; humidity sensing; smart packaging; microwave sensor; threshold sensor

Citation: Mulloni, V.; Marchi, G.; Gaiardo, A.; Valt, M.; Donelli, M.; Lorenzelli, L. Applications of Chipless RFID Humidity Sensors to Smart Packaging Solutions. *Sensors* **2024**, *24*, 2879. https://doi.org/10.3390/s24092879

Academic Editor: Simone Genovesi

Received: 20 February 2024
Revised: 4 April 2024
Accepted: 29 April 2024
Published: 30 April 2024

1. Introduction

As an alternative to traditional packaging, smart packaging today is a smart tool capable of interacting with the environment and creating a link with manufacturers and consumers while its functions far exceed the purpose of simply containing products [1]. Its market trend is continuously growing, and it is already estimated to be worth around several billion US dollars. The uses of smart packaging are widespread, from logistics and product traceability to safety control and waste reduction [2].

In particular, the food packaging market has been deeply transformed by the introduction of smart packaging [3]. This new technology can provide in real time important information on food products, reducing the amount of waste and the health risks. To implement these functions, internal or external electronic sensors and radio frequency identification (RFID) devices are usually exploited [4]. Electronic sensors can give accurate data for different kinds of parameters observed in food products, as in the case of electronic noses and electronic tongues. These types of sensors can sense both chemical and biological parameters and are rather selective and precise. However, their size and cost make them hard to embed in packaging, especially compared with the simplicity and low cost of a barcode as in traditional packaging solutions. Moreover, the typical electronic sensor needs to be powered, and this means that it has to be either wired or provided with a battery. Both approaches are difficult and expensive to implement in an efficient and low-cost food packaging solution.

On the other hand, RFID technology has progressed significantly in the past few years, and its identification and traceability capabilities make a major contribution to food safety and quality monitoring [5]. However, to implement smart packaging functions, the outstanding tracking functions typical of RFID systems must be coupled with sensing functionalities [6] to detect environmental parameters, such as the temperature, humidity and chemical composition, to address food quality and the risk of product deterioration. RFID tags can, in principle, integrate several types of sensors, providing identification and sensing in a contactless, wireless, passive and non-visual way at the same time, but this field of research is still in the explorative stage [5]. The usual approach to coupling identification and sensing functions is to monitor the impedance of the sensor, which depends on the parameter being sensed, together with the digital data from an IC chip, which are measured with a digital RFID reader. However, this complicates the reading system, and the two functions are often measured separately [6]. Passive RFID sensors require electromagnetic power to read and write the data stored on the chip. This power comes from the RFID reader and determines the reading range in UHF tags. This range decreases with the integration of a sensing device and is one of the most important limitations of RFID sensors. Chip-based RFID sensors guarantee high accuracy in the reading process but increase the complexity and cost of the RFID tag, making it less robust and convenient for integration in the package.

As an alternative and emerging solution, chipless RFID tags [7] have been proposed in the literature by several authors. The important advantage of this new technology is the absence of integrated circuits or chips in the tag. This important feature leads to simpler, low-cost, more robust and less power-demanding devices [8] compared with more traditional RFID technologies. Moreover, chipless RFID tags have better functionality in harsh environments and can have their support, like barcode tags, printed on using high-conductivity inks [9]. The tag structure is composed of several resonating structures, and much research is available for the design and encoding [10,11] strategies of the identification (ID) function in chipless RFIDs, as well as for the detection techniques. Given the proposed application, in this study, we focus on tags called backscattered chipless RFID tags, where the resonant elements provide the information directly through their spectral signatures, as opposed to retransmission-based tags [12], which require a transmitting and a receiving antenna or a microstrip line. The backscattered tags do not need additional elements; hence, the size of the tag is strongly reduced because it contains only the resonators.

While important results have been obtained for the coding capacity of the tag and maximization of the information density in chipless RFIDs for tracking and product identification applications [13], the integration of ID [14] and sensing functionalities [15] is still a topic of active and explorative research. The two functions are usually embedded in the structure of the tag, which is composed of a series of different resonators, some of which are used for ID coding, while others are used for sensing [8]. Even though a large variety of chipless RFID sensors has already been demonstrated in the literature for physical, chemical and mechanical sensing of different parameters [8,16], they are intrinsically more challenging than the ID geometries and are still in the early explorative stage regarding their implementation in marketable applications. The main obstacles to commercialization are the reduced information density compared with chipped RFID solutions and the limited reading distance, which is common among completely passive systems. Both challenges are currently highly active topics in chipless RFID research and development.

Furthermore, when integrating chipless RFID sensors into real-world smart packaging, challenges such as a low cost, environmental robustness [17], compatibility and data security must be addressed effectively [18]. These barriers highlight the need for standardized solutions that can ensure reliable operation in different packaging environments while maintaining cost efficiency and accurate detection performance (e.g., high sensitivity), as well as addressing privacy issues to gain consumer confidence [19]. Despite their potential, overcoming these challenges will be essential for the successful implementation of chipless RFID sensors in smart packaging applications.

In this contribution, we report the development of a chipless low-cost humidity sensor for wireless monitoring of moisture levels within packaged goods, which can be easily enabled with additional identification capabilities, and which is capable of detecting an extensive range of RH% with high sensitivity. The suitability and adaptation of the sensor for different packaging requirements is discussed, and the optimization of the sensor characteristics for each application in terms of sensitivity and operational humidity range is presented. In particular, the maximization of the sensitivity for quite low-humidity environments is useful for the packaging of perishable goods, where low levels of internal humidity are essential for proper conservation. In this case, in order to assure the package integrity and consequently the stability of an inert atmosphere inside the package, a threshold sensor for signaling the minimum amounts of humidity is the most convenient choice. Other applications may require more elaborate solutions, needing more frequent and precise monitoring of humidity levels inside the package. In this case, the sensor should be able to measure a wide range of ambient humidity while a high sensitivity in a specific subrange is not required. Intermediate or mixed solutions are also possible, depending on the characteristics of the goods to be monitored. A strategy to tailor the sensor performances while taking into account the needs of the different possible packaging applications will be presented and discussed.

2. Materials and Methods

The sensors were composed of a metallic resonator over a flexible 168 µm-thick low-loss substrate (Rogers 4350, Rogers Corp., Chandler, AZ, USA) covered by a 50 µm-thick Nafion NRE-212 membrane. This polymer was chosen because it is extremely sensitive to environmental humidity [20]. The geometry of the metallic structure was that of a square electric field-coupled (ELC) resonator [21,22] with a lateral dimension of 20 mm and was realized by microlithography and chemical etching. The geometry of the sensing structure was simulated and described in detail in [21,23], and it is schematically reported in Figure 1.

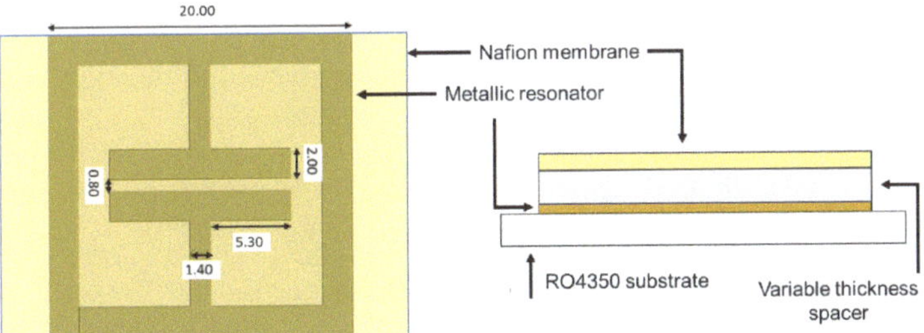

Figure 1. Structure and dimensions of the chipless sensors. All dimensions are in millimeters.

The measurements were performed on four samples with four different distances between the metallic surface and the Nafion membrane, namely 0 (contact), 150, 400 and 800 µm. The distance was determined using plastic spacers of low-loss material (polycarbonate) of a known thickness which were transparent to the electromagnetic field. The total thickness was additionally verified with a micrometer. The RF characterization was performed by using a dedicated apparatus with a custom-made gas-flow test chamber coupled with an Agilent E5061B ENA vector network analyzer. Gas sensing measurements were performed at room temperature (21 °C). The relative humidity (%RH) was monitored by a digital humidity sensor (Honeywell HIH-4000, 1.0% accuracy) at the chamber's exit. Here, %RH control was achieved by injecting a fixed fraction of the total dry air, namely 200 sccm (20% O_2 and 80% N_2), through mass flow controllers (MKSs) into a gas bubbler filled with deionized water to create the desired humidity conditions.

The circular antenna probe [9], with an internal diameter of 20 mm, was directly connected to the VNA outside the chamber while the sensing tag was inside, and the detection took place wirelessly, measuring the antenna return loss S11. In particular, the antenna was a commercial magnetic field probe, namely a model H20 magnetic probe (Signal Hound, Battle Ground, WA, USA) with a frequency band from 30 KHz up to 6 GHz. The probe presented a self-resonance value of around 1.36 GHz. The distance between the tag and the probe was 0.5 cm. The measurement set-up is shown in Figure 2.

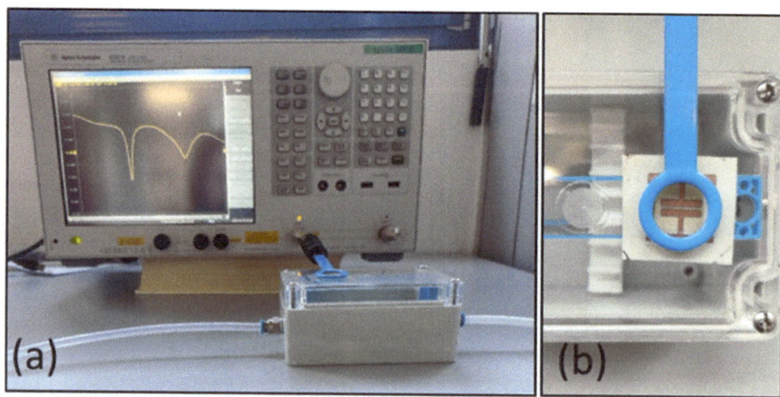

Figure 2. Set-up used for the measurements. (**a**) Measurement configuration showing the circular antenna probe connected to the vector network analyzer above the measurement chamber where the humidity is controlled. (**b**) Detailed image of the measurement chamber, showing the probe outside the chamber and the sensor inside the chamber.

3. Results and Discussion

The sensors investigated in this work were dual sensors, which means that two independent sensor parameters, the peak frequency and peak intensity, varied with the analyte concentration. A third parameter, the peak width, varied as well, but it could not be considered independent from the peak intensity, being almost inversely proportional to it. This behavior is physically related to the independent variation of both the real and imaginary part of the dielectric constant of the sensing material upon exposure to humidity, where the real part variation is principally related to the frequency shift, and the imaginary part is related to the change in the peak intensity [22]. For sensing purposes, this is an advantage because the sensor can be adapted to the application-specific needs in a more versatile way, and the simultaneous determination of two parameters increases the reliability and accuracy of the sensing process. This advantage has, however, an intrinsic drawback: the peak broadening and the decreased intensity set a limit to the humidity range that can be sensed, because the more sensitive the material is, the faster the weakening of the peak signal with increasing humidity is. This behavior is material-dependent, but the trend is typically the same for all humidity-sensitive materials because they incorporate water, which has a rather high real and imaginary part of the dielectric constant as the humidity increases. At a higher humidity, the resonant peak is therefore expected to shift to lower frequencies and decrease its intensity up to a limit that makes it no longer detectable. In this work, Nafion NRE-212 was used as the sensitive material. This material is a commercial polymeric membrane that has shown a great affinity for water. Moreover, it is mechanically and chemically very resistant and stable, and its absorption and desorption of water are quite fast [21,24], making it a rather convenient sensitive material for humidity detection.

In the case of the sensors evaluated in this work, the distance between the resonator and the sensitive material was varied to explore the effect of this parameter on the detectable humidity range and on the sensitivity. This was accomplished using a spacer of a variable thickness between the metallic resonator and the Nafion membrane to adapt the sensor to different humidity ranges and detection strategies. Four spacer thicknesses were selected, as reported in the previous section.

The four sensors were measured in the range of 0–70% RH between 1 and 2 GHz. The measured return loss intensities of the circular antenna probe are reported in Figure 3 at 0%, 25%, 50% and 70% relative humidity. The data are reported for the four spacer thicknesses, as indicated in the legend.

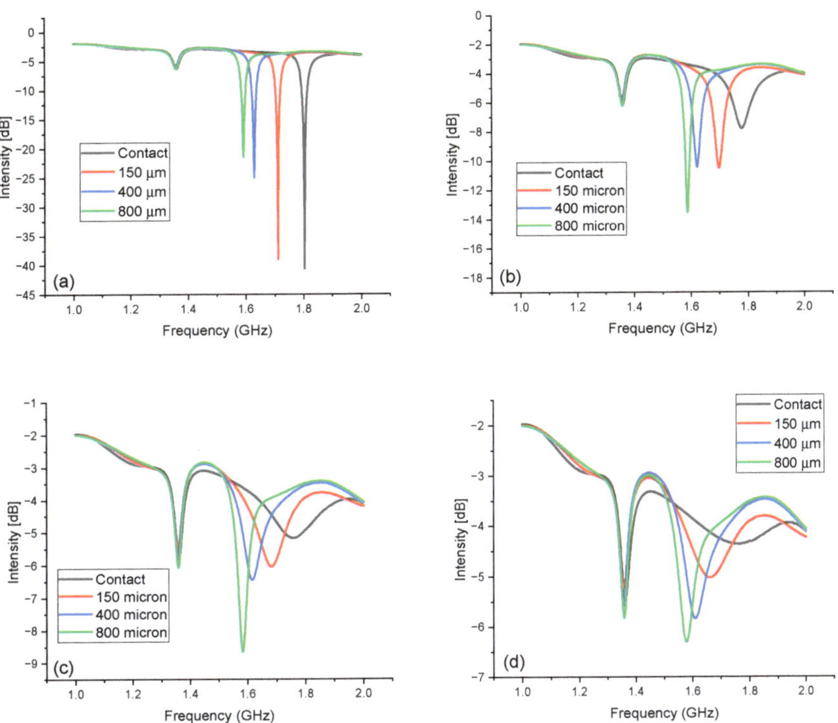

Figure 3. (**a**) Return loss intensity as a function of the frequency in the range of 1–2 GHz for the four resonators at (**a**) 0% RH. (**b**) 25% RH, (**c**) 50% RH and (**d**) 70% RH.

The peak intensity, bandwidth and frequency all varied markedly, going from 0 to 70% RH for all spacer values. The intensity variation is especially evident when considering the different vertical scales in the four plots of Figure 3. While in Figure 3a, the baseline signal seems almost negligible, in Figure 3d, the peak intensity is so low that the baseline signal is comparable with the peak intensities. In addition, the peak at 1.36 GHz, which was the internal resonance of the probe, did not vary appreciably with the humidity and spacer thickness and could be used to visualize the intensity variations and as a reference for sensor calibration. The intensity trend with the spacer thickness was reversed when going from 0% RH to 70% RH. The resonator with no spacer had the highest intensity at 0% RH but the lowest one at 70% RH, where it was quite broad and barely detectable. On the other hand, the 800 µm sensor had the lowest intensity at 0% RH but the highest one at 25, 50 and 70% RH. The 150 and 400 µm sensors had an intermediate behavior.

A more systematic variation of the peak intensity and frequency as a function of the %RH is reported in Figure 4. In order to better compare the different sensors, Figure 4a reports the frequency variations and not the frequency values, because the reference value of the sensors (0% RH) depended on the spacer thicknesses (see Figure 3a). Moreover, the frequency differences are shown only in the range where they had a linear trend with the %RH. It is worth noticing that the effects of the frequency shift were only on the second resonance peak, since the first peak was only due to the antenna characteristics (internal probe peak frequency). The data were plotted together with a linear fit of the experimental data, where the slope of the line represents the sensitivity of the sensor. The numerical values for the sensitivity are reported in Table 1 together with the standard error. The Pearson coefficients (R^2) for linearity are also reported. More detailed statistics are given in Supplementary Material S1. The range of linearity of the sensors was estimated by the last measured %RH value that gave an R^2 value larger than 0.99. Even though this is somehow an arbitrary and restrictive criterion, it should be noted that the peak broadening with increasing humidity also increased the error in the determination of the peak position, and this in turn increased the error in the sensitivity determination. The sensitivities reported in Table 1 show good values for all the spacer thicknesses, but they decreased when increasing the distance between the sensing material and the metallic resonator. The best values were given for the sensor with no spacer, where sensitivities of around 1 MHz/%RH were reached, while the sensor with a spacer of 800 µm showed a sensitivity which was about one fifth of this value. The highest values were better than those of other chipless RFID humidity sensors measured with a near-field probe [25,26]. The most sensitive microwave sensors exploit higher frequency ranges [27–30] and show higher shifts, but they require more sophisticated, large and expensive antenna systems, which are hardly suitable for applications like smart packaging. A more comprehensive list reporting the comparable existing literature is given in Table 2. For comparison, two examples of microstripped tags are also included [22,31]. They showed good performances compared with the ones measured with a near-field probe, but the detection was not wireless and required SMA connectors.

The peak intensity as a function of the %RH, reported in Figure 4b, was largest in the low humidity range, especially below 10% RH, for all the sensors investigated and did not have a linear behavior range. Below 10% RH, the peaks were quite sharp and well-defined, allowing for a very precise determination of the peak frequency. In the case of the sensor with no spacer and with the Nafion membrane in direct contact with the metallic resonator, the peak intensity and its variation were by far the largest for extremely low %RH values. In this case, when going from 0% to 5% RH, the intensity reduced to less than half, and the frequency shift was around 5 MHz. However, at higher %RH values, the intensity drop was less evident, and at approximately 20–30% RH, the signal approached a plateau, and the additional variation became difficult to detect.

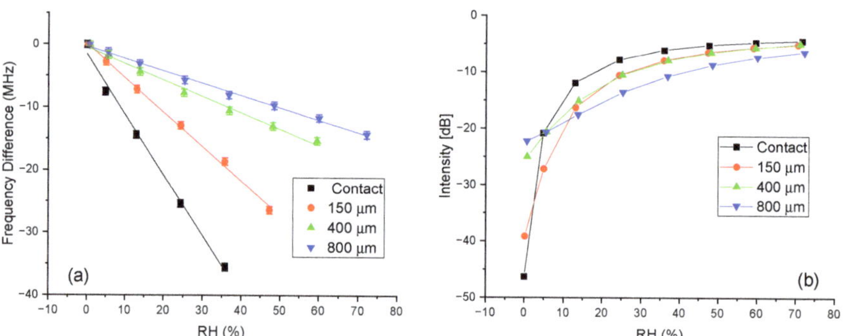

Figure 4. (**a**) Peak frequency variation as a function of the %RH in the range of 0–70% for the four resonators. The value at 0% RH was taken as reference. (**b**) Peak intensity as a function of the %RH in the range of 0–70% for the four resonators.

Table 1. Sensitivity (MHz/%RH) of the investigated sensors.

Spacer Thickness (μm)	Sensitivity (MHz/%RH)	Error (Standard Deviation)	Relative Error (%) [1]	Pearson Square (R^2)	Humidity Range (%)
0	−0.965	0.044	4.6%	0.99375	0–36
150	−0.547	0.014	2.5%	0.99756	0–47
400	−0.260	0.011	4.1%	0.9917	0–60
800	−0.195	0.006	2.8%	0.99529	0–72

[1] Calculated as the standard deviation/sensitivity ratio (%).

Table 2. Performance comparison of chipless sensors for humidity detection from the literature.

Ref.	Substrate	Sensitive Material	Tag Size (cm^2)	Resonator	Reading Range	%RH Range	Frequency Range (GHz)	Frequency Shift	Intensity [1] Variation
[25]	Photopaper	Poly-vinyl alcohol (PVA)	-	Interdigitated LC	2 cm: Near-field probe (NFP)	86–91	0.128–0.132	1.2 KHz/%RH	0.014 dB/%
[32]	Taconic TLX 0	PVA	1.5 × 0.68	ELC	30 cm: horn antennas	35–85	6–7	12 MHz/%RH	0.02 dB/%
[27]	Kapton HN	Substrate	5.4 × 5.4	Van-Atta reflectarray	5.5 m: horn antennas	10–75	25–35	40 MHz/%RH	Not reported
[29]	Kapton HN	Substrate	7.4 × 7.4	Van-Atta reflectarray	58 m: radar	30–75	24	-	0.2 dB/%
[26]	PET, paper	Substrate	4 × 4	Interdigitated LC	0.5 cm: NFP	20–90	0.12–0.3	0.37 MHz/%RH	Not reported
[31]	FR-4	SnO_2/graphene	5 × 2	4 C-shaped resonators	Microstrip (wired)	11–98	2.85–3.05	2.95 MHz/%RH	Negligible
[30]	Polycarbonate	Silicon nanowires	4 × 2	Rectangular loops	Not reported: horn antennas	74–98	3.295–3.33	1.5 MHz/%RH	-
[24]	DiClad	Nafion117	4 × 2	ELC	Microstrip	5–40	2.49–2.60	2.71 MHz/%RH	0.04 dB/%
This [2] work	Rogers 4350	Nafion 212	2 × 2	ELC	0.5 cm: NFP	0–36		0.965 MHz/%RH	1.12 dB%

[1] Average value over the %RH range reported. [2] Values for the contact sensor.

It should be noted that the values reported in Figure 4b include the baseline signal that can be estimated at approximately 3–4 dB, which means that at %RH values greater than 20%, the peak intensity signal was comparable with the baseline or less. The other three sensors had a similar behavior, but their detectable %RH range was larger and increased when increasing the spacer thickness. However, the corresponding drop at low %RH values also decreased. Furthermore, when comparing the intensity variation of the sensors reported in this work with those in Table 2, it is clear that the contact sensor reported in this study showed an intensity variation that exceeded all the other chipless sensors listed. This specificity enables the use of the intensity variation as the primary detection parameter or, alternatively, the use of both the frequency shift and intensity variation as sensing parameters to improve the reliability and sensitivity of the detection. Both of these approaches will be discussed in detail in Sections 3.1 and 3.2.

3.1. Threshold Humidity Sensor

The high sensitivity of the chipless RFID sensors in the low humidity range, coupled with remote signal detection, makes it particularly suitable for smart packaging integrity control in the case of a dry or inert atmosphere inside the package, which is common in the case of perishable goods. The remote monitoring proposed allows for verification of the integrity status of the package's internal atmosphere in a totally noninvasive manner even with non-transparent packages. Moreover, the simple integrity check does not require a precise determination of the humidity but only the detection of a value above a threshold. This can easily be determined by comparing the intensity at the reference frequency of the internal probe with the reference peak frequency (0% RH) of the sensor, requiring only a comparison of the signal at two discrete frequencies, greatly simplifying the detection mechanism and consequently the cost and complexity of the reader.

In Figure 5, the ratio between the intensity of the peak frequency at 0% RH and the probe peak intensity at 1.360 GHz is reported for all four resonators. The dashed horizontal lines are a guide to detect the %RH value, where this ratio is equal to one (black line) or two (red line). The trend is similar to what is reported in Figure 4b, but the ratio at a fixed frequency decayed faster with increasing humidity compared with the peak intensity because the effect of the peak shift added to the resonance broadening. The frequency and intensity of the probe peak remained remarkably constant within the experimental error of the measurement.

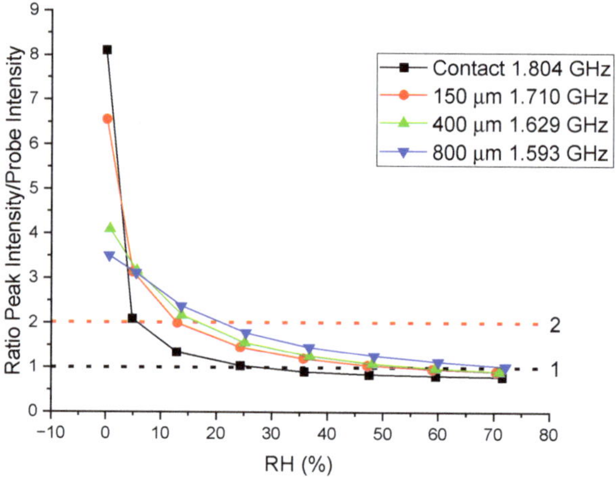

Figure 5. Intensity ratios of the reference peak frequency (0% RH) and the internal probe peak frequency for the four resonators investigated. The probe frequency is 1.360 GHz for all and does not vary with the humidity or spacer thickness of the sensor.

As can be seen from Figure 5, the ratio equal to one was reached only at a high humidity for all the sensors, with the sensor without a spacer crossing the black line at approximately 30%. The ratio equal to two is preferable because it gave a better differentiation among the sensors and a sharper drop near the threshold. The sensor with no spacer crosses the red line at 5%, the 150 μm sensor at 12%, the 400 μm sensor at 15% and the 800 μm sensor near 20% RH.

In order to have a threshold sensor capable of verifying the integrity of a package with an internal dry atmosphere, the sensor without a spacer can be the best choice because of its extremely high intensity drop and its low %RH threshold value. However, plastic packages are never totally hermetic, and a 5% humidity threshold could be too low because the sensor can detect a package integrity break when this is not real. Considering these tolerances, which depend on the specific package material and requirements, a sensor with a thin spacer can be, in this case, a more reasonable choice because the intensity drop is still high, but the threshold humidity can be tuned to higher %RH values.

3.2. Measurement of %RH inside the Package

When the purpose of smart packaging is not only to verify the package integrity but also to give precise information about the level of humidity inside the package, a threshold sensor is not enough, and a more accurate reading of the sensor signal is required. This may be the case for a non-hermetic package or packaged food that produces humidity during the degradation process. In this last case, a high humidity level may be an important indication of the food preservation status. In these conditions, the reading can still be made only at two discrete frequencies to reduce the detection complexity, but it must be much more precise, and the strong nonlinearity of the sensor response is an important problem because the sensitivity varies considerably with the %RH. This variation was extremely high for the contact sensor (%RH range: 0–36%) and the 150 µm sensor (%RH range: 0–47%). Therefore, the thicker spacer sensors are preferable for the usual ambient humidity values, which are seldom below 20%.

For improving the accuracy, the measurement of the frequency shift, which varies linearly, may provide a more reliable determination of the %RH levels compared with the intensity change, even if it needs a more complex reading system. The sensor without a spacer and the 150 µm sensor had the best sensitivities, but they were linear only up to 36 or 47 % RH. They are therefore the best choice only up to these humidity values. If a wide humidity range is required, then sensors with thick spacers must be used. In particular, the 800 µm sensor offered the widest range of linearity (%RH range: 0–70%) for the frequency variation, and its intensity trend approached linearity in the same range as well (see Figure 4b), indicating that the sensitivity did not vary much with the %RH for this parameter. This gives the opportunity to fully exploit the dual parameter detection, increasing the reliability and accuracy of the %RH determination. This not only improves the overall measurement error but can also prevent large errors due to incorrect calibration or sensor malfunctioning because it requires consistency between the peak frequency and intensity. In this case, the probe's internal resonance can be used as a reference for both the frequency and the intensity. The frequency difference between the resonator peak and the probe peak and their intensity ratio are reported in Figure 6, together with the correlation between the two parameters.

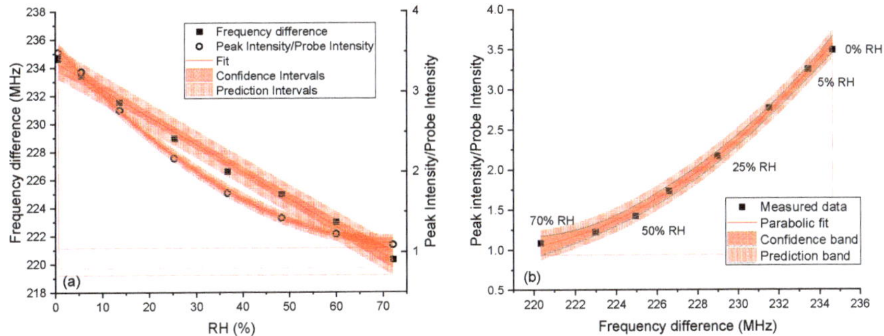

Figure 6. (**a**) Frequency difference (linear fit) and intensity ratio (parabolic fit) for the resonance peak of the 800 µm sensor relative to the probe peak at 1.36 MHz as a function of the %RH. (**b**) Correlation between frequency difference and intensity ratio (parabolic fit).

Together with the data fit, the intervals where the mean %RH value is located (confidence) or the future measurement value is located (prediction) with 95% probability are also reported. As expected, the frequency difference can be conveniently fitted with a linear trend, while the intensity ratio requires a parabolic fit. However, as can be seen from the confidence intervals in Figure 6a, the measurement of the intensity was more accurate than the frequency in the whole humidity range investigated. The fit parameters and the relevant statistics for the fitting functions are reported in Supplementary Material S1.

4. Conclusions

In this paper, we proposed the use of chipless RFID humidity sensors using Nafion 212 as the sensitive material for smart packaging of food items. A new methodology for tailoring the operating range of the sensor was presented based on the variation of the distance between the metallic resonator and the sensing material. Four sensors were proposed with different distances between the sensing material and the metallic resonator. Moreover, given the specificity of the detection mechanism, the proposed detection methodology can make use of two sensing parameters: the peak frequency shift and peak intensity variation. It was then shown how the combined use of the two parameters made the detection more precise and reliable.

Two different sensor applications were examined. In the first case, a threshold sensor for a package with a dry or controlled internal atmosphere was considered. In this case, a sensor able to detect extremely low humidity levels was the best choice to verify the package integrity, but an accurate determination of the %RH was not necessary. We found that the sensor with the sensing material in direct contact with the resonator showed an extremely high intensity variation at low humidity values and was therefore especially suited for this application. In this specific case, the detection at one or two frequencies was enough for integrity verification, and consequently, the reading system could be simplified.

In the second case, a non-hermetic package was examined. For this application, a more exact determination of the humidity is usually required over a wider span of %RH values. The best solution in this case was the sensor with a spacer of 800 μm. This sensor was worse in terms of frequency variation but covered a wider range of humidity. Moreover, it allowed the exploitation of two sensing parameters, the peak frequency and intensity, with the same measurement, increasing the reliability and accuracy of the %RH determination. Although the capabilities of the chipless RFID sensor developed in this work for the detection of different RH% values were successfully demonstrated, it is important to highlight that a laboratory-assembled configuration was required to perform the detection tests, which cannot be used as is for real applications. For the effective implementation of these sensors in smart packages, further research is needed, especially for chipless sensor signal reading technology, and this will be one of the directions for future work.

Supplementary Materials: The following supporting information can be downloaded at https://www.mdpi.com/article/10.3390/s24092879/s1. Contact Sensor: linear fit in Figure 2a, 150 μm Sensor: linear fit in Figure 2a, 400 μm Sensor: linear fit in Figure 2a, 800 μm Sensor: linear fit in Figure 2a, 800 μm Sensor: linear fit in Figure 6a, 800 μm Sensor: parabolic fit in Figure 6a, 800 μm Sensor: parabolic fit in Figure 6b.

Author Contributions: Conceptualization, V.M., A.G., G.M., M.V., M.D. and L.L.; methodology, V.M., A.G., G.M., M.V., M.D. and L.L.; software, V.M., M.V. and G.M.; validation, V.M., G.M. and A.G.; formal analysis, V.M. and G.M.; investigation, V.M., M.V., M.D., L.L. and G.M.; resources, M.D. and M.V.; data curation, V.M. and G.M.; writing—original draft preparation, V.M.; writing—review and editing, M.V., M.D., A.G., L.L. and G.M.; visualization, V.M. and G.M.; supervision, V.M. and M.D.; project administration, M.D. and L.L.; funding acquisition, M.D. and L.L. All authors have read and agreed to the published version of the manuscript.

Funding: This work was partially supported by the project "Smart ElectroMagnetic Environment in TrentiNo—SEME@TN", funded by the Autonomous Province of Trento (CUP: C63C22000720003).

Institutional Review Board Statement: Not applicable.

Informed Consent Statement: Not applicable.

Data Availability Statement: The data are available upon request from the corresponding author.

Acknowledgments: The authors wish to thank Antonio Orlando for the technical support with the experimental set-up for humidity control.

Conflicts of Interest: The authors declare no conflicts of interest.

References

1. Yam, K.L.; Takhistov, P.T.; Miltz, J. Intelligent Packaging: Concepts and Applications. *J. Food Sci.* **2005**, *70*, R1–R10. [CrossRef]
2. Müller, P.; Schmid, M. Intelligent Packaging in the Food Sector: A Brief Overview. *Foods* **2019**, *8*, 16. [CrossRef]

3. Chen, S.; Brahma, S.; Mackay, J.; Cao, C.; Aliakbarian, B. The Role of Smart Packaging System in Food Supply Chain. *J. Food Sci.* **2020**, *85*, 517–525. [CrossRef]
4. Han, J.; Ruiz-Garcia, L.; Qian, J.; Yang, X. Food Packaging: A Comprehensive Review and Future Trends. *Compr. Rev. Food Sci. Food Safe* **2018**, *17*, 860–877. [CrossRef]
5. Athauda, T.; Chandra Karmakar, N. Review of RFID-Based Sensing in Monitoring Physical Stimuli in Smart Packaging for Food-Freshness Applications. *Wirel. Power Transf.* **2019**, *6*, 161–174. [CrossRef]
6. Potyrailo, R.A.; Nagraj, N.; Tang, Z.; Mondello, F.J.; Surman, C.; Morris, W. Battery-Free Radio Frequency Identification (RFID) Sensors for Food Quality and Safety. *J. Agric. Food Chem.* **2012**, *60*, 8535–8543. [CrossRef]
7. Fathi, P.; Karmakar, N.C.; Bhattacharya, M.; Bhattacharya, S. Potential Chipless RFID Sensors for Food Packaging Applications: A Review. *IEEE Sens. J.* **2020**, *20*, 9618–9636. [CrossRef]
8. Mulloni, V.; Donelli, M. Chipless RFID Sensors for the Internet of Things: Challenges and Opportunities. *Sensors* **2020**, *20*, 2135. [CrossRef]
9. Marchi, G.; Zanazzi, E.; Mulloni, V.; Donelli, M.; Lorenzelli, L. Electromagnetic Modeling Strategy Supporting the Fabrication of Inkjet-Printed Chipless RFID Sensors. *IEEE J. Flex. Electron.* **2023**, *2*, 145–152. [CrossRef]
10. Hayati, M.; Majidifar, S.; Sobhani, S.N. Using a Hybrid Encoding Method Based on the Hexagonal Resonators to Increase the Coding Capacity of Chipless RFID Tags. *Int. J. RF Microw. Comput.-Aided Eng.* **2022**, *32*, e23474. [CrossRef]
11. Lalbakhsh, A.; Yahya, S.I.; Moloudian, G.; Hazzazi, F.; Sobhani, S.N.; Assaad, M.; Chaudhary, M.A. Hybrid Encoding Method for Radio Frequency Identification in the Internet of Things Systems. *IEEE Access* **2023**, *11*, 122554–122565. [CrossRef]
12. Bhuiyan, M.S.; Karmakar, N.C. An Efficient Coplanar Retransmission Type Chipless RFID Tag Based on Dual-Band McSrr. *PIER C* **2014**, *54*, 133–141. [CrossRef]
13. Mekki, K.; Necibi, O.; Boulejfen, N.; Larguech, S.; Alwadai, N.; Gharsallah, A. Development of a New 24-Bit High-Performance Chipless RFID Tag for Accurate Identification in IoT Systems. *IEEE Access* **2023**, *11*, 140943–140957. [CrossRef]
14. Herrojo, C.; Paredes, F.; Mata-Contreras, J.; Martín, F. Chipless-RFID: A Review and Recent Developments. *Sensors* **2019**, *19*, 3385. [CrossRef]
15. Ahmadihaji, A.; Izquierdo, R.; Shih, A. From Chip-Based to Chipless RFID Sensors: A Review. *IEEE Sens. J.* **2023**, *23*, 11356–11373. [CrossRef]
16. Zhang, J.; Tian, G.; Marindra, A.; Sunny, A.; Zhao, A. A Review of Passive RFID Tag Antenna-Based Sensors and Systems for Structural Health Monitoring Applications. *Sensors* **2017**, *17*, 265. [CrossRef]
17. Nadeem, M.; Habib, A.; Umair, M.Y. Chipless RFID Based Multi-Sensor Tag for Printed Electronics. *Heliyon* **2024**, *10*, e26494. [CrossRef]
18. Karmakar, N.C.; Amin, E.M.; Saha, J.K. *Chipless RFID Sensors*; John Wiley & Sons: Hoboken, NJ, USA, 2016; p. 274. [CrossRef]
19. Sarma, S.E.; Weis, S.A.; Engels, D.W. RFID Systems and Security and Privacy Implications. In *Cryptographic Hardware and Embedded Systems—CHES 2002*; Kaliski, B.S., Koç, Ç.K., Paar, C., Eds.; Springer: Berlin/Heidelberg, Germany, 2003; pp. 454–469. [CrossRef]
20. Mauritz, K.A.; Moore, R.B. State of Understanding of Nafion. *Chem. Rev.* **2004**, *104*, 4535–4586. [CrossRef]
21. Marchi, G.; Mulloni, V.; Manekiya, M.; Donelli, M.; Lorenzelli, L. A Preliminary Microwave Frequency Characterization of a Nafion-Based Chipless Sensor for Humidity Monitoring. In Proceedings of the 2020 IEEE Sensors, Rotterdam, The Netherlands, 25–28 October 2020; pp. 1–4. [CrossRef]
22. Marchi, G.; Mulloni, V.; Acerbi, F.; Donelli, M.; Lorenzelli, L. Tailoring the Performance of a Nafion 117 Humidity Chipless RFID Sensor: The Choice of the Substrate. *Sensors* **2023**, *23*, 1430. [CrossRef]
23. Mulloni, V.; Gaiardo, A.; Marchi, G.; Valt, M.; Vanzetti, L.; Donelli, M.; Lorenzelli, L. Sub-Ppm NO_2 Detection through Chipless RFID Sensor Functionalized with Reduced SnO_2. *Chemosensors* **2023**, *11*, 408. [CrossRef]
24. Marchi, G.; Mulloni, V.; Hammad Ali, O.; Lorenzelli, L.; Donelli, M. Improving the Sensitivity of Chipless RFID Sensors: The Case of a Low-Humidity Sensor. *Electronics* **2021**, *10*, 2861. [CrossRef]
25. Nabavi, S.; Anabestani, H.; Bhadra, S. A Printed Paper-Based RFID Tag for Wireless Humidity Sensing. In Proceedings of the 2022 IEEE Sensors, Dallas, TX, USA, 30 October–2 November 2022; pp. 1–4.
26. Feng, Y.; Xie, L. Low-Cost Printed Chipless RFID Humidity Sensor Tag for Intelligent Packaging. *IEEE Sens. J.* **2015**, *15*, 3201–3208. [CrossRef]
27. Hester, J.; Tentzeris, M. Inkjet-Printed Van-Atta Reflectarray Sensors: A New Paradigm for Long-Range Chipless Low Cost Ubiquitous Smart Skin Sensors of the Internet of Things. In Proceedings of the 2016 IEEE MTT-S International Microwave Symposium (IMS), San Francisco, CA, USA, 22–27 May 2016. [CrossRef]
28. Hester, J.; Tentzeris, M. Inkjet-Printed Flexible Mm-Wave Van-Atta Reflectarrays: A Solution for Ultralong-Range Dense Multitag and Multisensing Chipless RFID Implementations for IoT Smart Skins. *IEEE Trans. Microw. Theory Tech.* **2016**, *64*, 4763–4773. [CrossRef]
29. Henry, D.; Hester, J.; Aubert, H.; Pons, P.; Tentzeris, M. Long Range Wireless Interrogation of Passive Humidity Sensors Using Van-Atta Cross-Polarization Effect and 3D Beam Scanning Analysis. In Proceedings of the 2017 IEEE MTT-S International Microwave Symposium (IMS), Honolulu, HI, USA, 4–9 June 2017; p. 819.
30. Vena, A.; Perret, E.; Kaddour, D.; Baron, T. Toward a Reliable Chipless RFID Humidity Sensor Tag Based on Silicon Nanowires. *IEEE Trans. Microw. Theory Tech.* **2016**, *64*, 2977–2985. [CrossRef]

31. Tao, B.; Feng, L.; Miao, F.; Zang, Y. High Sensitivity Chipless RFID Humidity Sensor Tags Are Based on SnO_2/G Nanomaterials. *Vacuum* **2022**, *202*, 111126. [CrossRef]
32. Amin, E.M.; Karmakar, N.C. Development of a Low Cost Printable Humidity Sensor for Chipless RFID Technology. In Proceedings of the 2012 IEEE International Conference on RFID-Technologies and Applications (RFID-TA), Nice, France, 5–7 November 2012; pp. 165–170.

Communication

Point-of-Care Fluorescence Biosensing System for Rapid Multi-Allergen Screening [†]

Silvia Demuru *,[‡], **Hui Chai-Gao** [‡], **Yevhen Shynkarenko** [‡], **Nicola Hermann, Patricia-Daiana Boia, Peter Cristofolini, Bradley Petkus, Silvia Generelli, Samantha Paoletti, Stefano Cattaneo and Loïc Burr** *

Swiss Center for Electronics and Microtechnology (CSEM), 7302 Landquart, Switzerland;
hui.chai-gao@csem.ch (H.C.-G.); yevhen.shynkarenko@csem.ch (Y.S.); stefano.cattaneo@csem.ch (S.C.)
* Correspondence: silvia.demuru@csem.ch (S.D.); loic.burr@csem.ch (L.B.)
[†] This is an expanded research article based on the conference paper "Portable Fluorescence Biosensing System for Low-Cost, Quantitative, and Multiplexed Allergen Screening" that was presented at EUROSENSORS2023 conference, 10–13 September 2023 in Lecce, Italy.
[‡] These authors contributed equally to this work.

Abstract: With the steady increase in allergy prevalence worldwide, there is a strong need for novel diagnostic tools for precise, fast, and less invasive testing methods. Herein, a miniatured fluorescence-based biosensing system is developed for the rapid and quantitative detection of allergen-specific immunoglobulin-E. An antibody-based fluorescence assay in a microfluidic-patterned slide, combined with a custom-made portable fluorescence reader for image acquisition and user-friendly software for the data analysis, enables obtaining results for multiple allergens in just ~1 h with only 80 μL of blood serum. The multiplexed detection of common birch, timothy grass, cat epithelia, house dust mite, and dog epithelia shows quantitative IgE-mediated allergic responses to specific allergens in control serum samples with known total IgE concentration. The responses are verified with different control tests and measurements with a commercial fluorescence reader. These results open the door to point-of-care allergy screening for early diagnosis and broader access and for large-scale research in allergies.

Keywords: biosensor; allergy; allergen; IgE; antibody; microfluidics; point of care; fluorescence; optical reader; portable reader

Citation: Demuru, S.; Chai-Gao, H.; Shynkarenko, Y.; Hermann, N.; Boia, P.-D.; Cristofolini, P.; Petkus, B.; Generelli, S.; Paoletti, S.; Cattaneo, S.; et al. Point-of-Care Fluorescence Biosensing System for Rapid Multi-Allergen Screening. *Sensors* **2024**, *24*, 3280. https://doi.org/10.3390/s24113280

Academic Editors: Bruno Ando, Luca Francioso and Pietro Siciliano

Received: 16 April 2024
Revised: 8 May 2024
Accepted: 16 May 2024
Published: 21 May 2024

1. Introduction

The incidence of immune-mediated disorders such as allergies is on a consistent rise [1,2]. Yet, there remains a significant gap in our understanding regarding the interplay of various factors such as genetic predisposition, exposures during prenatal and postnatal life, and environmental influences in the initiation and progression of these conditions [3].

The current landscape of allergy diagnostics, while rich with historical significance and clinical utility, highlights a pressing need for innovation to enhance patient comfort and provide quantitative data for large-scale research and diagnostics [4]. Traditional methods such as the Skin Prick Test (SPT) and oral food challenges, although foundational in the identification and confirmation of allergen sensitivities, are often criticized for their invasiveness and the discomfort they may cause to patients [4–6]. Also, these techniques face limitations in their qualitative nature, which may not fully satisfy the requirements of modern, data-driven allergy research and personalized medicine approaches.

As the field progresses towards more sophisticated molecular diagnostics and precision treatments, there is a noticeable demand for developing less painful, more quantitatively precise methodologies [4,7]. This shift aims not only to improve patient experience but also to enable the detailed, large-scale data collection necessary for advancing our understanding of allergic diseases [8]. Innovative solutions are urgently needed to bridge the gap between traditional practices and the potential for targeted, data-rich diagnostic approaches, thus opening the door to more effective and less invasive detection [9–11]. Other methods of significant importance for allergy diagnostics is the measurement of specific immunoglobulin-E (sIgE) in serum by enzyme-linked immunosorbent assay (ELISA) and fluorescence enzyme immune assay (FEIA) [7]. While fluorescence-based assays are generally analyzed in standard laboratory settings, after the shipment of the blood samples, and using commercial microarray scanners, the long time needed for obtaining the results, their cost, and the size of the commercial scanners render these methods impractical for point-of-care settings, limiting broader and faster access to allergy diagnostic tools.

In this work, we present the development of a novel portable multiarray system tailored for specific allergen profiling. The multiarray system is engineered on microfluidic chips sized equivalently to standard microscopy slides and is integrated within a microfluidic channel featuring functionalized microstructures with recombinant proteins of allergens. To ensure the reliable detection of sIgE, an automated sample-on-chip processing system has been utilized, employing fluorescence-labeled antibodies to detect the signal from each pillar with different allergens. Moreover, to facilitate simple measurements of fluorescence signals and automated quantification of the sIgE responses, a compact, cost-effective, and rapid fluorescence reader has been developed. This reader is equipped with cross-platform software having a user-friendly interface, further enhancing its accessibility and utility.

This platform concept holds promise in facilitating rapid screening of allergen specific IgEs in various settings, such as doctor's offices, pharmacies, or university laboratories, for studying allergic disease causation and possible prevention strategies.

2. Materials and Methods

The microfluidic system, resembling a standard microscopy slide, was fabricated through injection molding of polycarbonate, with pillars of 300 μm in diameter and spaced 500 μm center to center in the same channel, with a channel width of 750 μm. The slides underwent coating with a dextran-based photo-linker polymer (OptoDex®, custom from CSEM, Landquart, Switzerland), serving for immobilizing allergens onto the micropillars, passivating the surface to mitigate non-specific bindings, and rendering the microfluidic channel hydrophilic.

Allergen recombinant proteins (including cat and dog epithelia, house dust mite, and common birch) and one allergen extract (timothy grass) were precisely dispensed onto the slides using a high-precision dispenser (sciDROP PICO, Scienion AG, Berlin, Germany) adding ~400 pL drop per pillar of protein solution, at an optimized concentration of 0.8 mg/mL in all cases, each allergen deposited six times for statistical analysis. All the recombinant proteins were purchased from Inbio (Cardiff, UK) and the allergen extract from BÜHLMANN Laboratories AG (Schönenbuch, Switzerland). Betula pendula (rBet v 1) was purified from Pichia pastoris culture by HPLC gel filtration; Felis domesticus (rFel d 1) from Pichia pastoris, clone N103Q, N-glycosylation site N103 mutated, was purified by multi-step column chromatography; house dust mite (rDer p 1) from Pichia pastoris (clone N52Q, N-glycosylation site N52 mutated) was purified by affinity chromatography; Canis familiar (rCan f 1) from Pichia culture supernatant using His-Trap chromatography. All recombinant proteins had a purity > 95% by SDS-PAGE. Finally, Timothy grass (BAG-G6) was an allergen extract (natural allergen) from Phleum Pratense. Also, Atto 647 Bioreagent (Sigma Aldrich, Buchs, Switzerland), was used to label BSA (Bovine serum albumin) for the fluorescence signal calibration, with concentrations from 10 to 5000 dye molecules/μm^2, each concentration deposited in triplicate for each slide.

For the control tests in each slide, human IgE (Abcam, Amsterdam, The Netherlands) and human IgG (Jackson ImmunoResearch Labs, West Grove, PA, USA), at a concentration IgE/BSA and IgG/BSA of 0.4/0.2 mg/mL, were added into the slide, each solution deposited in six pillars.

Subsequently, a photo-immobilization process was initiated using a UV chamber (2 min at 20 mW/cm^2, Beltron GmbH, Rödermark, Germany). With the process above, the slides with the immobilized proteins have long-term stability for at least 12 months when stored at 4 °C under vacuumed package (400 mbar nitrogen). Following lamination of a thin plastic foil (Simport T329-1, Simport-Saint-Mathieu-de-Beloeil, QC, Canada), control serum samples (PathTROL, TECOmedical AG, Sissach, Switzerland) were added into the fluidic slide channels to validate the efficacy of the multiarray. The control serum sample had a known total IgE concentration with verified presence of IgEs with specificity to the following allergens: timothy grass, cultivated rye (*S. cereale*), mugwort (*A. vulgaris*), house dust mite (*D. pteronyssinus*), house dust mite (*D. farinae*), cat epithelia, hazelnut, peanut, and milk.

Parallelized sample handling, accommodating up to 6 slides simultaneously, was achieved utilizing custom technology, with 80 µL of serum sample injected per slide with a peristaltic pump and incubated for ~30 min at 37 deg. Subsequent to sample deposition, detection antibodies (Mouse anti-hIgE) and AlexaFluor 647 Goat-Anti-mIgG (80 µL each, both from Life Technologies Europe, Zug, Switzerland) were introduced into the system and incubated for another ~45 min at 37 deg.

Measurement assessments were conducted utilizing both a commercial microarray reader (InnoScan 710, Innopsys, Carbonne, France) and a custom-made portable reader for a comparative analysis of their performance.

3. Results

3.1. Development of the Portable Fluorescence System

3.1.1. Microarray Assay

A microarray chip featuring over 300 micropillars was designed (Figure 1a). This chip comprises a meandering channel featuring an array of microstructures functionalized with allergen recombinant proteins, with immobilization facilitated through the Optodex chemistry. Upon introduction of patient serum onto the functionalized chip, there is a recognition between the sIgEs and the allergen forming a complex, and the subsequent reaction with fluorescence labeled anti-IgE antibodies reveals sIgEs through distinctive fluorescence responses from each micropillar (Figure 1b).

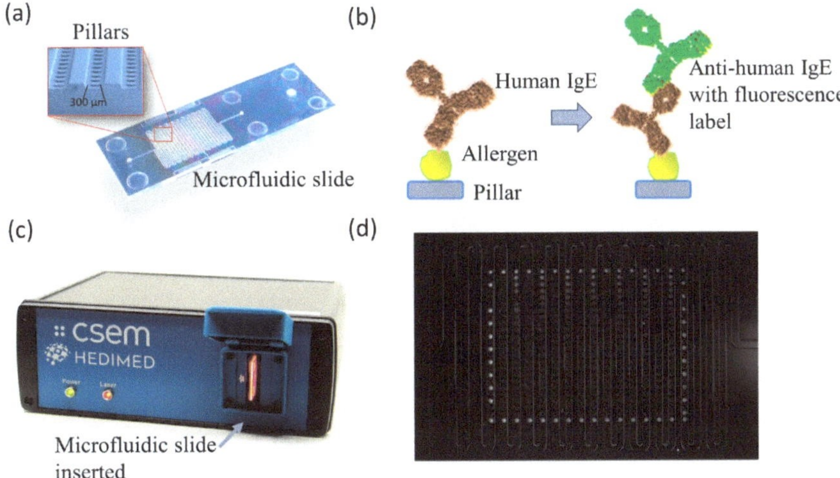

Figure 1. Portable system for allergen screening. (**a**) Image showing the microfluidic slide with a zoom on the micropillars; (**b**) simplified schematic of the biosensing assay on the slide; (**c**) image of the portable reader with a slide inserted; (**d**) the fluorescent signals on the micropillars with an image acquired with the portable reader and custom software.

3.1.2. Portable Reader

A portable reader was engineered for the fluorescence-based bio-detection (Figure 1c). Differing from commercial slide scanners that employ confocal scanning, this reader leverages fluorescence imaging, thereby reducing both size and cost while enhancing speed, albeit with a slightly reduced sensitivity (Figure 2). Key components of the reader include a camera, collection and beam-forming optics, a light source, emission and excitation filters, and a light trap (Figure 2a). A 642 nm fiber laser was selected as the light source, aligning with the spectral properties of the chosen dye Alexa-647. Light-forming optics ensure uniform sample illumination, while spectral filtering is accomplished through a combination of color glass and interferometric filters. Fluorescence emission was captured using a monochromatic CMOS camera featuring manual control for optimal imaging parameters. Notably, the portable reader boasts compact dimensions of $20 \times 16 \times 7$ cm^3, weighing a mere 1.5 kg, and comes at a fraction of the cost compared to conventional microarray readers, while still meeting the rigorous performance criteria essential for this assay.

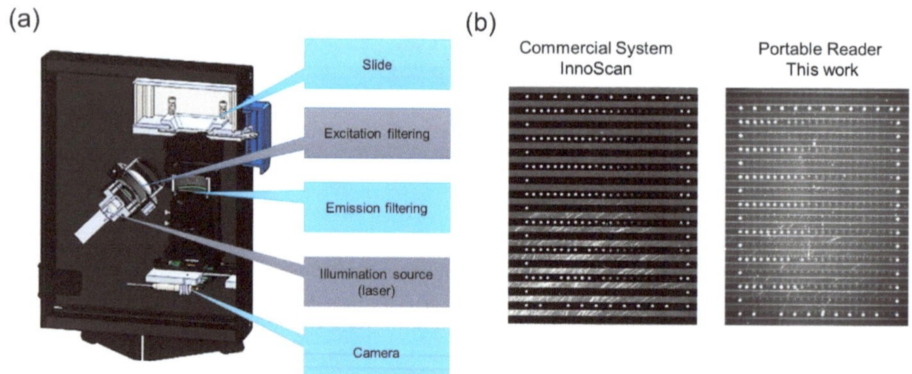

Figure 2. Portable reader components and comparison. (**a**) Schematic of the different components inside the developed portable reader; (**b**) Comparison of the data acquired by InnoScan and the portable reader, analyzing the microfluidic slide with different gradients of the fluorescent dye molecules Atto-BSA. Lower part shows correspondent line profile of the gradient from the lowest to the higher (24 points from about 10 to 5000 dye mol/μm^2).

Figure 2b shows the same slide read with the commercial InnoScan system and with the custom reader and correspondent line profiles. All 24 printed gradient points can be resolved with an InnoScan reader with 13 dye mol/μm^2 distinguishable and with a standard deviation of the background signal equal to 6.4 fluorescence intensity (n = 124 points) and a sensitivity of 0.84 ± 0.04 fluorescence intensity signal/(dye mol/μm^2), n = 5, giving a detection limit (LOD) calculated as $3\sigma/S$ (σ = standard deviation of the blank/S = sensitivity) equal to ~20 dye mol/μm^2. In our compact reader, with a higher background level and higher noise, only 15 out of 24 concentration points can be differentiated from the noise, reaching the lowest distinguishable point of 96 dye mol/μm^2, and with a blank signal variation equivalent to 76 fluorescence intensity (n = 124 points). Taking this into account, it can be calculated that the LOD of the portable device is $3\sigma/S$, with S equal to 1.1 ± 0.06 fluorescence intensity signal/(dye mol/μm^2), n = 5, so we have an LOD = ~200 dye mol/μm^2 (σ = standard deviation/S = sensitivity), hence 10X higher than the commercial system.

The measurement spectral range can be defined by the set of filters chosen for the Alexa-647 dye and the camera quantum efficiency; for our chosen components, it is ~670–1100 nm. Considering the dynamic range based on the normalized dye molecule concentration characterization, the LOD was estimated to be 200 dye mol/μm^2 as reported above, and the higher concentration could be measured by reducing the integration time and analog gain of the camera. So, theoretically, by using both factors, samples with dye concentration from ~200 dye mol/μm^2 up to about ~10,000,000 dye mol/μm^2 could be measured with the same system using only software control without any hardware changes. However, reaching a very high dye concentration does not make practical sense, and the current upper limit of the range used in the slides is 5000 dye mol/μm^2, which is still in the linear range. Then, the signal-to-noise level for the single measurement without variation in the integration time is estimated for a relevant signal level to be about SNR = ~130 (Signal/NoiseRMS = 10,000/76).

3.1.3. The Custom Software

The portable reader allows for full-slide imaging, markedly reducing readout times, as shown in Figures 1d and 2b with the pictures acquired from the portable reader with the custom software. The software has a locally hosted backend written in Python code, which is responsible for all the computational parts and data management. Also, we have developed a user-friendly frontend which is programmed using the software Flutter (Figure 3).

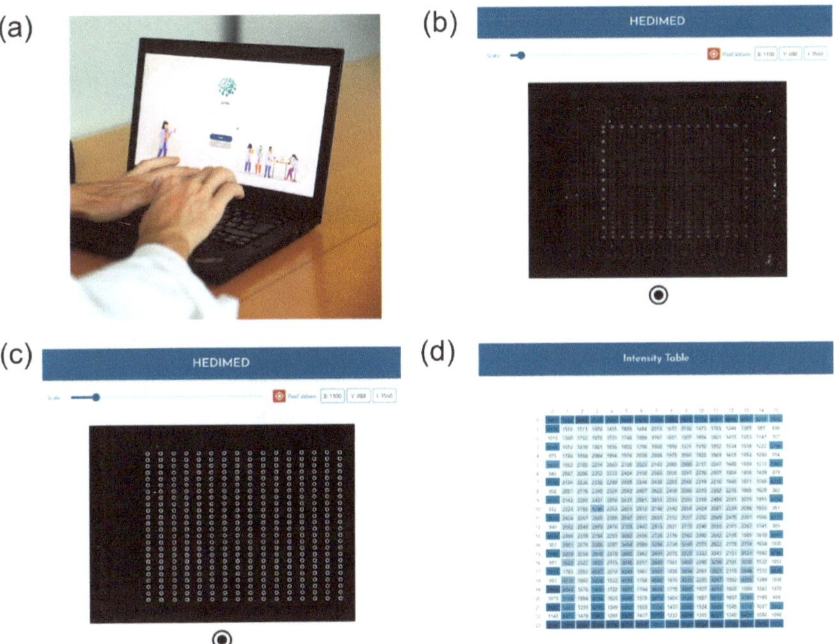

Figure 3. Images showing the custom software features. (**a**) Portable reader connected to the laptop with the program and login into the user interface; (**b**) the feature inside the program to acquire the image of the slide; (**c**) automated pillar recognition; (**d**) automated fluorescence data extraction and possibility to export in different Excel reports; the colors represent different fluorescence intensities.

Inside the software, you can login to your user profile (Figure 3a), acquire in about 1 s the image of the slide (Figure 3b), then automatically detect and identify the pillars in the slide (Figure 3c), and visualize and process the fluorescence intensity data extracted from the image in a table format (Figure 3d). Subsequently, the data can be exported and analyzed identically to the data extracted from commercial devices in a customizable excel sheet showing several resulting fluorescence intensity bar plots, as the ones reported in the data in the next paragraph. The fluorescence frame with high concentration of Atto/BSA dye enables the automated recognition of the area with the pillars.

The imagining area was optimized for our microfluidic slide to be about 2 cm $\times$ 1.5 cm. As for image acquisition speed, we acquire 10 images (~1 s); then, we take the averages of the images to reduce noise (basically instant, ~3 ms), and finally, we have the localization of points (~1 s) resulting in a total time of ~2 s. Considering the final software settings without any background subtraction and the rows of the microfluidic array from 1 to 8, the uniformity of the acquired signal can be estimated as the minimum background (I_{min}) intensity divided by the average value (I_{mean}), obtaining a percentage uniformity of $(I_{min})/(I_{mean}) = (1735/2317) \times 100 = 75\%$ ($n = 61$ for the mean). Considering the frame around the whole slide, we noticed a decrease in uniformity (65%, $n = 76$) possibly due to a non-uniform illumination, which could be corrected by hardware modification, software correction, or in the assay by performing a calibration with the dye molecule in both sides of the slide. In the reported data below, the results are all for allergens and calibration dye deposited in the first eight rows of the microfluidic array, hence with the 75% estimated uniformity.

3.2. Specific IgE Detection in Control Serum Samples

The performance of the portable reader was evaluated through comparison with the commercial confocal microarray reader after the allergen immobilization and incubation with control serum. Control serum samples with known total IgE concentration, equivalent to the minimum IgE concentration produced in case of the possible presence of an allergic condition such as allergic rhinitis, asthma, and atopic dermatitis [12,13], are used for the validation of our assay and reader. The images acquired with both readers can be seen in Figure 4a. The fluorescence signal obtained from the microarray system is depicted in Figure 4b. Utilizing human IgE positive control signals (represented by the brown bar) along with a negative control (human IgG, depicted by orange bar), the crosstalk threshold can be established. All the reported signals are <5 to 10% of variations between the six replicated with the same allergen in different pillars and the control signals in triplicates.

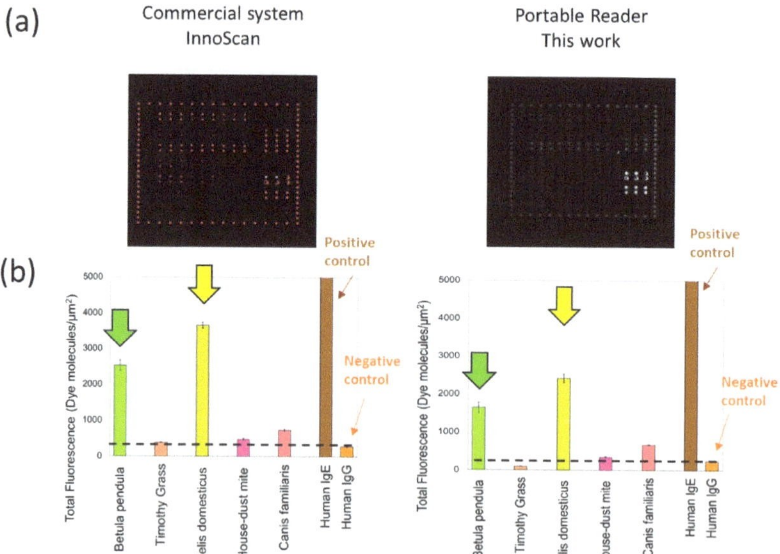

Figure 4. Results for specific IgE detection. (**a**) Optical image acquired by InnoScan and the portable reader, using a microfluidic slide with different recombinant allergens; (**b**) Relative fluorescence signals extracted and converted in dye molecules/μm^2 based on the internal Atto BSA dye calibration. Tested with control serum (274 IU/mL total IgE). The dashed lines indicate the signal over which there is a significant IgE signal, highlighting the maximum cross-sensitivity signal from the IgG negative control.

The results demonstrate the detection capability for different allergens, particularly common birch (Betula pendula, indicated by green bars) and cat epithelia (Felis domesticus, depicted by yellow bars) with both the readout systems. The signal for house dust mite (Der. pteronyssinus) and dog epithelia (Canis familiaris), while surpassing the IgE crosstalk signal, has a much lower signal indicating low specific IgEs present for these allergens in the assay. Specifically, for the commercial reader, the crosstalk signal for the commercial scanner is equal to 281 ± 12 dye mol/μm^2, and the signal for house-dust mite and Canis familiaris is $1.7\times$ and $2.6\times$ higher, respectively, and $9\times$ and $13\times$ higher for Betula pendula and Felis domesticus, respectively. In the case of the portable reader, the crosstalk signal extracted is of 239 ± 14 dye mol/μm^2, being similarly slightly higher for house-dust mite and Canis familiaris ($1.4\times$ and $2.8\times$, respectively) and significantly higher for Betula pendula and Felis domesticus ($6.9\times$ and $10.1\times$, respectively). This corroborates with control serum samples testing positive for cat and house dust mites, while common birch and Canis familiars, not included in the manufacturer's test, could not be directly confirmed. The reference test from the manufacturer can be requested online for the product PathTROL Allergy Control Sera-Level 2 (PW80201). It is possible that the very high specific IgE measured for common birch is due to a cross-reactivity with the specific IgEs for Cultivated rye [14], known to be present in the serum, and being in a similar birch and grass pollen family. The same is applied to different animal allergen, as also to Felis domesticus and Canis familiaris.

It is also important to remark that since there are appreciable total IgE levels in patients with allergen sensitization as well as those without, sIgE instead of total IgEs are the preferred method for determining the presence of allergic diseases [13]. Total IgE concentration is the addition of all the sIgE to the different allergens the individual has been exposed to; in non-allergic subjects, sIgE levels are very low and undetectable [7]. Thus, to identify the triggering antigen of allergic manifestations, one of the most common laboratory test requirements is the determination of sIgE concentration in serum, that can be performed in a point-of-care fashion with our device.

4. Discussion

The development of the portable fluorescence reader presents a viable option for point-of-care allergy diagnostics and broader biosensing applications. Its compact design (size $20\times$ smaller than standard commercial readers), lower cost ($40\times$ lower considering the prices of the current components used, ~1.5 kEUR), and rapid readout time ($60\times$ faster considering 1 min scan with Innoscan versus 1 s with the portable reader) make it highly accessible, with features including fast image acquisition and automated result computation enhancing usability. The signal uniformity in the whole slide can be easily improved by software correction or minor hardware and assay modification. The portable solution demonstrated significant IgE-sensitivity for specific allergen detection, with a performance comparable to the commercial system in terms of increased specific IgE signal over the IgG negative control signal. The system's adaptability to other fluorophores further underscores its versatility and potential for multi-allergen, multi-antibody, and multi-analyte detection in diverse clinical settings. The system also facilitates the parallel testing of up to 88 different allergens, considering the current protocol and number of replicates for control and internal calibration, thus presenting novel avenues for allergen screening directly within clinical settings. The commercial product ImmunoCAP Immuno Solid-Phase Allergen Chip (ISAC) by Thermo Fisher Scientific, after further development through the European Union (EU) project MeDALL, was reported to include, in 2014, a total of 176 allergens, of which 127 were recombinant allergens and the rest natural allergens, [15,16] including multiple allergens from different allergen sources such as latex, peanut, wheat, alder, olive, Plane Tree, Goosefoot, annual mercury, and many others. Similarly, these different allergens can be deposited in our developed microfluidic slide and used for the parallel multiplexed analysis. Finally, we would like to point out that as a secondary antibody for the immunodetection, the commercially available fluorescent-labeled anti-human IgE antibody is still limited by its low activity. The development of a high-active fluorescent-labeled anti-human IgE antibody conjugate could be carried out in the future with possible collaborations. Thus, the sensitivity of the assay can be also increased by using labeled anti-human IgE antibody directly instead of the mouse anti-human IgE/anti-mouse IgG system currently employed. Further tests with several serum samples and standardization of the sIgE signal output in µg, ng, or IU through collaboration with allergen diagnostic associations and doctors could enable the widespread use of this technology in clinical settings and for standardized research.

Author Contributions: Conceptualization, L.B., Y.S., H.C.-G., S.D., S.P. and S.C.; software development, N.H. and P.-D.B.; validation, L.B., S.D., H.C.-G., Y.S., P.C. and B.P.; formal analysis, L.B., S.D., Y.S. and H.C.-G.; writing—review and editing, S.D.; visualization, S.D., Y.S. and H.C.-G.; supervision, L.B., S.P. and S.C.; project administration, L.B., S.C., S.P., S.G. and S.D.; funding acquisition, L.B., S.G., S.P. and S.C. All authors have read and agreed to the published version of the manuscript.

Funding: This project has received funding from the European Union's Horizon 2020 research and innovation program under grant agreement No 874864 HEDIMED.

Institutional Review Board Statement: Not applicable.

Informed Consent Statement: Not applicable.

Data Availability Statement: The data that support the findings of this study are available upon reasonable request.

Acknowledgments: The authors would like to thank the HEDIMED Investigator Group for their support and suggestions on this research and potential use towards a better understanding of immune-mediated diseases.

Conflicts of Interest: The authors declare no conflicts of interest.

References

1. Huang, F.; Jia, H.; Zou, Y.; Yao, Y.; Deng, Z. Exosomes: An Important Messenger in the Asthma Inflammatory Microenvironment. *J. Int. Med. Res.* **2020**, *48*, 0300060520903220. [CrossRef] [PubMed]
2. Pawankar, R. Allergic Diseases and Asthma: A Global Public Health Concern and a Call to Action. *World Allergy Organ. J.* **2014**, *7*, 12. [CrossRef] [PubMed]
3. Laiho, J.E.; Laitinen, O.H.; Malkamäki, J.; Puustinen, L.; Sinkkonen, A.; Pärkkä, J.; Hyöty, H.; Anna, E.; Heikki, H.; Kalle, K.; et al. Exposomic Determinants of Immune-Mediated Diseases: Special Focus on Type 1 Diabetes, Celiac Disease, Asthma, and Allergies: The HEDIMED Project Approach. *Environ. Epidemiol.* **2022**, *6*, e212. [CrossRef] [PubMed]
4. World Allergy Organization. *White Book on Allergy*, 2011th ed.; Pawankar, R., Canonica, G.W., Holgate, S.T., Lockey, R.F., Eds.; World Allergy Organization (WAO): London, UK, 2011; pp. 1–225.
5. Nowak-Węgrzyn, A.; Assa'ad, A.H.; Bahna, S.L.; Bock, S.A.; Sicherer, S.H.; Teuber, S.S.; Adverse Reactions to Food Committee of the American Academy of Allergy, Asthma & Immunology. Work Group Report: Oral Food Challenge Testing. *J. Allergy Clin. Immunol.* **2009**, *123*, S365–S378. [CrossRef] [PubMed]
6. Bousquet, J.; Heinzerling, L.; Bachert, C.; Papadopoulos, N.G.; Bousquet, P.J. Practical Guide to Skin Prick Tests in Allergy to Aeroallergens. *Allergy Eur. J. Allergy Clin. Immunol.* **2012**, *67*, 18–24. [CrossRef] [PubMed]
7. Salazar, A.; Velázquez-soto, H.; Ayala-balboa, J.; Jiménez-martínez, M.C.; Salazar, A.; Velázquez-soto, H.; Ayala-balboa, J. *Allergen-Based Diagnostic: Novel and Old Methodologies with New Approaches*, 2017th ed.; Athari, S.S., Ed.; IntechOpen: London, UK, 2017.
8. Palareti, G.; Legnani, C.; Cosmi, B.; Antonucci, E.; Erba, N.; Poli, D.; Testa, S.; Tosetto, A. Comparison between Different D-Dimer Cutoff Values to Assess the Individual Risk of Recurrent Venous Thromboembolism: Analysis of Results Obtained in the DULCIS Study. *Int. J. Lab. Hematol.* **2016**, *38*, 42–49. [CrossRef] [PubMed]
9. Kesler, V.; Murmann, B.; Soh, H.T. Going beyond the Debye Length: Overcoming Charge Screening Limitations in Next-Generation Bioelectronic Sensors. *ACS Nano* **2020**, *14*, 16194–16201. [CrossRef] [PubMed]
10. Herbáth, M.; Papp, K.; Balogh, A. Applications of Antibodies in Microfluidics-Based Analytical Systems: Challenges and Strategies for Success Exploiting Fluorescence for Multiplex Immunoassays on Protein Microarrays. *J. Micromech. Microeng.* **2018**, *28*, 063001.
11. Chi, J.; Wu, Y.; Qin, F.; Su, M.; Cheng, N.; Zhang, J.; Li, C.; Lian, Z.; Yang, X.; Cheng, L.; et al. All-Printed Point-of-Care Immunosensing Biochip for One Drop Blood Diagnostics. *Lab Chip* **2022**, *22*, 3008–3014. [CrossRef] [PubMed]
12. Rengganis, I.; Rambe, D.S.; Rumende, C.M.; Abdullah, M. Total Serum IgE Levels among Adults Patients with Intermittent and Persistent Allergic Asthmas. *Med. J. Indones* **2018**, *27*, 279–283. [CrossRef]

13. Hamilton, R.G. *Allergy Clinical and Research Laboratory—Immunological Methods in the Diagnostic Allergy Clinical and Research Laboratory*, 6th ed.; Rose, N., Hamilton, R., Detrick, B., Eds.; John Wiley & Sons: Washington, DC, USA, 2002; pp. 883–890.

14. Vallier, P.; DeChamp, C.; Valenta, R.; Vial, O.; Deviller, P. Purification and Characterization of an Allergen from Celery Immunochemically Related to an Allergen Present in Several Other Plant Species. Identification as a Profilin. *Clin. Exp. Allergy* **1992**, *22*, 774–782. [CrossRef] [PubMed]

15. Tscheppe, A.; Breiteneder, H. Recombinant Allergens in Structural Biology, Diagnosis, and Immunotherapy. *Int. Arch. Allergy Immunol.* **2017**, *172*, 187–202. [CrossRef] [PubMed]

16. Lupinek, C.; Wollmann, E.; Baar, A.; Banerjee, S.; Breiteneder, H.; Broecker, B.M.; Bublin, M.; Curin, M.; Flicker, S.; Garmatiuk, T.; et al. Advances in Allergen-Microarray Technology for Diagnosis and Monitoring of Allergy: The MeDALL Allergen-Chip. *Methods* **2014**, *66*, 106–119. [CrossRef] [PubMed]

Article

Real-Time Monitoring of Odour Emissions at the Fenceline of a Waste Treatment Plant by Instrumental Odour Monitoring Systems: Focus on Training Methods

Christian Ratti, Carmen Bax *, Beatrice Julia Lotesoriere and Laura Capelli

Department of Chemistry, Materials and Chemical Engineering "Giulio Natta", Politecnico di Milano, Piazza Leonardo da Vinci 32, 20133 Milan, Italy; christian.ratti@polimi.it (C.R.); beatricejulia.lotesoriere@polimi.it (B.J.L.); laura.capelli@polimi.it (L.C.)
* Correspondence: carmen.bax@polimi.it

Abstract: Waste treatment plants (WTPs) often generate odours that may cause nuisance to citizens living nearby. In general, people are becoming more sensitive to environmental issues, and particularly to odour pollution. Instrumental Odour Monitoring Systems (IOMSs) represent an emerging tool for continuous odour measurement and real-time identification of odour peaks, which can provide useful information about the process operation and indicate the occurrence of anomalous conditions likely to cause odour events in the surrounding territories. This paper describes the implementation of two IOMSs at the fenceline of a WTP, focusing on the definition of a specific experimental protocol and data processing procedure for dealing with the interferences of humidity and temperature affecting sensors' responses. Different approaches for data processing were compared and the optimal one was selected based on field performance testing. The humidity compensation model developed proved to be effective, bringing the IOMS classification accuracy above 95%. Also, the adoption of a class-specific regression model compared to a global regression model resulted in an odour quantification capability comparable with those of the reference method (i.e., dynamic olfactometry). Lastly, the validated models were used to process the monitoring data over a period of about one year.

Keywords: e-nose; continuous monitoring; odour measurement; odour concentration; chemical sensors; gas sensing

Citation: Ratti, C.; Bax, C.; Lotesoriere, B.J.; Capelli, L. Real-Time Monitoring of Odour Emissions at the Fenceline of a Waste Treatment Plant by Instrumental Odour Monitoring Systems: Focus on Training Methods. *Sensors* **2024**, *24*, 3506. https://doi.org/10.3390/s24113506

Academic Editors: Bruno Ando, Luca Francioso and Pietro Siciliano

Received: 18 March 2024
Revised: 20 May 2024
Accepted: 21 May 2024
Published: 29 May 2024

1. Introduction

Nowadays, odour pollution is one of the main causes of public complaints to local authorities [1]. Several human activities, especially those related to waste treatment and disposal, can lead to the formation of unpleasant odours. Even though odour emissions are, in general, not harmful to human health, they often generate concerns among the people living in the proximity of these sources [2,3]. Therefore, odour is now considered a pollutant, and it has also been included in the Best Available Techniques (BAT) reference document for waste treatment plants (WTPs) among the pollutants to be monitored and controlled from this type of installation [4]. Odour monitoring, which is applied whenever an odour nuisance at a sensitive receptor is expected or has been substantiated, can be carried out using analytical methods or sensorial approaches [5]. Among analytical tools, Instrumental Odour Monitoring Systems (IOMSs), including electronic noses (e-noses), have been recently increasingly studied for air quality monitoring applications [6,7]. Several studies prove that, if properly trained, IOMSs are able to detect the presence or the absence of odours and discriminate between different odour emission sources [8,9]. The first studies describing the application of e-noses for the monitoring of odours at receptors for a limited duration date back to the early 2000s–2010s [10,11]. Recently, some scientific studies have also discussed the possibility of using e-noses for estimating odour concentration [12,13].

For more details on the applicability of e-noses for environmental monitoring, we refer to some very recent review papers on the topic [6,7]. Still, their application as real-time air quality monitoring tools for permanent installations is still far from being state-of-the-art. Nonetheless, in Italy, in the last 2–3 years, there have been some cases in which the Competent Authority has requested a WTP to realize a permanent installation of IOMS with the purpose of continuously analysing the ambient air and providing real-time identification of anomalous odour events. If properly working, this type of installation could act not just as an emission monitoring instrument but could represent an effective tool for odour mitigation, allowing the recognition of the causes of the odour events, thereby enabling the plant operators a prompt and targeted intervention and thus reducing the impact on the neighbourhoods.

As previously mentioned, despite its potentialities, the implementation of IOMS for continuous odour monitoring in real-life applications is still challenging [14], and is mainly hindered by some critical aspects related to prolonged outdoor use in critical conditions. Such critical aspects include sensors' drift over time and cross-sensitivity to atmospheric variables and non-odorous gases [15–17]. Cross-sensitivity towards temperature and humidity is particularly critical for environmental odour monitoring, because such parameters may vary within a rather wide range of values, thereby negatively affecting the sensors' outputs. As an example, typical relative humidity variations that may occur outdoors range from 20% to 100%, whereby some experimental studies have proven gas sensors typically used in e-nose systems (i.e., MOX sensors) to be sensitive to humidity variations of 1%.

For this reason, different humidity and temperature compensation strategies have been proposed in the literature [18] such as interference compensation models [19–21], methods based on separation models for environmental factors [22] and methods based on hardware optimization [23,24]. To the best of our knowledge, such solutions have been studied and implemented in laboratory conditions, while the application of compensation strategies in real scenarios is still very limited. Indeed, laboratory experiments are carried out in a controlled environment, whereas when operating outdoors, variable background conditions, as well as the presence of unknown interferents, may affect sensors' responses in an unpredictable way compared to what is observed in the lab, thus reducing the effectiveness of the compensation strategies developed in lab conditions.

In this context, this paper describes the implementation of a continuous odour emission monitoring system consisting of two e-noses at the fenceline of a WTP. Besides describing the experimental procedure involved in training and testing the IOMS in situ, the paper focuses on the development of a specific pre-treatment procedure to be applied to the monitoring data with the purpose of compensating oscillations of the e-nose signals due to temperature (T) and relative humidity (RH) variations. Two approaches, which differ in the definition of the reference baseline, have been investigated, and the efficacy of the two methods was evaluated in terms of classification and quantification performance achieved within specific field verification tests and compared with the ones obtained without applying any correction.

Furthermore, the paper reports the results of the IOMS monitoring carried out at the WTP fenceline for about one year. The frequency of odour events and the trend of the odour concentration at the monitoring sites were evaluated and correlated with other useful information regarding the plant operation, such as extraordinary maintenance activities, temporary shutdowns of the biogas upgrading section, etc., which could potentially be correlated with increased odour emissions, with the purpose of proving the possibility of using the IOMS network as an effective tool for odour management.

The novelty of this work is not related to the development of new algorithms, but rather to the whole procedure adopted for the realization of an effective odour monitoring system capable of providing accurate and useful information in a complex, real-life scenario.

2. Materials and Methods

2.1. Plant Description

The plant under investigation is a WTP that converts the organic fraction of municipal solid waste (OFMSW) into biomethane and high-quality compost (Figure 1). After pretreatments (plastic and sand separation), the OFMSW is mixed with water and transferred to anaerobic digestion tanks. Anaerobic digestion leads to the formation of biogas and digestate. The biogas is then refined to biomethane in an upgrading section, while the digestate is sent to an aerobic treatment section to produce high-quality compost.

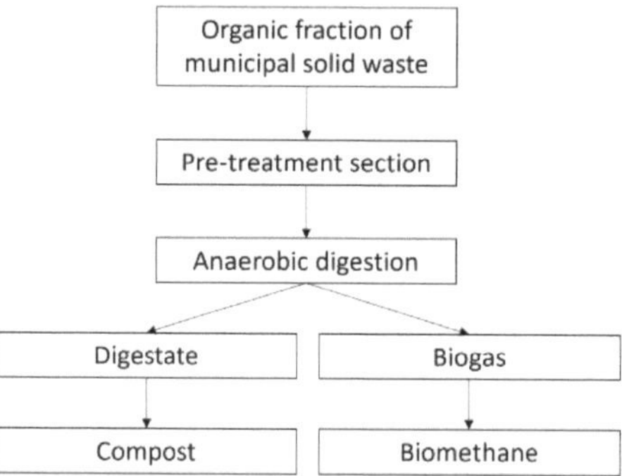

Figure 1. Plant operation scheme.

In order to prevent odour emissions, OFMSW storage and treatments are carried out in closed sheds operated under a slight vacuum and equipped with ventilators collecting exhaust air to a biofilter unit, which is the only conveyed emission of the plant. However, fugitive emissions from sheds cannot be completely neglected.

Thus, the following odour sources were considered in the study: the fugitive emissions from the sheds used for the storage and pre-treatment of the OFMSW and from the buildings used for the maturation and storage of compost, fugitive biogas emissions from the digestor and the upgrading section, and the conveyed emissions from the biofilters treating the air sucked from the buildings containing the OFMSW and the compost.

2.2. Monitoring System

For this study, as hardware, we used two identical e-noses WT1 (WT1 1179 and WT1 1180) commercialized by Ellona (Tolouse, France) and one weather station Vantage Pro2, commercialized by Davis Instruments Corporation (Hayward, Charlotte, NC, USA). Each WT1 e-nose is equipped with an array of 7 sensors, comprising 4 semiconducting metal oxides (MOX) sensors, 2 electrochemical sensors specific for the detection of hydrogen sulphide (H_2S) and ammonia (NH_3) and 1 photoionization detector (PID) calibrated with isobutylene. The instruments are also equipped with 2 sensors for monitoring temperature and relative humidity. The weather station Vantage Pro2 is equipped with a group of sensors including temperature and humidity, rainfall meter and anemometer, which offers the possibility to acquire meteorological data with a frequency of 15 min. All the data acquired by the e-noses and the weather station can be visualized and downloaded through the Ellona web interface for further elaboration.

Data processing was implemented externally from the Ellona software v. 2.4.2., in order to develop the compensation models required by the specific application, as described in the following sections.

The installation points for the e-noses have been defined based on the results of a parametric dispersion modelling study, which was carried out with the purpose of evaluating a preliminary correlation between odour concentration at the plant fenceline and the odour concentration expected at the nearest receptors. From this analysis, the most impacted receptors were identified and the selection of the instruments' installation points was defined to be representative of the emission plumes reaching the most impacted receptors most frequently.

Based on this preliminary study and other logistic considerations, it was decided to install the e-nose WT1 1180 at the Eastern boundary of the plant in the proximity of the biofilters and the entrance of the shed where the OFMSW is discharged and pre-treated before the anaerobic digestion process. The installation site of the WT1 1180 is also very close to the biogas upgrading section. The e-nose WT1 1179 was installed on the opposite side of the plant (Western boundary), i.e., on the rear of the waste receiving and treatment sheds, but close to the two digesters and the torch.

2.3. Electronic Nose Training

2.3.1. Data Acquisition

The main purpose of the e-nose training is to create a reference dataset (i.e., training set, TS) to be used for the implementation of models for classifying and quantifying the air analysed by the instruments during the monitoring phase. To carry this out, dedicated olfactometric campaigns are needed to characterize the plant odour emissions sources. Measurements should be carried out in different periods in order to account for both the intrinsic variability of the sources and the variability associated with the seasonality and different weather conditions. In this case, a total of 70 odour samples were collected directly at the odour sources of the plant as listed in Section 2.1 and analysed by dynamic olfactometry (EN 13725:2022) [25] to assess their odour concentration. Samples were collected over 5 olfactometric campaigns, conducted in different seasons between February 2022 and October 2022, and with different meteorological conditions.

In more detail, odour samples were collected in WTP sheds devoted to OFMSW treatment by means of a mechanical depression pump. Samples of exhaust gas from biofilters were collected by means of a static hood, which allows the isolation of a part of the biofilter surface and conveys the exhaust gas flux through the exit duct where the sample is taken by means of a mechanical depression pump. Finally, samples of fugitive biogas emissions were taken from the gas transfer pipeline from the digestor to the upgrading section, exploiting the overpressure of the gas flow for filling the gas sampling bag. In order to create a TS representative of conditions expected at the monitoring sites, characterized by lower concentrations than those measured at the emission sources, odour samples were diluted with odourless ambient air taken at the plant fenceline when odours were not perceivable by the operators.

After dilution, from the initial 70 samples collected directly at the odour sources, we obtained a total of 240 samples (Table 1), with odour concentrations ranging from about 20 ou_E/m^3 to about 1000 ou_E/m^3.

Table 1. Odour class and concentration range of the samples analysed by the electronic noses for the training and for the field tests for performance verification.

Class	Number of Diluted Samples Analysed	Odour Concentration Range [ou_E/m^3]
Air	33	20–50
Biogas	65	22–990
Biofilter	49	32–860
Sheds	93	27–950

In more detail, 196 samples collected during the first four campaigns were used to build the TS, whereas the 44 samples of the fifth campaign were used as external independent test sets for field performance verification, as described in Section 2.4 (Table 2).

Table 2. List of all the samples collected during the different campaigns and their division between training set and field test set.

	Training Set				Test Set
	22 February	**22 April**	**22 May**	**22 July**	**22 October**
Number of independent samples collected at the source	13	17	6	15	19
Number of diluted samples	47	65	25	59	44

Samples were presented to the e-nose directly in the field to account for background conditions at the monitoring sites and prevent bias in the TS. Odourless ambient air samples were also analysed to define the baseline of IOMS signals when no odour could be perceived. Sensor responses accounting for various analyses in the field were registered for the next steps of data processing.

2.3.2. Data Processing

The general scheme of the data processing procedure developed for this study is illustrated in Figure 2. The details of each step will be discussed in the following sub-sections. After an initial pre-treatment aimed at normalizing the data and compensating potential interferences, different features from the sensors' signal curves were extracted and selected by means of the Boruta algorithm [26]. Selected features were used as input for principal component analysis (PCA) to conduct an exploratory analysis [27,28]. Then, a two-step machine learning model was implemented. In the first phase, a Support Vector Machine (SVM) [29,30] classification model was built to predict the class of odours detected at the fenceline. In the second step, a class-specific regression model based on Support Vector Regression (SVR) [31] was developed to estimate the odour concentration. These algorithms were chosen because they proved to have a better performance on the specific dataset considered in this study compared to other classification algorithms (e.g., Random Forest and KNN) and other regression algorithms (e.g., PLS) [32]. Different approaches were compared to find out the optimal data processing procedure for the specific case, as will be described in the following sub-sections.

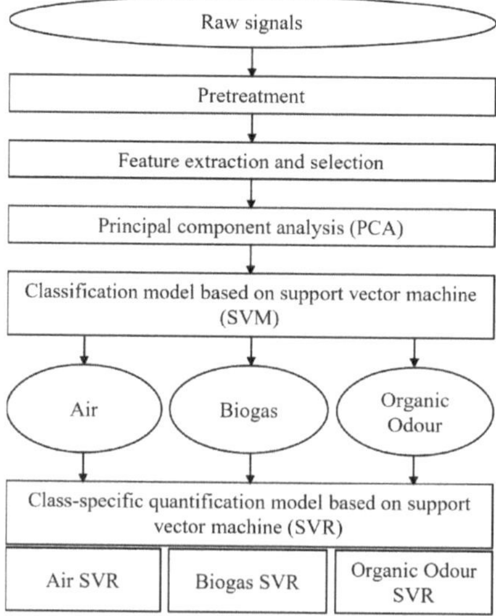

Figure 2. Data processing scheme resuming the different steps applied.

Pre-Treatment and Normalization of the Sensors' Resistances

As already mentioned, one of the main issues associated with the application of e-noses for environmental odour monitoring is their cross-sensitivity towards humidity and temperature variations. Such interference cannot be neglected in the case of outdoor installations, where considerable changes in temperature and humidity levels of the ambient air occur because of the day–night cycles, the season, and the variable meteorological conditions. All these variations cause alterations in e-nose sensor resistances comparable to those observed in the case of odour events. Thus, if not properly treated, they are likely to lead to misinterpretation of the monitoring results and overestimation of the odour detections.

For this reason, we decided to explore the possibility of applying a specific pre-treatment aimed at compensating interferences on e-nose sensor responses caused by temperature and humidity variations occurring during the monitoring, which could have been misclassified as odour events.

The proposed pre-treatment involves the normalization of raw sensor signals recorded during the monitoring versus a reference baseline, as reported in (1),

$$y_{norm} = \frac{y - y_{ref}}{y}, \tag{1}$$

where y_{norm} is the normalized value of resistance, y is the raw resistance value of the sensor and y_{ref} is the value of reference baseline. Two approaches for the definition of the reference baseline have been investigated: the first one uses as parameter y_{ref} as the mean sensor resistance value of the different sensors exposed to odourless ambient air, while the second approach is characterized by the implementation of a variable 'dynamic' y_{ref} obtained through a regression model, which correlates the value of resistance in non-odorous ambient air with the values of absolute humidity (Figure 3).

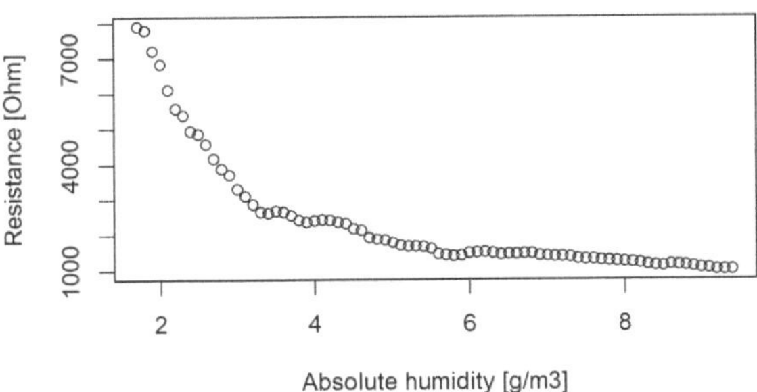

Figure 3. Example of sensor resistance variation exposed to non-odorous ambient air with different values of absolute humidity. The *x*-axis reports the value of absolute humidity in g/m^3, while the *y*-axis reports the resistance in Ohm.

Figure 3 shows, as an example, how the electrical resistance of one MOX sensor of the e-nose, exposed to odourless air, changes with the moisture content of the analysed air. In more detail, Figure 3 compares the MOX sensor baseline with the absolute humidity, expressed in g/m^3 and defined by (2),

$$AH = \frac{RH \, P_s}{R_w \, T \, 100}, \tag{2}$$

where AH is the absolute humidity, RH is the relative humidity, T is the temperature of the air expressed in K, R_w is the specific constant of gases for water vapour (equal to 461.5 J/kg/K), and P_s is the water vapour pressure evaluated using the empiric Equation (3) available on Perry's Chemical Engineers' Handbook [33],

$$\ln P_s = C1 + \frac{C2}{T} + C3 \ln T + C4 \, T^{C5} \tag{3}$$

where $C1$, $C2$, $C3$, $C4$, and $C5$ are empiric constants specific for the evaluation of the vapour pressure of water and P is the pressure expressed in Pa.

The visual inspection of the plot in Figure 3 points out the existence of a decreasing trend of the sensor's baseline with increasing absolute humidity, thereby suggesting the possibility of finding a correlation between these two variables through a regression model.

The regression model we chose consists of a polynomial regression based on a formula such as the one reported in (4),

$$y_{ref} = a_1 + a_2 AH + a_3 AH^2 + a_4 AH^3 + a_5 AH^4 + a_6 AH^5 \tag{4}$$

where the a coefficients are the empiric constants obtained from the regression model and AH is the absolute humidity value evaluated using the temperature and relative humidity registered by the instruments during the analysis. Regression models with different polynomial curves, up to the fifth grade, were tested for each sensor. The choice of the polynomial model was made mainly by a visual check by trying to select the polynomial curve with a lower degree that better approximates the sensor trend (e.g., first-grade polynomial in the case of a linear decreasing trend or third-degree polynomial in the case of an exponential-like trend).

It should be pointed out that this polynomial regression does not claim to have a physical meaning of the sensor's change in resistance in consideration of the chemical reactions occurring on the sensor's surface. However, since the interaction mechanism of water with the sensor surface is still not fully understood, especially in the case of complex mixtures with other chemical compounds, thus making mathematical modelling of the physical phenomena hardly applicable, the only possible way of "correcting" the effect of humidity on the sensors is to use an empirical regression based on the interpolation of experimental data obtained from sensor exposure to different humidity levels. It should be further noted that similar approaches (using polynomials) have been already proposed in the scientific literature [34] to correct for humidity and temperature effects on MOX sensors.

Feature Extraction and Selection

After normalization of the sensors' signals, feature extraction [35], calculated on a 5 min basis, was applied on the TS to extract the numerical parameters useful for the discrimination of the different odour emission sources considered. This step requires the implementation of mathematical operations on the sensor curves to obtain a multidimensional vector with informative values to be used as input for classification and quantification. In this case, 10 features were been extracted for each MOX sensor and 1 feature was extracted for each electrochemical sensor and for the PID, for a total of 43 features extracted (Table 3).

Table 3. List of the features extracted.

SMOX Sensors	Feature Name	Description
	A	Minimum of the resistance
	B	Exponential moving average (EMA)—area under the curve—alpha 0.05
	C	Exponential moving average (EMA)—minimum—alpha 0.05
	D	Exponential moving average (EMA)—area under the curve—alpha 0.1
	E	Exponential moving average (EMA)—minimum—alpha 0.1
	F	Difference between maximum and minimum of the resistance
	G	Minimum of the derivative of the resistance
	H	Area under the curve
	I	Mean of the resistance
Electrochemical Sensor and PID	**Feature Name**	**Description**
	L	Maximum of the sensor reading

To select the best features that entail useful information allowing the discrimination of a certain type of odour, the data processing procedure includes a step of feature selection. Different types of feature selection algorithms can be applied [36–38]. Here, we used Boruta [26], which is an algorithm based on the Random Forest classifier allowing us to rank the features based on their contribution to the performance of classification/regression. Compared to other algorithms, which are based on finding the smallest feature set giving the best classification result, Boruta allows us to identify all the features relevant to the classification purposes.

As a result, eleven features were selected for WT1 1180, and thirteen for WT1 1179, respectively. These features comprise the minimum of the resistance (A), the area under the curve and the minimum of the EMA with alpha 0.1 (D,E) [39], the difference between the maximum and minimum of the resistance (F), the minimum of the derivative of the resistance (G) and the maximum reading of the H_2S electrochemical sensor (L).

Exploratory Analysis

On the selected features, a principal component analysis (PCA) was applied to reduce data dimensionality and obtain a graphical visualization of the instrument's capability to discriminate the different odour classes (Appendix A). Also, the PCA allows us to identify possible odour samples acting as outliers, which are removed from the TS before the implementation of the classification and regression models. In this type of application, outliers are mainly attributable to sampling issues (e.g., contamination of the sampling lines, unpredictable variations in the sampled emission, etc.), which are not always clearly recognizable a priori, and thus are required to be identified through a graphic interpretation of the sample distribution in the so-called influence plot [28].

Classification and Regression Models

Not all the samples analysed (Section 2.3.1) were used for the implementation of the classification and regression models. As previously mentioned, the training models were developed using only 196 out of 240 samples, because the remaining 44 were collected later and left out as an external test set to assess the e-nose performance during operation on the field (as described in Section 2.4). Moreover, ca. 10% of the training samples were identified as outliers and removed to optimize the TS. Furthermore, the PCA allowed us to observe similarities between some of the odour classes considered, which made them hardly discriminable. Therefore, the final training model was built based on three classes: 'Air', which is representative of the condition of absence of odour; 'Biogas', which refers to potential fugitive emissions from the safety valves on the biodigesters; and 'Organic odour', which refers to the fugitive emissions from the sheds used for the receiving and pre-treatment of the organic wastes and the building dedicated to the maturation and storage of the compost and the biofilters.

The odour pattern registered by the e-nose is different depending on the odour class. As a consequence, samples belonging to different odour classes with the same odour concentration will provoke a different sensor response pattern, because of the different chemical composition of the odour classes. For this reason, the construction of an effective quantification model should be based on the identification of the odour class, first. Also, the identification of the odour class is of particular importance for the plant managers, because it enables the identification of the causes of the odour events detected at the fenceline.

The models implemented for classification and regression are based on the non-linear SVM (Support Vector Machine) with a Gaussian radial basis function (RBF) kernel [30]. As mentioned in the previous paragraphs, the score of the data processed by PCA is taken and used as input for the implementation of the classification models, whose outputs define the odour class of the gas analysed. In particular, for both e-noses, the first three principal components account for about 90% of the explained variance and so they were selected as input of the models.

In more detail, three classification models deriving from different data analysis procedures were compared: (1) Model CA—analysis of the raw sensor resistance without any pre-treatment; (2) Model CB—analysis of the sensor resistance after normalization of the signal as reported in Equation (1), with the parameter y_{ref} fixed at the general sensor resistance value of the different sensors exposed to odourless ambient air; (3) Model CC—analysis of the sensors resistance after normalization with absolute humidity compensation of the signal as reported in Equation (1), with a variable y_{ref} evaluated with the regression model expressed in Equation (4) at the absolute humidity measured during the analysis time window. The models obtained were optimized by means of internal 10 k-fold cross validation and to determine their performance an independent dataset was used for testing [40].

Regarding odour quantification, two different strategies were compared: the first approach (Model QA) is based on the application of a global regression model trained using a training set comprising all samples without differentiation of the odour class; instead, the second approach relies on a classification of the odour analysed prior to regression (Model QB). In this case, different regression models are built based on the number of odour classes defined in the TS (Figure 2). The aim of the comparison is to verify the assumption that the classification of the odour prior to its quantification can improve the accuracy of the odour concentration prediction.

The evaluation of the regression models' performance was conducted by comparing the odour concentration predicted on the independent test set with the concentration obtained by the reference method, i.e., dynamic olfactometry, and then by applying a statistical analysis based on the Bland–Altman (B&A) model [41,42].

2.4. Performance Verification in the Field

During the monitoring phase, it is important to carry out specific field tests to evaluate the reliability of the instruments' detections. Field performance testing is also one of the requirements of the recently published Italian technical norm UNI 11761:2023 [43] on instrumental odour measurement. In this case, sampling for field performance testing was carried out as for the training phase, but in a separate campaign after the definition of the training models: gas samples were withdrawn from the different odour emission sources considered, analysed by dynamic olfactometry (in accordance with EN 13725:2022), diluted, and then analysed by the e-noses to verify the instrument accuracy in recognizing the odour class and quantifying the odour concentration.

The odour classification capability was assessed in terms of the Matthews Correlation Coefficient (MCC), which is a better indicator compared to the global accuracy when the classes of interest are not balanced [44].

The Bland–Altman method was then applied to evaluate the odour quantification capability and compare the odour estimates by the e-noses with the reference method. This approach was chosen because dynamic olfactometry has a high measurement uncertainty (generally considered at about 50% [13]), which makes a comparison using traditional calibration methods based on the correlation coefficient (r) unsuitable [45]. For this reason, it is necessary to adopt "model comparison" methods, which enable us to compare outputs obtained from methods having comparable levels of uncertainty. The Bland–Altman method has been recently proposed as a suitable approach to provide quantitative information about the capability of IOMS to estimate odour concentrations [13].

To properly apply the Bland–Altman method, it is first necessary to check whether the differences in the values calculated with the two methods are normally distributed. This can be evaluated, for example, by applying a Shapiro–Wilk test [46] and verifying the resulting p-value. If the p-value is higher than 0.05, then the condition of normality cannot be rejected. It should be considered that, when dealing with odour concentration values, the agreement should be evaluated by comparing the differences in their logarithms.

The application of the Bland–Altman method provides the bias, which is the mean of the differences, and the Limits of Agreement (Lower LoA and Upper LoA), which define a range within which 95% of the differences between one measurement and the other are included.

Performance testing was mainly conducted with the purpose of comparing the different compensation strategies and the different regression models developed, finding the most suitable approach to develop a robust and accurate quantification model.

2.5. Monitoring Data Evaluation

After installation, the e-noses analysed the ambient air continuously, and the sensor resistance values were registered every 10 s. The monitoring phase lasted for about 1 year, resulting in a monitoring dataset of 73,685 observations.

These data were processed according to the data processing procedure described in Section 2.3.2. In more detail, data acquired continuously were divided into time intervals of 5 min to extract features from the sensor curves and process them with the classification and regression models developed. This means that the monitoring system produces an output every 5 min in terms of odour class detected and estimated odour concentration.

Monitoring data were analysed with the purpose of obtaining information about the detection of odours at the fenceline and their origin. Data were analysed first with the purpose of evaluating the odour impact in terms of the ratio between all the 'odour' detections (i.e., measurements classified differently than 'air') and the total number of measurements. Then, the relative detections of the different odour classes and the odour concentration trends over time were evaluated as well, with the purpose of identifying the most critical odour classes causing the majority and/or more intense odour events.

3. Results
3.1. Field Performance Testing
3.1.1. Odour Classification Performances

The different absolute humidity compensation models were evaluated by analysing the odour classification performances of the e-noses WT1 1179 and WT1 1180 when applying models.

The models were first validated (by internal cross-validation) and then tested by using an independent dataset as the test set. The confusion matrices resulting both from the validation and testing were assessed to compare the models' performances.

The results, expressed in terms of MCC values obtained with the different classification models for each electronic nose, are reported in Table 4.

Table 4. Comparison of the performance of different humidity compensation models for the electronic nose WT1 1179 and WT1 1180. The 'Validation' MCC is referred to the Matthews Correlation Coefficient evaluated on cross-validation, while the 'Test' MCC is referred to the same coefficient calculated on the independent test dataset used for testing.

	WT1 1179	
Model	**'Validation' MCC**	**'Test' MCC**
Model CA	0.86	0.85
Model CB	0.91	0.9
Model CC	0.91	0.95
	WT1 1180	
Model	**'Validation' MCC**	**'Test' MCC**
Model CA	0.55	0.69
Model CB	0.85	0.77
Model CC	0.88	1

It is possible to observe similar behaviours for both e-noses: the MCC values relevant to the models implemented without signal normalization (Model CA) perform the worst, while the models obtained after normalization and compensation with respect to the absolute humidity (Model CC) provide the best results. For the e-nose WT1 1179, the result performances obtained with the different models are not huge, with a minimum value of MCC obtained on the independent test set equal to 0.85 for Model CA and a maximum of 0.95 for Model CC. The e-nose WT1 1180, instead, shows greater variability in the performances obtained depending on the data processing procedure applied, ranging from 0.69 to 1 on the independent test set for Model CA and Model CC, respectively.

The results obtained highlight the importance of considering the interference of humidity and temperature in the response of the sensors in order to obtain reliable results in terms of odour class recognition.

The classification Model CC was further tested with an independent dataset. The results of this validation, reported as confusion matrixes in Figure 4a,b, show that, after the application of a suitable model for compensating the sensor responses considering the instantaneous absolute humidity value, the WT1 1179 and WT1 1180 are capable of properly discriminating the different odour classes with accuracy indexes of 97% [95% CI: 85–99%] and 100% [95% CI: 88–100%], respectively.

(a)

Reference

Prediction		Air	Biogas	Organic Odour
	Air	4	0	0
	Biogas	0	10	1
	Organic odour	0	0	19

(b)

Reference

Prediction		Air	Biogas	Organic Odour
	Air	4	0	0
	Biogas	0	10	0
	Organic odour	0	0	15

Figure 4. (**a**) WT1 1179 confusion matrix for odour classification resulting from testing with an independent dataset; (**b**) WT1 1180 confusion matrix for odour classification resulting from testing with an independent dataset.

3.1.2. Quantification Performances

Field testing was also applied to evaluate the capability of the e-noses to estimate the odour concentration, thereby comparing the odour concentration estimates from the instruments with the odour concentration values measured by dynamic olfactometry (according to EN 13725:2022).

The comparison was made by applying the Bland–Altman analysis, which can be used to estimate the agreement between two different methods [41,42]. The Shapiro–Wilk test was applied to both e-noses and for both models (Model QA and Model QB) obtaining a p-value higher than 0.05 in every case analysed, meaning that the condition of normal distribution of the differences cannot be rejected. Figure 5 reports the Bland–Altman plots relevant to Models QA and QB for the two e-noses, WT1 1179 and WT1 1180, respectively. The x-axis reports the mean of the logarithm of the odour concentrations, while the y-axis reports the differences in the logarithm of the odour concentrations. The plots further report the field measurements (black points), the bias (green dash-dot line), the Upper Limit of Agreement (LoA, blue dash-dot line), the Lower LoA (pink dash-dot line) and their 95% confidence intervals (delimited by the corresponding coloured dot lines). By looking at the plots in more detail (Figure 5), both Models QA and QB have their equality line (i.e., the line where the differences are equal to zero) contained in the 95% confidence interval of the bias, meaning that the bias is not significant.

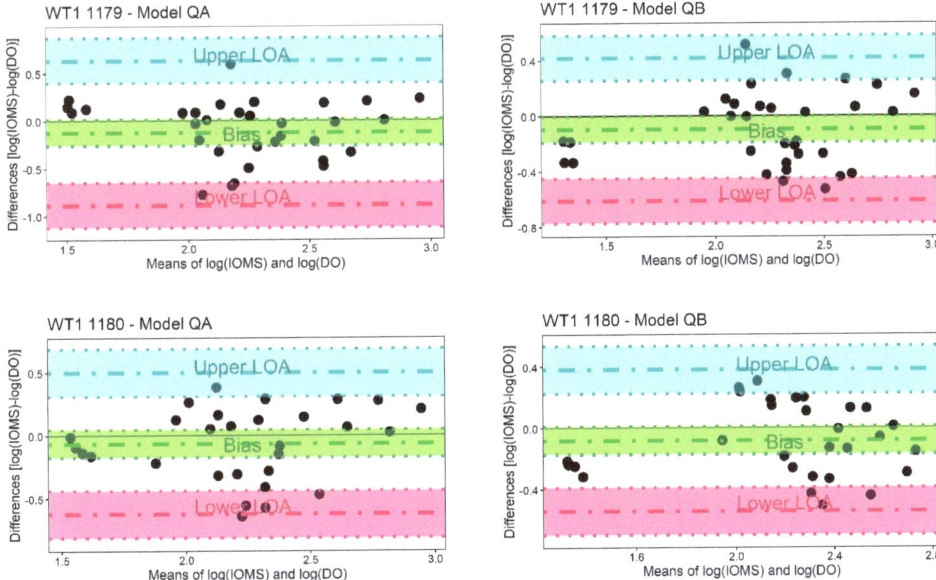

Figure 5. Bland–Altman plots obtained by applying the quantification Model QA and the quantification Model QB on the estimates of the electronic nose WT1 1179 and WT1 1180. In both plots, the bias is reported as a green dash-dot line, the lower limit of agreement as pink dash-dot line and the upper limit of agreement as blue dash-dot line. Each parameter is reported together with the corresponding 95% confidence interval, represented by the dotted lines of the same colour. The *x*-axis reports the mean of the logarithm of the concentration evaluated with the quantification model (IOMS) and the dynamic olfactometry (DO), while the *y*-axis reports their difference.

For an easier and more immediate comparison of the two models, the LoA is also expressed in Tables 5 and 6 in terms of multiplicative factors.

Table 5. Parameters of Bland–Altman model applied on Model QA and QB for the e-nose WT1 1179.

WT1 1179	Model QA		Model QB	
	Logarithmic Differences	Multiplicative Factors	Logarithmic Differences	Multiplicative Factors
Bias	−0.13		−0.1	
Upper LoA	0.63	4.27×	0.41	2.6×
Lower LoA	−0.89	0.13×	−0.61	0.24×

Table 6. Parameters of Bland–Altman model applied on Model QA and QB for the e-nose WT1 1180.

WT1 1180	Model QA		Model QB	
	Logarithmic Differences	Multiplicative Factors	Logarithmic Differences	Multiplicative Factors
Bias	−0.07		−0.08	
Upper LoA	0.5	3.13×	0.38	2.39×
Lower LoA	−0.63	0.24×	−0.55	0.28×

For the e-nose WT1 1179, the Lower and Upper LoA for Model QA are 0.13× and 4.27×, respectively, instead for Model QB they are equal to 0.24× and 2.6×. For the e-nose WT1 1180, these values are 0.24× and 3.13× for Model QA and 0.28× and 2.39× for Model QB. The outcomes of this analysis highlight a difference in the level of agreement between Models QA and QB for each e-nose. In more detail, Model QB results provide, for both e-noses, better estimates compared to the odour concentration values measured by the reference method.

To further compare the different models, we tried to assess the e-nose quantification performance adopting the same criteria for accuracy and for intermediate precision indicated in EN 13725:2022 for dynamic olfactometry.

According to the EN, to satisfy the accuracy criterion, the logarithm of the bias needs to be lower than 0.217 (Equation (5)), while the intermediate precision criteria are satisfied if the standard deviation(s) of the observations, in a non-logarithm form, multiplied by 1.96, is lower than 3 (Equation (6)), meaning that the factor that expresses the difference between two consecutive odour concentration measurements will not be larger than a factor of 3 in 95% of the cases.

$$\left| log_{10} bias \right| < 0.217 \tag{5}$$

$$1.96 * s < 3 \tag{6}$$

The same criteria were applied for the evaluation of the e-nose estimates, and the results are reported in Table 7. The accuracy criterion is satisfied by all the models for both e-noses since the absolute value of the logarithm of the bias is always lower than 0.217. On the other hand, the intermediate precision criterion is not satisfied by any of the models developed. Even if the criterion is not respected, comparing Models QA and QB, it is possible to observe that, for both electronic noses WT1 1179 and WT1 1180, Model QB is the one performing better, giving a factor closer to the limit value of 3. Based on these considerations, it was decided to adopt Model QB as a quantification model to predict the odour concentration of the ambient air analysed by the instruments at the plant fenceline.

Table 7. Results obtained from the Bland–Altman analysis for the quantification Model QA and QB applied on both electronic noses. For each model, the absolute value of the logarithm of the bias, the standard deviation of the difference in the measurements and the result of Equation (6) to check the compatibility with the intermediate precision criteria are reported.

Model	$\left| log_{10}(bias) \right|$	Standard Deviation	Intermediate Precision Criterion
WT1 1179—Model QA	0.13	2.41	4.72
WT1 1179—Model QB	0.1	1.82	3.57
WT1 1180—Model QA	0.07	1.91	3.74
WT1 1180—Model QB	0.085	1.71	3.35

The final data processing procedure applied and the different approaches considered have been resumed in Figure 6.

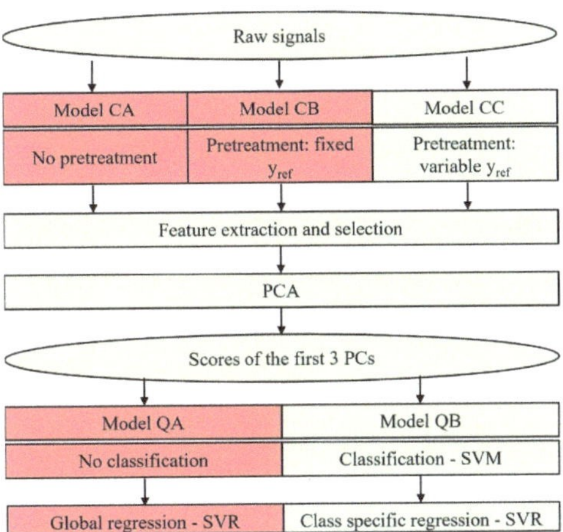

Figure 6. Detailed data processing scheme. The final data path flow are highlighted in green, while the discarded model approaches are coloured in red.

3.2. Monitoring Results

The signals registered by both e-noses from February 2022 to November 2022 were processed with the aim of obtaining a real-time qualitative and quantitative characterization of the ambient air at the plant fenceline. Based on the results discussed in the previous sections, we decided to adopt Model CC to compensate for humidity and assess the odour class, whereas the estimation of the odour concentration was obtained relying on Model QB (Figure 6).

The monitoring results have been then elaborated to assess the frequency of detection of the different odour classes considered at the instruments' installation sites.

Figure 7 reports the odour detection frequencies for both e-noses, WT1 1179 and WT1 1180. It can be observed that the WT1 1180 detected odour for about 11% of the monitoring period with a quasi-equal distribution between the two odour classes 'Biogas' and 'Organic odour' (5% and 6%, respectively). Instead, the WT1 1179 detected odour for about 7% of the monitoring period, with the odour events belonging principally to the 'Organic odour' class, which was recognized 6% of the time, compared to the odour class of 'Biogas', which was detected only ca. 1% of the time.

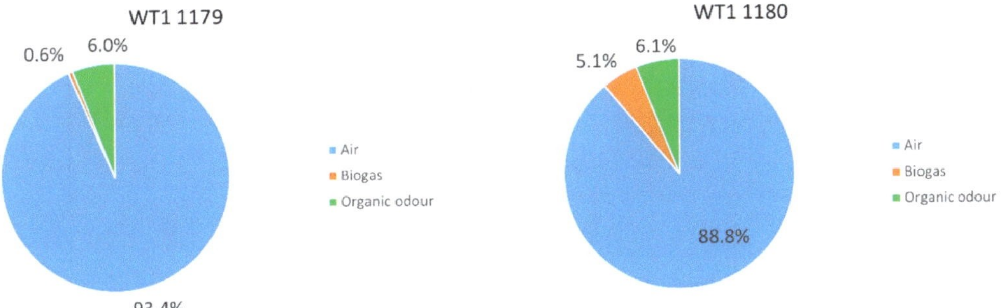

Figure 7. Odour detection frequency between the months of February and November 2022 for the electronic nose WT1 1179 and WT1 1180.

These differences can be explained by considering the different positions of the two e-noses. The WT1 1180 is installed in a more 'critical' point that allows the detection of odour emissions from the biofilters and the sheds where the OFMSW is stored and processed; moreover, the WT1 1180 is located also downwind the upgrading section, thus enabling the detection of fugitive emissions of biogas from that area, which turned out to be not negligible. Instead, the WT1 1179 is installed close to the biodigester and the torch, from which fugitive biogas emissions seem to be more limited.

The outputs of the regression models for WT1 1179 and WT1 1180 have been reported in Figures 8 and 9, showing the predicted odour concentration for the entire monitoring period. Looking at the results for the WT1 1179, the e-nose detected more frequent odour events attributed to the 'Organic odour' class; odour events related to 'Biogas' are rare, but they result in relatively high odour concentration peaks (up to ca. 1000 $[oue_{eq}/m^3]$). Conversely, the WT1 1180 detected odours attributed to the two classes 'Organic odour' and 'Biogas' with similar frequencies, and the relevant odour concentrations reached up to ca. 500 $[oue_{eq}/m^3]$.

What is particularly interesting from the point of view of the plant management is that the odour events recorded by the two e-noses, and especially those associated with high odour concentrations, could be often related to specific malfunctioning events or anomalous operating conditions of the plant. As an example, the increase in the frequency of the odour events registered by the e-nose WT1 1180 in October and November coincides with a malfunctioning of the ventilation system of the OFMSW storage buildings, which resulted in increased fugitive odorous emissions, because of the reduced capacity of keeping the shed's interior at a negative pressure. Moreover, the high concentration peaks of biogas detected by the e-nose WT1 1179 in October could be identified as related to extraordinary maintenance operations carried out on the pipelines transporting the biogas from the biodigester to the upgrading section.

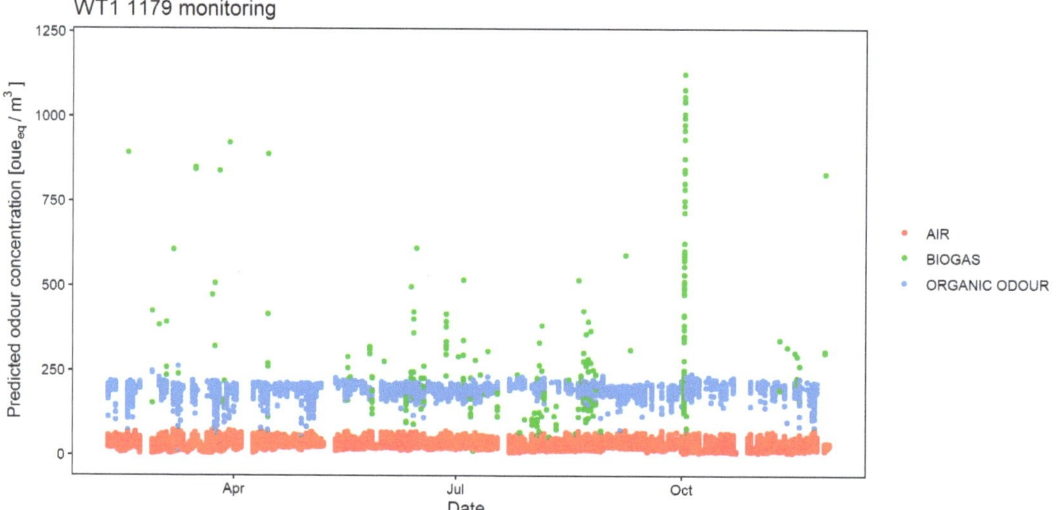

Figure 8. Odour class and relative odour concentration resulting from the ambient air analysis by the electronic nose WT1 1179 at the plant fenceline from February 2022 to November 2022.

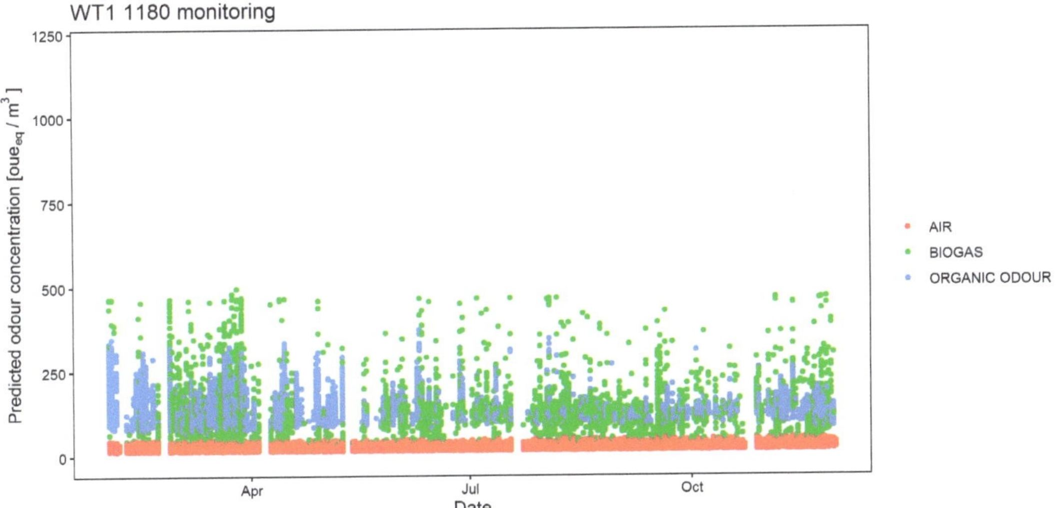

Figure 9. Odour class and relative odour concentration resulting from the ambient air analysis by the electronic nose WT1 1180 at the plant fenceline from February 2022 to November 2022.

4. Conclusions

The present study discusses the possibility of using e-noses to continuously monitor the odour emissions at the fenceline of a WTP by detecting the presence of odours, recognizing their provenance (odour source) and estimating the odour concentration. In more detail, this study focuses on the instrument training methods, thereby comparing the application of different training models.

First, the training involved extensive field data acquisition: 196 samples of known odour quality and concentration collected at the plant's main odour sources over four different olfactometric campaigns conducted in different seasons were analysed by the instruments, opportunely diluted to reproduce odour concentration levels that are likely to be detected at the plant fenceline. Furthermore, other 44 independent samples collected on a fifth campaign were used for testing.

Then, the work involved the implementation and the comparison of different models for the compensation of ambient air humidity and temperature, which are known as main interferents in the application of e-noses for environmental monitoring. Results show that the implementation of a suitable corrective model of sensor responses to compensate for changes in absolute humidity of the sample significantly improves the odour classification performances, passing from accuracies of ca. 70% to 100%. These observations further highlight the importance of including of a suitable compensation strategy whenever the e-nose is used outdoors, with variable atmospheric conditions.

Finally, in this study, we also studied and compared different models for the estimation of the odour concentration. In more detail, we wanted to confirm the hypothesis that a 'double-step' model, including first the identification of the odour class and then applying a regression model specific for the recognized odour class, would be more effective than a more simplified regression model without prior identification of the odour class. Indeed, the 'double-step' model proved to provide more accurate estimations, resulting in reduced factors for the Limits of Agreement evaluated by the Bland–Altman method. Moreover, quantification performances achieved by the trained electronic noses turned out to be comparable with the quality criteria fixed by EN 13725:2022 for the reference method, i.e., dynamic olfactometry.

As a conclusion, the study proves that, if properly trained, by including a suitable signal pre-treatment method for compensating atmospheric variables and by implementing specific quantification models based on a preliminary identification of the odour class for regression, e-noses can be very effective for environmental odour monitoring, achieving classification accuracies above 97% and quantification performances comparable to those considered acceptable by the reference standard for dynamic olfactometry.

Regarding the final application, the most important result consists of the realization of an effective odour emission monitoring system, enabling the real-time detection of odour events and the identification of their sources. This is a key feature for plant managers for identifying the possible causes of anomalous odour emissions and enabling prompt intervention, thus reducing the odour impact on the neighbourhoods.

Future studies should focus on the optimization of the classification and quantification models developed on a longer monitoring period, thereby accounting for the well-known problem of sensor drift over time and developing specific strategies to counteract it.

Author Contributions: Conceptualization, L.C. and C.B.; methodology, L.C., C.B., B.J.L. and C.R.; software, C.R.; validation, C.R. and C.B.; formal analysis, C.R. and C.B.; investigation, C.B., B.J.L. and L.C.; resources, L.C.; data curation, C.R.; writing—original draft preparation, C.R. and C.B.; writing—review and editing, C.B. and L.C.; visualization, C.R. and L.C.; supervision, L.C. and C.B.; project administration, C.B. and L.C.; funding acquisition, L.C. All authors have read and agreed to the published version of the manuscript.

Funding: This research has carried out within the MUSA—Multilayered Urban Sustainability Action—project, funded by the European Union—NextGenerationEU, under the National Recovery and Resilience Plan (NRRP) Mission 4 Component 2 Investment Line 1.5: Strenghtening of research structures and creation of R&D "innovation ecosystems", set up of "territorial leaders in R&D.

Institutional Review Board Statement: Not applicable.

Informed Consent Statement: Not applicable.

Data Availability Statement: Data are confidential.

Conflicts of Interest: The authors declare no conflicts of interest.

Appendix A

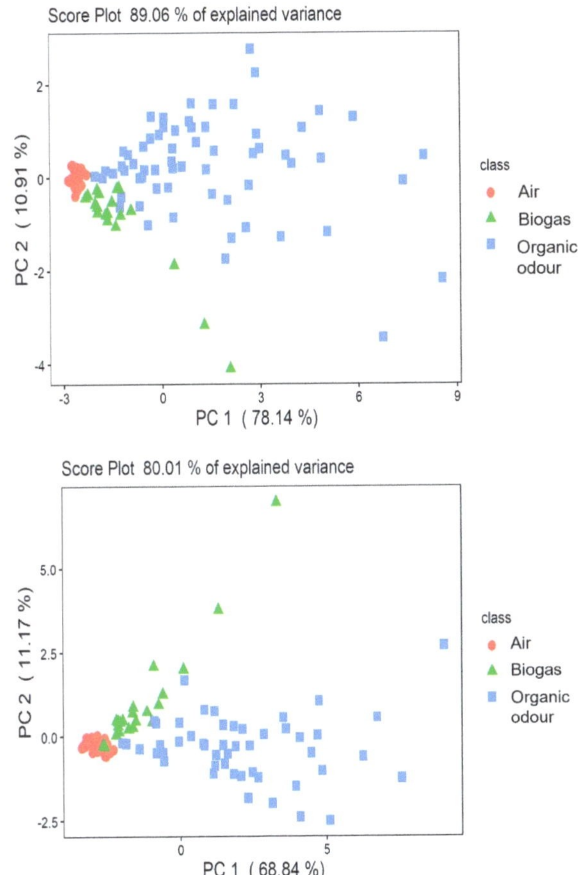

Figure A1. PCA score plot of the training set for the e-nose WT1 1179 (**above**) and WT1 1180 (**below**).

References

1. Henshaw, P.; Nicell, J.; Sikdar, A. Parameters for the assessment of odour impacts on communities. *Atmos. Environ.* **2006**, *40*, 1016–1029. [CrossRef]
2. Aatamila, M.; Verkasalo, P.K.; Korhonen, M.J.; Suominen, A.L.; Hirvonen, M.-R.; Viluksela, M.K.; Nevalainen, A. Odour annoyance and physical symptoms among residents living near waste treatment centres. *Environ. Res.* **2010**, *111*, 164–170. [CrossRef] [PubMed]
3. Sucker, K.; Both, R.; Bischoff, M.; Guski, R.; Krämer, U.; Winneke, G. Odor frequency and odor annoyance Part II: Dose-response associations and their modiWcation by hedonic tone. *Eur. Chemorecept. Res. Organ. Erlangen* **2008**, *81*, 683–694. [CrossRef] [PubMed]
4. Pinasseau, A.; Zerger, B.; Roth, J.; Canova, M.; Roudier, S. *Best Available Techniques (BAT) Reference Document for Waste Treatment*; Publications Office of the European Union: Luxembourg, 2018.
5. Bax, C.; Sironi, S.; Capelli, L. How Can Odors Be Measured? An Overview of Methods and Their Applications. *Atmosphere* **2020**, *11*, 92. [CrossRef]
6. Khorramifar, A.; Karami, H.; Lvova, L.; Kolouri, A.; Łazuka, E.; Piłat-Rożek, M.; Łagód, G.; Ramos, J.; Lozano, J.; Kaveh, M.; et al. Environmental Engineering Applications of Electronic Nose Systems Based on MOX Gas Sensors. *Sensors* **2023**, *23*, 5716. [CrossRef] [PubMed]
7. Jońca, J.; Pawnuk, M.; Arsen, A.; Sówka, I. Electronic Noses and Their Applications for Sensory and Analytical Measurements in the Waste Management Plants—A Review. *Sensors* **2022**, *22*, 1510. [CrossRef] [PubMed]

8. Nicolas, J.; Romain, A.C.; Wiertz, V.; Maternova, J.; André, P. Using the classification model of an electronic nose to assign unknown malodours to environmental sources and to monitor them continuously. *Sens. Actuators B Chem.* **2000**, *69*, 366–371. [CrossRef]

9. Blanco-Rodríguez, A.; Camara, V.F.; Campo, F.; Becherán, L.; Durán, A.; Vieira, V.D.; de Melo, H.; Garcia-Ramirez, A.R. Development of an electronic nose to characterize odours emitted from different stages in a wastewater treatment plant. *Water Res.* **2018**, *134*, 92–100. [CrossRef] [PubMed]

10. Sironi, S.; Capelli, L.; Céntola, P.; Del Rosso, R.; Il Grande, M. Continuous monitoring of odours from a composting plant using electronic noses. *Waste Manag.* **2007**, *27*, 389–397. [CrossRef]

11. Dentoni, L.; Capelli, L.; Sironi, S.; Del Rosso, R.; Zanetti, S.; Della Torre, M. Development of an Electronic Nose for Environmental Odour Monitoring. *Sensors* **2012**, *12*, 14363–14381. [CrossRef]

12. Bax, C.; Lotesoriere, B.J.; Capelli, L. Real-time Monitoring of Odour Concentration at a Landfill Fenceline: Performance Verification in the Field. *Chem. Eng. Trans.* **2021**, *85*, 19–24.

13. Burgués, J.; Doñate, S.; Deseada Esclapez, M.; Saúco, L.; Marco, S. Characterization of odour emissions in a wastewater treatment plant using a drone-based chemical sensor system. *Sci. Total. Environ.* **2022**, *846*, 157290. [CrossRef] [PubMed]

14. Oliva, G.; Zarra, T.; Pittoni, G.; Senatore, V.; Galang, M.; Castellani, M.; Belgiorno, V.; Naddeo, V. Next-generation of instrumental odour monitoring system (IOMS) for the gaseous emissions control in complex industrial plants. *Chemosphere* **2021**, *271*, 129768. [CrossRef] [PubMed]

15. Wang, C.; Yin, L.; Zhang, L.; Xiang, D.; Gao, R. Metal Oxide Gas Sensors: Sensitivity and Influencing Factors. *Sensors* **2010**, *10*, 2088–2106. [CrossRef] [PubMed]

16. Ziyatdinov, A.; Marco, S.; Chaudry, A.; Persaud, K.; Caminal, P.; Perera, A. Drift compensation of gas sensor array data by common principal component analysis. *Sens. Actuators B Chem.* **2010**, *146*, 460–465. [CrossRef]

17. Fonollosa, J.; Fernández, L.; Gutiérrez-Gálvez, A.; Huerta, R.; Marco, S. Calibration transfer and drift counteraction in chemical sensor arrays using Direct Standardization. *Sens. Actuators B Chem.* **2016**, *236*, 1044–1053. [CrossRef]

18. Liang, Z.; Tian, F.; Yang, S.X.; Zhang, C.; Sun, H.; Liu, T. Study on Interference Suppression Algorithms for Electronic Noses: A Review. *Sensors* **2018**, *18*, 1179. [CrossRef] [PubMed]

19. Nenova, Z.; Dimchev, G. Compensation of the Impact of Disturbing Factors on Gas Sensor Characteristics. *Acta Polytech. Hung.* **2013**, *10*, 97–111. [CrossRef]

20. Kashwan, K.R.; Bhuyan, M. Robust Electronic-Nose System with Temperature and Humidity Drift Compensation for Tea and Spice Flavour Discrimination. In Proceedings of the 2005 Asian Conference on Sensors and the International Conference on New Techniques in Pharmaceutical and Biomedical Research, Kuala Lumpur, Malaysia, 5–7 September 2005.

21. Yan, M.; Wu, Y.; Hua, Z.; Lu, N.; Sun, W.; Zhang, J.; Fan, S. Humidity compensation based on power-law response for MOS sensors to VOCs. *Sens. Actuators B Chem.* **2021**, *334*, 129601. [CrossRef]

22. Di Natale, C.; Martinelli, E.; D'Amico, A. Counteraction of environmental disturbances of electronic nose data by independent component analysis. *Sens. Actuators B Chem.* **2002**, *82*, 158–165. [CrossRef]

23. Emadi, T.A.; Shafai, C.; Freund, M.S.; Thomson, D.J.; Jayas, D.S.; White, N.D.G. Development of a polymer-based gas sensor—Humidity and CO_2 sensitivity. In Proceedings of the 2009 2nd Microsystems and Nanoelectronics Research Conference, Ottawa, ON, Canada, 13–14 October 2009; pp. 112–115. [CrossRef]

24. Wang, Y.; Zhou, Y. Recent Progress on Anti-Humidity Strategies of Chemiresistive Gas Sensors. *Materials* **2022**, *15*, 8728. [CrossRef]

25. CEN. EN 13725:2022; Stationary Source Emissions—Determination of Odour Concentration by Dynamic Olfactometry and Odour Emission Rate. CEN: Brussels, Belgium, 2022.

26. Kursa, M.B.; Rudnicki, W.R. Feature Selection with the Boruta Package. *J. Stat. Softw.* **2010**, *36*, 1–13. [CrossRef]

27. Abdi, H.; Williams, L.J. Principal component analysis. *Wiley Interdiscip. Rev. Comput. Stat.* **2010**, *2*, 433–459. [CrossRef]

28. Bro, R.; Smilde, A.K. Principal component analysis. *Anal. Methods* **2014**, *6*, 2812–2831. [CrossRef]

29. Bishop, C.M. *Neural Networks for Pattern Recognition*; Oxford University Press: Oxford, UK, 1995.

30. Cortes, C.; Vapnik, V.; Saitta, L. Support-Vector Networks Editor. *Mach. Learning* **1995**, *20*, 273–297. [CrossRef]

31. Smola, A.J.; Schölkopf, B.; Schölkopf, S. A tutorial on support vector regression. *Stat. Comput.* **2004**, *14*, 199–222. [CrossRef]

32. Panzitta, A.; Bax, C.; Lotesoriere, B.J.; Ratti, C.; Capelli, L. Realisation of a Multi-sensor System for Real-time Monitoring of Odour Emissions at a Waste Treatment Plant. *Chem. Eng. Trans.* **2022**, *95*, 139–144.

33. Green, D.W.; Southard, M.Z. *Perry's Chemical Engineers*; McGraw–Hill Education: New York, NY, USA, 2019.

34. Abdullah, A.N.; Kamarudin, K.; Kamarudin, L.M.; Adom, A.H.; Mamduh, S.M.; Juffry, Z.H.M.; Bennetts, V.H. Correction Model for Metal Oxide Sensor Drift Caused by Ambient Temperature and Humidity. *Sensors* **2022**, *22*, 3301. [CrossRef]

35. Yan, J.; Guo, X.; Duan, S.; Jia, P.; Wang, L.; Peng, C.; Zhang, S. Electronic Nose Feature Extraction Methods: A Review. *Sensors* **2015**, *15*, 27804–27831. [CrossRef]

36. Li, J.; Cheng, K.; Wang, S.; Morstatter, F.; Trevino, R.P.; Tang, J.; Liu, H. Feature selection: A data perspective. *ACM Comput. Surv.* **2017**, *50*, 94. [CrossRef]

37. Yan, K.; Zhang, D. Feature selection and analysis on correlated gas sensor data with recursive feature elimination. *Sens. Actuators B Chem.* **2015**, *212*, 353–363. [CrossRef]

38. Tang, J.; Alelyani, S.; Liu, H. *Feature Selection for Classification: A Review*; CRC Press: Boca Raton, FL, USA, 2014.

39. Muezzinoglu, M.K.; Vergara, A.; Huerta, R.; Rulkov, N.; Rabinovich, M.I.; Selverston, A.; Abarbanel, H.D.I. Acceleration of chemo-sensory information processing using transient features. *Sens. Actuators B Chem.* **2009**, *137*, 507–512. [CrossRef]
40. Marco, S. The need for external validation in machine olfaction: Emphasis on health-related applications Chemosensors and Chemoreception. *Anal. Bioanal. Chem.* **2014**, *406*, 3941–3956. [CrossRef] [PubMed]
41. Giavarina, D. Understanding Bland Altman analysis. *Biochem. Medica* **2015**, *25*, 141–151. [CrossRef] [PubMed]
42. Bland, J.M.; Altman, D.G. Measuring agreement in method comparison studies. *Stat. Methods Med. Res.* **1999**, *8*, 135–160. [CrossRef] [PubMed]
43. *UNI. UNI 11761:2023*; Emissioni e Qualità Dell'aria—Misurazione Strumentale Degli Odori Tramite IOMS (Instrumental Odour Monitoring Systems). UNI: Milano, Italy, 2023.
44. Chicco, D.; Jurman, G. The advantages of the Matthews correlation coefficient (MCC) over F1 score and accuracy in binary classification evaluation. *BMC Genom.* **2020**, *21*, 6. [CrossRef] [PubMed]
45. Burgués, J.; Esclapez, M.D.; Doñate, S.; Marco, S. RHINOS: A lightweight portable electronic nose for real-time odor quantification in wastewater treatment plants. *iScience* **2021**, *24*, 103371. [CrossRef]
46. Shapiro, S.S.; Wilk, M.B. An analysis of variance test for normality (complete samples). *Biometrika* **1965**, *52*, 591–611. [CrossRef]

 *sensors*

MDPI

Article

Stability Study of Synthetic Diamond Using a Thermally Controlled Biological Environment: Application towards Long-Lasting Neural Prostheses [†]

Jordan Roy [1], Umme Tabassum Sarah [1], Gaëlle Lissorgues [1], Olivier Français [1], Abir Rezgui [1], Patrick Poulichet [1], Hakim Takhedmit [1], Emmanuel Scorsone [2] and Lionel Rousseau [1,*]

[1] ESYCOM Laboratory for Electronics, Communication and Microsystems, CNRS UMR 9007, F-77454 Marne-la-Vallée, France; jordan.roy@esiee.fr (J.R.); sarah.ummetabassum@esiee.fr (U.T.S.); gaelle.lissorgues@esiee.fr (G.L.); olivier.francais@esiee.fr (O.F.); abir.rezgui@esiee.fr (A.R.); patrick.poulichet@esiee.fr (P.P.); hakim.takhedmit@univ-eiffel.fr (H.T.)

[2] Diamond Sensors Laboratory, CEA-LIST, F-91190 Gif-sur-Yvette, France; emmanuel.scorsone@cea.fr

* Correspondence: lionel.rousseau@esiee.fr

[†] Presented at the Eurosensors XXXIV, Lecce, Italy, 10–13 September 2023.

Abstract: This paper demonstrates, for the first time, the stability of synthetic diamond as a passive layer within neural implants. Leveraging the exceptional biocompatibility of intrinsic nanocrystalline diamond, a comprehensive review of material aging analysis in the context of in-vivo implants is provided. This work is based on electric impedance monitoring through the formulation of an analytical model that scrutinizes essential parameters such as the deposited metal resistivity, insulation between conductors, changes in electrode geometry, and leakage currents. The evolution of these parameters takes place over an equivalent period of approximately 10 years. The analytical model, focusing on a fractional capacitor, provides nuanced insights into the surface conductivity variation. A comparative study is performed between a classical polymer material (SU8) and synthetic diamond. Samples subjected to dynamic impedance analysis reveal distinctive patterns over time, characterized by their physical degradation. The results highlight the very high stability of diamond, suggesting promise for the electrode's enduring viability. To support this analysis, microscopic and optical measurements conclude the paper and confirm the high stability of diamond and its strong potential as a material for neural implants with long-life use.

Keywords: polycrystalline diamond; neural prosthesis; impedance spectroscopy; accelerated aging; passivated layer; electrodes; bio-interfaces

Citation: Roy, J.; Sarah, U.T.; Lissorgues, G.; Français, O.; Rezgui, A.; Poulichet, P.; Takhedmit, H.; Scorsone, E.; Rousseau, L. Stability Study of Synthetic Diamond Using a Thermally Controlled Biological Environment: Application towards Long-Lasting Neural Prostheses. *Sensors* **2024**, *24*, 3619. https://doi.org/10.3390/s24113619

Academic Editor: Silvio Sciortino

Received: 15 March 2024
Revised: 27 May 2024
Accepted: 1 June 2024
Published: 4 June 2024

1. Introduction

Implantable neuroprosthetic devices hold the potential to restore neurological function in people with disabilities. Experiments have shown that an array of microelectrodes inserted into the cortex can capture brain activity and trigger movements in prosthetic limbs [1,2] or elicit certain visual sensations through electrical stimulation [2,3]. In such applications, the durability and stability of the device are crucial in ensuring consistent functionality over time [4]. However, over the long term, the polymers used in the device may swell and allow moisture to seep in, leading to a degradation in performance and ultimately reducing its lifespan [5,6]. This has been shown as a major hindrance to the adoption of this technology for human health, especially considering the medical implications of open brain, spinal cord and peripheral nerve surgery regarding the use of soft implantable neuroprostheses, each showing specific constraints. While several electrode designs and materials exist, they all show limitations in their usable lifetime [7]. It is thus necessary to develop functional electrodes that are as biologically inert and long-lasting as possible, to reduce the need for replacement [8].

One solution is to envision an implant based on diamond material [9,10]. Nanocrystalline diamond is known as an advantageous material for electrochemistry and biosensing [11,12]. In the last decade, it has demonstrated impressive biocompatibility, making it one of the most promising materials for living tissue interfacing and particularly for neural interfacing [13]. Indeed, diamond, as a crystalline form of carbon, shows low toxicity towards living cells and offers great adherence to cells, facilitating its integration with tissue. Diamond also does not trigger inflammatory responses and has no natural oxides, a property that few other non-toxic materials share. Typical implant materials such as silicon, polyimide and parylene only share some of these properties, which impacts their performance over time [5,8,14–17].

Diamond's biocompatible nature has been thoroughly researched for many years [14,18–23], coupled with its semiconductor properties, allowing targeted functionalization [24]. This makes diamond an exceptional implant material.

Thus, in order to address the challenge of neuronal implants with a long lifetime (several years), a novel, full-diamond implant has been developed using a microfabrication process. It combines conductive diamond (boron-doped diamond (BDD)) for electrodes in contact with biological species and intrinsic diamond for full passivation (Figure 1) [25]. This passivation layer is embedded within a conductive layer (titanium nitride) in order to define the electrical access path and impede its resistivity.

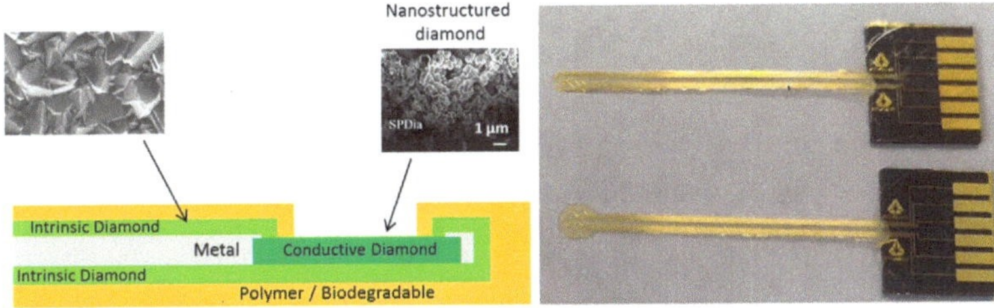

Figure 1. (**Left**) Schematic representation of our full-diamond electrode. (**Right**) Full-diamond implant.

This prosthesis shows promising properties in terms of biocompatibility and hermiticity thanks to the strong bonding between the intrinsic and conductive diamond, as well as its mechanical properties. Indeed, at micrometer scales, a thin layer of diamond shows appropriate flexibility for biological uses. Here, the thickness of the diamond implant is in the range of 3 µm. This structure is coated with a polymer to easily manipulate the device (total thickness around 15 µm) (Figure 1).

This implant has been tested on the visual cortices of rodents at the Paris Vision Institute and EPFL. The visual evoked potentials (EVP) were recorded after light stimulation of the contralateral eye. The results are very encouraging [26], showing suitable functionality for the stimulation or recording of neuronal activity.

However, the question of the long-term stability of the prosthesis requires further analysis. This paper focuses on this crucial point. In this context, the demonstration is based on the aging of an interdigitated electrode (IDE) with full-diamond insulating passivation, as a representative of the actual prosthesis, which is possible thanks to its full-diamond design.

The in vivo evaluation of electrodes requires highly skilled medical practitioners as well as the use of extensive facilities and equipment for a long period of time, namely the period for which it is necessary to evaluate the electrode's aging. In this particular case, it must be examined over a period of 10 years. This is not ethical or practical.

Thus, impedance monitoring is used here to evaluate changes in the diamond layer over time. For long-life testing, a thermally controlled biological environment has been designed and used for aging acceleration. For a comparative study, samples using a polymer instead of diamond have been used.

2. Materials and Methods Used for Longevity Test and Aging Monitoring

2.1. Passivation Layer as an Aging Reference

As the outer surfaces of the prosthesis are entirely covered in diamond, its stability can be derived from that of the diamond itself, using the same manufacturing process. An aging-specific design has been developed to evaluate the coating material's hermeticity in the test device. An IDE structure is used instead of a simple electrode as a test device to access the resistive and capacitive behavior over various frequencies through impedance analysis tools. A titanium nitride (TiN) IDE entirely coated (passivated) with intrinsic diamond, using the same fabrication process as for the previously described prosthesis, was fabricated.

Indeed, an electrode would show different impedance according to whether it is isolated from a conductive medium or not, allowing the evaluation of the insulation layer.

2.2. Design and Fabrication of Test Structures

In order to test the aging of the diamond coating and its insulating performance compared to classical polymers, two types of samples were created based on an IDE passivated with (i) intrinsic polycrystalline diamond or (ii) a polymer (SU8) (Figure 2).

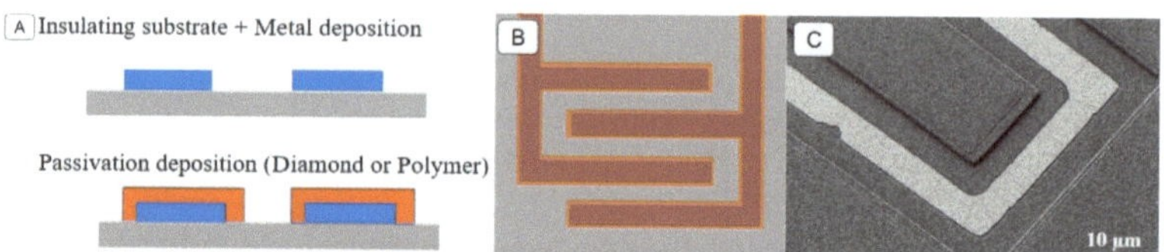

Figure 2. (**A**) Schematic representation of the test IDE electrode layers. (**B**) Schematic representation of the test electrode design. (**C**) Microscopic observation of a fabricated electrode.

The IDE was fabricated on silicon wafer with a 1.5 µm silicon dioxide layer. For the metallic layer, a titanium nitride (TiN) layer was sputtered on the wafer. By photolithography, the IDE was patterned and the TiN layer was etched in a liquid solution (TBR 19 Technic Inc., Cranston, RI, USA). Concerning the IDE with polymer passivation, we covered the TiN IDE with a 2 µm SU8 layer (SU8 2002).

For the IDE with diamond passivation, intrinsic diamond was grown on a diamond nano-seeded substrate with a microwave plasma-enhanced chemical vapor deposition (MPECVD) reactor (SDS6K) from the SEKI company (Tokyo, Japan). We grew the intrinsic diamond at the power of 3 kW at a pressure of 30 mTorr and a temperature of 780–900 °C, with a mixture of hydrogen and methane gas. In these conditions, we grew a 2 µm diamond layer. To grow diamond, it is necessary to prepare wafers by seeding diamond nanoparticles. After this step, we deposited an aluminium nitride (AlN) layer, and, by photolithography, this layer was patterned with AlN where it was desired to grow the diamond. After diamond growth, the entire AlN layer was etched in a liquid solution and the wafer was cleaned.

For the IDE design (Figure 2B), the finger parameters were as follows:

- SU8_100 and SU8_50 were IDEs with 100 or 50 fingers covered with SU8;
- Dmd_50 and Dmd-25 were IDEs with 50 or 25 fingers covered with intrinsic diamond;
- All IDEs had the same size—a finger width of 30 µm, with spacing of 20 µm, and a length of 4.66 mm.

The results obtained for the IDE are presented in Figure 2C. The thickness measurement of the passivating layer was 2 µm for diamond and an average of 2.5 µm for SU8, with a maximum value of 3 µm (Figure 3).

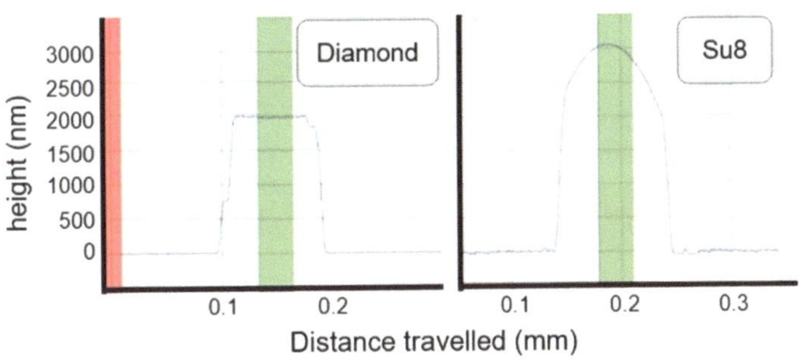

Figure 3. Fabricated layer profile for thickness measurements.

Sample parameters for labeling are presented in Table 1.

Table 1. Samples fabricated and their dimensions.

	SU8_100	SU8_50	Dmd_50	Dmd_25
Conductor		TiN (200 nm)		
Passivation material	Polymer	Polymer	Diamond	Diamond
Passivation thickness		2 µm to 3 µm		
IDE arrays	100	50	50	25

2.3. Aging Analysis

In chemical kinetics, Arrhenius' law is used to link the rate at which chemical interactions occur in relation to the temperature [27,28]. This method has been extensively used for many years to evaluate the aging of many objects, including electrodes [29].

Arrhenius' law postulates that higher temperatures increase the rate of the reaction, thus increasing the rate of electrode degradation. This relation can be used to calculate the coefficient of acceleration (k) in relation to the difference between the operation temperature and elevated temperature (T).

$$k = A \cdot e^{\frac{-E_a}{RT}} \tag{1}$$

where k: coefficient of acceleration, E_a: activation energy, A: frequency factor, R: gas constant, T: temperature in Kelvins.

A and E_a are empirical parameters determined through the comparative measurement of samples aged under different temperatures. Following previous work on the aging of bio-implants [29–31], the ASTM guidelines [32] provide values for the typical evaluation of such systems in the form of $k = Q^{\frac{\Delta T}{10}}$ with a conservative value of $Q = 2$. These values would give an acceleration factor of 19.03. Considering a chemically isolated aging medium, limited testing allowed the empirical determination of an acceleration factor of 12.9 in the conditions presented in this study.

2.4. Impedance Analysis

To evaluate the material's degradation, an analytical model of the device, based on the electrical behavior, was developed in relation to the electrode's geometry. The model was reduced to three passive components to reflect the main electric behaviors [33] of the device (Figure 4):

(i)   Rs represents the deposited metal's resistivity'
(ii)   Ce is the isolation between the conductors; since it includes the electrodes, it depends on the evolution of the electrode geometry;
(iii)   Rl represents the leakage currents through the substrates or between the electrodes.

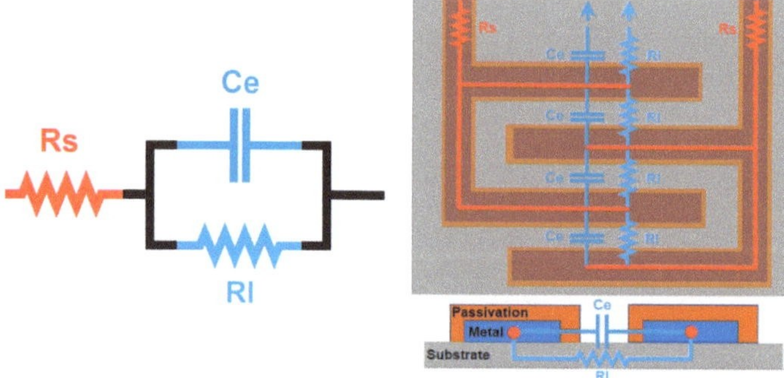

Figure 4. (**Left**) Simplified electrical model of the device. (**Right**) Schematic representation of the test pseudo-electrodes and their correlations with the model's parameters.

Considering Ce's expected capacitance, with a constant medium of permittivity ($\varepsilon_0\varepsilon_r$) and an equivalent distance between the electrodes (d), its variations are mainly due to changes in the IDE's surface area (A): $Ce = \varepsilon_0\varepsilon_r\frac{A}{d}$.

To take into account the statistical distribution of the capacitance over the IDE's surface, a constant phase element (CPE) has been introduced as a model for Ce. In this case, the impedance associated with Ce is $Zce = \frac{1}{Ce(j\omega)^\beta}$.

The value variation of β can be seen as a reflection of the charge distribution along the different electrodes and the surface roughness. For a better interpretation, Ce can be considered as a fractional capacitor, which offers a means to evaluate charge distributions as an indicator of the surface conductivity variation [34,35].

The described transfer function is defined in Equation (2):

$$Z(tot) = Rs + \frac{1}{\frac{1}{RI} + (j\omega Ce)^\beta} \tag{2}$$

The measurement setup is automated using a dedicated Python 3.7 script to allow the analysis of the changes in the parameters (impedance, temperature, time, etc.) over time. The data collected are then adapted using VBscript 5.6 and the parameters are finally extracted and plotted using Matlab R2021a. Lastly, fitting is automated and supervised using the least-squares method.

2.5. Aging Experimental Setup with Accelerating Behavior

To test the stability of the sample, an accelerating long-life setup has been fabricated. Figure 5 shows the setup based on an aluminum block (for thermal stability) placed on a hotplate. The block contains three tanks where a jar can be inserted. To ensure good thermal conductivity, each tank is filled with mineral oil. Each jar is filled with PBS solution and 6 samples can be placed in each one. To measure the temperature of the solution, thermocouples are inserted into the jar. The temperature of the PBS is maintained at 78 °C to ensure the acceleration of the aging according to the law of Arrhenius [29,36,37], as seen previously.

The whole system is placed inside a Styrofoam box to protect it from the environment and ensure good thermal insulation from the outside. To check the integrity of the sample, impedance measurements are conducted regularly (every hour) with a large frequency range, from 100 Hz to 10 MHz. The results are coupled with an analytical model for the monitoring of the implants' aging over time. The system also allows the stimulation of the electrode strip via current injection. The current stimulation mimics the behavior of the implant during artificial spiking activity (induced by stimulation).

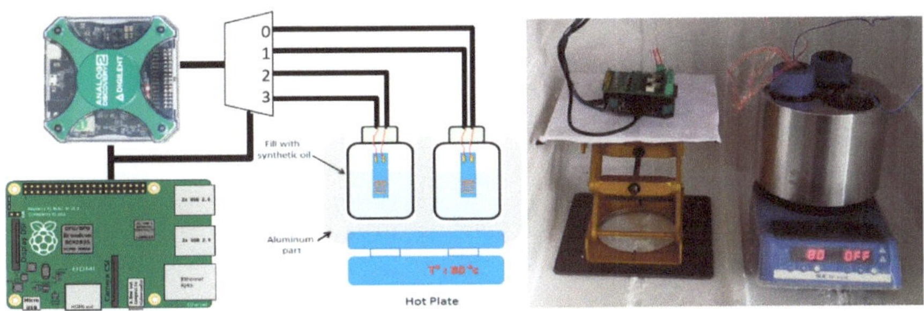

Figure 5. (**Left**) Schematic representation of the test setup. (**Right**) Test bench setup with samples.

3. Results

The collected data consist of four samples: two, denoted SU8_100 and SU8_50 (Table 1), are associated with the polymer passivation layer, and the two others concern the diamond passivation layer, labeled Dmd_50 and Dmd_25 (see Table 1). The data include a total of 7009 impedance spectra (modulus and phase) over 100 frequency points logarithmically spaced between 100 Hz and 15 MHz, with a measurement occurring every hour for each sample. This represents continuous recording for 39 weeks under accelerated aging conditions.

Considering that there are 7009 measurements for each electrode, data analysis is required for the interpretation of the results. Thus, for a better understanding, further visual analysis is performed on a single frequency point considered representative of neural activity recording: 1 kHz. Parameter evolution is extracted using automated supervised least-squares fitting.

3.1. Stability Comparison

To evaluate the stability of the pseudo-electrodes over time, the evolution of each sample's impedance modulus and phase at 1 kHz is considered, as in Figure 6 (dotted lines refer to diamond; continuous lines refer to the SU8 polymer). The values presented are normalized compared to the first measurement. Our tests show different aging changes and potential failure modes for each material. At first consideration, the diamond passivation layer shows higher insulation properties, represented by the mean |Z| value over time, presented in Table 2. The diamond passivation also shows more stable behavior compared to the SU8 layer, which presents some failure, which is visible as a high modulus and phase variation, potentially indicating the occurrence of a distinct failure.

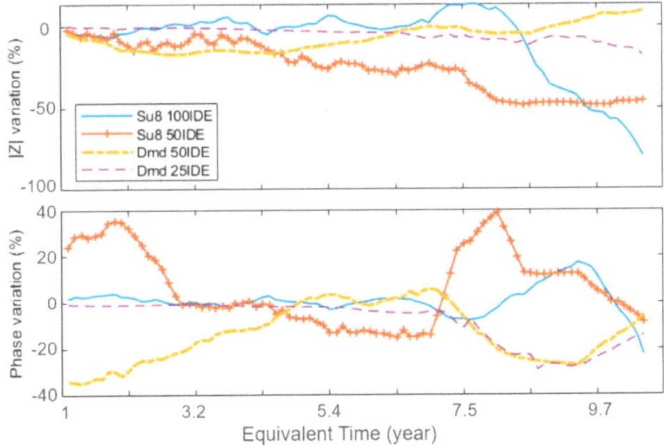

Figure 6. Extracted impedance variation (modulus and phase) for each sample design at 1 kHz over time.

Table 2. Sample parameters' evolution over time, including initial transient period.

	SU8_100	**SU8_50**	**Dmd_50**	**Dmd_25**
Passivation	Polymer	Polymer	Diamond	Diamond
Mean $|Z|$ @1 kHz	5.2 kΩ	6 kΩ	15 kΩ	35 kΩ
Module variation	±51.1%	±27.6%	±14.1%	±9.8%
Phase variation	±24.8%	±29.4%	±14.3%	±9%
Rs variation	±104.4%	±135.4%	±47.9%	±34.3%
Rl variation	±58%	±115.8%	±28%	±37%
Ce variation	±71%	±136.4%	±24.9%	±35.5%
Stabilized Ce	1.15 μF	524 nF	540 nF	356 nF
Equivalent stabilized Ce surface	14.2 μm²	6.5 μm²	6.7 μm²	4.4 μm²
β variation	±12.1%	±14.2%	±2.9%	±5.2%
−30% functionality (equivalent time)	6.4 years	8.1 years	⊘	⊘

The diamond sample's phase shows a synchronous variation after an equivalent aging time of 7.5 years of comparable amplitude, indicating a common cause. This effect seems to be correlated to the acceleration of the degradation of the SU8 samples, which may have altered the PBS medium.

To evaluate the reproducibility, several sets of diamond electrodes have been separately aged for a longer period, to extract the statistical variance from the process. The results can be observed in Figure 7. Although these samples have not been monitored over time, their final impedance values at 1 kHz after being aged together show two main aging behaviors: one set increased in modulus, as the Dmd_50 sample previously showed, while the second set progressively decreased, as Dmd_25 did. Further studies are needed to analyze the reasons for such differences as no obvious pattern emerged. One sample failed with a −64% modulus variation, which may have been due to an error in this particular sample's preparation as the wire connections were found to be exposed to the medium. This sample has, nevertheless, been retained in our data.

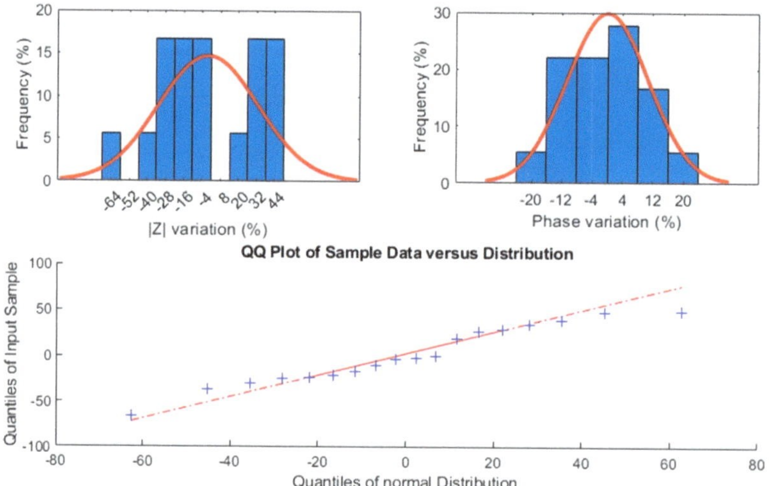

Figure 7. Impedance variations of diamond electrodes over 23 years of equivalent aging time on 6 sets of 3 electrodes.

3.2. Model's Parameters' Evolution

For each material and spectrum, the extraction of the model parameter values was performed, and this allowed the monitoring over time of the degradation occurring within the samples. Figure 8 shows the side-by-side comparison of the Rs, Rl, Ce and β parameters' evolution throughout the experimental duration.

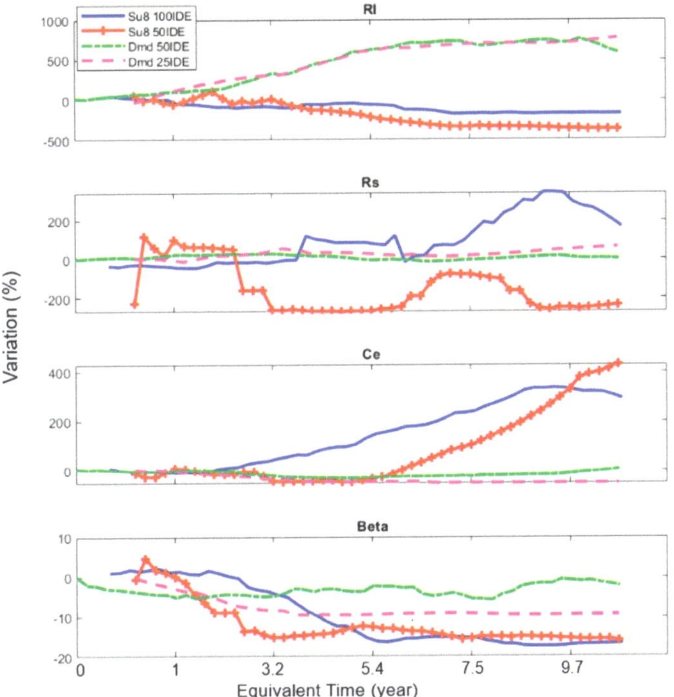

Figure 8. Evolution of the variations in the extracted model parameters for the SU8 and diamond samples as a function of the equivalent time.

The capacitance of polymer samples tends to increase over time, which may be due to an increase in their surface, eventually becoming porous, which reflects the degradation of the passivation layer. Rl and β seem to offer some indications on this matter, as the current leakage increases for the polymer layer (Rl decreases), while the charge distribution (β) shows imperfect capacitive behavior (β decreased). β is also related to the roughness of the surface; a decrease in β over time can be associated with an increase in the roughness value. As such, it is inferred that, over time and successive measurements, the polymer passivation tends to degrade and leak.

Comparatively, after an initial stabilization period, the diamond samples show relatively stable capacitance values over time. This is confirmed by the stability of the β parameter, showing very little variation over time, also confirming the stability of the surface electrode capacitance Ce. The Rl value tends here to increase at first until reaching a plateau, in synchronization with Ce. This indicates that the diamond passivation starts in a less conductive state before stabilizing, meaning an increase in its electrical insulation behavior. It can be assumed that the diamond layer is in a more hydrogenated state after growth, which stabilizes as the hydrogen reacts [38,39].

This comparison, across more than nine months of experiments under accelerated aging conditions, highlights the very good stability of diamond as a passivation layer compared to the polymer material (SU8). Indeed, its capacitance and charge distribution show high stability over time, indicating no loss in functionality, besides an initial settling time due to the fabrication process. This confirms the potential of using diamond as an electrically stable layer for biomedical devices.

3.3. Microscopic Observation

To correlate the previously shown analytical results, a sample analysis has been performed, starting with profilometer measurements (Figure 9), which can be compared with Figure 3. The diamond samples retain their original shape, indicating no loss in structural integrity, while the SU8 samples show distortion and a diminished profile, indicating the clear instability of the passivation layer.

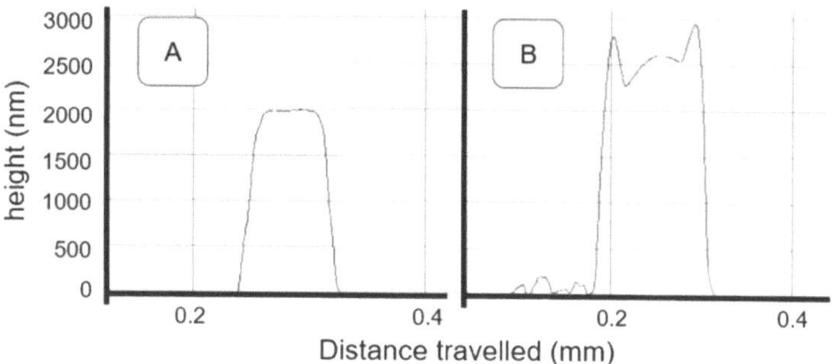

Figure 9. Thickness profile comparison before and after aging. (**A**) Dmd_50 after aging. (**B**) SU8_50 after aging.

Several more observations have been made using both microscopic and SEM observation, as shown in Figure 10, where the comparison before and after aging shows a clear passivation failure for SU8. The diamond layer has retained its original shape, further confirming its structural integrity, although some undefined particles appear lodged between the IDE's elements. These particles are supposed to originate from the SU8 sample's degraded layers, as previously mentioned during the analysis. The high density of these particles on the SU8 samples, and near their degraded parts in particular, further increase our confidence in their origin, without leading to a conclusion (Figures 10 and 11).

Figure 10. Microscopic and SEM observation of IDE's surface before and after aging. (**A**) SU8_50 before aging. (**B**) SU8_50 after aging. (**C**) Dmd_50 before aging. (**D**) Dmd_50 after aging.

A further analysis of the SU8 passivation failures shows the progress of the localized structural decohesion of the passivation, as seen in Figure 11. This delamination seems to further progress until the SU8 layer's structure breaks, exposing the inner layers. These effects accumulate over the electrode's surface, explaining the relatively high impedance variations compared to diamond, as discussed previously. For the diamond samples, no delamination was optically found.

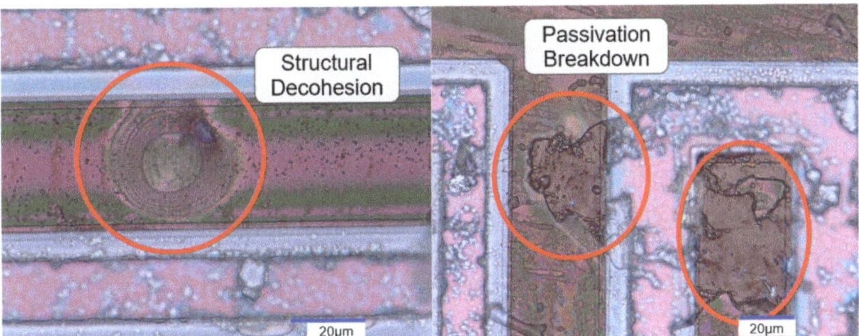

Figure 11. Optical microscopic observation of SU8_50's surface after aging with a focus on different failure types.

4. Discussion

This work serves to demonstrate that the diamond material will be the next-generation material for long-term medical implants. With its intrinsic properties, carbon interface, lack of oxidation, high stability and biocompatibility, the diamond material shows great promise for the medical field. To validate these assumptions, this study of IDE structures provides a complete analysis of the aging characteristics of the material and an analysis on a simple device. This choice was made to allow a fine comparison between the bioimpedance measurements over time under specific aging conditions and to obtain analytical models expressing several significant parameters, such as capacitance, resistance or charge variations. Our analytical model, focusing on a fractional capacitor, provides clear insights into the surface conductivity variation, and the presented results highlight the superior stability of diamond as a passivation layer compared to polymer materials like SU8.

The next step will be to complete the aging setup with additional varying conditions, such as temperature ramps, pH cycling or mechanical stress, with samples in separate media to avoid potential cross-contamination. A deeper analytical model coupled with micro-photography monitoring could allow the extraction and validation of more parameters, paving the way towards a deeper understanding of the material's properties and improving our initial predictions. Our methodology has proven its efficiency and will be extended to converge towards an improved and fully characterized neuro-implant.

Furthermore, to validate the attractiveness of diamond materials for medical applications, a comparison between different materials is planned, such as silicon nitride or silicon dioxide as a passivation layer in an accelerated long-life setup, as well as validating the stability of conductive diamond as an interface for sensing, while recording or stimulating neuronal activity. These studies will result in a full-diamond implant that combines conductive and intrinsic diamond for long-life use in medical devices. Thanks to this development based on diamond materials, the production of medical implants that are stable throughout a patient's life can be envisaged.

The sample's fabrication is based on a completely new manufacturing process based on MPCVD diamond growth on 4-inch wafers and is fully compatible with standard cleanroom microfabrication and commercial equipment [26]. Therefore, the production of diamond-based neural implants will be directly transferable, and previous projects have already conducted some diamond implant manufacturing [40]. Furthermore, today's microelectronics sector is investing widely in this material for the development of new applications in MEMS sensors [41] or high-energy devices [42,43] as examples. Thus, in the near future, it is expected to become possible to easily produce diamond layers on larger wafers, thus multiplying the production output of such devices.

Introducing a new material for medical application requires many years of research and regulatory processes; as such, it is necessary to use animal experiments to validate the innocuity of the material. In the European project NEUROCARE, several preliminary studies have been performed with diamond samples in animals to validate their biocompatibility [44], but it will be necessary to continue these experiments in a future project to accelerate the transfer of this technology to the medical field.

5. Conclusions

This study aimed to assess, for the first time, the long-term stability and aging characteristics of intrinsic polycrystalline diamond as a biocompatible coating material for in vivo implants.

This investigation was based on an electric impedance analysis of diamond-passivated IDE samples compared to polymer-coated IDE samples. The results obtained demonstrated evolving impedance patterns over time, revealing information about the behavior of the test devices and therefore the materials. The key parameters concerning electrical insulation were clearly stable for the diamond compared to the polymer, where failures appeared during the experiments. Accelerated aging over an equivalent time of 12 years with thermal activation was used to obtain these results.

The impedance analysis of isolated test electrodes, coupled with the developed analytical model, provides valuable insights into the material efficiency of full-diamond electrodes at a low cost and with ease of use. The consideration of Ce as a fractional capacitor has allowed the evaluation of the charge distributions as indicators of the surface conductivity variation and behavior, offering a better understanding of diamond's aging dynamics while used as a passivation layer.

This work demonstrates the strong interest in diamond materials for the in vivo long-term use of neuronal implants.

Author Contributions: Conceptualization, L.R., G.L., H.T. and E.S.; methodology, L.R., O.F., U.T.S. and J.R.; software, O.F., A.R. and J.R.; validation, L.R. and O.F.; formal analysis, J.R.; investigation, L.R., U.T.S., G.L., H.T. and E.S.; resources, L.R. and A.R.; data curation, J.R.; writing—original draft preparation, L.R.; writing—review and editing, O.F., P.P., G.L., A.R. and J.R.; visualization, J.R.; supervision, L.R., O.F., A.R., G.L., H.T. and E.S.; project administration, L.R.; funding acquisition, L.R. All authors have read and agreed to the published version of the manuscript.

Funding: This research was funded by the European Research Council (ERC), ERC starting grant 2017 (758700—NEURODIAM), Université Gustave Eiffel, CNRS, ESYCOM, UMR 9007 CNRS/Université Gustave Eiffel, F-77454 Marne-la-Vallée, France.

Institutional Review Board Statement: Not applicable.

Informed Consent Statement: Not applicable.

Data Availability Statement: Data are contained within the article.

Conflicts of Interest: The authors declare no conflicts of interest. The funders had no role in the design of the study; in the collection, analyses, or interpretation of the data; in the writing of the manuscript; or in the decision to publish the results.

References

1. Lorach, H.; Galvez, A.; Spagnolo, V. Walking Naturally After Spinal Cord Injury Using a Brain–spine Interface. *Nature* **2023**, *618*, 126–133. [CrossRef]
2. Blackrock Neurotech | Empowered by Thought. Available online: https://blackrockneurotech.com/ (accessed on 29 April 2024).
3. Liu, X.; Chen, P.; Ding, X.; Liu, A.; Li, P.; Sun, C.; Guan, H. A Narrative Review of Cortical Visual Prosthesis Systems: The Latest Progress and Significance of Nanotechnology for the Future. *Ann. Transl. Med.* **2022**, *10*, 716. [CrossRef] [PubMed]

4.   Eitutis, S.T.; Tam, Y.C.; Roberts, I.; Swords, C.; Tysome, J.R.; Donnelly, N.P.; Axon, P.R.; Bance, M.L. Detecting and Managing Partial Shorts in Cochlear Implants: A Validation of Scalp Surface Potential Testing. *Clin. Otolaryngol.* **2022**, *47*, 641–649. [CrossRef] [PubMed]

5.   Hassler, C.; Boretius, T.; Stieglitz, T. Polymers for Neural Implants. *J. Polym. Sci. Part B Polym. Phys.* **2011**, *49*, 18–33. [CrossRef]

6.   Onken, A.; Schütte, H.; Wulff, A.; Lenz-Strauch, H.; Kreienmeyer, M.; Hild, S.; Stieglitz, T.; Gassmann, S.; Lenarz, T.; Doll, T. Predicting Corrosion Delamination Failure in Active Implantable Medical Devices: Analytical Model and Validation Strategy. *Bioengineering* **2022**, *9*, 10. [CrossRef]

7.   Lacour, S.P.; Courtine, G.; Guck, J. Materials and Technologies for Soft Implantable Neuroprostheses. *Nat. Rev. Mater.* **2016**, *1*, 1–14. [CrossRef]

8.   Cho, Y.; Park, S.; Lee, J.; Yu, K.J. Emerging Materials and Technologies with Applications in Flexible Neural Implants: A Comprehensive Review of Current Issues with Neural Devices. *Adv. Mater.* **2021**, *33*, 2005786. [CrossRef]

9.   Malisz, K.; Świeczko-Żurek, B.; Sionkowska, A. Preparation and Characterization of Diamond-like Carbon Coatings for Biomedical Applications—A Review. *Materials* **2023**, *16*, 3420. [CrossRef]

10.   Hrdlička, V.; Matvieiev, O.; Navrátil, T.; Šelešovská, R. Recent Advances in Modified Boron-doped Diamond Electrodes: A Review. *Electrochim. Acta* **2023**, *456*, 142435. [CrossRef]

11.   Purcell, E.K.; Becker, M.F.; Guo, Y.; Hara, S.A.; Ludwig, K.A.; McKinney, C.J.; Monroe, E.M.; Rechenberg, R.; Rusinek, C.A.; Saxena, A.; et al. Next-Generation Diamond Electrodes for Neurochemical Sensing: Challenges and Opportunities. *Micromachines* **2021**, *12*, 128. [CrossRef]

12.   Devi, M.; Vomero, M.; Fuhrer, E.; Castagnola, E.; Gueli, C.; Nimbalkar, S.; Hirabayashi, M.; Kassegne, S.; Stieglitz, T.; Sharma, S. Carbon-based Neural Electrodes: Promises and Challenges. *J. Neural Eng.* **2021**, *18*, 041007. [CrossRef]

13.   Garrett, D.J.; Tong, W.; Simpson, D.A.; Meffin, H. Diamond for Neural Interfacing: A Review. *Carbon* **2016**, *102*, 437–454. [CrossRef]

14.   Nistor, P.A.; May, P.W. Diamond thin films: Giving biomedical applications a new shine. *J. R. Soc. Interface* **2017**, *14*, 20170382. [CrossRef]

15.   Barrese, J.C.; Rao, N.; Paroo, K.; Triebwasser, C.; Vargas-Irwin, C.; Franquemont, L.; Donoghue, J.P. Failure mode analysis of silicon-based intracortical microelectrode arrays in non-human primates. *J. Neural Eng.* **2013**, *10*, 066014. [CrossRef] [PubMed]

16.   Tintelott, M.; Schander, A.; Lang, W. Understanding Electrical Failure of Polyimide-based Flexible Neural Implants: The Role of Thin Film Adhesion. *Polymers* **2022**, *14*, 3702. [CrossRef] [PubMed]

17.   Xie, X.; Rieth, L.; Williams, L.; Negi, S.; Bhandari, R.; Caldwell, R.; Sharma, R.; Tathireddy, P.; Solzbacher, F. Long-term reliability of Al_2O_3 and Parylene C bilayer encapsulated Utah electrode array based neural interfaces for chronic implantation. *J. Neural Eng.* **2014**, *11*, 026016. [CrossRef]

18.   King, E.J.; Yoganathan, M.; Nagelschmidt, G. The effect of diamond dust alone and mixed with quartz on the lungs of rats. *Br. J. Ind. Med.* **1958**, *15*, 92–95. [CrossRef]

19.   Vaijayanthimala, V.; Cheng, P.Y.; Yeh, S.H.; Liu, K.K.; Hsiao, C.H.; Chao, J.I.; Chang, H.C. The long-term stability and biocompatibility of fluorescent nanodiamond as an in vivo contrast agent. *Biomaterials* **2012**, *33*, 7794–7802. [CrossRef] [PubMed]

20.   Yu, S.J.; Kang, M.W.; Chang, H.C.; Chen, K.M.; Yu, Y.C. Bright fluorescent nanodiamonds: No photobleaching and low cytotoxicity. *J. Am. Chem. Soc.* **2005**, *127*, 17604–17605. [CrossRef]

21.   Grodzik, M.; Sawosz, F.; Sawosz, E.; Hotowy, A.; Wierzbicki, M.; Kutwin, M.; Jaworski, S.; Chwalibog, A. Nano-nutrition of chicken embryos. The effect of in ovo administration of diamond nanoparticles and L-glutamine on molecular responses in chicken embryo pectoral muscles. *Int. J. Mol. Sci.* **2013**, *14*, 23033–23044. [CrossRef]

22.   Zakrzewska, K.E.; Samluk, A.; Wierzbicki, M.; Jaworski, S.; Kutwin, M.; Sawosz, E.; Chwalibog, A.; Pijanowska, D.G.; Pluta, K.D. Analysis of the cytotoxicity of carbon-based nanoparticles, diamond and graphite, in human glioblastoma and hepatoma cell lines. *PLoS ONE* **2015**, *10*, e0122579. [CrossRef] [PubMed]

23.   Xiao, X.; Wang, J.; Liu, C.; Carlisle, J.A.; Mech, B.; Greenberg, R.; Guven, D.; Freda, R.; Humayun, M.S.; Weiland, J.; et al. In vitro and in vivo evaluation of ultrananocrystalline diamond for coating of implantable retinal microchips. *J. Biomed. Mater. Res. Part B Appl. Biomater.* **2006**, *77*, 273–281. [CrossRef] [PubMed]

24.   Koizumi, S.; Teraji, T.; Kanda, H. Phosphorus-doped chemical vapor deposition of diamond. *Diam. Relat. Mater.* **2000**, *9*, 935–940. [CrossRef]

25.   Fan, B.; Rusinek, C.A.; Thompson, C.H. Flexible, Diamond-based Microelectrodes Fabricated Using the Diamond Growth Side for Neural Sensing. *Microsyst. Nanoeng.* **2020**, *6*, 1–12. [CrossRef]

26.   Wilfinger, C.; Zhang, J.; Nguyen, D.; Degardin-Chicaud, J.; Bergonzo, P.; Picaud, S.; Borda, E.; Ghezzi, D.; Scorsone, E.; Lissorgues, G.; et al. *In Vivo Recording of Visually Evoked Potentials with Novel Full Diamond Ecog Implants*; Elsevier: Amsterdam, The Netherlands, 2023. [CrossRef]

27.   Arrhenius, S.A. Über die Dissociationswärme und den Einfluß der Temperatur auf den Dissociationsgrad der Elektrolyte. *Z. Phys. Chem.* **1889**, *4*, 96–116. [CrossRef]

28.   Arrhenius, S.A. Über die Reaktionsgeschwindigkeit bei der Inversion von Rohrzucker durch Säuren. *Z. Phys. Chem.* **1889**, *4*, 226–248. [CrossRef]

29.   Hukins, D.W.; Mahomed, A.; Kukureka, S.N. Accelerated aging for testing polymeric biomaterials and medical devices. *Med. Eng. Phys.* **2008**, *30*, 1270–1274. [CrossRef] [PubMed]

30. Patel, P.R.; Zhang, H.; Robbins, M.T.; Nofar, J.B.; Marshall, S.P.; Kobylarek, M.J.; Kozai, T.D.; Kotov, N.A.; Chestek, C.A. Chronic in vivo stability assessment of carbon fiber microelectrode arrays. *J. Neural Eng.* **2016**, *13*, 066002. [CrossRef] [PubMed]

31. Green, R.A.; Hassarati, R.T.; Bouchinet, L.; Lee, C.S.; Cheong, G.L.; Yu, J.F.; Dodds, C.W.; Suaning, G.J.; Poole-Warren, L.A.; Lovell, N.H. Substrate dependent stability of conducting polymer coatings on medical electrodes. *Biomaterials* **2012**, *33*, 5875–5886. [CrossRef]

32. Standard Guide for Accelerated Aging of Sterile Barrier Systems for Medical Devices. Available online: https://www.astm.org/f1980-16.html (accessed on 29 April 2024).

33. Randles, J. Kinetics of Rapid Electrode Reactions. *Discuss. Faraday Soc.* **1947**, *1*, 11–19. [CrossRef]

34. Elwakil, A.S. Fractional-order Circuits and Systems: An Emerging Interdisciplinary Research Area. *IEEE Circuits Syst. Mag.* **2010**, *10*, 40–50. [CrossRef]

35. Tripathy, M.C.; Behera, S. Modelling and Analysis of Fractional Capacitors. *Int. J. Eng. Appl. Sci. (IJEAS)* **2015**, *2*, 29–32.

36. Logan, S.R. The Origin and Status of the Arrhenius Equation. *Int. J. Eng. Appl. Sci. (IJEAS)* **1982**, *59*, 279. [CrossRef]

37. Janting, J.; Theander, J.G.; Egesborg, H. On Thermal Acceleration of Medical Device Polymer Aging. *IEEE Trans. Device Mater. Reliab.* **2019**, *19*, 313–321. [CrossRef]

38. Shiryaev, A.A.; Grambole, D.; Rivera, A.; Herrmann, F. On the Interaction of Molecular Hydrogen with Diamonds: An Experimental Study Using Nuclear Probes and Thermal Desorption. *Diam. Relat. Mater.* **2007**, *16*, 1479–1485. [CrossRef]

39. Łoś, S.; Fabisiak, K.; Paprocki, K.; Szybowicz, M.; Dychalska, A.; Spychaj-Fabisiak, E.; Franków, W. The Hydrogenation Impact on Electronic Properties of p-Diamond/n-Si Heterojunctions. *Materials* **2021**, *14*, 6615. [CrossRef] [PubMed]

40. Wilfinger, C.A. Fabrication of Full Soft Diamond Implants for Functional Rehabilitation. Ph.D. Thesis, Gustave Eiffel University, Champs-sur-Marne, France, 2023. NNT: 2023UEFL2025.

41. Liao, M. Progress in Semiconductor Diamond Photodetectors and MEMS Sensors. *Funct. Diam.* **2021**, *1*, 29–46. [CrossRef]

42. Chen, Z.; Fu, Y.; Kawarada, H.; Xu, Y. Microwave Diamond Devices Technology: Field-effect Transistors and Modeling. *Int. J. Numer. Model. Electron. Netw. Devices Fields* **2021**, *34*, e2800. [CrossRef]

43. Choi, U.; Kwak, T.; Han, S.; Kim, S.-W.; Nam, O. High Breakdown Voltage of Boron-doped Diamond Metal Semiconductor Field Effect Transistor Grown on Freestanding Heteroepitaxial Diamond Substrate. *Diam. Relat. Mater.* **2022**, *121*, 108782. [CrossRef]

44. Cottance, M. Contribution to the Development of Neuro-Electric Interfaces. Ph.D. Thesis, Paris-Est University, Paris, France, 2014. NNT: 2014PEST1105.

Article

Highly Sensitive and Selective Detection of L-Tryptophan by ECL Using Boron-Doped Diamond Electrodes †

Emmanuel Scorsone *, Samuel Stewart and Matthieu Hamel

Université Paris-Saclay, CEA, List, F-91120 Palaiseau, France; samuel.stewart@cea.fr (S.S.)
* Correspondence: emmanuel.scorsone@cea.fr
† This paper is an extended version of our paper published in Proceedings of the XXXV EUROSENSORS Conference, Lecce, Italy, 10–13 September 2023.

Abstract: L-tryptophan is an amino acid that is essential to the metabolism of humans. Therefore, there is a high interest for its detection in biological fluids including blood, urine, and saliva for medical studies, but also in food products. Towards this goal, we report on a new electrochemiluminescence (ECL) method for L-tryptophan detection involving the in situ production of hydrogen peroxide at the surface of boron-doped diamond (BDD) electrodes. We demonstrate that the ECL response efficiency is directly related to H_2O_2 production at the electrode surface and propose a mechanism for the ECL emission of L-tryptophan. After optimizing the analytical conditions, we show that the ECL response to L-tryptophan is directly linear with concentration in the range of 0.005 to 1 µM. We achieved a limit of detection of 0.4 nM and limit of quantification of 1.4 nM in phosphate buffer saline (PBS, pH 7.4). Good selectivity against other indolic compounds (serotonin, 3-methylindole, tryptamine, indole) potentially found in biological fluids was observed, thus making this approach highly promising for quantifying L-tryptophan in a broad range of aqueous matrices of interest.

Keywords: boron-doped diamond; electrochemiluminescence; L-tryptophan

Citation: Scorsone, E.; Stewart, S.; Hamel, M. Highly Sensitive and Selective Detection of L-Tryptophan by ECL Using Boron-Doped Diamond Electrodes. *Sensors* **2024**, *24*, 3627.
https://doi.org/10.3390/s24113627

Academic Editor: Rene Kizek

Received: 15 May 2024
Revised: 27 May 2024
Accepted: 1 June 2024
Published: 4 June 2024

1. Introduction

L-tryptophan is an amino acid that is essential to the metabolism of humans [1]. It is involved in the biosynthesis of proteins, but it also plays other important roles, along with its metabolites such as serotonin, in neurophysiological functions [2,3]. L-tryptophan can only be obtained through food intake, and the recommended daily dose for adults is estimated to be between 250 mg and 425 mg [4,5]. Low levels of the amino acid could be compensated for by the consumption of L-tryptophan-based dietary supplements as a possible strategy to help in the treatment of disorders such as depression, insomnia, obesity, or Parkinson disease [6,7]. High levels of L-tryptophan may also be harmful for the central nervous system and have been suggested to accelerate aging and neurodegeneration processes [8–10]. Therefore, there is a high interest in the detection of this important amino acid in biological fluids, including blood, urine, and saliva, in medical studies, in particular in the area of neuroscience and oncology [11–13]. A variety of analytical methods have been reported for the determination of L-tryptophan, which include for instance high-performance liquid chromatography (HPLC) [14–16] and colorimetric and spectrofluorometric methods [17–20]. These methods require expensive equipment and trained personnel and are time-consuming. Therefore, there is a need for alternative solutions that are low-cost, rapid, accurate and easy to use. In this context, a range of electrochemical sensors has been investigated with promising results [11,21–25]. Electrogenerated chemiluminescence (ECL) is also attractive here because it offers potentially higher sensitivity than voltammetric methods and good selectivity. ECL methods usually take one of two different approaches. In the first approach, L-tryptophan's ability to enhance or quench the luminescence of certain metal complexes is used. For example, tryptophan has been utilized as a luminescence enhancer of the Ru(bpy)$_3^{2+}$ and KMnO$_4$ system [26]. A very low limit of detection (LOD), around 10 nM, was achieved; however, selectivity was found to be poor in this case. In particular, an increase in luminescence observed from other amino acids like histamine were present. A similar approach was also used as an end HPLC detector in order to distinguish between the two different enantiomers (L-tryptophan and D-tryptophan) down to the picomolar level [27]. Moreover, the ECL quenching ability of tryptophan has also been extensively investigated, in particular its effect on iridium complexes [28–30]. This relies on indole being subjected to electrochemical oxidation, which reduces the generation of electrogenerated excited states and quenches ECL emission of the iridium complex. While this method has been shown to be more selective than the signal-enhancing technique, it was found to display a non-linear response, meaning that quantitative measurements may be limited. In the second approach, L-tryptophan may be detected using ECL through its direct oxidation into an excited intermediate state that luminesces itself. This phenomenon was firstly reported by Sakura et al., who found that tryptophan would luminesce when oxidized on a platinum electrode at 0.78 V in a 0.1 M NaOH solution and in the presence of a bromating agent [31]. However, this method required very high concentrations of tryptophan to be present (>1 μM), and so it was not deemed to be useful for routine analysis. This process has recently been refined, first by Chen et al., who found that upon the addition of hydrogen peroxide, both indole and tryptophan would luminesce in a basic solution (0.03 M NaOH and 0.1 M KCl) [32]. Using this method, they were able to achieve an LOD of both indole and tryptophan around 100 nM. Since then, further attempts to improve the sensitivity of this method have been investigated, most notably by Zholudov et al., who tested the use of another oxidizing coreactant, namely tetraphenylborate (TPB), with tryptophan in aqueous solutions on glassy carbon electrodes [33]. Although this method did work in near-neutral conditions, meaning that it was good for biological assays, ECL emission was rather weak, with an LOD of 300 nM. Thus, hydrogen peroxide has been demonstrated to be the best coreactant so far for assisting the electro-oxidation of tryptophan. Moreover, boron-doped diamond (BDD) electrodes are known for their exceptional electrochemical properties, including a wide potential window in aqueous media > 3 V, low double-layer capacitance, high stability and low adsorption properties [34–38]. Such electrodes have recently shown to be capable of opening new perspectives in ECL analysis [39]. In this paper, we investigate the feasibility and performance of electrogenerated hydrogen peroxide at the surface of BDD electrodes as a new efficient way to detect L-tryptophan in aqueous solutions without the need for adding further co-reactant to the analytical solution.

2. Materials and Methods

2.1. Chemicals

L-tryptophan, luminol, phosphate buffer saline (PBS) tablets, tryptamine, indole, serotonin, and 3-methylindole were all purchased from Sigma-Aldrich, Saint Quentin Fallavier, France and were of analytical grade. Deionized water was obtained from a Direct-Q UV 3 water purification system (Merck Millipore, Guyancourt, France).

2.2. Electrochemical Cell

A three-electrode system was used, including lab-grown boron-doped diamond (BDD) electrodes as both working and counter electrodes, and a platinum wire as pseudo-reference electrode. Polycristalline BDD was grown by plasma enhanced chemical vapor deposition (PE-CVD) on a highly conductive 4-inch <100> silicon wafer in a SekiDiamond AX6500 diamond growth reactor. The growth parameters were 1% methane in hydrogen at a pressure of 40 Torr and microwave power 3.5 kW. Trimethylboron was added to the gas phase as a source of boron dopant. The thickness of the resulting diamond film was ca. 800 nm with a doping level of 2 × 10^{21} boron atom.cm^{-3}, as determined by SIMS measurements. A scanning electron microscopy (SEM) image of the surface of the BDD layer is shown in Figure S1. Electrodes were fabricated using a 10×10 mm^2 (working electrode, WE) or 15×15 mm^2 (counter electrode, CE) section of BDD. Electrical contact was taken through the silicon backside using copper tape. Moreover, electrodes from cut sheets of 0.128 mm foil gold and 1 mm foil glassy carbon (Sigma-Aldrich, Saint Quentin Fallavier, France) were fabricated for comparison with BDD.

2.3. ECL Measurements

The two electrodes (working and counter), along with a Pt quasi-reference wire electrode, were then placed inside a 10 mL custom three-dimensional (3D)-printed cell fitted with an optical glass window, so that the worker electrode was facing the optical detector. When measuring total light emission, electrochemical measurements were performed using a potentiostat Autolab PGSTAT128N and the photon emission was measured with a PDM03-9107-USB photomultiplier tube (ET Enterprises, Uxbridge, UK) using the following parameters: acquisition time 500 ms, high voltage 950 V. All measurements were carried out in the dark. Spectral data were obtained by placing the electrochemical cell into a Fluoromax 4P spectrofluorometer (Horiba Jobin Yvon, Paris, France). In this latter case, a portable PalmSens EMStat4 potentiostat was used for electrochemical excitation. ECL intensities were recorded at different fixed wavelengths with a slit size of 25 nm (i.e., ± 12.5 nm). For example, point '350 nm' was measured from 337.5 to 362.5 nm, and so on.

3. Results

3.1. In Situ Generation of Hydrogen Peroxide on BDD Electrode

BDD electrodes are particularity suitable for in situ generation of reactive oxygen species (ROS) in solution, due to their large potential window in aqueous solutions and low adsorption properties. ROS are a product of oxygen reduction reactions (ORRs) from dissolved oxygen and/or water, either through oxidation or reduction reactions. The simplest of these ROS is the superoxide anion radical, which is formed as dissolved oxygen gains an electron. This is often the first step towards the production of other molecules such as hydrogen peroxide [40–42]. The production of hydrogen peroxide on electrode surfaces may be assessed using the luminol molecular probe. ECL emission from luminol is enhanced in the presence of both the superoxide radical cation and H_2O_2, but at high cathodic overpotential ECL enhancement from H_2O_2 is predominant [43]. Figure 1 shows the evolution of ECL intensity of luminol on either gold, glassy carbon (GC), or BDD electrodes as cyclic voltammetry was performed starting from 0 V to a lower cathodic potential and then back to 1 V vs. Ag I AgCl, at a scan rate of 0.1 V.s^{-1}. The lower potentials tested started from 0 V down to -3 V vs. Ag I AgCl in 0.5 V increments. As the resulting signals span several orders of magnitude, the results are displayed in a logarithmic scale. A typical voltammogram and resulting ECL signal recorded in such conditions is shown in Figure S2 for the BDD electrode. When no reduction potential is applied (lower potential = 0 V), both the GC and Au electrodes perform better in the standard luminol/O$_2$ system. As the lower potential is decreased to -1 V, there is a clear increase in ECL efficiency for all electrodes. This increase is less significant for Au electrodes, which start producing superoxide radicals at -1 V, where a reduction peak is observed on the voltammogram. However, the electrolysis of water occurs, producing hydrogen gas that will compete with the formation of ROS. The GC electrode features the largest increase in luminescence at this point. The generation of hydrogen peroxide seems to be maximal on GC electrodes from -1 V vs. Ag I AgCl. BDD has also a large increase in ECL emission, starting from -0.5 V vs. Ag I AgCl. However, as observed from ECL during the reduction reaction, the formation of H_2O_2 is not fully optimized until after -1.5 V due to the lower inner sphere reaction rate of BDD electrodes. When using a lower potential below -1 V for the Au and GC electrodes, steady ECL emission is reached as the generation of ROS is limited due to the high surface adsorption and the lower potential window in aqueous solution. The BDD electrode, however, continues to increase its luminescence as the lower voltage is decreased to the –3 V limit set, thus suggesting much higher efficiency of H_2O_2 generation on this electrode.

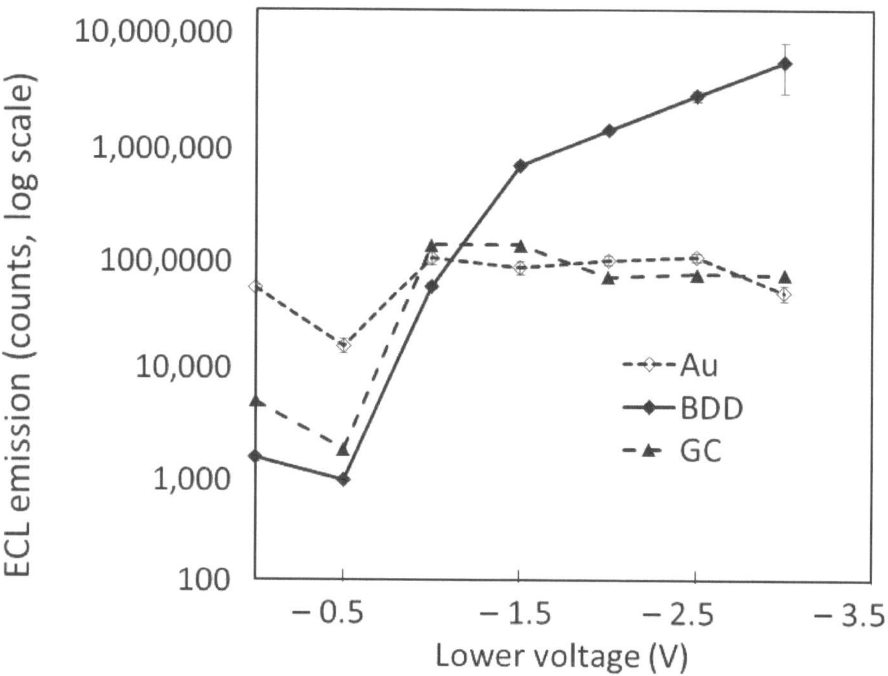

Figure 1. Lower applied voltage plotted against the corresponding oxidation ECL peak emission during the oxidation of luminol in log10 scale. Three electrodes, Au (short dash line) GC (long dash line), and BDD (continuous line), were tested, starting at 0 V before then sweeping to the lower potential and finally sweeping to 1 V, at a scan rate of 0.1 V.s^{-1} in a 0.1 M PBS solution with 1 µM of luminol (n = 3).

3.2. ECL Reaction in the Presence of L-Tryptophan

Simply oxidizing L-tryptophan on a BDD electrode in neutral conditions does not result in any luminescence, suggesting that the conditions that will produce an excited intermediate are not met. Thus, successive application of cathodic and then anodic polarization on BDD electrodes was examined in the presence of 500 nM tryptophan in 0.1 M PBS solution by cyclic voltammetry (Figure 2a). The appearance of an ECL signal is observed upon anodic polarization at approx. 0.9 V vs. Ag|AgCl. This occurs only after the reduction potential applied during CV reaches at least approximately −1.5 V vs. Ag|AgCl. According to previous experiments with luminol, this corresponds to the potential from which a large increase in the production of H_2O_2 occurs. This signal was found to gradually increase as a reduction potential was decreased further, down to −3.3 V vs. Ag|AgCl (Figure S3). However, potentials below −3 V were found to have significantly lower reproducibility between five replicate measurements. After this voltage, it can be assumed that other reactions, like the evolution of hydrogen gas at the cathode, disrupt the production of ROS that aid in enhancing the ECL signal. Similarly, the ECL signal was found to gradually increase while increasing the upper oxidation potential up to 1.7 V vs. Ag|AgCl. Above 1.5 V, reproducibility was also affected, hence an optimum value of 1.5 V vs. Ag|AgCl was considered. The effects of scan rate on the ECL signal emitted from L-tryptophan were also examined while scanning from 0 V to −3 V and then back to +1.5 V vs. Ag|AgCl in a solution of 100 nM L-tryptophan in 0.1 M PBS (Figure S4). Here it was found that ECL emission increased almost linearly from 10 to 250 mV.s^{-1}.

After ca. 300 mV.s^{-1}, a plateau seemed to be reached, where the ECL signal intensity remains constant. Therefore, 250 mV.s^{-1} was found to be optimal. Taking into account the CV optimum values as determined above, a calibration curve was drawn in the range 0–1 μM L-tryptophan in 0.1 M PBS (Figure 2b). ECL measurements over 5 replicates demonstrated good reproducibility with the percentage relative standard deviation (%RSD) ranging from typically 3–9%. Calibration graph showed excellent linearity across the whole range (0–1000 nM) with a linear fit coefficient of determination of $R^2 = 0.99\%$. LOD (Blank + $3 \times$ SD) and LOQ (Blank + $10 \times$ SD) of 0.4 and 1.4 nM were achieved in PBS solution, respectively.

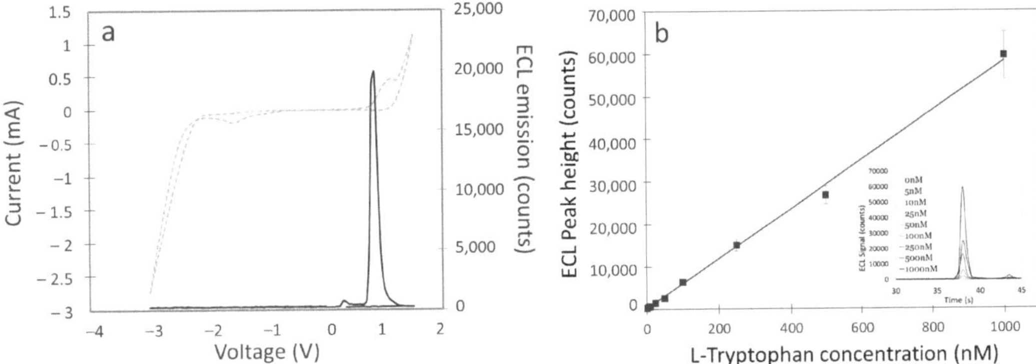

Figure 2. (**a**) CV (dashed line) and ECL (solid line) emission from 500 nM L-tryptophan in 0.1 M PBS solution. CV scan was 0 V $\rightarrow -3$ V $\rightarrow 1.5$ V $\rightarrow 0$ V, at a scan rate 0.25 V.s^{-1}. (**b**) Resulting calibration curve for L-tryptophan obtained in the same experimental conditions (inset: ECL peak emission for each concentration of tryptophan measured ($0, 5, 10, 25, 50, 100, 250, 500, 1000$ nM) in 0.1 M PBS (pH 7.4) on BDD electrodes after successive reduction and oxidation).

In order to establish the reaction mechanism and strengthen the hypothesis that the generation of hydrogen peroxide is responsible for the ECL of L-tryptophan, spectral data can provide valuable information. L-tryptophan is known to fluoresce at ca. 350 nm [44] when excited by UV light. However, all ECL measurements have shown maximum emission at different wavelengths, meaning that it generally goes through a chemical transformation before becoming ECL emitter. A spectrum with a peak at around 550 nm has been observed in the cases where a bromating agent is used as a coreactant. This is known to be associated with the chemiluminescence of the indole ring with the addition of a hydroperoxide group [45]. Moreover, L-tryptophan ECL can also produce luminescence at around 450 nm, which is said to be due to the formation of a dioxetane group [31]. In our study, ECL emission maximum intensity from L-tryptophan was recorded using the same electrochemical conditions as used for calibration in a spectrofluorimeter at different wavelengths in the range 350–700 nm with a slit size of 25 nm. The graph presented in Figure 3a is a reconstruction of the light emission recorded in these conditions according to wavelength. A maximum emission at ca. 425 nm is observed, suggesting that it is through the formation of the dioxetane intermediate that the ECL of L-tryptophan is generated. Dioxetane is common to ECL reactions and is known to be enhanced through the presence of ROS. Moreover, the same reaction was examined on the four different electrode materials (Au, Pt, GC and BDD) in 500 μM L-tryptophan solution. A range of lower potentials were tested ($0, -0.5, -1, 1.5, -2, -2.5, -3$ V) for each of the electrodes. It appeared that only the BDD electrode was able to generate an ECL signal from tryptophan in these conditions (Figure 3b). This unique ECL reaction to BDD suggests again that it is favored by the production of H_2O_2 produced at the surface of the diamond electrode and made available to ECL reaction through low adsorption at the electrode surface. Thus, a reaction mechanism is proposed in Figure 4. It is based partially on the work already performed by Sakura et al. and Chen et al., who examined the ECL reactions of tryptophan/H_2O_2/Br$^-$

and tryptophan/H_2O_2 systems, respectively [31,32]. During electro-oxidation, H_2O_2 is generated through several steps of reduction and protonation through proton exchange with water. The tryptophan molecule firstly goes through electro-oxidation at 0.9 V and then is further oxidized through interaction with H_2O_2 to a hydroperoxide tryptophan intermediate. From this, a dioxetane group is formed, which is then protonated and rearranged to form the excited intermediate.

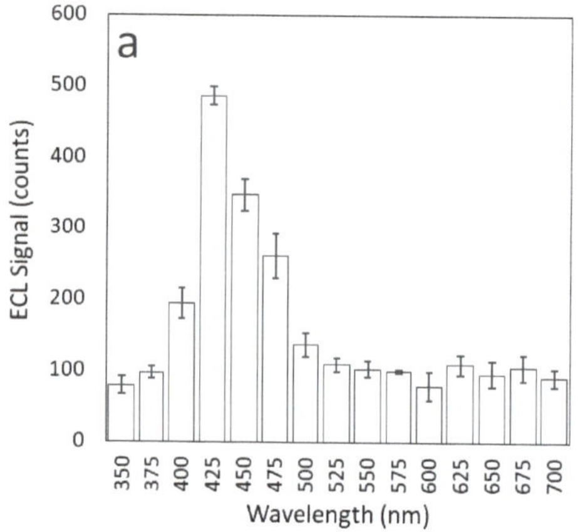
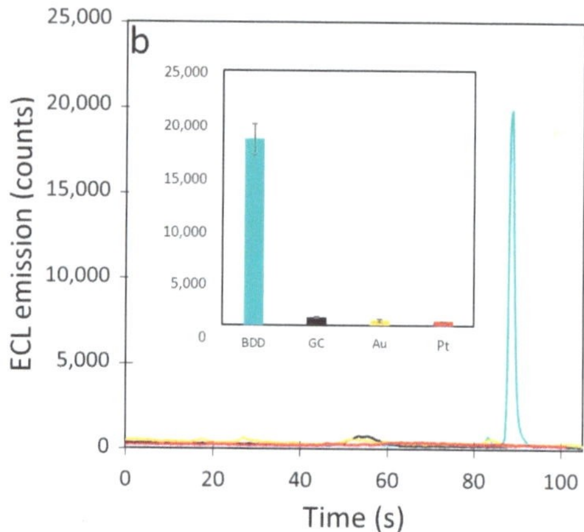

Figure 3. (**a**) ECL signal intensity versus wavelength from BDD electrode and (**b**) ECL emission plotted against time for four electrodes (yellow: Au, red: Pt, black: GC and turquoise: BDD) recorded with the PMT. In both cases a solution containing 0.1 M PBS (pH 7.4) and 500 nM tryptophan was used; CV scan was 0 V $\rightarrow$ -3 V $\rightarrow$ 1.5 V $\rightarrow$ 0 V, at a scan rate 0.1 V.s^{-1}.

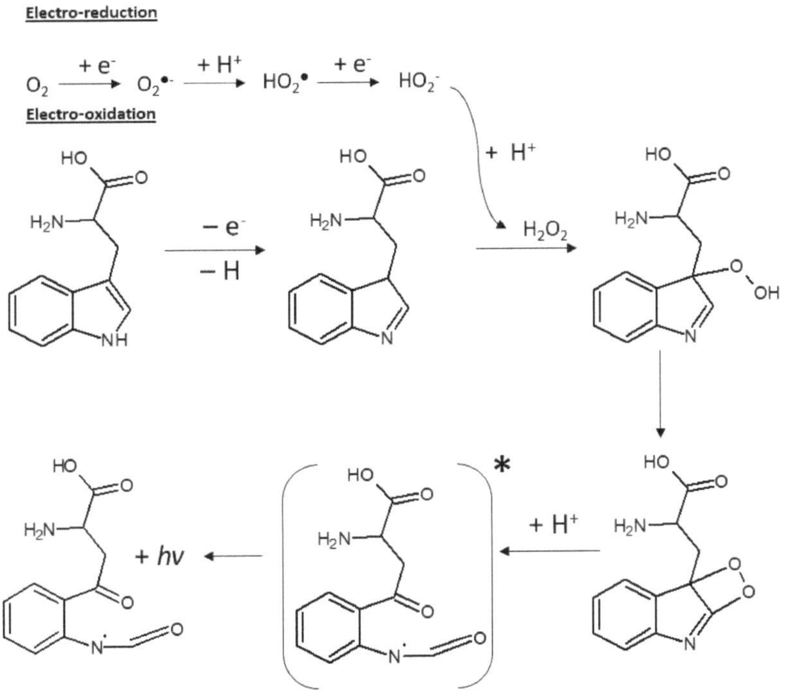

Electro-reduction

$$O_2 \xrightarrow{+\,e^-} O_2{}^{\bullet-} \xrightarrow{+\,H^+} HO_2{}^{\bullet} \xrightarrow{+\,e^-} HO_2{}^-$$

Electro-oxidation

Figure 4. Proposed mechanism for the ECL of L-tryptophan at BDD electrode using successive electro-reduction and electro-oxidation.

ECL presents the advantage of potentially being a very selective method, as only certain types of molecules are able to undergo this reaction. By examining the list of known ECL emitters, the most likely interfering species that occur naturally in human blood would be other polyaromatic compounds featuring specifically an indole ring. Four indolic compounds that have been reported to produce ECL emission under certain conditions were measured: (i) tryptamine, an indole ring with a 2-aminoethyl group substitution (ii) indole, the base indole ring, (iii) 3-methylindole, an indole ring with a methyl group substitution and finally a (iv) serotonin with a 2-aminoethyl and alcohol group substitution. 10 μM of each of the compounds were examined on BDD electrodes using the same conditions presented above. The ECL peak heights from each of the chemicals tested are presented in Figure 5. None of the compounds except for tryptophan gave a significant ECL signal over the blank.

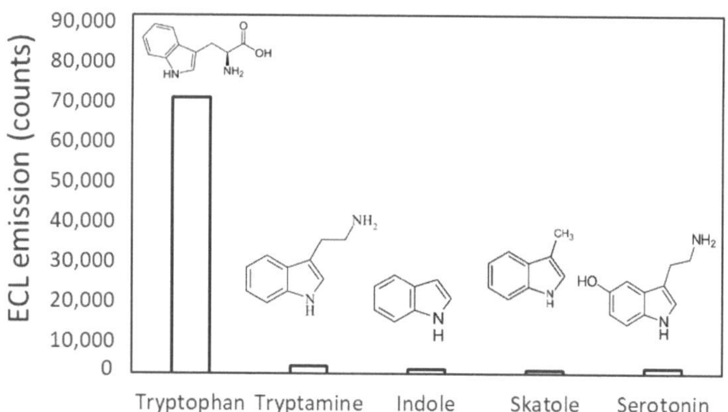

Figure 5. ECL maximum peak emission from 10 µM of five different indolic compounds (tryptophan, tryptamine, indole, skatole (3-methylindole) and serotonin) on BDD electrodes after successive reduction and oxidation. CV scan was 0 V → −3 V → 1.5 V → 0 V, at a scan rate 0.1 V.s^{-1}.

4. Discussion

We have demonstrated that ECL emission from L-tryptophan can be achieved in aqueous solutions at the surface of BDD electrodes through successive reduction and oxidation processes. This unique reaction does not take place when other electrode materials such as gold or glassy carbon are used. Also, it appears to be highly specific to L-tryptophan in our experimental conditions among the few indole derivatives tested. The chemiluminescence of indole and some derivatives induced by electrogenerated superoxide ion on glassy carbon electrode in aprotic solvent was investigated by Okajima and co-workers [46]. They showed that superoxide ions produced at the electrode's surface from dissolved oxygen could act as proton acceptors to indole compounds carrying hydrogen at the *N*-position. The so-formed hydroperoxy radical may proceed to the chemical oxidation of indole derivatives having an electron releasing group at the 3-position, leading to the formation of a 1,2-dioxetane intermediate through radical-radical coupling. A similar reaction has been observed in aqueous solution, e.g., on N-centered radicals derived from tryptophan at pH 10, when the latter radicals are produced by pulse radiolysis [47]. In this case, hydroperoxy radicals are produced from dissolved oxygen and water. However, the reaction described by Okajima and co-workers has been observed to be completely inhibited in acetonitrile upon the addition of water, quenching for instance the chemiluminescence of 3-methylindole [48]. This is thus explained here by the absence of indole radical formation in the presence of water (proton donor). Those observations suggest that the hydroperoxy radical could play a role in the chemiluminescence of tryptophan in water in some adequate pH conditions, given that an indole radical can be formed. This is supported by the fact that, according to Mahé and coworkers, hydroperoxy radicals are indeed mostly adsorbed on GC and Au surfaces and only made available in solution from BDD electrodes. Nevertheless, we have shown that at the high overpotentials used, hydrogen peroxide is the main species produced on BDD upon cathodic polarization. Therefore, we assume that H$_2$O$_2$ is the main contributor to tryptophan chemiluminescence in our study. When reversing polarization, an indole-derivative radical is formed on tryptophan and all interferences under test through electrochemical oxidation on the *N*-position [49]. Except for serotonin, all indole derivatives under testing have a similar structure but for the functional group in the 3-position. Hence, selectivity toward tryptophan can only be due to the nature of this group. At pH = 7.4, tryptophan is a zwitterion in which the amino group is protonated and the carboxylic acid is deprotonated. Thus we assume that, by contrast with the other interference species, the carboxylic group act as a proton acceptor, allowing interaction of hydrogen peroxide with the indole radical at the 3-position, leading to the dioxetane-like intermediate and subsequent excited state. We have investigated this reaction at physiological pH, which correspond to the pH of interest in many biological assays. However, the reaction is expected to be highly pH-dependent. A thorough investigation of ECL emission according to pH would certainly help with our understanding of the reaction mechanisms involved in the detection of tryptophan in water at BDD electrodes.

5. Conclusions

We have reported on a new electrochemiluminescence (ECL) method for L-tryptophan detection involving the in situ production of hydrogen peroxide at the surface of boron-doped diamond (BDD) electrodes. In analytical terms, this means that the reaction can be achieved without the need for an additional coreactant in solution. An ECL mechanism has been proposed based on spectroscopic data that involves the generation of a dioxetane group and subsequent excited state. A low limit of detection in the order of 0.4 nM has been observed, which to our knowledge is the lowest reported to date. Moreover, the detection has been shown to be rather selective against other indolic compound that may be found in particular in biological samples. Altogether, this makes this approach attractive for L-tryptophan detection in biological fluids or food products.

Supplementary Materials: The following supporting information can be downloaded at: https://www.mdpi.com/article/10.3390/s24113627/s1, Figure S1: Typical scanning electron microscope (SEM) images of the surface of the BDD electrodes (worker or counter electrodes); Figure S2: Typical response of BDD electrode to a 0.1 M PBS solution containing 1 μM luminol. CV scan was 0 V → −2.5 V → 0 V → 0 V, at a scan rate 0.1 V/s. Figure S3: Influence of the lower cathodic potential on the ECL emission of 500 nM L-tryptophan in 0.1 M PBS. Scan 0 V → x V → 1.5 V → 0 V at a scan rate of 0.1 V.s^{-1}. Figure S4: Influence of scan rate on the ECL emission of 500 nM L-tryptophan in 0.1 M PBS. Scan 0 V → −3 V → 1.5 V → 0 V. Figure S5: ECL peak emission for each concentration of tryptophan measured (0, 5, 10, 25, 50, 100, 250, 500, 1000 nM) in 0.1 M PBS (pH 7.4) on BDD electrodes after successive reduction and oxidation.

Author Contributions: Conceptualization, S.S. and E.S.; methodology, S.S.; validation, S.S.; formal analysis, S.S.; investigation, S.S.; data curation, S.S. and E.S.; writing—original draft preparation, E.S. and S.S.; writing—review and editing, E.S., S.S. and M.H.; supervision, M.H. All authors have read and agreed to the published version of the manuscript.

Funding: This research received no external funding.

Institutional Review Board Statement: Not applicable.

Informed Consent Statement: Not applicable.

Data Availability Statement: The datasets generated from the current study are available from the corresponding author on reasonable request.

Conflicts of Interest: The authors declare no conflicts of interest.

References

1. Palego, L.; Betti, L.; Rossi, A.; Giannaccini, G. Tryptophan biochemistry: Structural, nutritional, metabolic, and medical aspects in humans. *J. Amino Acids* **2016**, *2016*, 8952520. [CrossRef] [PubMed]
2. Hoglund, E.; Øverli, Ø.; Winberg, S. Tryptophan metabolic pathways and brain serotonergic activity: A comparative review. *Front. Endocrinol.* **2019**, *10*, 435368. [CrossRef] [PubMed]
3. Cervenka, I.; Agudelo, L.Z.; Ruas, J.L. Kynurenines: Tryptophan's metabolites in exercise, inflammation, and mental health. *Science* **1979**, *357*, eaaf9794. [CrossRef] [PubMed]
4. Young, V.R. Adult Amino Acid Requirements: The Case for a Major Revision in Current Recommendations. *J. Nutr.* **1994**, *124* (Suppl. 8), 1517S–1523S. [CrossRef] [PubMed]
5. World Health Organization. *Protein and Amino Acid Requirements in Human Nutrition: Report of a Joint WHO/FAO/UNU Expert Consultation*; WHO Press: Geneva, Switzerland, 2007; p. 150.
6. Kałużna-Czaplińska, J.; Gątarek, P.; Chirumbolo, S.; Chartrand, M.S.; Bjørklund, G. How important is tryptophan in human health? *Crit. Rev. Food Sci. Nutr.* **2019**, *59*, 72–88. [CrossRef] [PubMed]
7. Zhou, Z.Y.; He, L.C.; Mao, Y.; Chai, W.S.; Ren, Z.Q. Green preparation and selective permeation of D-Tryptophan imprinted composite membrane for racemic tryptophan. *Chem. Eng. J.* **2017**, *310*, 63–71. [CrossRef]

8. Chouraki, V.; Preis, S.R.; Yang, Q.; Beiser, A.; Li, S.; Larson, M.G.; Weinstein, G.; Wang, T.J.; Gerszten, R.E.; Vasan, R.S.; et al. Association of amine biomarkers with incident dementia and Alzheimer's disease in the Framingham Study. *Alzheimer's Dement.* **2017**, *13*, 1327–1336. [CrossRef]

9. van der Goot, A.T.; Zhu, W.; Vazquez-Manrique, R.P.; Seinstra, R.I.; Dettmer, K.; Michels, H.; Farina, F.; Krijnen, J.; Melki, R.; Buijsman, R.C.; et al. Delaying aging and the aging associated decline in protein homeostasis by inhibition of tryptophan degradation. *Proc. Natl. Acad. Sci. USA* **2012**, *109*, 14912–14917. [CrossRef] [PubMed]

10. Steinhart, H. *L-Tryptophan: Current Prospects in Medicine and Drug Safety*; Walter de Gruyter: Berlin, Germany, 1994.

11. Majidi, M.R.; Omidi, Y.; Karami, P.; Johari-Ahar, M. Reusable potentiometric screen-printed sensor and label-free aptasensor with pseudo-reference electrode for determination of tryptophan in the presence of tyrosine. *Talanta* **2016**, *150*, 425–433. [CrossRef]

12. Török, N.; Tanaka, M.; Vécsei, L. Searching for Peripheral Biomarkers in Neurodegenerative Diseases: The Tryptophan-Kynurenine Metabolic Pathway. *Int. J. Mol. Sci.* **2020**, *21*, 9338. [CrossRef]

13. Mandarano, M.; Orecchini, E.; Bellezza, G.; Vannucci, J.; Ludovini, V.; Baglivo, S.; Tofanetti, F.R.; Chiari, R.; Loreti, E.; Puma, F. Kynurenine/Tryptophan Ratio as a Potential Blood-Based Biomarker in Non-Small Cell Lung Cancer. *Int. J. Mol. Sci.* **2021**, *22*, 4403. [CrossRef] [PubMed]

14. Lomenova, A.; Hroboňová, K. Application of achiral–chiral two-dimensional HPLC for separation of phenylalanine and tryptophan enantiomers in dietary supplement. *Biomed. Chromatogr.* **2021**, *35*, e4972. [CrossRef] [PubMed]

15. Lian, W.; Ma, D.J.; Xu, X.; Chen, Y.; Wu, Y.L. Rapid high-performance liquid chromatography method for determination of tryptophan in gastric juice. *J. Dig. Dis.* **2012**, *13*, 100–106. [CrossRef] [PubMed]

16. DeVries, J.W.; Koski, C.M.; Egberg, D.C.; Larson, P.A. Comparison between a spectrophotometric and a high-pressure liquid chromatography method for determining tryptophan in food products. *J. Agric. Food Chem.* **1980**, *28*, 896–898. [CrossRef] [PubMed]

17. Sathya, V.; Srinivasadesikan, V.; Ming-Chang, L.; Padmini, V. Highly sensitive and selective detection of tryptophan by antipyrine based fluorimetric sensor. *J. Mol. Struct.* **2023**, *1272*, 134241. [CrossRef]

18. Jafari, M.; Tashkhourian, J.; Absalan, G. Chiral recognition of tryptophan enantiomers using chitosan-capped silver nanoparticles: Scanometry and spectrophotometry approaches. *Talanta* **2018**, *178*, 870–878. [CrossRef] [PubMed]

19. Wang, S.-D.; Xie, L.-X.; Zhao, Y.-F.; Wang, Y.-N. A dual luminescent sensor coordination polymer for simultaneous determination of ascorbic acid and tryptophan. *Spectrochim. Acta Part A* **2020**, *242*, 118750. [CrossRef] [PubMed]

20. Pundi, A.; Chang, C.-J.; Chen, Y.-S.; Chen, J.-K.; Yeh, J.-M.; Zhuang, C.-S.; Lee, M.-C. An aniline trimer-based multifunctional sensor for colorimetric Fe3+, Cu2+ and Ag+ detection, and its complex for fluorescent sensing of L-tryptophan. *Spectrochim. Acta Part A* **2021**, *247*, 119075. [CrossRef] [PubMed]

21. Rajalakshmi, K.; Abraham John, S. Sensitive and selective determination of l-tryptophan at physiological pH using functionalized multiwalled carbon nanotubes–nanostructured conducting polymer composite modified electrode. *J. Electroanal. Chem.* **2014**, *734*, 31–37. [CrossRef]

22. Ensafi, A.A.; Karimi-Maleh, H.; Mallakpour, S. Simultaneous Determination of Ascorbic Acid, Acetaminophen, and Tryptophan by Square Wave Voltammetry Using N-(3,4-Dihydroxyphenethyl)-3,5-Dinitrobenzamide-Modified Carbon Nanotubes Paste Electrode. *Electroanalysis* **2012**, *24*, 666–675. [CrossRef]

23. Ratautaite, V.; Brazys, E.; Ramanaviciene, A.; Ramanavicius, A. Electrochemical sensors based on L-tryptophan molecularly imprinted polypyrrole and polyaniline. *J. Electroanal. Chem.* **2022**, *917*, 116389. [CrossRef]

24. He, Q.; Liu, J.; Feng, J.; Wu, Y.; Tian, Y.; Li, G.; Chen, D. Sensitive Voltammetric Sensor for Tryptophan Detection by Using Polyvinylpyrrolidone Functionalized Graphene/GCE. *Nanomaterials* **2020**, *10*, 125. [CrossRef] [PubMed]

25. Szunerits, S.; Coffinier, Y.; Galopin, E.; Brenner, J.; Boukherroub, R. Preparation of Boron Doped Diamond Nanowires and Their Application for Sensitive Electrochemical Detection of Tryptophan. *Electrochem. Commun.* **2010**, *12*, 438–441. [CrossRef]

26. Lin, Z.; Chen, X.; Cai, Z.; Li, P.; Chen, X.; Wang, X. Chemiluminescence of Tryptophan and Histidine in $Ru(Bpy)_3^{2+}$-$KMnO_4$ Aqueous Solution. *Talanta* **2008**, *75*, 544–550. [CrossRef] [PubMed]

27. Chikura, K.; Kirisawa, M. Notes Electrochemiluminescence Determination of D-, L-Tryptophan Using Ligand-Exchange High-Performance Liquid Chromatography. *Anal. Sci.* **1991**, *7*, 971–973. [CrossRef]

28. Chen, K.; Schmittel, M. An Iridium (III)-Based Lab-on-a-Molecule for Cysteine/Homocysteine and Tryptophan Using Triple-Channel Interrogation. *Analyst* **2013**, *138*, 6742–6745. [CrossRef] [PubMed]

29. Wu, F.; Tong, B.; Zhang, Q. Application of a New Iridium Complex as a Chemiluminescence Reagent for the Determination of Tryptophan. *Anal. Sci.* **2011**, *27*, 529–533. [CrossRef]

30. Zhu, S.; Lin, X.; Ran, P.; Xia, Q.; Yang, C.; Ma, J.; Fu, Y. A Novel Luminescence Functionalized Metal-Organic Framework Nanoflowers Electrochemiluminesence Sensor via "on-off" System. *Biosens. Bioelectron.* **2017**, *91*, 436–440. [CrossRef]

31. Sakura, S. Chemiluminescence of Tryptophan Enhanced by Electrochemical Energy. *Electrochim. Acta* **1992**, *37*, 2731–2735. [CrossRef]

32. Chen, G.N.; Lin, R.E.; Zhao, Z.F.; Duan, J.P.; Zhang, L. Electrogenerated Chemiluminescence for Determination of Indole and Tryptophan. *Anal. Chim. Acta* **1997**, *341*, 251–256. [CrossRef]

33. Zholudov, Y.; Lysak, N.; Snizhko, D.; Reshetniak, O.; Xu, G. Electrochemiluminescence Analysis of Tryptophan in Aqueous Solutions Based on Its Reaction with Tetraphenylborate Anions. *Analyst* **2020**, *145*, 3364–3369. [CrossRef]

34. Macpherson, J.V. A practical guide to using boron doped diamond in electrochemical research. *Phys. Chem. Chem. Phys.* **2015**, *17*, 2935–2949. [CrossRef] [PubMed]
35. Yence, M.; Cetinkaya, A.; Ozcelikay, G.; Kaya, S.I.; Ozkan, S.A. Boron-Doped Diamond Electrodes: Recent Developments and Advances in View of Electrochemical Drug Sensors. *Crit. Rev. Anal. Chem.* **2022**, *52*, 1122–1138. [CrossRef] [PubMed]
36. Sarakhman, O.; Švorc, L. A Review on Recent Advances in the Applications of Boron-Doped Diamond Electrochemical Sensors in Food Analysis. *Crit. Rev. Anal. Chem.* **2022**, *52*, 791–813. [CrossRef] [PubMed]
37. Freitas, J.M.; Oliveira, T.C.; Munoz, R.A.A.; Richter, E.M. Boron Doped Diamond Electrodes in Flow-Based Systems. *Front. Chem.* **2019**, *7*, 190. [CrossRef] [PubMed]
38. Deng, Z.; Zhu, R.; Ma, L.; Zhou, K.; Yu, Z.; Wei, Q. Diamond for antifouling applications: A review. *Carbon* **2022**, *196*, 923–939. [CrossRef]
39. Irkham; Fiorani, A.; Einaga, Y. Electrogenerated Chemiluminescence at Diamond Electrode. In *Diamond Electrodes, Fundamentals and Applications*; Einaga, Y., Ed.; Springer Nature: Singapore, 2022; p. 119. [CrossRef]
40. Drogui, P.; Elmaleh, S.; Rumeau, M.; Bernard, C.; Rambaud, A. Hydrogen Peroxide Production by Water Electrolysis: Application to Disinfection. *J. Appl. Electrochem.* **2001**, *31*, 877–882. [CrossRef]
41. Chávez, E.I.; De Rosa, C.; Martínez, C.A.; Hernández, J.M.P. On-Site Hydrogen Peroxide Production at Pilot Flow Plant: Application to Electro-Fenton Process. *Int. J. Electrochem. Sci.* **2013**, *8*, 3084–3094. [CrossRef]
42. Guillet, N.; Roué, L.; Marcotte, S.; Villers, D.; Dodelet, J.P.; Chhim, N.; Vin, S.T. Electrogeneration of Hydrogen Peroxide in Acid Medium Using Pyrolyzed Cobalt-Based Catalysts: Influence of the Cobalt Content on the Electrode Performance. *J. Appl. Electrochem.* **2006**, *36*, 863–870. [CrossRef]
43. Mahé, É.; Bornoz, P.; Briot, E.; Chevalet, J.; Comninellis, C.; Devilliers, D. A Selective Chemiluminescence Detection Method for Reactive Oxygen Species Involved in Oxygen Reduction Reaction on Electrocatalytic Materials. *Electrochim. Acta* **2013**, *102*, 259–273. [CrossRef]
44. Royer, C.A. Fluorescence Spectroscopy. *Protein Stab. Fold. Theory Pract.* **1995**, *40*, 65–89. [CrossRef] [PubMed]
45. Sugiyama, N.; Yamamoto, H.; Omote, Y.; Akutagawa, M. The Chemiluminescence of Indole Derivatives. IV. Correlation between Chemiluminescence and Structure of 2, 3-Dimethylindoles. *Bull. Chem. Soc. Jpn.* **1968**, *41*, 1917–1921. [CrossRef]
46. Okajima, T.; Ohsaka, T. Chemiluminescence of indole and its derivatives induced by electrogenerated superoxide ion in acetonitrile solutions. *Electrochem. Acta* **2002**, *47*, 1561–1565. [CrossRef]
47. Fang, X.; Jin, F.; Jin, H.; von Sonntag, C. Reaction of the superoxide radical with the N-centered radical derived from N-acetyltryptophan methyl ester. *J. Chem. Soc. Perkin Trans.* **1998**, *2*, 259–263. [CrossRef]
48. Stewart, S.; Scorsone, E.; Prunier, A.; Hamel, M. Novel ECL method for the determination of skatole in porcine adipose tissue. *Anal. Chem.* **2022**, *94*, 6403–6409. [CrossRef]
49. Goyal, R.N.; Kumar, N.; Singhal, N.K. Oxidation chemistry and biochemistry of indole and effect of its oxidation product in albino mice. *Bioelectrochem. Bioenerg.* **1998**, *45*, 47–53. [CrossRef]

Article

Electrochemical Diffusion Study in Poly(Ethylene Glycol) Dimethacrylate-Based Hydrogels

Eva Melnik [1,*], **Steffen Kurzhals** [1], **Giorgio C. Mutinati** [1], **Valerio Beni** [2] and **Rainer Hainberger** [1]

[1] Molecular Diagnostics, AIT Austrian Institute of Technology GmbH, 1210 Vienna, Austria; steffen.kurzhals@ait.ac.at (S.K.); giorgio.mutinati@ait.ac.at (G.C.M.); rainer.hainberger@ait.ac.at (R.H.)

[2] Bioelectronics and Organic Electronics, Smart Hardware, Digital Systems, RISE Research Institutes of Sweden, 60233 Norrköping, Sweden; valerio.beni@ri.se

* Correspondence: eva.melnik@ait.ac.at

Abstract: Hydrogels are of great importance for functionalizing sensors and microfluidics, and poly(ethylene glycol) dimethacrylate (PEG-DMA) is often used as a viscosifier for printable hydrogel precursor inks. In this study, 1–10 kDa PEG-DMA based hydrogels were characterized by gravimetric and electrochemical methods to investigate the diffusivity of small molecules and proteins. Swelling ratios (*SRs*) of 14.43–9.24, as well as mesh sizes ξ of 3.58–6.91 nm were calculated, and it was found that the *SR* correlates with the molar concentration of PEG-DMA in the ink (*MCI*) (SR = 0.1127 × MCI + 8.3256, R^2 = 0.9692) and ξ correlates with the molecular weight (M_w) (ξ = 0.3382 × M_w + 3.638, R^2 = 0.9451). To investigate the sensing properties, methylene blue (MB) and MB-conjugated proteins were measured on electrochemical sensors with and without hydrogel coating. It was found that on sensors with 10 kDa PEG-DMA hydrogel modification, the DPV peak currents were reduced to 8% for MB, 73% for MB-BSA, and 23% for MB-IgG. To investigate the diffusion properties of MB(-conjugates) in hydrogels with 1–10 kDa PEG-DMA, diffusivity was calculated from the current equation. It was found that diffusivity increases with increasing ξ. Finally, the release of MB-BSA was detected after drying the MB-BSA-containing hydrogel, which is a promising result for the development of hydrogel-based reagent reservoirs for biosensing.

Keywords: hydrogels; electrochemical sensors; diffusivity study; methylene blue; MB-conjugated proteins

Citation: Melnik, E.; Kurzhals, S.; Mutinati, G.C.; Beni, V.; Hainberger, R. Electrochemical Diffusion Study in Poly(Ethylene Glycol) Dimethacrylate-Based Hydrogels. *Sensors* 2024, 24, 3678. https://doi.org/10.3390/s24113678

Academic Editors: Hai-Feng (Frank) Ji, Bruno Ando, Pietro Siciliano and Luca Francioso

Received: 14 March 2024
Revised: 17 May 2024
Accepted: 23 May 2024
Published: 6 June 2024

Correction Statement: This article has been republished with a minor change. The change does not affect the scientific content of the article and further details are available within the backmatter of the website version of this article.

1. Introduction

Biosensors enable the rapid and sensitive monitoring of analytes in human body fluids and are thus highly promising in realizing rapid tests and point-of-care devices [1]. Electrochemical sensors are reliable transducers, can be produced cost-effectively using roll-to-roll screen printing, and are suitable for a wide range of analytes in various medical and environmental applications [2–5]. The assays used for detection usually differ only marginally from the assays used in gold standard methods such as microarrays, lateral flow tests, or enzyme-linked immunosorbent assays [6–11]. This is because of the high selectivity and specificity of the established and well-researched assays, which, however, may require sample preparation steps (e.g., target analyte extraction, sample dilution) and target analyte preparation steps (e.g., the addition of chemicals, proteins, enzymes, or antibodies). In this respect, well-designed microfluidic systems, such as those produced by injection molding, micromilling, PDMS punching, or roll-to-roll compatible lithographic processes [12,13], support the automation of sample handling. To make microfluidics operational for various applications in molecular diagnostics, the integration of reagents is a fundamental challenge. Various concepts have been pursued for this purpose, such as blister bags, freeze-dried pellets, functionalized beads, sponges, paper strips, biopolymers, and hydrogels [14–18]. For the large-scale production of microfluidics, automated pick-and-place tools, lamination systems, and spotting and printing technologies have been developed for the assembly process [14].

Among these techniques, spotting methods are particularly suitable because they are also used for the functionalization of biosensor systems [19–22] and are compatible with roll-to-roll manufacturing processes [13]. Moreover, they can be used for spotting liquid reagents in a microfluidic chamber. In order to increase the shelf life of the reagents, they are commonly embedded in biopolymers or hydrogel networks [23–26]. In contrast to polymers, hydrogel networks are insoluble and specifically release only the desired substance. This is a decisive advantage, as it prevents a change in viscosity and thus the flow and reaction properties. In addition to their role as reservoirs, hydrogels also play the role of protective and filter layers in biosensing [27–30], e.g., to prevent nonspecific blood cell binding on the sensor surface [22,30].

In this study, poly(ethylene glycol) di-methacrylate (PEG-DMA)-based hydrogels are investigated in combination with electrochemical sensors. The main focus lies in the diffusion properties of biomolecules through hydrogel films attached to the sensor element in order to use the hydrogels as reagent reservoirs and/or molecular filters. Despite various other methods described in the literature [31–34], we investigate the diffusion of biomolecules using electrochemical (EC) screen-printed sensors. To better understand the diffusion properties, PEG-DMA-based inkjet printable hydrogel inks with 5.4% (w/w) 1–10 kDa PEG-DMA are formulated using di(ethylene glycol) vinyl ether (DEGVE) as a monomer and lithium phenyl (2,4,6-trimethylbenzoyl) phosphinate (LAP) as a photoinitiator [22].

Figure 1 depicts the workflow of this study. It starts with the modification of the EC screen-printed sensor with hydrogel (Figure 1A). The characterization of the hydrogels via a gravimetric method enables the calculation of the swelling ratio, the polymer content, the water content, the average molecular weight of polymer chains between two crosslinks ($\overline{M}_c$), and the mesh size (ξ) [33,35]. For the EC measurement, the hydrogel is then overlayed with a methylene blue (MB) or MB-conjugate protein solution (Figure 1B). As proteins, bovine serum albumin (BSA) and mouse IgG are chosen because they are good representatives for bioanalytical assays [14,18]. The measurement is immediately started after applying the MB-(conjugate), and the diffusion is monitored using differential pulse voltammetry (DPV) (Figure 1C). In order to understand the diffusion properties, the following experiments are carried out: (i) measurement of MB(-conjugates) on sensors without hydrogel, (ii) measurement of MB(-conjugates) on sensors coated with a 10 kDa PEDG-DMA hydrogel, (iii) measurement of MB(-conjugates) on sensors coated with different molecular weight (MW) PEG-DMA hydrogels, and (iv) measurement of MB-BSA on sensors coated with wet and dry hydrogels.

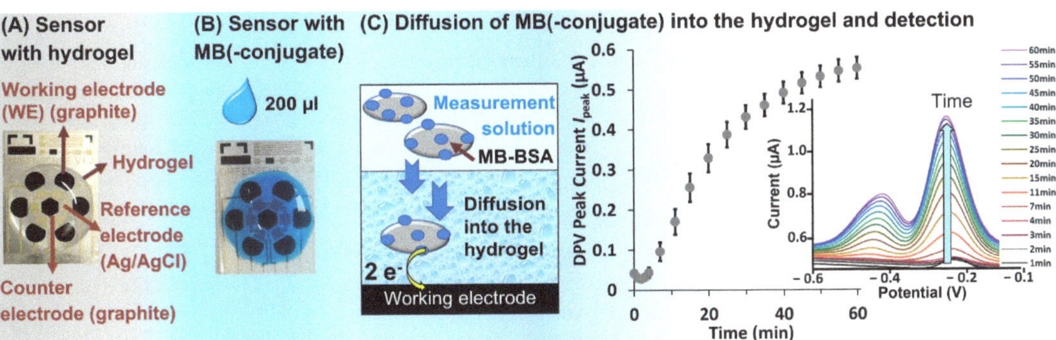

Figure 1. (**A**) Sensors covered with a hydrogel are overlayed with (**B**) MB(-conjugate) solution for (**C**) electrochemical diffusion monitoring with DPV.

2. Materials and Methods

2.1. Materials

Mouse IgG, bovine serum albumin (BSA), hexaamine ruthenium (III) chloride [Ru(NH$_3$)$_6$]Cl$_3$, Poly(ethylene glycol) dimethacrylate (PEG-DMA) 10 kDa, 2 kDa, and 1 kDa, di(ethylene glycol) vinyl ether (DEGVE), lithium phenyl (2,4,6-trimethylbenzoyl) phosphinate (LAP), and Dimethyl sulfoxide (DMSO), as well as the chemicals for the physiological phosphate-buffered saline buffer referred to as PBS buffer (10 mmol/L Phosphate, 137 mmol/L NaCl, 2.7 mmol/L KCl; pH 7.3), were purchased from Sigma Aldrich Europe (St. Gallen, Switzerland).

PEG-DMA 5 kDa was purchased from Biopharma PEG Scientific Inc., Watertown, NY, USA. PEG-DMA 3.4 kDa from Alfa Aesar (Haverhill, MA, USA) and the dialysis membrane 3.5 kDa Side-A-Lyser from Thermo Fisher Scientific were purchased from VWR Austria (Vienna, Austria). The methylene-blue N-hydroxy succinimide ester (MB-NHS) for labeling the biomolecules was purchased from Biotium, Fremont, MA, USA.

2.2. Electrochemical Sensors and Measurement Setup

Screen-printed electrochemical (EC) sensor arrays with graphite working electrodes (WE) were produced by RISE Research Institutes of Sweden (Digital Systems, Smart Hardware, Bio- and Organic Electronics, Norrköping, Sweden) according to the AIT design elaborated in the GREENSENSE Project (European Union's Horizon 2020 research and innovation program under Grant Agreement No. 761000). The screen-printed sensor array consists of six graphite working electrodes (WEs) surrounding a central counter electrode (CE) and a common silver/silver chloride reference electrode (RE) on a PET substrate (Figure 2A). This configuration enabled the measurement of six values simultaneously (WEs = 6). The working electrode area was determined by using a microscope on ten sensors at different positions on the screen-printed sheet (see Section 3.1). For the electrochemical characterization, a 1000 μM hexaamine ruthenium(III) chloride [Ru(NH$_3$)$_6$]Cl$_3$ solution in 50 mmol/L NaCl or MB- and MB-conjugates (see Section 2.6) were used. As the electrochemical measurement method, DPV was chosen with the following parameters: equilibration time t_{equ} = 8 s, initial potential E_i = -0.6 V, final potential E_f = -0.1 V, step potential E_s = 0.005 V, interlevel potential E_p = 0.05 V, step time t_p = 0.005 s, and scan rate SR = 0.5 V/s. A PalmSens MUX8 R2 potentiostat (Figure S1) was used for the DPV measurements. The monitoring measurements of MB(-conjugates) were carried out for up to 160 min (one DPV per minute, or per five minutes), and the DPV peak currents I_{peak} were recorded. From the I_{peak}, the diffusivity (also called diffusion coefficient) D of the electrochemical species can be calculated according to Equation (1) [35].

$$I_{peak} = \frac{nFAD^{\frac{1}{2}}C}{\pi^{\frac{1}{2}}t_p^{\frac{1}{2}}}\left(\frac{1-\sigma}{1+\sigma}\right), \; with \; \sigma = e^{\frac{2FE_p}{2RT}} \tag{1}$$

In Equation (1), n is the number of electrons (2 for methylene blue, and 2 multiplied by the degree of labeling for BSA or IgG, which are 5.4 and 3.8, respectively (see Section 2.6)), F is the Faraday constant (96,485 C $\times$ mol^{-1}), A is the working electrode area (see Section 3.1), C is the concentration of the analyte, R is the gas constant, and T the temperature in Kelvin (293.15 K = 20 °C).

2.3. Hydrogel Preparation

The hydrogel inks were fabricated by mixing 30 μL of 166 mg/mL PEG-DMA (1–10 kDa) in water with 60 μL di(ethylene glycol) vinyl ether (DEGVE) and 2 μL of a lithium phenyl (2,4,6-trimethylbenzoyl) phosphinate (LAP) solution (10 mg LPA per 100 μL in a 1:1 mixture of ethanol and ultrapure water). The PEG-DMA weight percent was kept constant at 5.4 % (w/w) in each ink by decreasing the molar concentration (μmol/L) with respect to the increase in molecular weight of PEG-DMA (see Section 3.2). This strategy was chosen to fulfill the following requirements: (a) the inks should have a viscosity of

<3.0 cP (to be applicable later in a printing process) and (b) the hydrogels should have good solidity after crosslinking. Preliminary tests showed that the inks prepared with 5.4% (w/w) of 1, 2, 3.4, and 10 kDa PEG-DMA composition fulfilled these criteria. No hydrogel with good solidity could be prepared with 5.4% (w/w) of 5 kDa PEG-DMA. As a consequence, only the 1, 2, 3.4, and 10 kDa hydrogel could be investigated in this study.

For the hydrogel preparation on the EC-sensor, 80 µL of the inks were applied and UV-crosslinked at a wavelength of 365 nm (1 J/cm^2) using the UVP crosslinker CL-3000 (Analytik Jena US, Jena, Germany). The crosslinking step forms acrylate radicals that are highly reactive and can bind to untreated organic surfaces. This process is so strong that even screen-printed graphite-based sensor surfaces can be modified with hydrogels without any pre-treatment. This process led to an approximately 2 mm thick wet hydrogel layer on the sensor surface, which was washed two times with PBS buffer (1 × 10 min, 1 × 15 min) (Figure 1A). The hydrogel layers were tested in wet (Figure 1A) and dry states.

2.4. Gravimetric Hydrogel Characterization

The gravimetric analysis of hydrogels allows for calculating parameters that characterize the mesh structure of the hydrogel [36]. Therefore, hydrogel structures were fabricated, as described in Section 2.3, and gravimetric analyses were performed using a sensitive scale. For each PEG-DMA ink variation, a set of three sensors was prepared with hydrogel coatings, and the weights and volumes were determined. The variables used for this analysis are also listed in Table S1.

For a better understanding of water uptake and, therefore, of the hydrophilic properties of the hydrogel, the swelling ratio (SR) and water content of the hydrogel was calculated. The weight of the screen-printed sensor (W_{sens}), the weight of the washed hydrogel on the sensor (W_{wash}), and the weight of the dried hydrogel on the sensor after drying for 24 h at room temperature (W_{dry}) were determined. The weight of the swollen hydrogel (W_{sh}) was calculated by $W_{sh} = W_{wash} - W_{sens}$ and the weight of the polymer (W_p) by $W_p = W_{dry} - W_{sens}$. The SR was finally calculated according to the equation $SR = W_{sh}/W_p$. The weight of the water in the hydrogel (W_w) was determined according to the equation $W_w = W_{wash} - W_{dry}$.

Another interesting parameter is the number-average molecular weight of the polymer chains between crosslinks ($\overline{M}_c$). Using this parameter in combination with other variables, the mesh size ξ of the hydrogel can be calculated. As $\overline{M}_c$ and ξ allow for different hydrogels to be compared with each other, they are the primary objective of this calculation. The $\overline{M}_c$ can be calculated using Equation (2) [37].

$$\frac{1}{\overline{M}_c} = \frac{2}{\overline{M}_n} - \frac{\tilde{v}/V_1\left[ln(1 - v_{2,s}) + v_{2,s} + \chi v_{2,s}^2\right]}{v_{2,r}\left[\left(\frac{v_{2,s}}{v_{2,r}}\right)^{\frac{1}{3}} - \frac{v_{2,s}}{v_{2,r}}\right]} \tag{2}$$

In Equation (2), $\overline{M}_n$ is the number-average molecular weight of PEG-DMA, $\tilde{v}$ is the specific volume of the polymer ($\tilde{v} = 1/\rho_p$ with ρ_p being the density of the polymer of 1.1 kg × L^{-1}), V_1 is the molar volume of the solvent (18 × 10^{-3} L × mol^{-1} for water), χ is the Flory–Huggin's polymer–solvent interaction parameter (0.495 for PEG-water system) [37], $v_{2,s}$ is the volume fraction of the swollen gel, and $v_{s,r}$ is the volume fraction of the relaxed gel (gel after crosslinking). The parameter $v_{2,s}$ is given by $v_{2,s} = V_p/V_s$, where V_p is the volume of the polymer and V_s the volume of the gel. The parameter $v_{2,r}$ is given by $v_{2,s} = V_p/V_r$, where V_r is the relaxed volume of the gel. From $\overline{M}_c$, the mesh size ξ can be calculated using Equation (3) [38]

$$\xi = v_{2,s}^{-1/3} * C_n^{\frac{1}{2}}\left(\frac{2\overline{M}_c}{M_r}\right)^{\frac{1}{2}} * l \tag{3}$$

In Equation (3), C_n is the rigidity factor of the polymer (4 for PEG) [39], M_r is the molecular weight of repeating units (44×10^{-3} kg/mol for PEG), and l is the carbon–carbon bond length (0.154 nm).

2.5. Hydrogel Characterization with Scanning Electron Microscopy

For SEM imaging analyses, 10 kDa PEG-DMA hydrogels were processed on a screen-printed Ag layer to provide an electrical contact to the sample. Next, lyophilization was performed in a Christ Alpha 2–4 LCSplus freeze dryer. The freeze dryer was set, and the pre-cooling was switched on. The settings are listed in Table 1. During the warm-up of the system, two of the sensors were shock-frozen in liquid nitrogen. Subsequently, they were transferred into the freeze drier, and the main drying process was performed overnight. After lyophilization, the sensors were stored under vacuum until further use. The sputtering of a gold/palladium mixture was deliberately omitted, as otherwise, the fine structure of the hydrogel would no longer have been visible (Figure S2).

Table 1. Settings for the lyophilization.

	Freeze	Warm-Up	Main Drying	Post-Drying
Time (min)	90	30	Setting: ∞	20
Temperature (°C)	−40	−50	−50	25
Vacuum (mbar)			0.015	0.015
Pressure (mbar)			off	off

A ZEISS Gemini SEM was used to image the hydrogel layers after drying in vacuum (in the desiccator) and after lyophilization. SEM imaging was performed with a SE2 secondary electron detector at an electron high tension (EHT) voltage of 5 kV.

2.6. Conjugation of Biomolecules with Methylene Blue

To obtain MB-BSA (or MB-IgG) conjugates, 300 µL of a 5 mg/mL BSA or IgG solution in physiological PBS and 121 µL for BSA (53 µL for IgG) of a 5 mg/mL solution of MB-NHS in DMSO were mixed and incubated for 1.5 h at room temperature. The MB-NHS/protein solutions were then dialyzed in a 3.5 kDa Side-A-Lyser against physiological PBS buffer at 4 °C for 12 h. Then, the liquids were transferred to a 1 mL volumetric flask and filled up to 1 mL using physiological PBS buffer. By using UV-VIS absorption measurements, a degree of labeling (DOL) of 5.4 could be detected for MB-BSA and 3.8 for MB-IgG.

3. Results

3.1. Sensor Characterization

The area of the working electrode was determined on ten sensors by a microscope image analysis, which comprised sixty working electrodes. The measured areas showed good homogeneity of 2.13 ± 0.05 mm^2. Functionality and reproducibility were tested with a 1000 µmol/L hexaamine ruthenium(III) chloride [Ru(NH$_3$)$_6$]Cl$_3$ solution in 50 mmol/L NaCl by performing DPV (Figure 2C) on nine sensors (six working electrodes each) taken from different positions on the screen-printed sheet (Figure 2B). The average peak current amounted to 16.44 ± 0.61 µA.

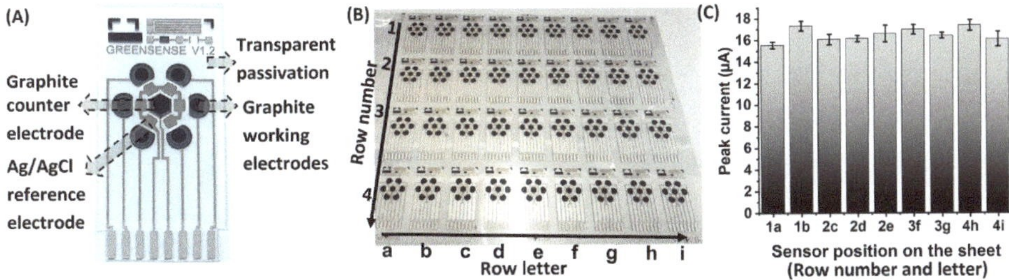

Figure 2. (**A**) Graphite sensor, (**B**) screen-printed sheet, and (**C**) peak currents measured on nine sensors of the screen-printed sheet.

3.2. Gravimetric Hydrogel Characterization

Hydrogels with different PEG-DMA molecular weights were prepared as described in Section 2.3. Gravimetric analyses were performed according to Section 2.4. The swelling ratio, the number-average molecular weight of the polymer chain between the crosslinks, $\overline{M}_c$ (g/mol), and the mesh size ξ (nm) were calculated for each PEG-DMA hydrogel. Table 2 summarizes the results. (A more detailed list is given in Table S2).

Table 2. Values of the swelling ratio, the number-average molecular weight of the polymer chain between the crosslinks $\overline{M}_c$ (g/mol), and mesh size ξ (nm).

Molecular Weight of PEG-DMA, M_w (kDa)	Molar Concentration in the Ink, MCI (µmol/L)	Swelling Ratio, $SR = W_{sh}/W_p$	Number-Average Molecular Weight of Polymer Chain between CrossLinks, $\overline{M}_c$ (g/mol)	Mesh Size, ξ (nm)
1	54.1	14.43	407.37	3.58
2	27.1	11.66	630.38	4.46
3.4	15.9	9.53	900.61	5.15
10	5.4	9.24	1238.22	6.91

It was found that the higher the molecular weight of PEG-DMA, the lower the degree of swelling (SR). This may seem surprising at first glance. However, as shown in Table 2, the molar concentration in the ink (MCI) decreases with the increasing molecular weight of PEG-DMA. Therefore, there are more dimethacrylate crosslink points available in the 1 kDa PEG-DMA ink than in the 10 kDa PEG-DMA ink, which affects the swelling ratio. Linear fitting revealed that the two quantities correlate via the linear equation $SR = 0.1127 \times MCI + 8.3256$ with an $R^2 = 0.9692$ (Figure S3A), which proves the plausibility of the assumption. The swelling ratio does not correlate linearly with the molecular weight of PEG-DMA (M_w) (Figure S3B). However, a clear linear correlation between the mesh size and the molecular weight of PEG-DMA was found with $\xi = 0.3382 \times M_w + 3.638$, with an $R^2 = 0.9451$ (Figure S3D).

3.3. Hydrogel Characterization with Scanning Electron Microscopy

To better understand the 3D structure of the PEG-DMA hydrogel, SEM images of the 10 kDa PEG-DMA hydrogel were taken after lyophilization and after drying in the vacuum chamber, which reflects the wet and dry structure of the hydrogel. At magnifications <600, the lyophilized hydrogel shows an extended network structure that appears to be interrupted by polymer walls, in contrast to the vacuum-dried hydrogel. At magnifications of around 3000, only the vacuum-dried hydrogel shows a clear pore structure with a pore diameter of about 1 µm. With a magnification of 10,000, a nano-pore structure becomes visible on the polymer wall of the lyophilized hydrogel, although this is very difficult to resolve with the SEM (see Figure 3).

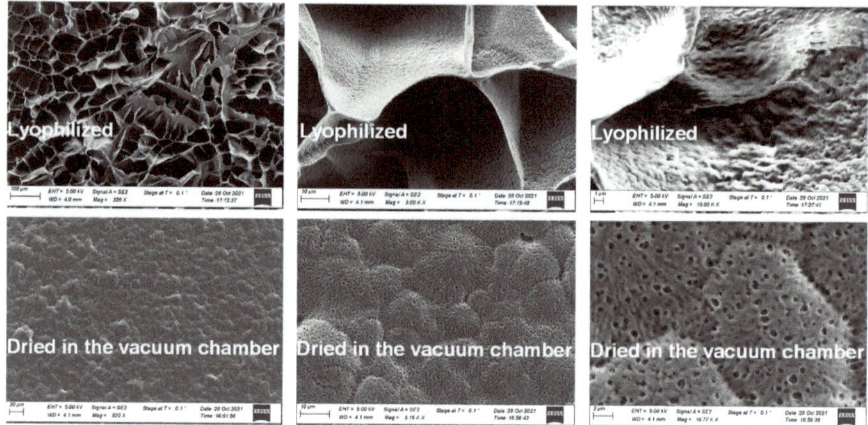

Figure 3. SEM images of vacuum-dried and lyophilized hydrogel structures.

3.4. MB-Conjugate Measurement without Hydrogel

The MB-conjugation leads to a redox-labeling of the biomolecules. In the first experiment, different concentrations of MB(-conjugates) in physiological PBS buffer were measured on sensors without a hydrogel layer. It was found that when the DPV I_{peak} values (at 0 min) for MB-BSA and MB-IgG conjugates are normalized by dividing them by the degree of labeling (DOL), comparable results are obtained for both conjugates. This indicates that conjugated MB molecules equally contribute to the current output (Figure 4).

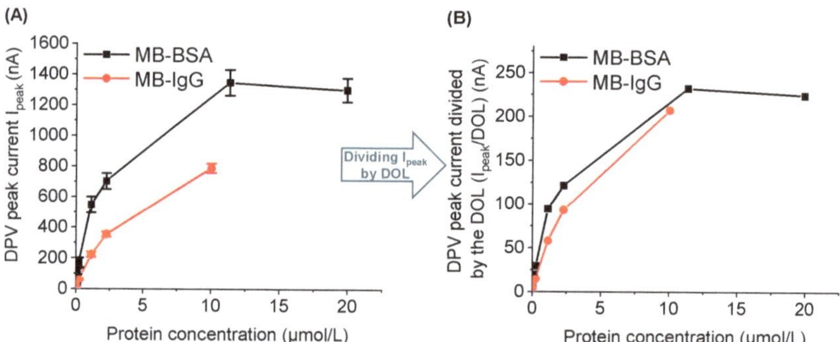

Figure 4. (**A**) Concentration-dependent measurement of MB-BSA and MB-IgG on unmodified sensors. (**B**) Normalization of the DPV peak current I_{peak} with the DOL of 5.8 for MB-BSA and 3.8 for MB-IgG.

3.5. Response of MB-Conjugates of Sensors with and without Hydrogel

The unmodified sensors and the sensors modified with a 10 kDa PEDG-DMA hydrogel were measured with MB(-conjugates) for 60 min to allow the signals to equilibrate (see Figure S4). A DPV measurement was performed every minute for the first five minutes of the test and every five minutes thereafter. The DPV I_{peak} current values after 60 min were plotted versus the concentration of MB. Figure 5A shows the concentration-dependent measurement results of methylene blue (MB) on EC sensors with and without a hydrogel coating. The DPV Ipeak values were found to have decreased to 8% for MB (Figure 5A), to 73% for MB-BSA (Figure 5B) and to 23% for MB-IgG (Figure 5B) compared to the results of the unmodified sensor. Furthermore, the DOL normalization of the DPV I_{peak} currents (see Section 3.4) obtained from the sensors with the hydrogel coating did not lead to comparable results for MB-BSA and MB-IgG. This indicates different transport properties of BSA and

IgG in the hydrogel. However, the results demonstrate a good migration property of MB(-conjugates) through the ~2 mm thick wet hydrogel.

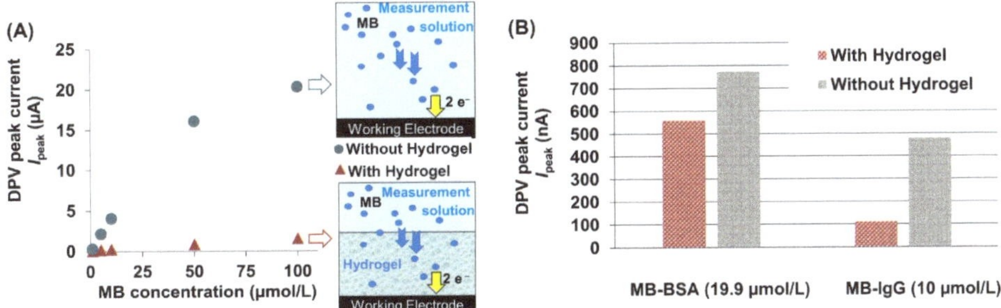

Figure 5. (**A**) Detection of MB with and without a hydrogel coating. (**B**) Detection of MB-BSA (19.9 µmol/L) and MB-IgG (10 µmol/L) with and without a hydrogel coating.

3.6. Hydrgel Characteriazion with Different PEG-DMA Molecular Weights

Hydrogels with different MW PEG-DMAs (1, 2, 3.4, and 10 kDa) were compared regarding the diffusion properties for the MB(-conjugates). For this purpose, 50 µmol/L MB, 22.7 µmol/L MB-BSA, and 10 µmol/L MB-IgG were measured on the sensors modified with the different hydrogel layers.

The hydrogel layers were used directly after washing with PBS, without an intermediate drying step. The measurements were started after the application of the MB(-conjugate) solutions and performed for 60 min.

The diffusivity of the MB(-conjugates) was calculated using Equation (1), and the peak currents after 60 min on the sensor with or without hydrogel were used.

Table 3 shows the results of the calculated diffusivity. Measurements without hydrogel result in electrochemical diffusivity values for MB, MB-BSA, and MB-IgG with a deviation from the values reported in the literature of 6.74×10^{-10} [40], 5.9×10^{-11} [32,33,41], and 4.0×10^{-11} m^2 × s^{-1} [33,41], respectively. This deviation might be explained by the fact that the literature values were calculated values using the hydrodynamic radius of the molecules and did not consider the migration properties under the given conditions (temperature, buffer solution composition) or alternated migration properties of the loaded MB(-conjugates) in electrochemical fields. Furthermore, the diffusivity of the MB-conjugates is significantly reduced in the case of hydrogel modification on the sensor (see Table 3). For a better comparison of the current values, the peak currents were normalized with respect to the MB concentration in µM (for MB, $I_{norm} = I_{peak}/$(MB concentration in µM), for MB-BSA and MB-IgG, $I_{norm} = I_{peak}/$(DOL × protein concentration in µM)) (Figure S5). To compare the maximum current and the equilibrium times, a Langmuir fit was performed using the equation $I_{norm}(t) = I_{norm\_max} \times t/(k_t + t)$, where current I_{peak_max} is the maximum equilibration current, t is the measurement time of each data point, and k_t is the equilibrium rate constant (Figure S5, Table S3). The highest I_{norm_max} was found for MB, followed by MB-BSA and MG-IgG (see Table 3). Figure 6A shows a plot of I_{norm_max} versus diffusivity and Figure 6B shows diffusivity versus mesh size ξ, where a clear correlation can be observed.

Table 3. Results of the comparison of different PEG-DMA hydrogels for the sensing of MB(-conjugates).

PEG-DMA (kDa)	Methylene Blue		MB-BSA		MB-IgG	
	Diffusivity ($m^2 \times s^{-1}$)	$I_{norm\_max} \pm sd$, (nA/(µM MB))	Diffusivity ($m^2 \times s^{-1}$)	$I_{norm\_max} \pm sd$, (nA/(µM MB))	Diffusivity ($m^2 \times s^{-1}$)	$I_{norm\_max} \pm sd$, (nA/(µM MB))
no hydrogel	1.6×10^{-8}		4.8×10^{-12}		1.5×10^{-11}	
1	3.3×10^{-11}	17.8 ± 0.2	7.4×10^{-13}	5.0 ± 0.2	3.7×10^{-12}	2.6 ± 0.1
2	2.2×10^{-11}	15.1 ± 0.6	2.0×10^{-12}	6.5 ± 0.4	4.8×10^{-13}	2.6 ± 0.1
3.4	6.1×10^{-11}	26.4 ± 1.1	1.8×10^{-12}	7.2 ± 0.9	6.4×10^{-13}	5.1 ± 0.7
10	8.3×10^{-11}	29.7 ± 1.2	2.5×10^{-12}	9.0 ± 0.8	7.5×10^{-12}	3.7 ± 0.1

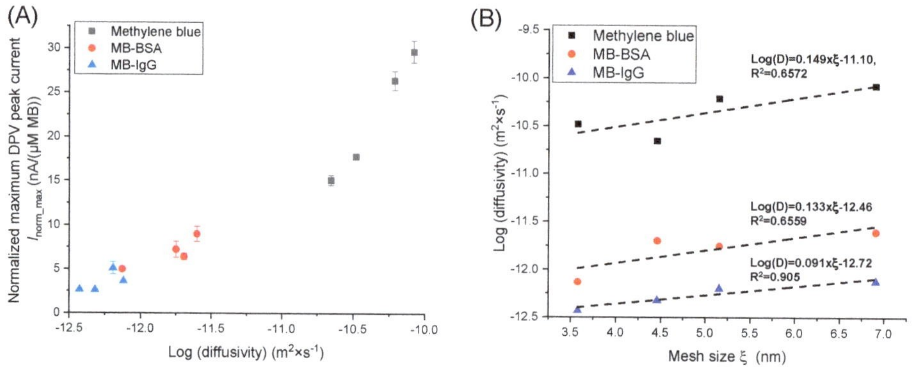

Figure 6. Plots of (**A**) I_{norm_max} versus diffusivity and (**B**) diffusivity versus mesh size ξ.

3.7. MB-BSA Diffusion into and out of the Wet Hydrogel and of the Dry Hydrogel

For the fabrication of hydrogel reservoirs, the hydrogel structure must be filled with a reagent. After the drying procedure, this reagent should be able to migrate out of the hydrogel into the supernatant solution. To test this ability, 10 kDa PEG-DMA hydrogels were fabricated on EC sensors, and the MB-BSA diffusion into and out of a wet hydrogel and a dry hydrogel was investigated.

Figure 7A shows the EC measurement in the wet hydrogel, where, in the first 60 min, MB-BSA migrated from the top solution (22.7 µmol/L MB-BSA, 200 µL) into the hydrogel until it was evenly distributed in the gel (noticeable by the fact that all six WE showed a stable current) and I_{peak} reached the value of 282.96 ± 16.80 nA. Subsequently, the MB-BSA solution was replaced with PBS buffer (200 µL). From this point onwards, the current decreased again as MB-BSA migrated out of the hydrogel, and thus, the redox-active MB moved away from the sensor surface. At minute 90, a new equilibrium was reached with an average I_{peak} of 165.45 ± 11.25 nA, and the top solution was exchanged again with fresh PBS buffer (200 µL). After 130 min, an equilibrium with 93.12 ± 14.46 nA was reached. After changing the buffer again, the current only dropped to 63.93 ± 22.35 nA at minute 158. A further buffer change did not lead to a further significant drop in current (at 175 min, 65.55 ± 20.34 nA). If this current value is calculated as a percentage of the current after diffusion of MB-BSA in the wet hydrogel, it can be said that around 22 % of the MB-BSA remains in the hydrogel and cannot be washed out any further. Subsequently, the PBS buffer was replaced with an MB-BSA solution, and again, the migration of MB-BSA into the hydrogel could be monitored until minute 241 (I_{peak} 361.24 ± 36.48 nA), which proves the reversibility of the diffusion process.

For the fabrication of hydrogel reservoirs, the prior drying of the hydrogel is important to avoid a dilution of the reagent with the water in the gel. Because of this fact, the hydrogel was dried in vacuum and subsequently, the migration of MB-BSA into the dry hydrogel layer was monitored by EC measurements. Figure 7B depicts the measurement. After the application of the MB-BSA solution (22.7 µmol/L, 200 µL), the signal increased and the hydrogel swelled. After 60 min, the current was 2.32 ± 0.60 µA (n = 6), with a quite high standard deviation, and thus, in a similar µA range as the current without the hydrogel (1.30 ± 0.08 µA for 19.9 µmol/L, see Figure 4A). After the saturation of the hydrogel with MB-BSA, the top solution was removed, and the hydrogel was dried. Drying leads to the inclusion of the reagents in the dry hydrogel polymer structure. The inclusion of proteins in the dense polymeric network of the dry hydrogel can also stabilize their 3D structure [42–44] After the drying step, the hydrogel network was overlayed with PBS buffer (200 µL). Figure 7C shows the plot of the current. At minute 30, an equilibrium was reached with an average I_{peak} of 1.07 ± 0.26 µA, and the top solution was again exchanged with fresh PBS buffer (200 µL). The reduction in the I_{peak} current can be explained by the dilution effect of MB-BSA. The hydrogel can only take up 128.5 µL of the MB-BSA solution (Table S2, 10 kDa W_w), but it becomes diluted with 200 µL PBS buffer in the hydrogel. Then, 60 min after the PBS buffer application, an equilibrium with 0.82 ± 0.23 µA was reached. After changing the buffer again, the current only dropped to 0.40 ± 0.35 µA by minute 90. Another buffer change further reduced the current to 0.28 ± 0.26 µA, which corresponds to 12% of the MB-BSA current after filling and 26% of the current after the first addition of PBS buffer. The latter indicates a reduced mobility of MB-BSA in the dried hydrogel network.

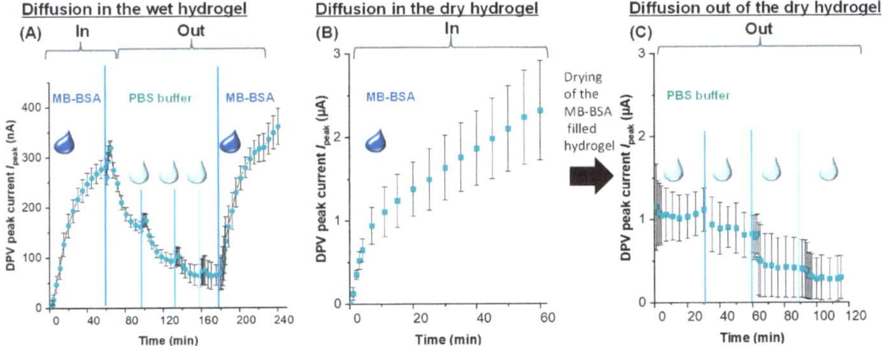

Figure 7. Diffusion study of MB-BSA in the wet (**A**), in the dry hydrogels (**B**), and out of the dry hydrogel (**C**).

4. Discussion

The diffusion properties of biomolecules are an important parameter and must be considered in the production of hydrogels for biosensor applications. In our study, we approached this topic by preparing and characterizing different printable PEG-DMA-based hydrogels. Gravimetric and electrochemical characterization methods were used and finally brought into relation. The gravimetric method showed that the swelling ratio (*SR*) can be adjusted primarily via the molar concentration of the PEG-DMA compound in the hydrogel ink. This is crucial and must be considered if either excessive swelling should be prevented in a microfluidic channel or excessive swelling of the hydrogels is desired, e.g., if hydrogels are used as valves in a microfluidic channel.

The calculated number-average molecular weight of the polymer chain between the crosslinks ($\overline{M_c}$) and mesh sizes ξ are in good agreement with values from the literature [39], where a PEG-DMA mixture was used for the preparation of hydrogels.

Therefore, the addition of DEGVE to hydrogel inks only has the purpose of ensuring good printability and minimally influences the formation of the network structure.

The number-average molecular weight of the polymer chain $\overline{M_c}$ and the mesh size ξ provide information about the strength of the hydrogel and enable the estimation of the diffusion properties of the hydrogel, as demonstrated further below. Several experiments were systematically conducted to investigate the diffusion properties. First, MB(-conjugates) were measured on sensors without the hydrogel. It was found that the degree of labeling (DOL) influences the level of the measured current, and the normalization of the currents enables the direct comparison of MB-conjugates with different DOLs.

First, this result is decisive because it demonstrates the accuracy of the method of labeling the biomolecules. Moreover, this result also shows that the methylene blue molecules bound to the biomolecules contribute equally to the resulting DPV peak current. This is particularly important for the calculation of the diffusivity from the electrochemical DPV equation (Equation (1)) for the number of electrons per mole of MB(-conjugate) (n), as we will discuss further below.

Second, the sensors coated with a 10 kDa PEG-DMA-based hydrogel showed that hydrogel coatings on sensors lead to a reduction in DPV peak currents to 8% for MB, 73% for MB-BSA, and 23% for MB-IgG with respect to the measurements without the hydrogel. This is probably because fewer molecules reach the electrode surface in the presence of a hydrogel coating, leading to limited diffusion. As the sensing kinetics depends significantly on the thickness of the hydrogel layer, the thickness of the hydrogel layer must also be considered when producing filter layers over sensors to avoid a limitation in the sensitivity of the sensor.

Third, sensors coated with PEG-DMA hydrogels of different molecular weights (M_w) show that a higher M_w leads to higher equilibrium currents and faster diffusion. As the mesh sizes do not change following the diffusion of MB-BSA and MB-IgG into the hydrogel network, the calculated values from the independent gravimetric method can be correlated with the diffusivity of these molecules, again demonstrating the plausibility of the chosen methods. It was found that the diffusivity increases for MB-BSA, MB-IG, and MB with increasing mesh sizes. This can be explained by increased molecule mobility due to a reduced interaction between the molecules and the polymer chains of the hydrogel network.

Fourth, wet and dry hydrogels were used to study the diffusion of MB-BSA. It was found that up to 88 % of the MB-BSA introduced into dried hydrogels could be released from the hydrogel. Therefore, the hydrogel inks investigated in this study have a good potential to be used as a reagent reservoir.

In future studies, the use and long-term stability of hydrogels as reagent reservoirs on sensors and in microfluidics will be investigated. Spotting and dispensing methods, as already demonstrated for the measurement of lactate on microneedles [22], will be investigated for printing in microfluidics.

Supplementary Materials: The following supporting information can be downloaded at: https://www.mdpi.com/article/10.3390/s24113678/s1, Figure S1: Measurement setup: (1) PalmSens multiplexer MUX8-R2 (PSTrace software 5.7), (2) connector, (3) sensor, and (4) computer; Table S1: Variables of the gravimetric analysis; Table S2: Determined values from gravimetric hydrogel characterization; Figure S2: SEM images of vacuum-dried and lyophilized hydrogel structures; Figure S3: Correlation between (A) the swelling ratio and the molar concentration in the ink, (B) the swelling ratio and the molecular weight of PEG-DMA (kDa), (C) mesh size and the molar concentration in the ink, and (D) mesh size and the molecular weight of PEG-DMA (kDa); Figure S4: DPV measurement of electrochemical sensors of MB-BSA and MB-IgG (A) without and (B) with the hydrogel (10 kDa PEG-DMA as crosslinker) over 60 min. During the first five minutes, a DPV measurement was performed every minute and afterward, every five minutes; Figure S5: Normalized DPV peak currents (nA/μM MB) of MB, MB-BSA, and MB-IgG measurements on 1, 2, 3.4, and 10 kDa PEG-DMA hydrogel-modified sensors over 60 min. Curves were fitted with a Langmuir fitting function; Table S3: Values of the equilibrium rate constant k_f.

Author Contributions: Conceptualization, E.M., G.C.M. and R.H.; data curation, E.M.; formal analysis, E.M. and S.K.; funding acquisition, E.M., G.C.M., R.H. and V.B.; investigation, E.M., S.K. and V.B.; methodology, E.M., G.C.M. and R.H.; project administration, E.M., G.C.M. and R.H.; supervision, E.M., G.C.M. and R.H.; validation, E.M.; visualization, E.M. and S.K.; writing—original draft, E.M.;

writing—review and editing, E.M., S.K., G.C.M., R.H. and V.B. All authors have read and agreed to the published version of the manuscript.

Funding: This work received funding from the Austrian Research Promotion Agency (FFG) under the HydroChip2 (grant no. 883914) and the Predict project (grant no. 870027) as well as from the European Union's Horizon 2020 research and innovation program under Grant Agreement No. 761000 (GREENSENSE).

Institutional Review Board Statement: Not applicable.

Informed Consent Statement: Not applicable.

Data Availability Statement: The data are contained within this article and Supplementary Materials.

Acknowledgments: Special thanks to Alice K. Pschenitschnigg (AIT) for support in the measurements for this publication. Thanks also to Magdalena Wegerer for performing the SEM imaging. Special thanks also to Kathrin Freitag (RISE), Jessica Åhlin (RISE), and Jan Strandberg (RISE) for the manufacturing and quality control of the sensor arrays.

Conflicts of Interest: The authors declare no conflicts of interest.

References

1. Kim, E.R.; Joe, C.; Mitchell, R.J.; Gu, M.B. Biosensors for healthcare: Current and future perspectives. *Trends Biotechnol.* **2023**, *41*, 374–395. [CrossRef]
2. Gonzalez-Macia, L.; Morrin, A.; Smyth, M.R.; Killard, A.J. Advanced printing and deposition methodologies for the fabrication of biosensors and biodevices. *Analyst* **2010**, *135*, 845. [CrossRef]
3. Ronkainen, N.J.; Halsall, H.B.; Heineman, W.R. Electrochemical biosensors. *Chem. Soc. Rev.* **2010**, *39*, 1747–1763. [CrossRef] [PubMed]
4. da Silva, E.T.S.G.; Souto, D.E.P.; Barragan, J.T.C.; Giarola, J.D.F.; De Moraes, A.C.M.; Kubota, L.T. Electrochemical Biosensors in Point-of-Care Devices: Recent Advances and Future Trends. *ChemElectroChem* **2017**, *4*, 778–794. [CrossRef]
5. Couto, R.A.S.; Lima, J.L.F.C.; Quinaz, M.B. Recent developments, characteristics and potential applications of screen-printed electrodes in pharmaceutical and biological analysis. *Talanta* **2016**, *146*, 801–814. [CrossRef]
6. Schrattenecker, J.D.; Heer, R.; Hainberger, R.; Fafilek, G. Impedimetric IgG-Biosensor with In-Situ Generation of the Redox-Probe. *Proceedings* **2017**, *1*, 534. [CrossRef]
7. Schrattenecker, J.D.; Heer, R.; Melnik, E.; Maier, T.; Fafilek, G.; Hainberger, R. Hexaammineruthenium (II)/(III) as alternative redox-probe to Hexacyanoferrat (II)/(III) for stable impedimetric biosensing with gold electrodes. *Biosens. Bioelectron.* **2018**, *127*, 25–30. [CrossRef] [PubMed]
8. Nimse, S.B.; Sonawane, M.D.; Song, K.-S.; Kim, T. Biomarker detection technologies and future directions. *Analyst* **2016**, *141*, 740–755. [CrossRef] [PubMed]
9. Ricci, F.; Adornetto, G.; Palleschi, G. A review of experimental aspects of electrochemical immunosensors. *Electrochim. Acta* **2012**, *84*, 74–83. [CrossRef]
10. Thoeny, V.; Melnik, E.; Asadi, M.; Mehrabi, P.; Schalkhammer, T.; Pulverer, W.; Maier, T.; Mutinati, G.C.; Lieberzeit, P.; Hainberger, R. Detection of breast cancer-related point-mutations using screen-printed and gold-plated electrochemical sensor arrays suitable for point-of-care applications. *Talanta Open* **2022**, 100150. [CrossRef]
11. Thoeny, V.; Melnik, E.; Huetter, M.; Asadi, M.; Mehrabi, P.; Schalkhammer, T.; Pulverer, W.; Maier, T.; Mutinati, G.C.; Lieberzeit, P.; et al. Recombinase polymerase amplification in combination with electrochemical readout for sensitive and specific detection of PIK3CA point mutations. *Anal. Chim. Acta* **2023**, *1281*, 341922. [CrossRef]
12. Faustino, V.; Catarino, S.O.; Lima, R.; Minas, G. Biomedical microfluidic devices by using low-cost fabrication techniques: A review. *J. Biomech.* **2016**, *49*, 2280–2292. [CrossRef]
13. Toren, P.; Smolka, M.; Haase, A.; Palfinger, U.; Nees, D.; Ruttloff, S.; Kuna, L.; Schaude, C.; Jauk, S.; Rumpler, M.; et al. High-throughput roll-to-roll production of polymer biochips for multiplexed DNA detection in point-of-care diagnostics. *Lab Chip* **2020**, *20*, 4106–4117. [CrossRef]
14. Deng, J.; Jiang, X. Advances in Reagents Storage and Release in Self-Contained Point-of-Care Devices. *Adv. Mater. Technol.* **2019**, *4*, 1800625. [CrossRef]
15. Smith, S.; Sewart, R.; Becker, H.; Roux, P.; Land, K. Blister pouches for effective reagent storage on microfluidic chips for blood cell counting. *Microfluid. Nanofluid* **2016**, *20*, 163. [CrossRef]
16. Xu, J.; Wang, J.; Su, X.; Qiu, G.; Zhong, Q.; Li, T.; Zhang, D.; Zhang, S.; He, S.; Ge, S.; et al. Transferable, easy-to-use and room-temperature-storable PCR mixes for microfluidic molecular diagnostics. *Talanta* **2021**, *235*, 122797. [CrossRef]
17. Zirath, H.; Schnetz, G.; Glatz, A.; Spittler, A.; Redl, H.; Peham, J.R. Bedside Immune Monitoring: An Automated Immunoassay Platform for Quantification of Blood Biomarkers in Patient Serum within 20 Minutes. *Anal. Chem.* **2017**, *89*, 4817–4823. [CrossRef]

18. Zhao, Z.; Al-Ameen, M.A.; Duan, K.; Ghosh, G.; Lo, J.F. On-chip porous microgel generation for microfluidic enhanced VEGF detection. *Biosens. Bioelectron.* **2015**, *74*, 305–312. [CrossRef]

19. Kirk, J.T.; Fridley, G.E.; Chamberlain, J.W.; Christensen, E.D.; Hochberg, M.; Ratner, D.M. Multiplexed inkjet functionalization of silicon photonic biosensors. *Lab Chip* **2011**, *11*, 1372. [CrossRef]

20. Abe, K.; Hashimoto, Y.; Yatsushiro, S.; Yamamura, S.; Bando, M.; Hiroshima, Y.; Kido, J.; Tanaka, M.; Shinohara, Y.; Ooie, T.; et al. Simultaneous Immunoassay Analysis of Plasma IL-6 and TNF-α on a Microchip. *PLoS ONE* **2013**, *8*, e53620. [CrossRef]

21. Melnik, E.; Muellner, P.; Mutinati, G.C.; Koppitsch, G.; Schrank, F.; Hainberger, R.; Laemmerhofer, M. Local functionalization of CMOS-compatible Si 3 N 4 Mach-Zehnder interferometers with printable functional polymers. *Sens. Actuators B Chem.* **2016**, *236*, 1061–1068. [CrossRef]

22. Kurzhals, S.; Melnik, E.; Plata, P.; Cihan, E.; Herzog, P.; Felice, A.; Bocchino, A.; O'Mahony, C.; Mutinati, G.C.; Hainberger, R. Detection of lactate via amperometric sensors modified with direct electron transfer enzyme containing PEDOT:PSS and hydrogel inks. *IEEE Sens. Lett.* **2023**, *7*, 1–4. [CrossRef]

23. Lee, J.; Ko, J.H.; Lin, E.-W.; Wallace, P.; Ruch, F.; Maynard, H.D. Trehalose hydrogels for stabilization of enzymes to heat. *Polym. Chem.* **2015**, *6*, 3443–3448. [CrossRef]

24. Panescu, P.H.; Ko, J.H.; Maynard, H.D. Scalable Trehalose-Functionalized Hydrogel Synthesis for High-Temperature Protection of Enzymes. *Macromol. Mater. Eng.* **2019**, *304*, 1800782. [CrossRef]

25. Mancini, R.J.; Lee, J.; Maynard, H.D. Trehalose glycopolymers for stabilization of protein conjugates to environmental stressors. *J. Am. Chem. Soc.* **2012**, *134*, 8474–8479. [CrossRef]

26. Zhang, B.; Yao, H.; Qi, H.; Zhang, X.-L. Trehalose and alginate oligosaccharides increase the stability of muscle proteins in frozen shrimp (*Litopenaeus vannamei*). *Food Funct.* **2020**, *11*, 1270–1278. [CrossRef]

27. Lesch, A.; Cortés-Salazar, F.; Amstutz, V.; Tacchini, P.; Girault, H.H. Inkjet printed nanohydrogel coated carbon nanotubes electrodes for matrix independent sensing. *Anal. Chem.* **2015**, *87*, 1026–1033. [CrossRef]

28. Melnik, E.; Strasser, F.; Muellner, P.; Heer, R.; Mutinati, G.C.; Koppitsch, G.; Lieberzeit, P.; Laemmerhofer, M.; Hainberger, R. Surface Modification of Integrated Optical MZI Sensor Arrays Using Inkjet Printing Technology. *Procedia Eng.* **2016**, *168*, 337–340. [CrossRef]

29. Bauer, M.; Duerkop, A.; Baeumner, A.J. Critical review of polymer and hydrogel deposition methods for optical and electrochemical bioanalytical sensors correlated to the sensor's applicability in real samples. *Anal. Bioanal. Chem.* **2023**, *415*, 83–95. [CrossRef]

30. Shafique, H.; de Vries, J.; Strauss, J.; Jahromi, A.K.; Moakhar, R.S.; Mahshid, S. Advances in the Translation of Electrochemical Hydrogel-Based Sensors. *Adv. Healthc. Mater.* **2023**, *12*, e2201501. [CrossRef]

31. Tokuyama, H.; Nakahata, Y.; Ban, T. Diffusion coefficient of solute in heterogeneous and macroporous hydrogels and its correlation with the effective crosslinking density. *J. Membr. Sci.* **2020**, *595*, 117533. [CrossRef]

32. Johnson, E.M.; Berk, D.A.; Jain, R.K.; Deen, W.M. Hindered diffusion in agarose gels: Test of effective medium model. *Biophys. J.* **1996**, *70*, 1017–1023. [CrossRef] [PubMed]

33. Zhou, Y.; Li, J.; Zhang, Y.; Dong, D.; Zhang, E.; Ji, F.; Qin, Z.; Yang, J.; Yao, F. Establishment of a Physical Model for Solute Diffusion in Hydrogel: Understanding the Diffusion of Proteins in Poly(sulfobetaine methacrylate) Hydrogel. *J. Phys. Chem. B* **2017**, *121*, 800–814. [CrossRef] [PubMed]

34. Axpe, E.; Chan, D.; Offeddu, G.S.; Chang, Y.; Merida, D.; Hernandez, H.L.; Appel, E.A. A Multiscale Model for Solute Diffusion in Hydrogels. *Macromolecules* **2019**, *52*, 6889–6897. [CrossRef] [PubMed]

35. Deffo, G.; Nde Tene, T.F.; Medonbou Dongmo, L.; Zambou Jiokeng, S.L.; Tonleu Temgoua, R.C. *Differential Pulse and Square-Wave Voltammetry as Sensitive Methods for Electroanalysis Applications*; Elsevier: Oxford, UK, 2024; pp. 409–417. [CrossRef]

36. Caccavo, D.; Cascone, S.; Lamberti, G.; Barba, A.A. Hydrogels: Experimental characterization and mathematical modelling of their mechanical and diffusive behaviour. *Chem. Soc. Rev.* **2018**, *47*, 2357–2373. [CrossRef] [PubMed]

37. Bray, J.C.; Merrill, E.W. Poly(vinyl alcohol) hydrogels. Formation by electron beam irradiation of aqueous solutions and subsequent crystallization. *J. Appl. Polym. Sci.* **1973**, *17*, 3779–3794. [CrossRef]

38. Hickey, A.S.; Peppas, N.A. Mesh size and diffusive characteristics of semicrystalline poly(vinyl alcohol) membranes prepared by freezing/thawing techniques. *J. Membr. Sci.* **1995**, *107*, 229–237. [CrossRef]

39. Lin, S.; Sangaj, N.; Razafiarison, T.; Zhang, C.; Varghese, S. Influence of physical properties of biomaterials on cellular behavior. *Pharm. Res.* **2011**, *28*, 1422–1430. [CrossRef] [PubMed]

40. Selifonov, A.A.; Shapoval, O.G.; Mikerov, A.N.; Tuchin, V.V. Determination of the Diffusion Coefficient of Methylene Blue Solutions in Dentin of a Human Tooth using Reflectance Spectroscopy and Their Antibacterial Activity during Laser Exposure. *Opt. Spectrosc.* **2019**, *126*, 758–768. [CrossRef]

41. Merrill, E.W.; Dennison, K.; Sung, C. Partitioning and diffusion of solutes in hydrogels of poly(ethylene oxide). *Biomaterials* **1993**, *14*, 1117–1126. [CrossRef]

42. Gil, M.S.; Cho, J.; Thambi, T.; Giang Phan, V.H.; Kwon, I.; Lee, D.S. Bioengineered robust hybrid hydrogels enrich the stability and efficacy of biological drugs. *J. Control. Release* **2017**, *267*, 119–132. [CrossRef] [PubMed]

43. Simon, D.; Obst, F.; Haefner, S.; Heroldt, T.; Peiter, M.; Simon, F.; Richter, A.; Voit, B.; Appelhans, D. Hydrogel/enzyme dots as adaptable tool for non-compartmentalized multi-enzymatic reactions in microfluidic devices. *React. Chem. Eng.* **2019**, *4*, 67–77. [CrossRef]

44. Davari, N.; Bakhtiary, N.; Khajehmohammadi, M.; Sarkari, S.; Tolabi, H.; Ghorbani, F.; Ghalandari, B. Protein-Based Hydrogels: Promising Materials for Tissue Engineering. *Polymers* **2022**, *14*, 986. [CrossRef] [PubMed]

Article

3D Printed Hydrogel Sensor for Rapid Colorimetric Detection of Salivary pH

Magdalena B. Łabowska [1,*], Agnieszka Krakos [2] and Wojciech Kubicki [2]

[1] Department of Mechanics, Materials and Biomedical Engineering, Faculty of Mechanical Engineering, Wroclaw University of Science and Technology, Smoluchowskiego 25, 50-371 Wroclaw, Poland

[2] Department of Microsystems, Faculty of Electronics, Photonics and Microsystems, Wroclaw University of Science and Technology, Janiszewskiego 11/17, 50-372 Wroclaw, Poland; agnieszka.krakos@pwr.edu.pl (A.K.); wojciech.kubicki@pwr.edu.pl (W.K.)

* Correspondence: magdalena.labowska@pwr.edu.pl

Abstract: Salivary pH is one of the crucial biomarkers used for non-invasive diagnosis of intraoral diseases, as well as general health conditions. However, standard pH sensors are usually too bulky, expensive, and impractical for routine use outside laboratory settings. Herein, a miniature hydrogel sensor, which enables quick and simple colorimetric detection of pH level, is shown. The sensor structure was manufactured from non-toxic hydrogel ink and patterned in the form of a matrix with 5 mm × 5 mm × 1 mm individual sensing pads using a 3D printing technique (bioplotting). The authors' ink composition, which contains sodium alginate, polyvinylpyrrolidone, and bromothymol blue indicator, enables repeatable and stable color response to different pH levels. The developed analysis software with an easy-to-use graphical user interface extracts the R(ed), G(reen), and B(lue) components of the color image of the hydrogel pads, and evaluates the pH value in a second. A calibration curve used for the analysis was obtained in a pH range of 3.5 to 9.0 using a laboratory pH meter as a reference. Validation of the sensor was performed on samples of artificial saliva for medical use and its mixtures with beverages of different pH values (lemon juice, coffee, black and green tea, bottled and tap water), and correct responses to acidic and alkaline solutions were observed. The matrix of square sensing pads used in this study provided multiple parallel responses for parametric tests, but the applied 3D printing method and ink composition enable easy adjustment of the shape of the sensing layer to other desired patterns and sizes. Additional mechanical tests of the hydrogel layers confirmed the relatively high quality and durability of the sensor structure. The solution presented here, comprising 3D printed hydrogel sensor pads, simple colorimetric detection, and graphical software for signal processing, opens the way to development of miniature and biocompatible diagnostic devices in the form of flexible, wearable, or intraoral sensors for prospective application in personalized medicine and point-of-care diagnosis.

Keywords: pH sensor; hydrogel ink; bioplotting; 3D printing; hydrogel sensor; colorimetric detection; salivary diagnostics

Citation: Łabowska, M.B.; Krakos, A.; Kubicki, W. 3D Printed Hydrogel Sensor for Rapid Colorimetric Detection of Salivary pH. *Sensors* **2024**, *24*, 3740. https://doi.org/10.3390/s24123740

Academic Editors: Bruno Ando, Luca Francioso and Pietro Siciliano

Received: 25 April 2024
Revised: 9 May 2024
Accepted: 17 May 2024
Published: 8 June 2024

1. Introduction

The interest in biocompatible polymer materials has dynamically increased in the biomedical and biopharma sectors in the last few years [1–3]. Among them, hydrogels provide unique physiochemical properties, including relatively good mechanical stability, high liquid absorption, and insolubility, allowing versatile applications where biocompatibility and fabrication feasibility are critical [4,5]. Hydrogel layers have been commonly used as smart, pain-relieving wound patches [6,7], as well as precise structural scaffolds for cell culture and tissue engineering [8,9]. On the other hand, the responsive character of hydrogels, showing notable changes in their rheology, resistivity, or color tunability when exposed to variable external environments, make them a perfect tool for biosensing applications [10,11], including pH measurements.

Recent advances in additive manufacturing techniques have provided new possibilities for the fabrication of hydrogel-based devices [12,13]. Hydrogel inks of different content can be printed automatically, rapidly, and more easily in the process of bioplotting. This technique enables the formation of repeatable and more advanced hydrogel structures compared to manual methods [14,15]. As mentioned above, hydrogel matrices find application in cell culture, ensuring the 3D spheroid growth of microbials [16,17]. These microfluidic-based hydrogel platforms provide a passive physiological cellular environment and also become novel integrated perfusion systems, with drugs encapsulated in a polymer protective cover [18]. These approaches are interesting and commonly available alternatives to standard lab-on-chip technologies [19], which are based on more demanding and advanced silicon, glass, and PDMS microengineering.

Bioplotting of hydrogel matrices also opens the way for dedicated, low-cost, and biodegradable sensor platforms. According to the state of the art [20–22], considerable attention has been paid to hydrogel biosensors and their possible role as new diagnostic tools for modern and personalized medicine. In this regard, hydrogel matrices can be used as stimuli-responsive materials, as immobilization substrates for sensing molecules, and as protection surfaces. However, the main focus has recently been on the development of simple and non-toxic biosensors that allow the real-time monitoring of chemical biomarkers in human fluids in a wide detection range [23–25]. Such point-of-care devices can be used in a disposable way, for single ex vivo use, and in a constant wearable mode. As much effort has recently been put to provide non-invasive medical diagnostics, the monitoring of potential health disfunctions based on tears, sweat, or saliva can provide a real revolution in biopharma strategies.

Saliva, also called a "mirror of the body", is a valuable diagnostic specimen that can directly reflect many pathological states of the human body condition [26]. Due to the ease of sample collection, storage, and acceptable sensitivity, saliva evaluation can constitute an important alternative to standard blood examination, which without the need for qualified personnel appears to be a perfect solution for home care devices [27]. Variations in saliva pH levels can be an indication of a number of health issues, including systemic illnesses, metabolic disorders, and malfunctioning salivary glands. For instance, too high or too low saliva pH can be a sign of various health issues, whereas too low pH can encourage gum disease and tooth decay. Salivary diagnostics may provide rapid insight into both intraoral disfunctions (e.g., periodontal diseases), as well as serious systemic disorders (e.g., diabetes mellitus, cancer). According to the most current WHO guides, intraoral testing can be a chance for developing countries to provide inexpensive and easy-to-use medical diagnostics, notably improving the rapid profiling of health conditions in underserved groups [28,29]. However, the general need for accurate and low-cost biosensors, produced using available and simple fabrication methods, is a key issue that requires global interest and novel development strategies.

A recent trend toward the use of biocompatible hydrogel matrices that are sensitive to local environmental changes (e.g., pH, temperature, gases) and the presence of specific biomarkers (e.g., glucose, urea, antigens, drugs) in combination with microfluidics and microsystem techniques can provide new solutions of both extraoral and intraoral biosensors. Accompanied by rapid bioplotting technology, a new generation of portable salivary diagnostic systems can be developed. Thanks to the unique properties of hydrogels, different sensing methods are available, strictly related to their composition and application. The most popular are hydrogel sensors based on colorimetry, conductance, and piezoresistivity. Utilizing hydrogel sensors for determining changes in pH in saliva is useful to assess the condition of the oral mucosa and detect metabolic disorders, dysfunctional salivary glands, candida, and even systemic diseases like diabetes, gastroesophageal reflux, cancer, or autoimmune diseases [30].

For example, Wen et al. [31] proposed a polyacrylamide structure for the detection of ammonia based on the color change of the hydrogel induced by the pH value. A similar approach was presented in the work of Tang et al. [32], where a composition of polyvinyl alcohol and sodium alginate was used as a functional layer of the hydrogel. Lee et al. [33] and Wu et al. [34] also presented an interesting solution to enable the rapid detection of glucose using agarose and methacrylate gelatin (GelMA). Apart from the color change of hydrogels exposed to water solution, including different proportions of glucose, a visible shrinkage in the hydrogel geometry was observed in such structures [34].

Conductance detection, as well as piezoresistivity changes, are in turn typically used for ethanol [35,36] and drug detection [37]. The literature review shows that the hydrogels based on gelatin, polyacrylamide, and bisacrylamide are the most common in this regard. Nevertheless, in these cases, the hydrogel sensor fabrication procedures may seem rather complicated, often requiring multistep protocols as well as specialized equipment and laboratories [25,38]. Moreover, in the case of exclusively bisacrylamide structures, these solutions are considered toxic [39], thus the potential application of such sensors for near-oral diagnostics is rather controversial.

Based on the aforementioned issues, the preparation of a hydrogel sensor with total biocompatibility and simplified manufacturing steps is not trivial and needs further investigation. In this work, a hydrogel structure for pH assessment toward salivary diagnostics utilizing rapid colorimetry detection is shown. Non-toxic and polymer inks of unique but simple and available composition were prepared, based on sodium alginate and polyvinylpyrrolidone, for simplified and automated bioplotting of the small-scale sensor matrices. Dedicated software was developed to quickly and easily process and analyze the color tunability of the hydrogel sensor exposed to pH changes. The hydrogel structures proposed in this work were investigated in terms of mechanical properties (tensile tests) and sensing capabilities in response to defined saliva samples (raw or with additives). The results of the experiments confirmed fine mechanical durability and appropriate operational stability of the sensor, as well as quick response to pH changes, which in combination with simple software-based signal detection reveal the prospective application of the device in future point-of-care diagnostics.

2. Materials and Methods

2.1. Fabrication of the Hydrogel pH Sensor

Hydrogel pads were bioplotted on a dedicated substrate, which contained a matrix of 25 square cavities, $5 \times 5 \times 1$ mm^3 each (Figure 1a). The matrix was used to enable easy and repeatable capture of the image for testing various pH samples at the same conditions. However, the sensing hydrogel layer may be plotted onto other surfaces and in a more sophisticated pattern on demand. The substrate was manufactured using a multi-jet 3D printing technique (Projet 3500 SD Max, 3D Systems, Rock Hill, SC, USA) using photocurable and biodegradable VisiJet M3 Crystal construction material (3D Systems, Rock Hill, SC, USA) and VisiJet S300 support material (3D Systems, Rock Hill, SC, USA). The substrate was treated to obtain repeatable surface properties, according to the post-processing protocol described elsewhere [40]. Post-processing included an air heating process followed by an oil bath at 65 °C, a water-based detergent bath in an ultrasonic cleaner, and a final washing step with isopropyl alcohol (IPA).

The hydrogel ink solution was composed of 5% (v/v) sodium alginate (Sigma Aldrich, Saint Louis, MO, USA), 1% (v/v) polyvinylpyrrolidone (Sigma Aldrich, Saint Louis, MO, USA), and 0.1% (v/v) bromothymol blue pH indicator (Warchem, Zakręt, Poland). The chemical characteristics and their impact on the output hydrogel structure were taken into consideration during the choice of ink composition and the proportions of the ingredients. Physical characteristics, including the sensor material's strength, permeability, absorption capacity, and stability under various environmental conditions, were also taken into account. The water-based mixture was prepared by stirring (400 RPM) at 60 °C for 60 min using a laboratory hot plate with a magnetic stirrer. Sensor pads were filled with a repeatable dose of a hydrogel ink utilizing a commercially available 3D printer (BioX, Cellink, San Diego, CA, USA) with the process parameters as follows: $p = 20$ kPa, v = 4 mm/s, and T = 30 °C

(Figure 1b). Next, a 4% (v/v) calcium chloride solution (Warchem, Zakręt, Poland) was gently pipetted into the pads to enable cross-linking of the hydrogel structures in 10 min.

(a)

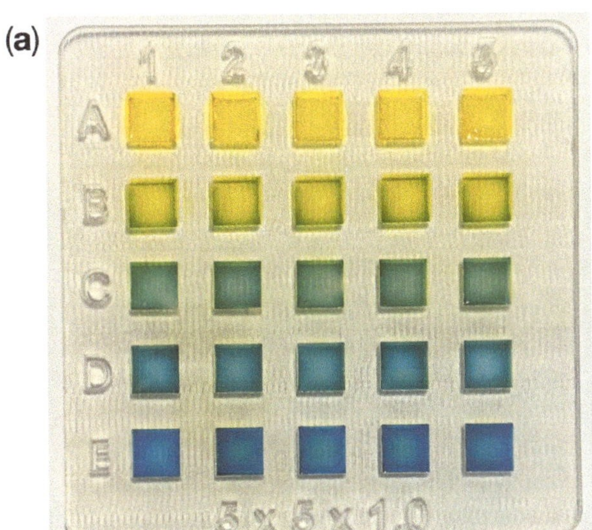

(b)

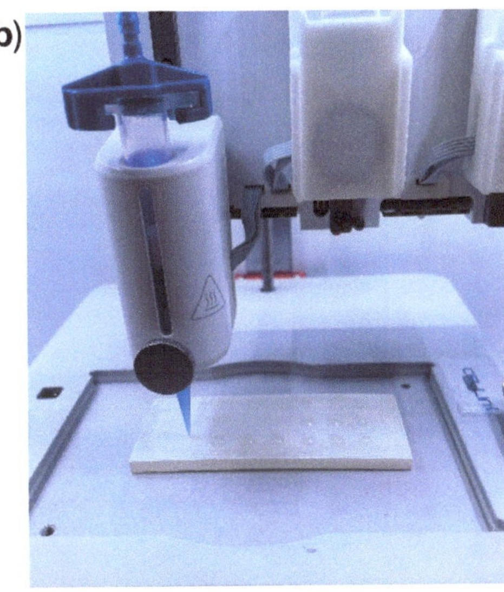

Figure 1. Preparation of the hydrogel pH sensors: (**a**) 3D printed substrate (outer dimensions: $54 \times 51 \times 1.1$ mm^3) containing a matrix of 25 bioplotted 5×5 mm^2 hydrogel sensor pads, (**b**) hydrogel pattern bioplotting process utilizing a BioX Cellink printer.

2.2. Mechanical Properties of the Hydrogel Structures

Due to the high water content in their structure, hydrogels pose relatively poor mechanical properties compared to other materials. The mechanical durability of hydrogel materials depends mainly on the fluid content, as well as the strength of the bonds between the polymer chains of the structure, which are directly affected by the cross-linking process. Appropriate durability is essential to ensure the resilience of thin layers, preventing them from tearing during use and enabling prolonged storage capability. Furthermore, the use of additive manufacturing technology ensures precise deposition of layers of equal thickness, ensuring repeatable results. To assess the mechanical durability of the hydrogel layers, the tensile strength and compression tests were performed using the MultiTest-i-1 instrument (Mecmesin, Slinfold, United Kingdom). The apparatus was equipped with the ILC-S strain gauge sensor with a measuring range of 100 N, operating at a constant velocity of 5 mm/min (Figure 2). Tensile tests were conducted for both pure hydrogel structures (including sodium alginate and polyvinylpyrrolidone), as well as structures containing the pH indicator (bromothymol blue). Cuboidal hydrogel samples with dimensions $5 \times 0.7 \times 50$ mm^3 were fabricated on a 3D BioX bioplotter to ensure replicability. Mechanical tests were also conducted for hydrogel structures subjected to various pH environments to verify their potential shrinkage or swelling under different ambient conditions. In this case, a compression test was conducted on $5 \times 5 \times 5$ mm^3 hydrogel cuboids with a pH indicator included in the structures, placed in different pH solutions for 10 min. Three pH values (extremes and typical for saliva: pH 3.5, 9.0, and 6.5) were used for examination. Based on instrument force indication, Young's modulus (E) and compressive modulus of elasticity (E_c) values for the hydrogel samples were calculated.

Afterward, statistical analysis was performed with the Statistica package (version 13, TIBCO Software Inc., Palo Alto, CA, USA) to visualize the maximum, mean, and minimum values of Young's modulus and the compressive modulus of elasticity for each group of samples in box-and-whisker chart form.

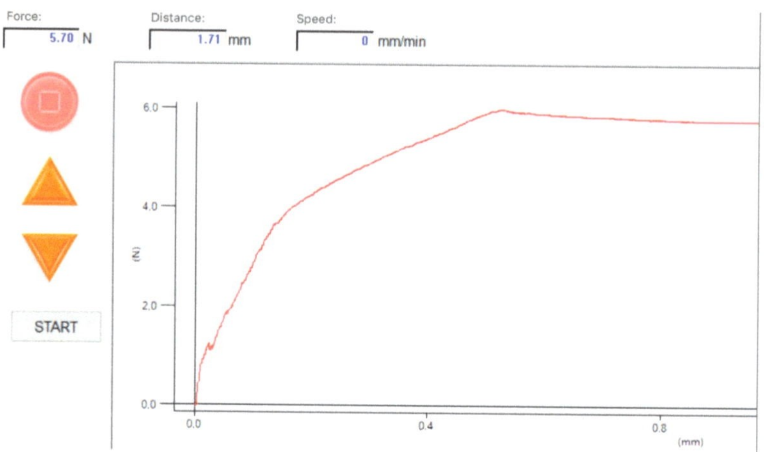

Figure 2. Tensile strength of the hydrogel matrices. Emperor Force software v1.18-408 for visualization of the maximum tensile strength.

2.3. Colorimetric Detection—Software-Enhanced Analysis of the Sensor Responses

The response of the hydrogel sensor was obtained by first capturing a color image of the matrix using a 24 MP CMOS camera. Next, the image was digitally processed by dedicated software with a graphical user interface (LabVIEW, National Instruments), developed based on our previous experience with optical signal processing [41,42]. As the color of the individual pad changed proportionally to the change of the pH level, the software-based algorithm was used to extract red (R), green (G), and blue (B) components from individual regions of interest (ROIs) of the image. The software enables manual or programmed selection of ROIs, and the measured data is presented in both graphical and tabularized forms (Figure 3). The numerical components of RGB plots are stored and may be compared, enabling measurement of pH values of the samples, as well as preparation of the calibration curve for the pH pattern (e.g., when the composition of the ink is changed). Due to the applied algorithm, the colorimetric detection is rapid and simple and can be performed for numerous pads in parallel. The shape of the ROIs can also be adapted to other patterns of the sensor pads.

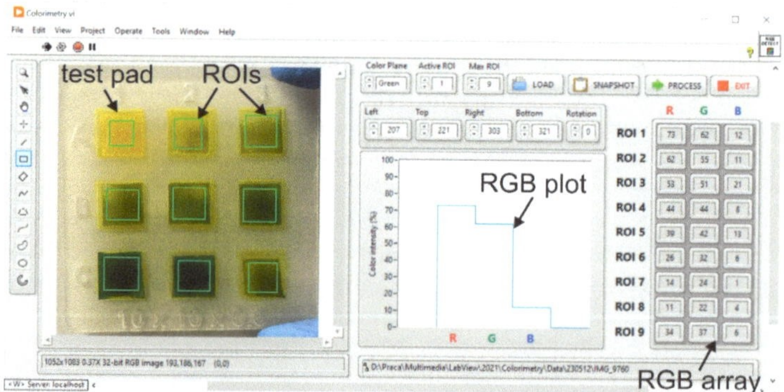

Figure 3. Front panel of the developed LabVIEW application for image colorimetry. RGB plots are automatically processed for selected rectangular ROIs selected in the image of the sensor matrix.

2.4. Calibration Curve for Hue Values and Tests of the Hydrogel Sensor

All experiments were carried out under stabilized laboratory conditions at 22 °C and 35% humidity. The sensor calibration curve was obtained for a pH range of 3.5–9.0 with a

0.5 step. The calibration curve was determined based on the hue value, which is derived by the ratio of the dominant wavelength to other wavelengths in the color, indicated by its position (in degrees) on the RGB color wheel: H = [0°, 360°]. The hue value can be calculated based on the RGB components from Formula (1).

$$(A)\ H = 60° \times \left[(G' - B') / (R' - B') \right] \text{ for R} \geq \text{G} \geq \text{B} \tag{1}$$

$$(B)\ H = 60° \times \left[2 - (R' - B') / (G' - B') \right] \text{ for G} > \text{R} \geq \text{B}$$

$$(C)\ H = 60° \times \left[2 + (B' - R') / (G' - R') \right] \text{ for G} \geq \text{B} > \text{R}$$

$$(D)\ H = 60° \times \left[4 - (G' - R') / (B' - R') \right] \text{ for B} > \text{G} > \text{R}$$

$$(E)\ H = 60° \times \left[4 + (R' - G') / (B' - G') \right] \text{ for B} > \text{R} \geq \text{G}$$

$$(F)\ H = 60° \times \left[6 - (B' - G') / (R' - G') \right] \text{ for R} \geq \text{B} > \text{G}$$

where R′G′B′ value are calculated from:

$$R' = (R\ value) / 255$$
$$G' = (G\ value) / 255$$
$$B' = (B\ value) / 255$$

Next, a ready-to-use hydrogel sensor was tested with solutions of artificial saliva for medical and dental research (Pickering Labs, Mountain View, CA, USA) and its mixtures with beverages: lemon juice, black tea, green tea, bottled water, tap water, and coffee. The concentrations of the beverages examined included 100% (pure beverage) and their mixtures with artificial saliva in proportions of 50%, 20%, and 10%. Furthermore, the examination involved sensor response time (pH-based color change) as a function of time. Therefore, colorimetric data were collected every 30 s within the first 5 min after the application of the test sample, and additionally after 10 and 15 min.

3. Results and Discussion

3.1. Mechanical Properties of the Bioplotted Hydrogel Matrices

Mechanical strength tests of hydrogel structures based on sodium alginate and polyvinylpyrrolidone showed a Young's modulus (E) value of 731 ± 110 kPa, whereas the addition of the bromothymol blue pH indicator slightly increased the value to 765 ± 128 kPa (Figure 4a). Based on the change in pH of the environment, sodium alginate-based hydrogels react by shrinking or swelling, which directly affects the strength of the polymer's chemical bonds. This relationship may affect the degree of response of the pH indicator incorporated into the hydrogel structure. Compressive modulus of elasticity (E_C) values increase with pH increases, therefore E_C for pH 3.5 is 12.9 ± 2 kPa, for pH 6.5 is 18.8 ± 1 kPa, and for pH 9.0 is 24.7 ± 1.8 kPa (Figure 4b).

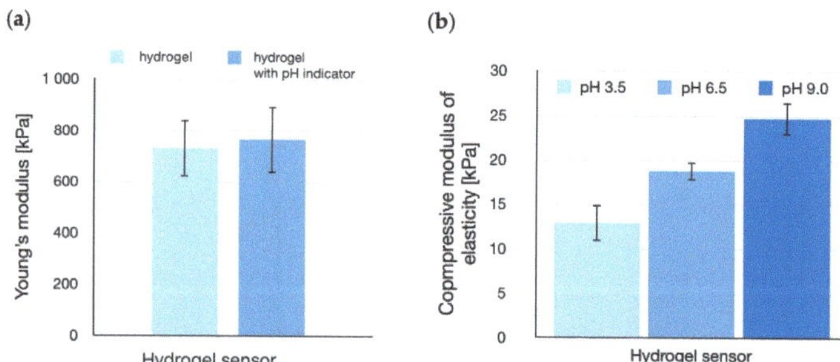

Figure 4. Tensile strength tests of the hydrogel sensor. (**a**) Young's modulus (E) of hydrogel and hydrogel with pH indicator added (N = 10), (**b**) comparison of the compressive modulus of elasticity (E_c) of hydrogel exposed to different pH (3.5, 6.5, 9.0) (N = 10).

3.2. Calibration of the Colorimetric Hydrogel Sensor and pH Salivary Diagnostics

In Figure 5a, the sensor calibration curve is presented. As shown, hydrogel pads containing bromothymol blue pH indicator provided appropriate responses to the defined pH solutions. Ranging from 3.5 to 9.0 with 0.5 step, visible color tunability of the hydrogel pads can be observed 30 s after measurement begins. Observation by the naked eye suggests the color change from light green for acidic solutions to dark green for alkaline solutions (Figure 5a), but the RGB analysis clearly indicated an increase in the sum of the green and red components (resulting in a yellow hue) in the first case and a blue component in the latter, which confirms the proper response of the bromothymol blue indicator [43]. Moreover, software-enhanced analysis of the responses of the sensors was conducted and presented in Figure 5b in the form of the hue values as a function of pH values. The tendency of the components R and G to decrease with the simultaneous increase of the B component is consistent with the decimal code tables [44], in which R and G achieved maximum values for the yellow color, and minimum for the blue one. These results confirmed the appropriate operation of the hydrogel sensor and suggested further research with artificial salivary samples.

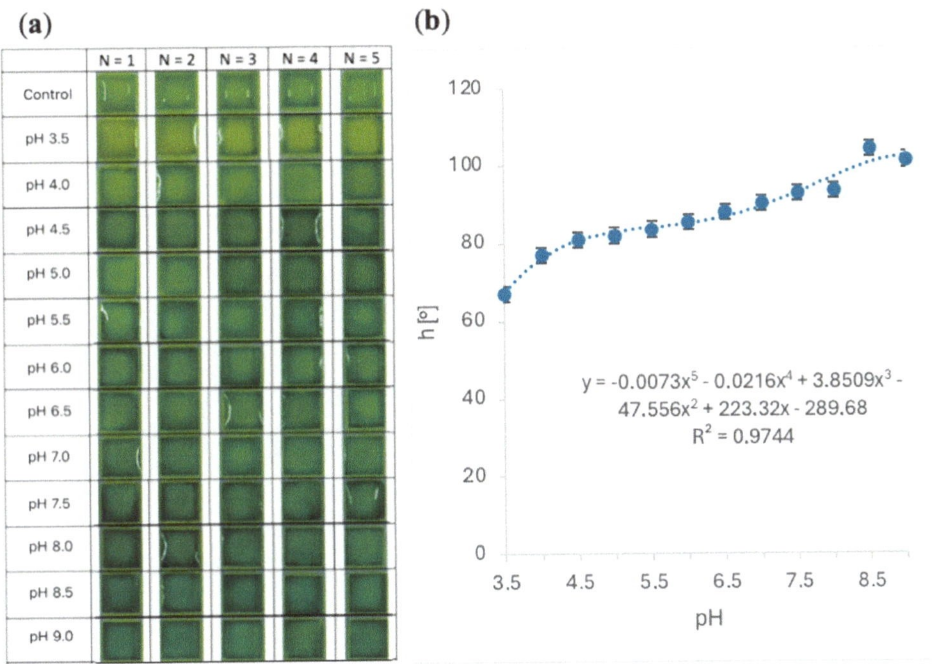

Figure 5. Colorimetry response of hydrogel sensor to pH values in the range 3.5 to 9.0. (**a**) Visible color tunability of hydrogel sensors after 30 s of various pH application, (**b**) calibration curve for hue values as a function of pH with fitted polynomial curve after 30 s of various pH application (mean values, N = 5).

The investigation shows repeatability of measurements, where this is noticeable in the triangular plot for the hydrogel sensor control group (Figure 6a). The sensor shows an increase in the values of the color components in the time function, slight for acidic solutions (pH 3.5, Figure 6b) and noticeable for neutral solutions (pH 7.0, Figure 6c) and alkaline solutions (pH 9.0, Figure 6d). An acidic reaction causes shifts toward the R component, for neutrals toward the G component, and for alkalis toward the B component.

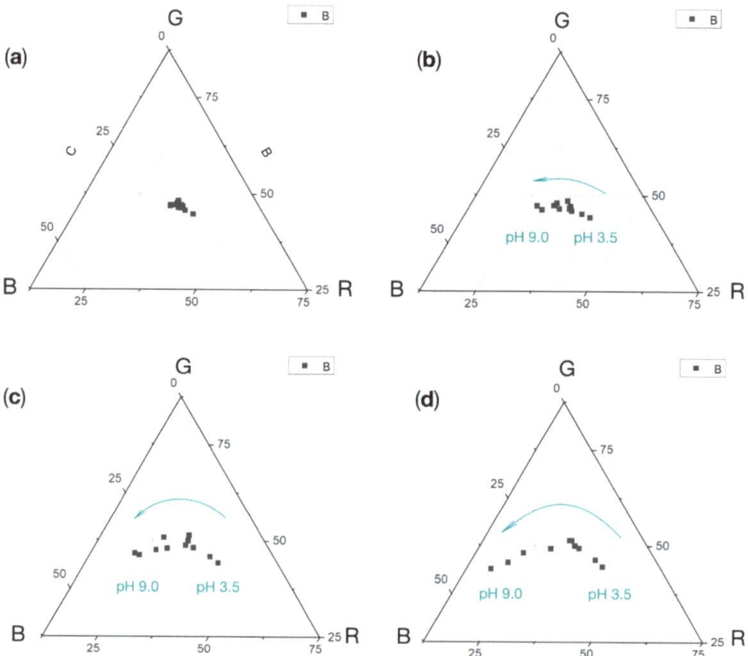

Figure 6. Triangle charts of RGB components for color changes for various pH 3.5–9.0 (mean values, N = 5) at (**a**) 0 min, (**b**) 1 min, (**c**) 5 min, (**d**) 15 min.

The stability of the sensor is shown using a radar plot (Figure 7), where similar results can be seen in the analysis of the RGB components from the beginning of the measurement to 4 min with a slight upward trend toward the G component and a slight downward trend toward R component. Measurements after 4 min show larger changes in component values, which may be the result of a chemical reaction of the indicator or evaporation of water from the hydrogel structure of the sensor.

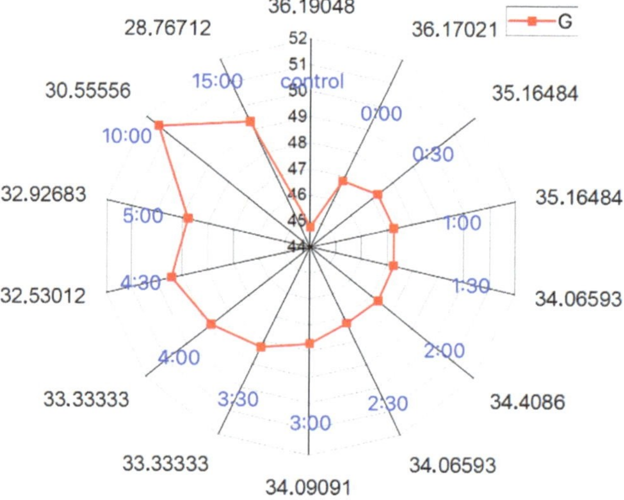

Figure 7. Radar chart of hydrogel sensor responses at 0–15 min (example for ph 6.5 solution)—stability of the hydrogel sensor (mean values, N = 5).

3.3. Measure of Salivary pH and pH of Saliva Mixed with Beverages

In this paper, the hydrogel sensor response to color changes under the influence of the pH of synthetic saliva, beverages (100%), and a mixture of saliva and beverages in ratios of 50%, 20%, and 10% were examined. Prior to the experiments, artificial saliva and saliva with the beverage mixture were measured with a pH meter. The pH values for each beverage were as follows: artificial saliva—pH 6.0, lemon juice—pH 2.3, coffee—pH 4.2, black tea—5.0, green tea—pH 5.5, tap water—7.1, bottled water—pH 7.3. The pH measurements of each beverage for hue values were plotted onto the calibration curve (Figure 8). At the same time, the pH values of the artificial saliva and the beverage mixture changed due to the pH value of the saliva. The values of the RGB components of the color-changing hydrogel under the influence of a mixture of beverages and saliva are shown in triangular charts in Figure 9 for 50% concentration.

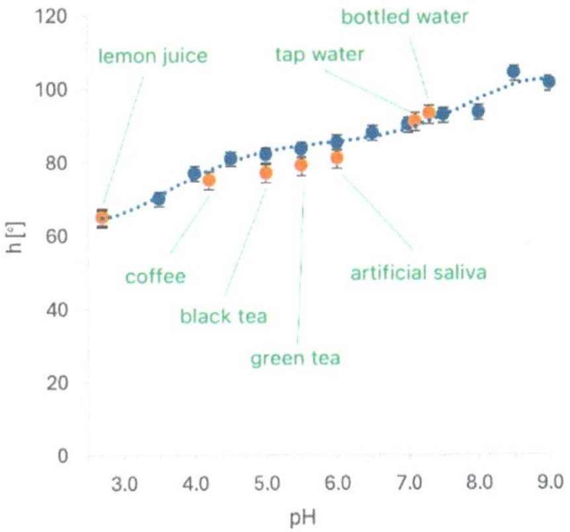

Figure 8. Hue values of beverages as a function of pH after 30 s of measurement (mean values, N = 5).

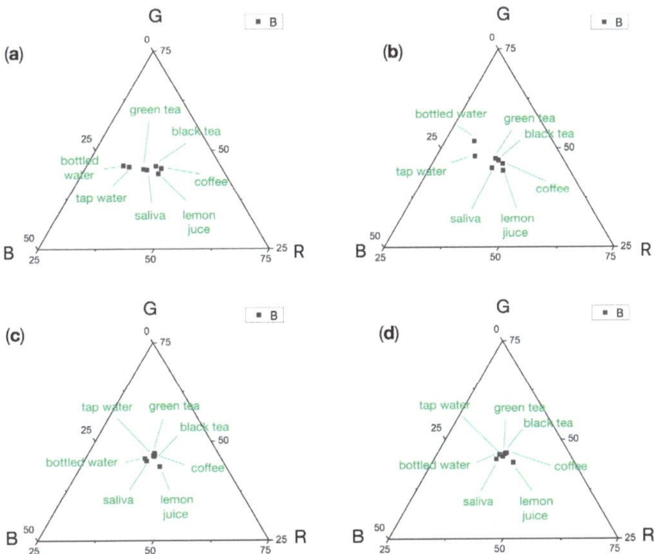

Figure 9. Triangle chart of the RGB component for hydrogel sensor responses to saliva and saliva-beverages mixtures after 15 min of application (mean values, N = 5); (**a**) beverages, 100% concentration; (**b**) beverages and saliva mixtures, 50% concentration; (**c**) beverages and saliva mixtures, 20% concentration; (**d**) beverages and saliva mixtures, 10% concentration.

Analysis of RGB components shows noticeable changes in the presence of beverages in saliva at a concentration of 50%. Minor correlations were observed for the concentration of 20% and 10% beverage content in saliva. While more acidic reactions of the substances tested tended to shift toward the R and G components (successively, mixtures with lemon juice, coffee, black tea, green tea, and pure artificial saliva), for the neutral reaction, a trend of increase toward the B component was observed (values close to pH 7 for bottled water and tap water).

4. Conclusions

The research carried out in this study showed the possibility of fabrication using additive manufacturing technologies of a colorimetric hydrogel sensor based on sodium alginate and polyvinylpyrrolidone with the addition of a pH indicator. The results showed the possibility of a fast (30 s) and reproducible response to pH range (3.5 to 9.0) with simultaneous high sensor stability. To confirm the durability of the bioplotted hydrogel sensor, the mechanical properties of the sensor were tested, which showed the strength of the thin films, reaching above 700 kPa of Young's modulus value. For the purpose of the sensor validation, dedicated software with an easy-to-use graphical user interface was developed to extract the R(ed), G(reen), and B(lue) components of the color image of the hydrogel pads and evaluate the pH value. After a sensor calibration, it was validated using samples of artificial saliva and its mixtures with beverages (lemon juice, coffee, black and green tea, bottled and tap water). As a result, a clear response to acidic and alkaline solutions of the sensor was observed, revealed in the form of substantial color change, visible also with the naked eye. However, consistent with the recent literature reports, the influence of different reactions on the hydrogel structure can lead to shrinkage of the hydrogel (in acidic conditions) or swelling of the hydrogel (in neutral and alkaline environments). This can, in turn, affect the rate of response, relax the polymer chains upon swelling, and release the pH indicator from the material structure. On that basis, further research to enhance the sensor reliability and provide the best pH indications is needed. Attention should also be paid to the rate of evaporation of water from the hydrogel structure, and methods should be developed to enable the preservation or storage of the sensor after its production to prevent the hydrogel from drying out.

Author Contributions: Conceptualization, methodology, validation, data curation, writing—review and editing: M.B.Ł., A.K. and W.K.; software, formal analysis, resources, visualization, supervision, project administration, funding acquisition: W.K.; investigation: M.B.Ł.; writing—original draft preparation, A.K. All authors have read and agreed to the published version of the manuscript.

Funding: This research was funded by the European Union's Horizon 2020 research and innovation programme under the Marie Skłodowska-Curie grant agreement No. 872370, and the Polish Ministry of Education and Science programme entitled the International Co-financed Projects (PWM, 5089/H2020/2020/2, 2019–2024).

Institutional Review Board Statement: Not applicable.

Informed Consent Statement: Not applicable.

Data Availability Statement: The raw data supporting the conclusions of this article will be made available by the authors on request.

Conflicts of Interest: The authors declare no conflicts of interest. The funders had no role in the design of the study; in the collection, analyses, or interpretation of data; in the writing of the manuscript; or in the decision to publish the results.

References

1. Kasai, R.D.; Radhika, D.; Archana, S.; Shanavaz, H.; Koutavarapu, R.; Lee, D.Y.; Shim, J. A review on hydrogels classification and recent developments in biomedical applications. *Int. J. Polym. Mater. Polym. Biomater.* **2022**, *72*, 1059–1069. [CrossRef]
2. Parhi, R. Cross-linked hydrogel for pharmaceutical applications: A review. *Adv. Pharm. Bull.* **2017**, *7*, 515–530. [CrossRef] [PubMed]

3. Sánchez-Cid, P.; Jiménez-Rosado, M.; Romero, A.; Pérez-Puyana, V. Novel trends in hydrogel development for biomedical applications: A review. *Polymers* **2022**, *14*, 3023. [CrossRef] [PubMed]
4. Ahmed, E.M. Hydrogel: Preparation, characterization, and applications: A review. *J. Adv. Res.* **2015**, *6*, 105–121. [CrossRef] [PubMed]
5. Gun'ko, V.; Savina, I.; Mikhalovsky, S. Properties of water bound in hydrogels. *Gels* **2017**, *3*, 37. [CrossRef] [PubMed]
6. Guo, C.; Wang, Y.; Song, H.; Li, W.; Kong, Q.; Wu, Y. A pain reflex-inspired hydrogel for refractory wound healing. *Mater. Des.* **2022**, *221*, 110986. [CrossRef]
7. Liang, Y.; He, J.; Guo, B. Functional Hydrogels as Wound Dressing to Enhance Wound Healing. *ACS Nano* **2021**, *15*, 12687–12722. [CrossRef]
8. Szymczyk-Ziółkowska, P.; Łabowska, M.B.; Detyna, J.; Michalak, I.; Gruber, P. A review of fabrication polymer scaffolds for biomedical applications using additive manufacturing techniques. *Biocybern. Biomed. Eng.* **2020**, *40*, 624–638. [CrossRef]
9. Akther, F.; Little, P.; Li, Z.; Nguyen, N.-T.; Ta, H.T. Hydrogels as artificial matrices for cell seeding in microfluidic devices. *RSC Adv.* **2020**, *10*, 43682–43703. [CrossRef]
10. Zhang, X.; Xu, X.; Chen, L.; Zhang, C.; Liao, L. Multi-responsive hydrogel actuator with photo-switchable color changing behaviors. *Dye Pigment.* **2020**, *174*, 108042. [CrossRef]
11. Park, J.S.; Choi, J.S.; Han, D.K. Platinum nanozyme-hydrogel composite (PtNZHG)-impregnated cascade sensing system for one-step glucose detection in serum, urine, and saliva. *Sens. Actuators B Chem.* **2022**, *359*, 131585. [CrossRef]
12. Zhao, C.; Lv, Q.; Wu, W. Application and prospects of hydrogel additive manufacturing. *Gels* **2022**, *8*, 297. [CrossRef] [PubMed]
13. Ligon, S.C.; Liska, R.; Gurr, M.; Mu, R.; Gmbh, H.B.F.D.; Bleiche, A.D.R.D.-L. Polymers for 3D Printing and Customized Additive Manufacturing. *Chem. Rev.* **2017**, *117*, 10212–10290. [CrossRef] [PubMed]
14. Zhang, R.; Larsen, N.B. Stereolithographic hydrogel printing of 3D culture chips with biofunctionalized complex 3D perfusion networks. *Lab Chip* **2017**, *17*, 4273–4282. [CrossRef] [PubMed]
15. Pan, B.; Shao, L.; Jiang, J.; Zou, S.; Kong, H.; Hou, R.; Yao, Y.; Du, J.; Jin, Y. 3D printing sacrificial templates for manufacturing hydrogel constructs with channel networks. *Mater. Des.* **2022**, *222*, 111012. [CrossRef]
16. Krakos, A.; Cieślak, A.; Hartel, E.; Łabowska, M.; Kulbacka, J.; Detyna, J. 3D bio-printed hydrogel inks promoting lung cancer cell growth in a lab-on-chip culturing platform. *Microchim. Acta* **2023**, *190*, 349. [CrossRef] [PubMed]
17. Simitian, G.; Virumbrales-Muñoz, M.; Sánchez-de-Diego, C.; Beebe, D.J.; Kosoff, D. Microfluidics in vascular biology research: A critical review for engineers, biologists, and clinicians. *Lab Chip* **2022**, *22*, 3618–3636. [CrossRef]
18. Zhao, Z.; Wang, Z.; Li, G.; Cai, Z.; Wu, J.; Wang, L.; Deng, L.; Cai, M.; Cui, W. Injectable Microfluidic Hydrogel Microspheres for Cell and Drug Delivery. *Adv. Funct. Mater.* **2021**, *31*, 2103339. [CrossRef]
19. Sun, H.; Liu, Z.; Hu, C.; Ren, K. Cell-on-hydrogel platform made of agar and alginate for rapid, low-cost, multidimensional test of antimicrobial susceptibility. *Lab Chip* **2016**, *16*, 3130–3138. [CrossRef]
20. Sun, X.; Agate, S.; Salem, K.S.; Lucia, L.; Pal, L. Hydrogel-Based Sensor Networks: Compositions, Properties, and Applications—A Review. ACS Appl. *Bio Mater.* **2021**, *4*, 140–162. [CrossRef]
21. Herrmann, A.; Haag, R.; Schedler, U. Hydrogels and Their Role in Biosensing Applications. *Adv. Healthc. Mater.* **2021**, *10*, 2100062. [CrossRef] [PubMed]
22. Buenger, D.; Topuz, F.; Groll, J. Hydrogels in sensing applications. *Prog. Polym. Sci.* **2012**, *37*, 1678–1719. [CrossRef]
23. Erfkamp, J.; Guenther, M.; Gerlach, G. Enzyme-Functionalized Piezoresistive Hydrogel Biosensors for the Detection of Urea. *Sensors* **2019**, *19*, 2858. [CrossRef] [PubMed]
24. Mir, M.; Ali, M.N.; Zahoor, A.; Smith, P.J. Comparison of conductometric behavior of two hydrogel based sensing polymeric composites in the physiological pH range. *Biomed. Phys. Eng. Express* **2018**, *4*, 65017. [CrossRef]
25. Luo, S.; Wang, R.; Wang, L.; Qu, H.; Zheng, L. Breath alcohol sensor based on hydrogel-gated graphene field-effect transistor. *Biosens. Bioelectron.* **2022**, *210*, 114319. [CrossRef] [PubMed]
26. Arunkumar, S.; Arunkumar, J.S.; Burde, K.N.; Shakunthala, G.K. Developments in diagnostic applications of saliva in oral and systemic diseases—A comprehensive review. *J. Sci. Innov. Res.* **2014**, *3*, 372–387. [CrossRef]
27. Lakshmi, K.R.; Nelakurthi, H.; Kumar, A.S.; Rudraraju, A. Oral fluid-based biosensors: A novel method for rapid and noninvasive diagnosis. *Indian J. Dent. Sci.* **2017**, *9*, 60–66. [CrossRef]
28. Eaton, K.A. The platform for better oral health in Europe-report of a new initiative. *Community Dent. Health* **2012**, *29*, 131–133. [CrossRef] [PubMed]
29. WHO. A guide to aid the selection of diagnostic tests. *Bull. World Health Organ.* **2017**, *95*, 639–645. [CrossRef]
30. Tavakoli, J.; Tang, Y. Hydrogel Based Sensors for Biomedical Applications: An Updated Review. *Polymers* **2017**, *9*, 364. [CrossRef]
31. Wen, G.Y.; Zhou, X.L.; Tian, X.Y.; Xie, R.; Ju, X.J.; Liu, Z.; Chu, L.Y. Smart hydrogels with wide visible color tunability. *NPG Asia Mater.* **2022**, *14*, 29. [CrossRef]
32. Tang, Q.; Hu, J.; Li, S.; Lin, S.; Tu, Y.; Gui, X. Colorimetric hydrogel indicators based on polyvinyl alcohol/sodium alginate for visual food spoilage monitoring. *Food Sci. Technol.* **2022**, *57*, 6867–6880. [CrossRef]
33. Lee, T.; Kim, C.; Kim, J.; Seong, J.B.; Lee, Y.; Roh, S.; Cheong, D.Y.; Lee, W.; Park, J.; Hong, Y.; et al. Colorimetric Nanoparticle-Embedded Hydrogels for a Biosensing Platform. *Nanomaterials* **2022**, *12*, 1150. [CrossRef] [PubMed]
34. Wu, M.; Zhang, Y.; Liu, Q.; Huang, H.; Wang, X.; Shi, Z.; Li, Y.; Liu, S.; Xue, L.; Ley, Y. A smart hydrogel system for visual detection of glucose. *Biosens. Bioelectron.* **2019**, *142*, 111547. [CrossRef] [PubMed]

35. Erfkamp, J.; Guenther, M.; Gerlach, G. Hydrogel-based piezoresistive sensor for the detection of ethanol. *J. Sens. Sens. Syst.* **2018**, *7*, 219–226. [CrossRef]
36. Peng, H.-Y.; Wang, W.; Gao, F.H.; Lin, S.; Ju, X.J.; Xie, R.; Liu, Z.; Faraj, Y.; Chu, L.-Y. Smart Hydrogel Gratings for Sensitive, Facile, and Rapid Detection of Ethanol Concentration. *Ind. Eng. Chem. Res.* **2019**, *58*, 17833–17841. [CrossRef]
37. Ghorbanizamani, F.; Moulahoum, H.; Celik, E.G.; Timur, S. Ionic liquid-hydrogel hybrid material for enhanced electron transfer and sensitivity towards electrochemical detection of methamphetamine. *J. Mol. Liq.* **2022**, *361*, 119627. [CrossRef]
38. Meng, L.; Meng, P.; Zhang, Q.; Wang, Y. Fast screening of ketamine in biological samples based on molecularly imprinted photonic hydrogels. *Anal. Chim. Acta* **2013**, *771*, 86–94. [CrossRef] [PubMed]
39. Bisht, G.; Zaidi, M.G.H.; Biplab, K.C. In vivo Acute Cytotoxicity Study of Poly(2-amino ethyl methacrylate-co-methylene bis-acrylamide) Magnetic Composite Synthesized in Supercritical CO_2. *Macromol. Res.* **2018**, *26*, 581–591. [CrossRef]
40. Podwin, A.; Dziuban, J. Modular 3D printed lab-on-a-chip bio-reactor for the biochemical energy cascade of microorganisms. *J. Micromech. Microeng.* **2017**, *27*, 104004. [CrossRef]
41. Łabowska, M.B.; Krakos, A.; Kubicki, W. 3D bioprinted hydrogel sensor towards rapid salivary diagnostics based on pH colorimetric detection. *Proceedings* **2023**, *97*, 160. [CrossRef]
42. Walczak, R.; Kubicki, W.; Dziuban, J. Low cost fluorescence detection using a CCD array and image processing for on-chip gel electrophoresis. *Sens. Actuators B* **2017**, *240*, 46–54. [CrossRef]
43. Arafat, M.T.; Mahmud, M.D.; Wong, S.Y.; Li, X. PVA/PAA based electrospun nanofibers with pH-responsive color change using bromothymol blue and on-demand ciprofloxacin release properties. *J. Drug Deliv. Sci. Technol.* **2021**, *61*, 102297. [CrossRef]
44. Karolin, M.; Meyyapan, T. RGB Based Secret Sharing Scheme in Color Visual Cryptography. *Int. J. Adv. Res. Comput. Commun. Eng.* **2015**, *4*, 151–155.

Article

A Low-Cost Sensing Solution for SHM, Exploiting a Dedicated Approach for Signal Recognition

Bruno Andò [1], Danilo Greco [1,*], Giacomo Navarra [2] and Francesco Lo Iacono [2]

[1] Department of Electrical Electronic and Computer Engineering (DIEEI), University of Catania, 95123 Catania, Italy; bruno.ando@unict.it

[2] Department of Engineering and Architecture, Kore University of Enna, 94100 Enna, Italy; giacomo.navarra@unikore.it (G.N.); francesco.loiacono@unikore.it (F.L.I.)

[*] Correspondence: danilo.greco@phd.unict.it

Abstract: Health assessment and preventive maintenance of structures are mandatory to predict injuries and to schedule required interventions, especially in seismic areas. Structural health monitoring aims to provide a robust and effective approach to obtaining valuable information on structural conditions of buildings and civil infrastructures, in conjunction with methodologies for the identification and, sometimes, localization of potential risks. In this paper a low-cost solution for structural health monitoring is proposed, exploiting a customized embedded system for the acquisition and storing of measurement signals. Experimental surveys for the assessment of the sensing node have also been performed. The obtained results confirmed the expected performances, especially in terms of resolution in acceleration and tilt measurement, which are 0.55 mg and 0.020°, respectively. Moreover, we used a dedicated algorithm for the classification of recorded signals in the following three classes: noise floor (being mainly related to intrinsic noise of the sensing system), exogenous sources (not correlated to the dynamic behavior of the structure), and structural responses (the response of the structure to external stimuli, such as seismic events, artificially forced and/or environmental solicitations). The latter is of main interest for the investigation of structures' health, while other signals need to be recognized and filtered out. The algorithm, which has been tested against real data, demonstrates relevant features in performing the above-mentioned classification task.

Keywords: structural health monitoring; sensing system; embedded architecture; time–frequency analysis; signals classification

Citation: Andò, B.; Greco, D.; Navarra, G.; Lo Iacono, F. A Low-Cost Sensing Solution for SHM, Exploiting a Dedicated Approach for Signal Recognition. *Sensors* **2024**, *24*, 4023. https://doi.org/10.3390/s24124023

Academic Editor: Filippo Ubertini

Received: 15 May 2024
Revised: 13 June 2024
Accepted: 17 June 2024
Published: 20 June 2024

1. Introduction

The continuous monitoring of buildings and structures is strategic, particularly with the aim of health assessment and preventive maintenance planning, especially in seismic areas [1]. Actually, the possibility of identifying potential risks associated with aging, as well natural hazards and environmental factors, is essential for disaster prevention [2].

Structural health monitoring (SHM) aims to provide a robust and effective approach to obtaining valuable information on structural conditions and supporting methodologies for the identification and, under specific conditions, localization of potential damages. Comprehensive reviews on SHM are available in [3–6], while their specific use in bridge monitoring is addressed in [7]. SHM can be implemented at either a global or local level [1]. Global monitoring is very convenient as it allows for assessing the overall behavior of the entire structure by measuring dynamic responses through the use of sensors (e.g., accelerometers or tiltmeters). On a different scale, the analysis of local cracks or fatigues requires other techniques like nondestructive testing (NDT).

SHM can be approached through two main strategies: discrete monitoring and continuous monitoring of health-related quantities. Discrete monitoring involves highly accurate measurements, typically sparse in time and performed only at specific locations. This approach lacks both time-continuous and spatially dense monitoring due to its requirement for accurate and costly instrumentation [8,9]. Conversely, the use of low-cost sensing nodes, often based on micro-electromechanical sensors (MEMS), can be conveniently adopted for the implementation of solutions allowing for the continuous monitoring of structural health. The development of a multi-sensor node for applications in the field of structural monitoring is addressed in [10], along with its assessment by real seismic signals.

Although these systems will show lower accuracy with respect to high-cost instrumentation, they offer the possibility of realizing sensor networks for time-continuous and spatially dense monitoring. This approach is valuable for the implementation of early warning systems (EWSs), allowing for the timely detection of anomalous structural behaviors [11,12]. It clearly emerges that EWSs are hence strategic in implementing monitoring tasks in many contexts, such as schools, hospitals, public or private buildings, and civil infrastructures (e.g., bridges). The commonly adopted methodology to fix the number and the location of sensing nodes often relies on the knowledge of reliable structural models, which could allow for optimizing the sensor network and identifying the most relevant monitoring locations [1]. Also, introducing redundancy in the number and location of monitoring nodes allows for improving the system's information reliability. Optimizing data transmission strategies is also strategic for wireless sensor networks in general and, particularly, for EWSs involving the use of dense network architectures [13].

It is important to bear in mind that data provided by monitoring systems must be appropriately processed to extract relevant features, which are crucial for feeding dedicated algorithms aimed at assessing structural health and identifying the nature and potential locations of damage [14]. To such an aim, different approaches can be adopted. As an example, in [15], signal decomposition implemented through discrete wavelet transform (DWT) is proposed as a suitable approach to filtering out exogenous sources coming from the environment while highlighting main seismic sources. The development of a low-cost multi-sensor node, compliant with the early warning system for SHM and aimed to measure inertial vibrations and tilt, is reported in [16]. Also, this work presents a dedicated signal processing to assess the sensing node performance in the time and frequency domains. In [17], a continuous wavelet transform (CWT) operator has been introduced with the aim of analyzing the time–frequency content of signals provided by the sensing node investigated in [16]. In particular, cross-correlation and coherence operators have been used to validate the system performances against reference instrumentation.

This paper represents a follow-up with respect to the work presented in [15–17]. In particular, the aim of this work is twofold: a deep assessment of the multi-sensor node performance [16] and the development of a dedicated methodology to discriminate among different kinds of signals: (i) noise floor (NF), being mainly related to the intrinsic noise of the sensing system; (ii) exogenous sources (ES), not correlated to the dynamic behavior of the structure; and (iii) structural responses (SR), i.e., the response of the structure to external stimuli, such as seismic events and artificially forced and/or environmental solicitations. The latter is of main interest for the investigation of structures' health, while other signals need to be recognized and filtered out.

Main novelties introduced by this paper concern the following:

- More detailed device assessment is provided by highlighting its behavior under different kinds of input stimulations.
- The use of dedicated key performance indexes aims to compare the behavior of the sensor node with respect to the ground truth. In particular, in addition to the traditional root-mean-square error, a time-correlation index and a Wavelet-correlation index have been introduced;
- The dedicated methodology for the classification of different signals is provided by the sensing system, using a threshold approach supported by the receiving operating curve (ROC) theory.

The main outcomes of the proposed approach are related to the following:

- There is a possibility to adopt an embedded sensing node for the implementation of continuous SHM; actually, the low-cost feature of this MEMS-based device, compared with highly accurate instrumentation usually adopted to perform discrete monitoring, enables the development of distributed EWSs.
- The use of CWT-based operators, with respect to discrete wavelet transform, shows a more accurate frequency sampling, thus providing a more reliable analysis of transients in the output signal.
- The methodology is implemented to separate exogenous dynamics from signals of interest (i.e., structural response), which are potentially useful to feed paradigms aimed at extracting health-related quantities and, consequently, tracking short- and long-term behaviors of the monitored structure. It must be considered that the last task is outside of the purpose of this work.

The activity presented through this work has been developed in the framework of the "HCH LowCost GeoEngineering Check" project, which aims to develop suitable methodologies for the continuous monitoring of buildings. The strategy adopted is based on the use of sensors and advanced tools for data processing, exploiting both standard instrumentation and low-cost sensing nodes, such as the one investigated in the following.

A brief description of the low-cost multi-sensor node architecture is reported in Section 2, along with the approach proposed for the classification of different classes of signals. Section 3 reports results related to the assessment of the sensing architecture and the validation of the classification algorithm. To achieve the former aim, the node is prodded by periodical stimuli with different frequencies. Moreover, tests using signals recorded during real events are also performed. During these experiments, reference instruments are used for the sake of comparison along with dedicated metrics. The classification algorithm, which has been tested against real data, demonstrates relevant features in performing the above-mentioned classification task.

2. Materials and Methods

2.1. Brief Description of the Multi-Sensor Node

Although the developed sensing platform has already been presented in [15,16], in the following, a brief description of its architecture and main functionalities is provided. Considering the addressed application and the need for a low-cost solution, the main specifications required for the tilt and acceleration measurements are as follows:

- Acceleration: range ± 1.5 g, resolution 0.005 g;
- Tilt: range $\pm 10°$, resolution $0.02°$;
- Frequency range: $0.5 \div 20$ Hz.

As schematized in Figure 1, the sensing node consists of the following main parts [16,17]:

- The triaxial MEMS inclinometer SCL3300-D01-PCB by Murata, with working range set to $\pm 10°$ full-scale and the tilt resolution of $0.0055°$/LSB, with a $0.001 °/\sqrt{Hz}$ noise density;
- The MEMS triaxial accelerometer SCA3300-D01-PCB by Murata, set to an operating range of ± 1.5 g and using a 70 Hz 1st-order low-pass filter;
- The embedded module RT1062 Teensy 4.0 series, based on a ARM® Cortex®-M7 MPU at 600 MHz;
- The DS3234 real-time clock;
- The GPS MODULE—COPERNICUS II DIP;
- A micro-SD card for data storing;
- The HC-05 Bluetooth Bee Master Slave 2 in1 module;
- A lithium EEMB 3.7 V 2000 mAh 103,454 battery.

A real view of the device installed in a real environment is shown in Figure 2 [16]. The adopted sampling rate is 200 Hz, which is compliant with the addressed frequency range. The node can be connected to a PC through a USB interface and/or a Bluetooth protocol. Also, a graphical user interface (GUI) has been realized [16], allowing for a complete management of the node, including data transmission and visualization. Actually, a full set of commands are implemented in order to control and set the node operation. In particular, the implemented working modes allow for continuous or event-triggered data storing/transmission. In the latter case, "activation thresholds" are settable by the GUI.

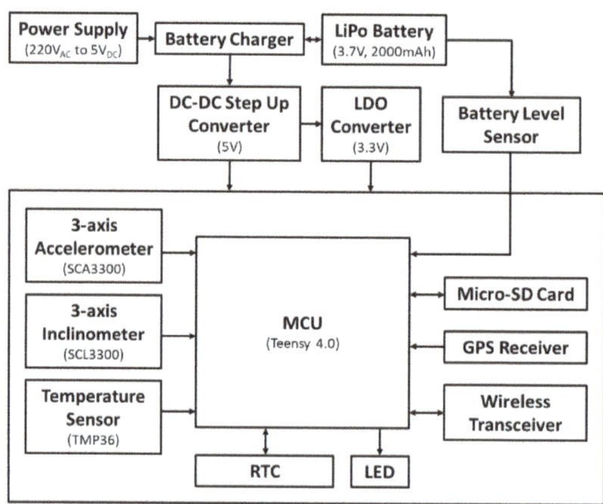

Figure 1. Schematization of the multi-sensor node [17].

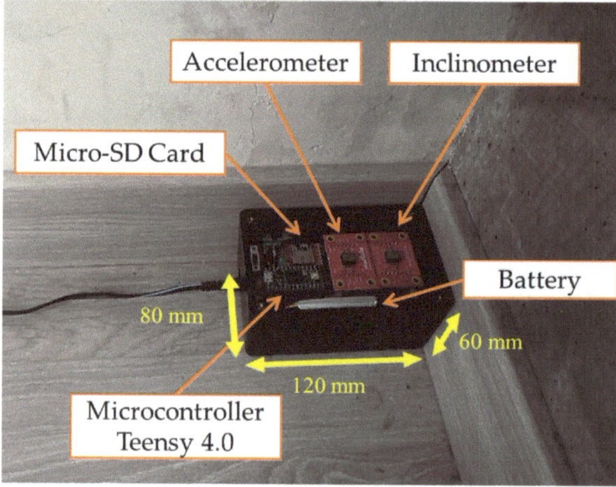

Figure 2. The sensor node [16] installed in a real environment.

2.2. The Classification Algorithm

The main aim of the developed algorithm is the detection and classification of different classes of signals: noise floor (NF), exogenous sources (ES), and structural responses (SR). In the following, the proposed classification approach is described, while the results obtained by its experimental assessment are given in Section 3.2.

The first task accomplished toward the development of the classification methodology has been the collection of a dedicated dataset, including the following kinds of signals:

- NF recorded by the sensor node.
- ES acquired by exposing the sensor node to environment-related solicitation, such as human walking, home appliances, and pulse-like dynamics.
- SR whose strength must be compliant with the target sensitivity of the developed low-cost node.
- Since during the experimental survey, it was not possible to observe SR generated by natural events showing strength compliant with the sensor node specifications, signals recorded by a reference accelerometer positioned close to the sensor platform

were used. Such signals were selected by considering the occurrence of low-strength seismic events. Dedicated post-processing was then implemented, with the aim of adapting the strength of acquired signals to values compliant with the characteristics of the sensor node.

Typical examples of the acceleration trends for the case of the above-mentioned signals are reported in Figure 3, along with their frequency contents. In particular, the CWT operator has been used to investigate the time–frequency characteristics of considered signals.

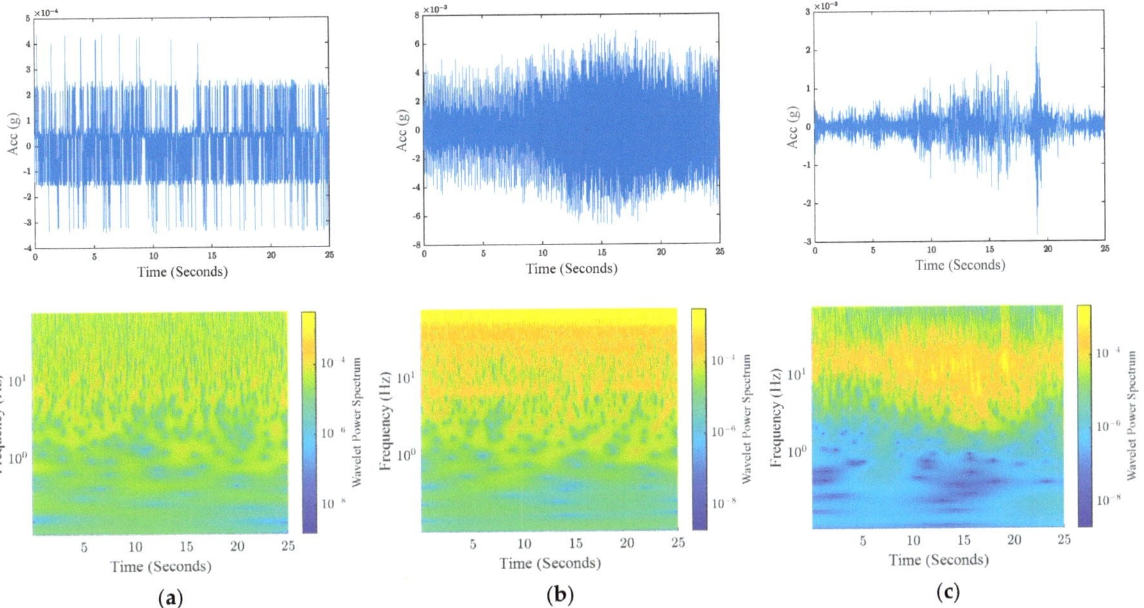

Figure 3. Time evolution of typical signals and their time–frequency representation: (**a**) NF, (**b**) ES, (**c**) SR.

Table 1 reports the number of patterns for each class of the collected dataset. Each pattern consists of a time window of 10 s of the acceleration module. The dataset is divided into two parts: 100 patterns for each class to be used as the test dataset, while the remaining patterns (setting dataset) are used for the identification of the classification model.

Table 1. Number of patterns for each class.

	NF	ES	SR
Number of patterns	500	447	432

To develop a feature-based classification methodology, the root-means-square (RMS) value of the acceleration module, calculated across the 10 s time window, is used. This can be considered as a key feature conveying a strategic piece of information related to the signal strength in the considered time interval. Figure 4 shows the distribution of RMS values for the whole set of considered patterns. As can be observed by such distribution and the detail embedded in Figure 4, it is not possible to find a separation threshold allowing for a simple clustering of different kinds of signals. To better substantiate the above statement, the receiving operating characteristic (ROC) theory is used [18].

To such aim, the following sensitivity (*Se*) and specificity (*Sp*) quantities are used [18]:

$$Se = \frac{TP}{TP + FN} \tag{1}$$

$$Sp = \frac{TN}{TN + FP} \qquad (2)$$

where standard definitions of true positive (*TP*), true negative (*TN*), false positive (*FP*), and false negative (*FN*) are used [18].

Figure 5 shows trends of *Se* and *Sp* as a function of the considered threshold (*Th*), for the two cases discriminating NF from other sources and SR from ES.

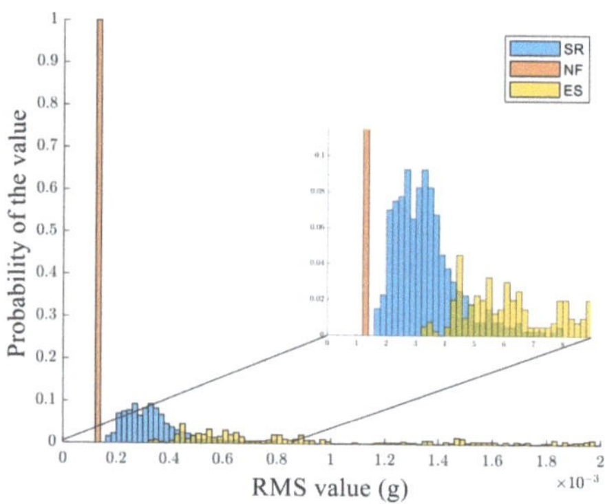

Figure 4. Distribution of RMS values for the whole set of considered patterns. The detailed view aims to show the superposition of patterns belonging to different classes.

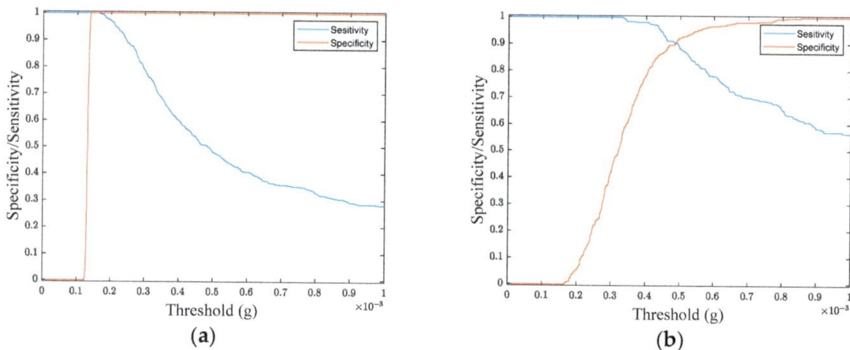

Figure 5. Sensitivity (*Se*) and specificity (*Sp*) values as a function of the considered threshold for the two cases discriminating (**a**) NF from other sources, (**b**) SR from ES.

Optimal thresholds (Th_{op}), allowing for the implementation of separation tasks, are estimated by minimizing the distance of *Se* and *Sp* from the condition of complete separation (*Se* = 1 and *Sp* = 1), through the following expression [18]:

$$\left(1 - Se\left(Th_{op}\right)\right)^2 + \left(1 - Sp\left(Th_{op}\right)\right)^2 = min\left(\left(1 - Se(Th)\right)^2 + \left(1 - Sp(Th)\right)^2\right) \qquad (3)$$

In case condition (3) is achieved for different threshold values belonging to an interval, the mean value calculated across such interval is used as the optimal threshold.

As can be observed by the achieved results (Figure 5b), the classification task is worth improving. Further outcomes of this analysis are reported and discussed in Section 3.2.

On the basis of the above results, in order to extract other relevant features to effectively implement the classification task, a deeper investigation of the dataset has been performed. As evidenced by Figure 3, the three classes of signals show different characteristics in the frequency domain. In particular, NF shows a wide band energy spectrum, while ES and SR present opposite behaviors. The first one has most of its energy located at higher frequencies, whereas the latter is bounded at lower frequencies. The last statement is also supported by the literature, in which the typical frequency range of structural responses to natural/forced solicitation is mainly below 20 Hz, especially for tall buildings [19–21].

On the basis of the above findings, the whole dataset has been processed by a low-pass (LP) and a high-pass (HP) filter, both with a cutoff frequency of 20 Hz. RMS values of filtered data, calculated across the 10 s time window, namely, RMS_{LP} and RMS_{HP}, represent the new features to be used for implementing the classification task.

As can be observed from Figure 6, the distribution of LP-filtered data demonstrates the possibility for clustering the NF class, while the same applies to the ES class in the case of HP data.

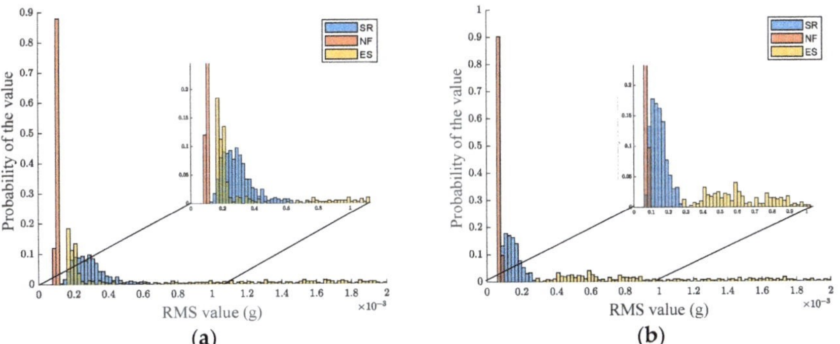

Figure 6. Distribution of RMS values for the whole set of considered patterns. The detailed view aims to show an overlap of patterns belonging to different classes. (**a**) Low-pass filtered signals; (**b**) high-pass filtered signals.

As a first step, optimal thresholds, which allow for implementing the following separation tasks, have been estimated:

- Tr_{LP}: to separate NF from ES and SR, by LP-filtered data;
- Tr_{HP}: to separate ES from NF and SR, by HP-filtered data.

Figure 7 shows *Se* and *Sp* values for the above tasks as a function of separation thresholds, Tr_{LP} and Tr_{HP}. The following optimal thresholds have been estimated: $Tr_{LP} = 1.38 \times 10^{-4}$ g, $Tr_{HP} = 2.69 \times 10^{-4}$ g.

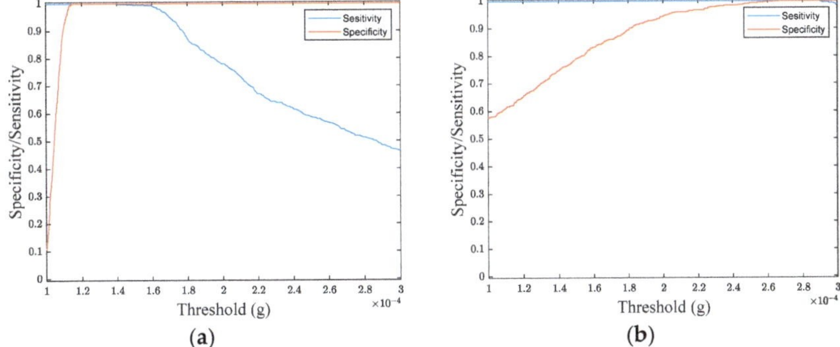

Figure 7. Sensitivity (*Se*) and specificity (*Sp*) values as a function of the considered threshold for the following tasks: (**a**) separation of NF from other sources by LP data and (**b**) separation of ES from other sources by HP data.

Considering all the above statements, the following classification rules have been identified:

$$Class = \begin{cases} NF: & RMS_{LP} \leq Tr_{LP} \\ SR: & \begin{cases} RMS_{LP} > Tr_{LP} \\ RMS_{HP} < Tr_{HP} \end{cases} \\ ES: & RMS_{HP} \geq Tr_{HP} \end{cases} \tag{4}$$

The above rules are used to implement the algorithm shown in Figure 8, aimed at the real-time classification of data acquired by the sensing platform. Although compliant with its deployment in the adopted microcontroller platform, for the sake of convenience, at this stage, the classification algorithm is implemented in MATLAB R2022b.

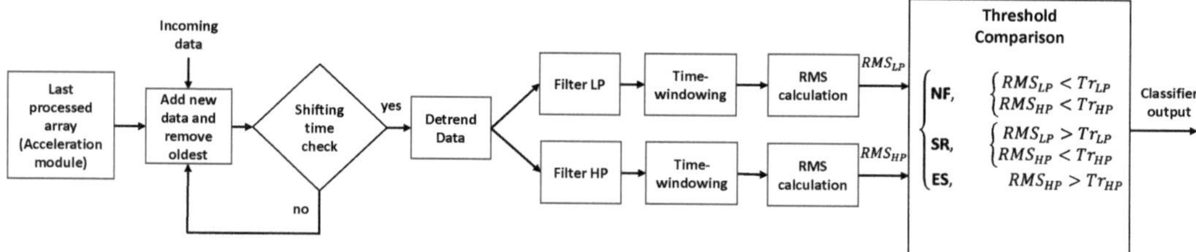

Figure 8. The real-time classification algorithm.

The algorithm operates in time windows of 10 s, extracting the classification features from the time evolution of the acceleration module. Since the algorithm is expected to work continuously on real-time data sequences, a storing array is used, which is continuously updated with new data. As soon as a time increment of 1 s is achieved (condition checked by "shifting time check" routine), a new pattern containing the last 10 s of the time sequence is released and processed in order to extract features and estimate its class of belonging.

Summarizing, each pattern is processed as follows:

- A zero-order de-trend is accomplished, thus removing the effect of signal offset;
- The de-trended signal is then processed by LP and HP filters;
- RMS_{LP} and RMS_{HP} values are calculated;
- The set of rules (4) is applied to estimate the class of belonging for the considered pattern.

Results obtained by the above-presented classification algorithm are discussed in Section 3.2.

3. Results and Discussion

3.1. Assessment of the Low-Cost Sensing Node

In this section, results of experimental surveys performed to validate the developed sensing node are reported. For the sake of convenience, it must be considered that the device resolution has already been investigated in [16], leading to the results shown in Table 2.

Table 2. Results of the resolution analysis for acceleration and tilt measurements [16].

	X	Y	Z
Acceleration (mg)	0.30	0.30	0.55
Tilt (°)	0.016	0.016	0.020

In order to test the system response to acceleration stimuli, the experimental setup shown in Figure 9a is used, equipped with the APS 129 HF ELECTRO-SEIS® long-stroke vibration exciter, available at the Experimental Dynamics Laboratory of the L.E.D.A. Research Institute at the University of Enna "Kore", Enna, Italy [22]. The sensing platform is fixed to the moving platform of the shaker. The system, after a settling time, performs 10 reliable periods at the desired frequency and amplitude. The reference value is provided by a QA-700 accelerometer by Honeywell, Charlotte, NC, USA, whose main characteristics are as follows:

- Operating range: ±30 g;
- Bias: <8 mg;
- One-year composite repeatability: <1200 µg;
- Temperature sensitivity: <70 µg/°C;
- Intrinsic noise: <7 µg rms (0–10 Hz), 70 µg rms (10–500 Hz).

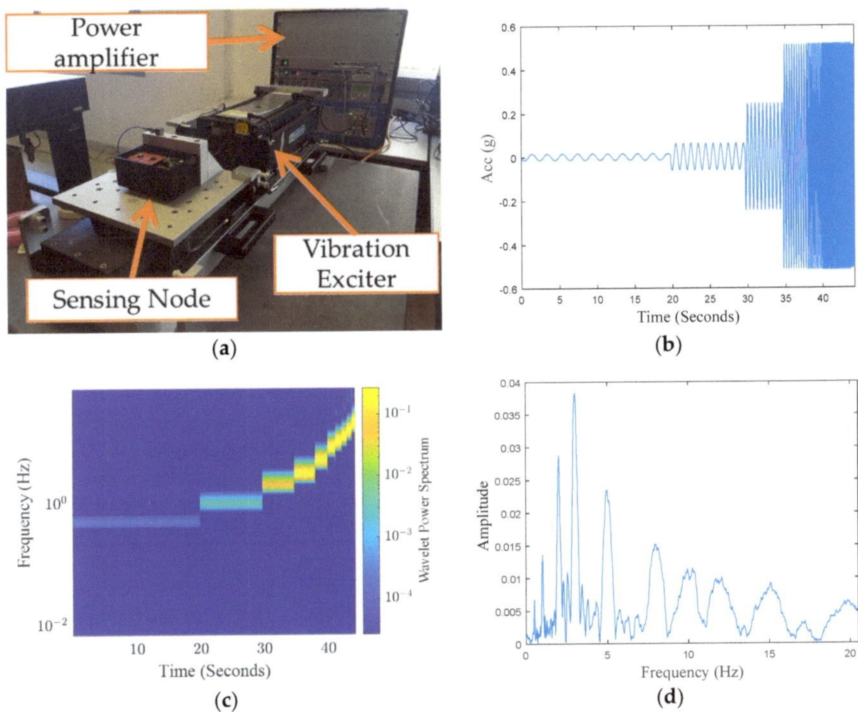

Figure 9. Vibration exciter test. (**a**) Setup adopted during the test along the *X*-axis. (**b**) Time series of the concatenated time window of 10 periods for each frequency [17]. (**c**) Wavelet analysis of the corresponding time window [17]; (**d**) Discrete Fourier transform of the concatenated signal.

Characteristics of stimulation signals are reported in Table 3 in terms of their nominal frequency and amplitude values. Figure 9b shows the concatenation of 10 periods of the signal recorded by the sensing node for each considered frequency [17]. Results obtained by the wavelet analysis, shown in Figure 9c [17], as well as the discrete Fourier transforms shown in Figure 9d, confirm the compliance of the system response to the applied stimulus.

Table 3. Nominal value of frequency, f, and the amplitude, A, of applied signals.

f (Hz)	0.5	1.0	2.0	3.0	5.0	8.0	10.0	12.0	15.0	20.0
A (g)	0.015	0.060	0.241	0.510	0.510	0.510	0.509	0.510	0.510	0.510

In order to quantify such performances, the following index has been defined:

$$\delta_V = 100 * \frac{1}{n} * \sum_{i=1}^{n} \left| \frac{V_1 - V_2}{V_1} \right| \tag{5}$$

where V states for the amplitude, A, or the frequency, f, of the applied stimulus.

Moreover, the repeatability, assessing the system performances in the acceleration domain, is estimated as 3 times the standard deviation of peak values distribution. The results obtained for above defined indexes, under different operating conditions, are shown in Figure 10. The δ_A, for the three axes, is limited to under 1% in most of the investigated frequencies, with the exception of a few cases where its value reaches 5%. The δ_f values show that the frequency discrepancy is extremely low (less than 0.03%) for each solicitation frequency and for each axis. The repeatability has an upward trend with the frequency, which rises after 10 Hz, being, however, confined below 4.5% of the nominal values.

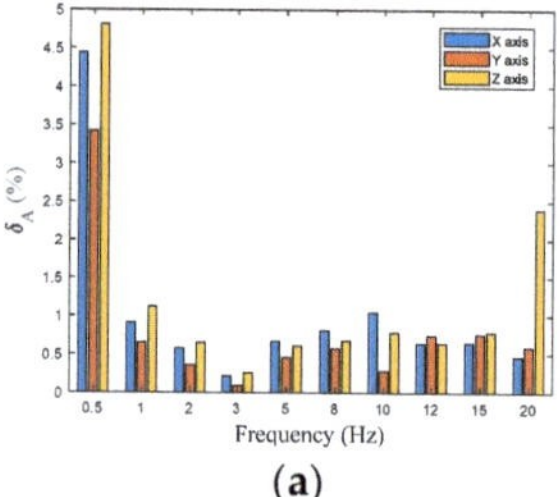
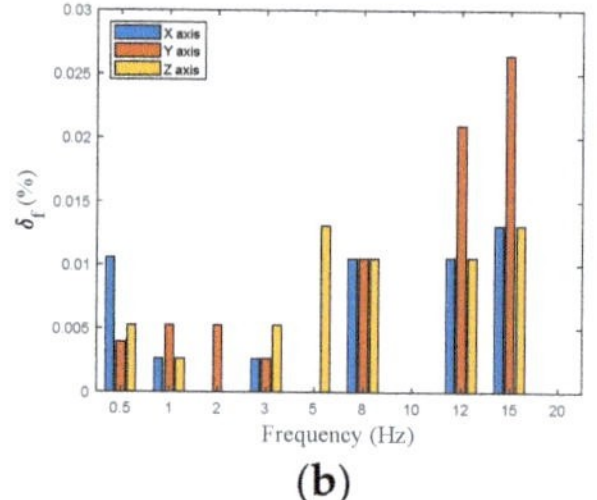
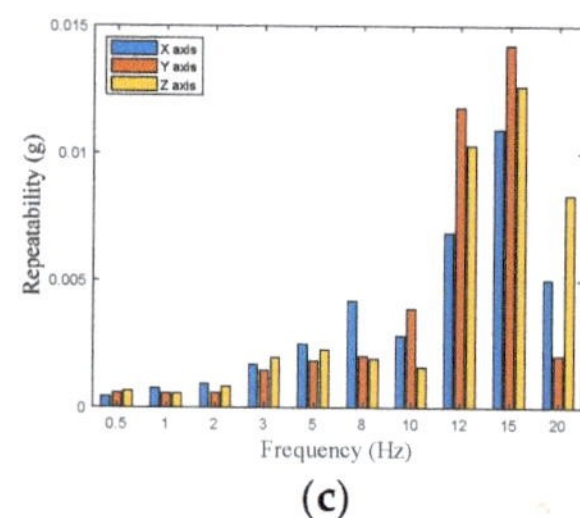

(a)   (b)   (c)

Figure 10. Vibration exciter test. Bar chart of indexes value under different operating conditions for each axis: (**a**) δ_A, (**b**) δ_f, (**c**) repeatability.

The next set of experiments is dedicated to testing the system response to dynamics imposed by using a controlled moving platform. The sensing node is hooked at the center of the vibrating platform, as shown in Figure 11. The vibrating platform has 6 degrees of freedom, with ± 1.5 g horizontal acceleration range, ± 1 g vertical acceleration range, and an RT3-S real-time digital control system, exploiting a 2 kHz control loop in position, velocity, and acceleration. A detailed description of the shaking table facility is reported in [23,24]. For the sake of comparison and validation, a dedicated reference system has been used, which consists of eight MEMS DC accelerometers (model 3711B1110G by PCB Piezotronics, Depew, NY, USA) placed in the actuator of the vibrating platform (two along the *X*-axis, two along the *Y*-axis, and four along the *Z*-axis), whose main specifications are as follows:

- Range ± 10 g;
- Frequency range 0–1.0 kHz;
- Nonlinearity $\leq 1\%$;
- Transverse sensitivity $\leq 3\%$.

Figure 11. Setup employed during the tests with the vibrating platform that reports the axis orientation of the sensing system.

In particular, the response of the sensing node to the following signals is observed:

- Frequency sweep test: an acceleration stimulus, ranging from 0.5 Hz to 20 Hz, applied along the X, Y, and Z-axes;
- Tilt test: a periodic tilt, ranging from $-2.5°$ to $2.5°$, applied for two different frequency values;
- Seismic test: a typical seismic signal.

The results obtained for the frequency sweep test are reported in Figure 12, which shows the output from both the reference system and the sensing platform. Figure 13 shows the wavelet power spectrum of the above signals [25]. In particular, to achieve this goal, the synchrosqueezing wavelet is used to narrow the frequency distribution in each time instant. The time–frequency analysis clearly demonstrates the coherence with the adopted frequency sweep stimulus.

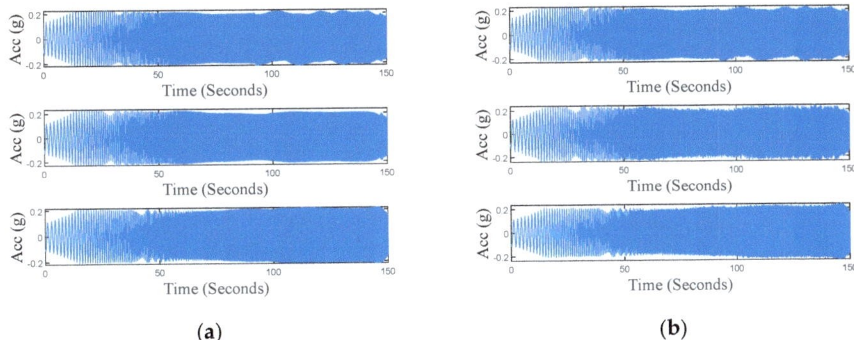

Figure 12. Time series along the X-axis (**top**), Y-axis (**center**), and Z-axis (**bottom**), in the frequency sweep test, recorded by (**a**) reference instrumentation; (**b**) the sensing platform.

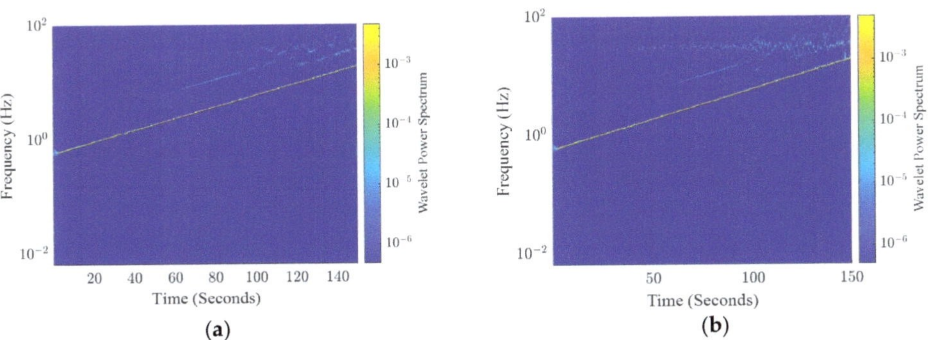

Figure 13. Wavelet analysis for the frequency sweep test. (**a**) Wavelet power spectrum of the reference instrumentation signals; (**b**) wavelet power spectrum of the sensing platform signals.

The following indexes are used in order to assess the performance of the sensing platform:

$$\zeta = 100 * \sqrt{\frac{\sum_{i=1}^{n}\left(V_1 - V_2\right)^2}{\sum_{i=1}^{n}\left(V_1\right)^2}} \tag{6}$$

$$R_{V_1,V_2}(l) = \begin{cases} \frac{\sum_{i=0}^{n-|l|-1} V_1(i)*V_2(i+l)}{n}, & l < 0 \\ \frac{\sum_{i=0}^{n-|l|-1} V_1(i)*V_2(i+l)}{n}, & l \geq 0 \end{cases} \tag{7}$$

$$R_{norm} = \frac{R_{1,2}(0)}{\sqrt{R_{V_1,V_1}(0)R_{V_2,V_2}(0)}} \tag{8}$$

where

- ζ estimates the percentage error with respect to the nominal values on the whole signal;
- V_1, V_2 are signals recorded by the reference system and the sensing node, respectively;
- n is the number of samples;
- $R_{V_1,V_2}(l)$ is the cross-correlation between the two signals, V_1 and V_2, as a function of the lag (l);
- R_{norm} is the normalized cross-correlation between the reference and the acquired signals, defined in $l = 0$.

Moreover, the following index, C_{norm}, representing the normalized cross-correlation in the time–frequency domain, is used to assess the similarity of the time–frequency content of the sensor output against the reference signal:

$$
C_{norm} = \frac{\Sigma_{xy}\left(\left(C_1(x,y) - \overline{C_1}\right)\left(C_2(x,y) - \overline{\overline{C_2}}\right)\right)}{\sqrt{\Sigma_{x,y}\left(C_1(x,y) - \overline{C_1}\right)^2 \Sigma_{x,y}\left(C_2(x,y) - \overline{C_2}\right)^2}}
\tag{9}
$$

where

- C_1 and C_2 are two matrixes containing the wavelet coefficients;
- $\overline{C_1}$ and $\overline{C_2}$ are mean values of elements belonging to the above matrixes.

The results obtained for performance indexes (6), (8), and (9) are reported in Table 4. High values of R_norm (close to 1.0) for the three axes prove the capability of the sensing platform to follow the dynamics of the solicitation in the whole investigated frequency range. Values of C_norm are over 0.9, which demonstrates the coherence in the frequency domain of the two signals. The ζ values provide a quantification of the difference in magnitude between the two signals, which, on average, is about 8.5%. As can be observed, the Z and Y axes show higher ζ values with respect to the X-axis, most probably due to the presence of noise superimposed to the output signal along these directions. Further investigations will be dedicated to identifying possible strategies aimed at reducing the effect of such influencing quantities on the system performance.

Table 4. Performance indexes estimated for the frequency sweep test.

	X-Axis	*Y*-Axis	*Z*-Axis
ζ	2.76%	9.99%	12.86%
R_{norm}	0.99	0.99	0.99
C_{norm}	0.97	0.92	0.92

Concerning the tilt test, the vibrating platform generates a series of oscillations at fixed frequencies (0.2 Hz and 0.5 Hz) in the range of $\pm 2.5°$. The test is performed through the X and Y axes (as defined in Figure 11) of the sensor node, which is intended to measure a quasi-static tilt. Figure 14 shows the time series provided by both the reference system and the sensing platform. Performance indexes (6), (8), and (9) calculated for this test are reported in Table 5. As can be observed, estimated values for ζ are lower than 4%, and values of R_{norm} and C_{norm} are close to 1 for both considered axes. Such results demonstrate the capability of the sensor node to follow the imposed solicitation.

During the last test on the accelerometer, the vibrating platform was used to reproduce the ground motion recorded during a real earthquake. Particularly, the tree components of the Kobe earthquake, recorded in Takatori, Japan, on 16 January 1995, were used. This event was characterized by high horizontal and vertical ground motion ($M_w = 6.9$). The test aimed to validate the sensing platform response to this kind of realistic complex solicitation (rich magnitude and frequency content). Figure 15 shows output signals of both the reference system and the sensing platform, for the three axes. From the wavelet analysis, shown in Figure 16, it is possible to observe that the energy of the wavelet in the X and Y axes is mainly located in the low-frequency range, while the response along the Z-axis shows relevant energy content for higher frequency. Table 6 reports performance indexes (6), (8), and (9) calculated for this test. Values of R_{norm} are close to 1.0 for the three axes, thus validating the capability of the node to measure input with rich frequency content. Values obtained for the index ζ, along the X and Y axes are in line with the previous test, resulting in values below than 6%. Values of C_{norm} for the X and Y axes confirm a suitable coherence in the time–frequency domain between the sensing node and the reference system, while highlighting lower coherence for the Z-axis. The worst performances observed for the Z-axis are most probably due to the presence of high-frequency noise along this direction.

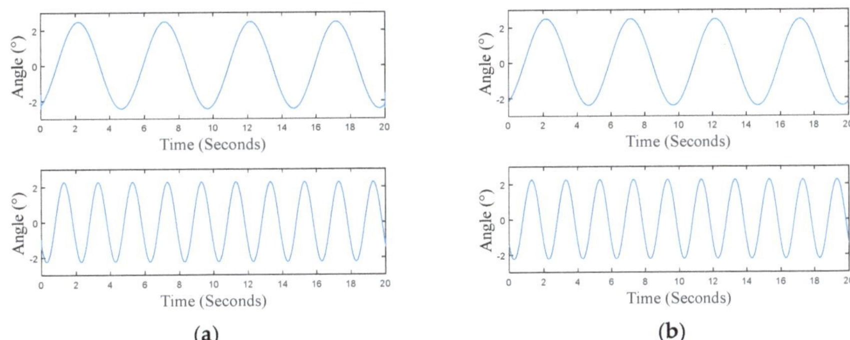

Figure 14. Time series recorded during the tilt test, in the case of the node rotation applied around the Y-axis. **Top**: 0.2 Hz; **bottom**: 0.5 Hz. (**a**) Reference signals; (**b**) sensing platform signals.

Table 5. Performance indexes estimated for the tilt test.

	0.2 Hz		0.5 Hz	
	Y-Axis	**X-Axis**	**Y-Axis**	**X-Axis**
ζ	1.66%	1.14%	2.15%	3.85%
R_{norm}	0.99	0.99	0.99	0.99
C_{norm}	0.99	0.99	0.99	0.94

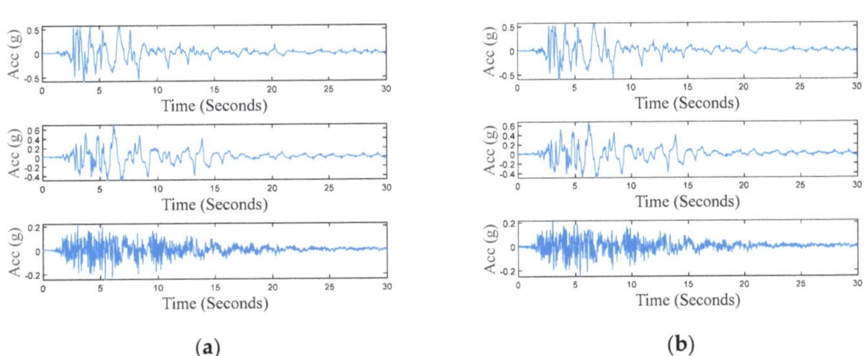

Figure 15. Time series of the seismic test along the X-axis (**top**), Y-axis (**center**), and Z-axis (**bottom**), recorded by (**a**) reference instrumentation; (**b**) the sensing platform.

Table 6. Performance indexes estimated for the seismic test.

	X-Axis	**Y-Axis**	**Z-Axis**
ζ	3.77%	5.05%	19.45%
R_{norm}	0.99	0.99	0.98
C_{norm}	0.98	0.97	0.84

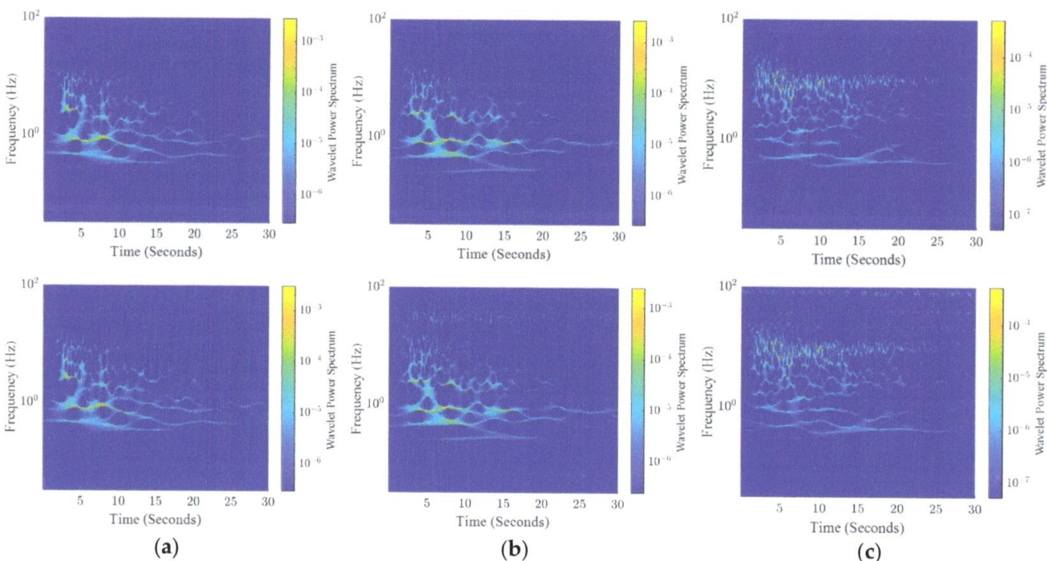

Figure 16. Wavelet power spectrum of the seismic test recorded by reference instrumentation (**top**) and the sensing platform (**bottom**), along: (**a**) *X*-axis, (**b**) *Y*-axis; (**c**) *Z*-axis.

3.2. Outcomes of the Classification Algorithm

Results obtained by applying the algorithm presented in Section 2.2 are shown in Figure 17 where blue symbols represent the expected class for each pattern and the red symbols the output of the classification procedure. For the sake of completeness, Figure 17a reports results obtained by the classification algorithm exploiting RMS values of the raw acceleration module and thresholds defined in Figure 5 where misclassifications of SR and ES patterns are clearly observable. Figure 17b shows results obtained by the algorithm, shown in Figure 8 and exploiting rules (4), which allows for overcoming the misclassification issue. To support this finding, Table 7 shows the confusion matrix obtained for the classification algorithm shown in Figure 8 for the setting and test datasets. As can be observed, the obtained results confirm the suitability of the classification methodology.

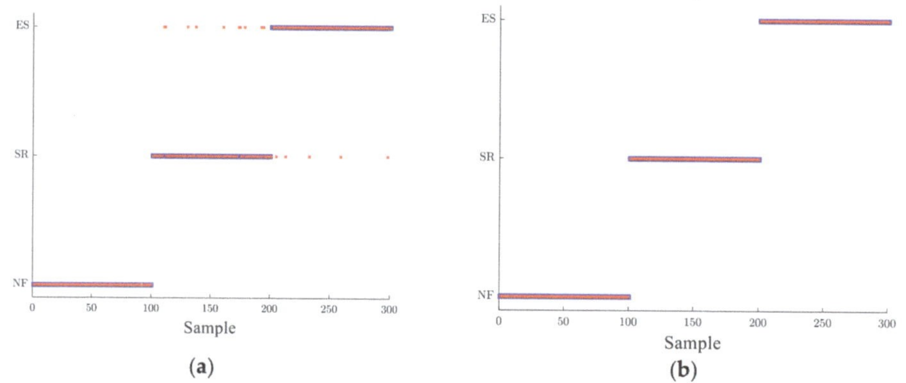

Figure 17. Validation test: expected data (blue symbols) and estimated data (red symbols). (**a**) Algorithm based on RMS values of the raw acceleration module; (**b**) algorithm shown in Figure 8 and exploiting rules (4).

Table 7. Confusion Matrix obtained for the classification algorithm shown in Figure 8 for the setting and test datasets.

			Classification Output		
			NF	SR	ES
Expected Output	**Setting dataset**	NF	400	0	0
		SR	0	400	0
		ES	0	0	400
	Test dataset	NF	100	0	0
		SR	0	100	0
		ES	0	0	100

4. Conclusions

In this paper, the development of a low-cost sensing platform for applications in the field of SHM is presented, along with a signal processing approach allowing for the analysis and classification of different kinds of dynamics recorded by the sensor node. The latter has shown suitable performances in both acceleration and tilt measurements.

In particular, tests performed by forcing periodic acceleration along the three axes, at 10 different frequencies in the operating range (0–20 Hz), revealed amplitude discrepancies of less than 1% in most of the frequencies tested, with the exception of a few cases where its value reached 5%.

Performing a frequency sweep from 0.5 to 20.0 Hz, the value of the correlation between the signal provided by the reference instrumentation and the signal recorded by the node was very close to 1, while the difference in magnitude between the two signals was, on average, about 8.5%.

The tilt test performed in the range of $\pm 2.5°$ has shown a discrepancy lower than 4%. Finally, the test exploiting the reproduction of a real earthquake has demonstrated suitable performance of the sensing platform when solicited by realistic time series.

A dedicated approach for the classification of different kinds of signals recorded by the sensing node has been introduced and assessed. In particular, the algorithm developed allows us to separate three classes of signals, namely, noise floor, structural response, and exogenous signals. The algorithm has been tested by using a dedicated dataset, and the results obtained allow for affirming the excellent performance of the methodology developed.

Future activities will be dedicated to implementing the deployment of the preprocessing and classification algorithm in the embedded architecture of the sensor node. Moreover, a large experimental survey will be performed by recording long-lasting time series, thus extending the available dataset. If required, the procedure adopted to set classification thresholds will be iteratively applied in order to adapt the set of estimated rules to a wide set of real observations.

Moreover, efforts will be also dedicated to identifying possible strategies aimed at reducing the effect of influencing quantities on the system performances.

Author Contributions: Conceptualization, methodology and development B.A.; formal analysis and validation D.G.; investigation D.G., G.N. and F.L.I. All authors have read and agreed to the published version of the manuscript.

Funding: This work is supported by Action: 1.1.5 POR FESR 2014-2020, project name "HCH LowCost GeoEngineering Check", Project code: G89J18000710007.

Data Availability Statement: The data presented in this study are available: https://doi.org/10.608 4/m9.figshare.26065126.

Conflicts of Interest: The authors declare no conflicts of interest.

References

1. Structural Health Monitoring: An Overview. In *Structural Health Monitoring with Application to Offshore Structures*; World Scientific: Singapore, 2019.
2. Fabbrocino, G.; Rainieri, C. Some remarks on the seismic safety management of existing health facilities. In Proceedings of the 15th World Conference on Earthquake Engineering (WCEE), Lisbon, Portugal, 24–28 September 2012.
3. Scuro, C.; Fusaro, P.A. Structural Health Monitoring Systems: An Overview. In Proceedings of the 2022 IEEE International Workshop on Metrology for Living Environment (MetroLivEn 2022), Cosenza, Italy, 25–27 May 2022; pp. 232–236. [CrossRef]
4. Chen, H.P.; Ni, Y.Q. Introduction to Structural Health Monitoring. In *Structural Health Monitoring of Large Civil Engineering Structures*; John Wiley & Sons Ltd.: Hoboken, NJ, USA, 2018.
5. López-Castro, B.; Haro-Baez, A.G.; Arcos-Aviles, D.; Barreno-Riera, M.; Landázuri-Avilés, B. A Systematic Review of Structural Health Monitoring Systems to Strengthen Post-Earthquake Assessment Procedures. *Sensors* **2022**, *22*, 9206. [CrossRef] [PubMed]
6. Warsi, Z.H.; Irshad, S.M.; Khan, F.; Shahbaz, M.A.; Junaid, M.; Amin, S.U. Sensors for Structural Health Monitoring: A Review. In Proceedings of the 2019 Second International Conference on Latest trends in Electrical Engineering and Computing Technologies (INTELLECT), Karachi, Pakistan, 13–14 November 2019; pp. 1–6.
7. Saidin, S.S.; Jamadin, A.; Abdul Kudus, S.; Mohd Amin, N.; Anuar, M.A. An Overview: The Application of Vibration-Based Techniques in Bridge Structural Health Monitoring. *Int. J. Concr. Struct. Mater.* **2022**, *16*, 69. [CrossRef]
8. Brownjohn, J.M.W. Structural health monitoring of civil infrastructure. *Philos. Trans. R. Soc. A* **2007**, *365*, 589–622. [CrossRef] [PubMed]
9. Feng, L.; Zhou, B.; Sun, X.; Liang, Z.; Xie, H. A study on a cRIO-based monitoring and safety early warning system for large bridge structures. In Proceedings of the International Conference on Measuring Technology and Mechatronics Automation, Changsha, China, 13–14 March 2010; pp. 361–364.
10. Andò, B.; Baglio, S.; Pistorio, A. A distributed monitoring systems for structural early warning. In Proceedings of the 2015 2015 IEEE International Workshop on Measurements & Networking, Coimbra, Portugal, 12–13 October 2015; pp. 1–5. [CrossRef]
11. Wang, J.; Fu, Y.; Yang, X. An integrated system for building structural health monitoring and early warning based on an Internet of things approach. *Int. J. Distrib. Sens. Netw.* **2017**, *13*, 1550147716689101. [CrossRef]
12. Waidyanatha, N. Towards a typology of integrated functional early warning systems. *Int. J. Crit. Infrastruct.* **2010**, *6*, 31–51. [CrossRef]
13. Wang, D.; Liao, W. Wireless transmission for health monitoring of large structures. *IEEE Trans. Instrum. Meas.* **2006**, *55*, 972–981. [CrossRef]
14. Malekloo, A.; Ozer, E.; AlHamaydeh, M.; Girolami, M. Machine learning and structural health monitoring overview with emerging technology and high-dimensional data source highlights. *Struct. Health Monit.* **2022**, *21*, 1906–1955. [CrossRef]
15. Andò, B.; Baglio, S.; Pistorio, A. A low cost multi-sensor system for investigating the structural response of buildings. *Ann. Geophics* **2018**, *61*, 1–14. [CrossRef]
16. Andò, B.; Castorina, S.; Graziani, S.; Greco, D.; Manenti, M.; Pistorio, A. An Embedded Multi-Sensor Architecture for Applications. In Proceedings of the Structural Health Monitoring, 2023 IEEE Sensors Applications Symposium (SAS), Ottawa, ON, Canada, 18–20 July 2023; pp. 1–5. [CrossRef]
17. Andò, B.; Greco, D.; Navarra, G. A Low-Cost Solution and Continuous Wavelet Transform Analysis for Structural Health Monitoring. *Proceedings* **2024**, *97*, 38. [CrossRef]
18. Nahm, F.S. Receiver operating characteristic curve: Overview and practical use for clinicians. *Korean J. Anesthesiol.* **2022**, *75*, 25–36. [CrossRef] [PubMed]
19. Moisidi, M.; Vallianatos, F.; Gallipoli, M.R. Assessing the main frequencies of modern and historical buildings using ambient noise recordings: Case studies in the historical cities of Crete (Greece). *Heritage* **2018**, *1*, 12. [CrossRef]
20. Brownjohn, J.M.; De Stefano, A.; Xu, Y.L.; Wenzel, H.; Aktan, A.E. Vibration-based monitoring of civil infrastructure: Challenges and successes. *J. Civ. Struct. Health Monit.* **2011**, *1*, 79–95. [CrossRef]
21. Valašková, V.; Papán, D. Theoretical frequency analysis of the historical building. *Procedia Eng.* **2015**, *111*, 821–827. [CrossRef]
22. Costanza, A.; Fertitta, G.; D'Anna, G.; Yang, W.; Iacono, F.L.; Navarra, G.; Patanè, D. A study for the selection of a calibration system for seismic sensors. *Quad. Geofis.* **2022**, *175*, 1–38. [CrossRef]
23. Iacono, F.L.; Navarra, G.; Oliva, M.; Cascone, D. Experimental investigation of the dynamic performances of the LEDA shaking tables system. In Proceedings of the AIMETA 2017—Proceedings of the 23rd Conference of the Italian Association of Theoretical and Applied Mechanics, Salerno, Italy, 4–7 September 2017; Volume 5, pp. 897–915.
24. Fossetti, M.; Iacono, F.L.; Minafò, G.; Navarra, G.; Tesoriere, G. A new large scale laboratory: The LEDA research centre (Laboratory of Earthquake engineering and Dynamic Analysis). In Proceedings of the International Conference on Advances in Experimental Structural Engineering (AESE 2017), Pavia, Italy, 6–8 September 2017; Volume 2017, pp. 699–717. [CrossRef]
25. Torrence, C.; Compo, G.P. A practical guide to wavelet analysis. *Bull. Am. Meteorol. Soc.* **1998**, *79*, 61–78. [CrossRef]

Article

Wearable Sensor Node for Safety Improvement in Workplaces: Technology Assessment in a Simulated Environment

Fabrizio Formisano [1,*]**, Michele Dellutri** [2]**, Ettore Massera** [1]**, Antonio Del Giudice** [1]**, Luigi Barretta** [3] **and Girolamo Di Francia** [1]

[1]   ENEA, 80055 Portici, Italy; ettore.massera@enea.it (E.M.); antonio.delgiudice@enea.it (A.D.G.); girolamo.difrancia@enea.it (G.D.F.)
[2]   STMicroelectronics, 95121 Catania, Italy; michele.dellutri@st.com
[3]   STMicroelectronics, 80022 Arzano, Italy; luigi.barretta@st.com
*   Correspondence: fabrizio.formisano@enea.it

Abstract: Personal protective equipment (PPE) has been universally recognized for its role in protecting workers from injuries and illnesses. Smart PPE integrates Internet of Things (IoT) technologies to enable continuous monitoring of workers and their surrounding environment, preventing undesirable events, facilitating rapid emergency response, and informing rescuers of potential hazards. This work presents a smart PPE system with a sensor node architecture designed to monitor workers and their surroundings. The sensor node is equipped with various sensors and communication capabilities, enabling the monitoring of specific gases (VOC, CO_2, CO, O_2), particulate matter (PM), temperature, humidity, positional information, audio signals, and body gestures. The system utilizes artificial intelligence algorithms to recognize patterns in worker activity that could lead to risky situations. Gas tests were conducted in a special chamber, positioning capabilities were tested indoors and outdoors, and the remaining sensors were tested in a simulated laboratory environment. This paper presents the sensor node architecture and the results of tests on target risky scenarios. The sensor node performed well in all situations, correctly signaling all cases that could lead to risky situations.

Keywords: wearable; smart personal protective equipment (PPE); safety; particulate matter (PM); gas exposure; IoT; sensor node; indoor outdoor localization

Citation: Formisano, F.; Dellutri, M.; Massera, E.; Giudice, A.D.; Barretta, L.; Di Francia, G. Wearable Sensor Node for Safety Improvement in Workplaces: Technology Assessment in a Simulated Environment. *Sensors* **2024**, *24*, 4993. https://doi.org/ 10.3390/s24154993

Academic Editor: Lorenzo Scalise

Received: 13 June 2024
Revised: 17 July 2024
Accepted: 20 July 2024
Published: 1 August 2024

1. Introduction

According to the INAIL (Italian National Institute for Occupational Accident Insurance) annual report in 2022 [1], the number of reports of accidents at work submitted to the Institute in the year 2022 was 703,432, representing an increase of 139 thousand cases compared with 2021. There were 1208 reports of fatal accidents, a decrease of 15.2% compared with 2021. The reports of pathologies of professional origin rose by 18,164 (+25.1%). The occurrence of accidents in the workplace by mode is shown in Table 1; the most accidents were the result of falling from heights.

Even one fatal accident at work is too much; this is why research and improvement in this field are important and desirable. With the recent improvements in the reliability of cyber–physical systems, new applications can be conceived to make workplaces safer than they were before. WAs workers can be continuously monitored in workplaces utilizing modern intelligent systems, it is possible to prevent and warn workers of risks. Smart PPE devices are cyber–physical systems that incorporate communication, elaboration, and sensing components that are deeply intertwined to pursue a specific task. This work presents a smart PPE which, together with a communication infrastructure and a remote server for visualization and alarm, contributes to the mitigation of risks related to the duties of workers and eventually improves the overall working conditions in workplaces. The smart PPE device oversees some quantities in the environment in which the worker is operating to recognize and prevent uncontrollable or hazardous situations. There are studies in the literature that cover the topic of the design of smart PPE [2–4]. In [2], the authors proposed a system that measures some parameters such as humidity, temperature, acceleration, and gas and exploits artificial intelligence to optimize choices in terms of safety at work. Here, sensors were incorporated in the helmet and the belt of the worker. However, the system designed in [2] does not address the problem of locating the worker in the working environment. The smart PPE presented in this work introduces an RTLS (real-time locating system) for indoor locations and a GPS for outdoor locations that lay the groundwork for an augmented reality remote assistant project covered in future works. In [3], a specific application is covered, namely the mitigation of risks in the case of workers who operate in construction sites where the main threat is represented by height. In [4], the authors presented a device that can monitor the concentration of carbon dioxide present in the air together with physiological sensors applied on the body of the worker; all this information is used to assess the worker's state and, in case of anomaly, issue an alarm. The novelty in the present work, compared to the state of the art, is that the proposed solution offers an extensive sensor array and includes locating capabilities for both indoor and outdoor environments in a compact form factor. The node is equipped with a GPS sensor and an RTLS, the combined data of which are fed back to the control center with precise information about the position of the worker. Starting from that knowledge, in case of emergency, the rescuers can be informed of the exact position of the worker. The sensor node puts together a large mix of features, retaining reduced dimensions and low consumption. The features, among the others presented in the continuation of this article, include gas monitoring (VOC, CO_2, CO, O_2), air quality monitoring by means of a PM (particulate matter) sensor, a temperature and humidity sensor; in addition, the proposed device is equipped with a gyroscope, accelerometer, RTLS, GPS, LORA communication, and BLE. This work improves the preceding versions of the sensor node [5,6]. Some important improvements were made to achieve a compact form factor and improve portability. Low-power policies were introduced to improve battery endurance. A new localization system was implemented using two different localizing technologies, GPS and RTLS. Finally, a new experimental sensor array was integrated into the new node. Section 2 describes the details of the architecture, the sensors, and the devices used. Then, in Section 3, the results are shown along with the position testing. Section 4 is dedicated to the discussion of the results. Finally, conclusions and future work are reported in Section 5.

Table 1. Fatal accidents by mode of occurrence.

Height Fall	Object Fall	Loss Control Vehicles	Contact with Moving Objects	Vehicle Start	Contact with Washing Machines	Other
32.5%	16.8%	14.6%	6.8%	6.3%	5.5%	17.5%

2. Materials and Methods

A sensor node was developed within an IoT architecture, enabling remote hazard detection in both indoor and outdoor environments. This architecture utilizes a real-time locating system (RTLS) combining ultra-wideband (UWB) and global positioning system (GPS) technologies (Figure 1). The figure depicts the entire system, including the sensor node (described in the next paragraph), an LoRaWAN™-based wireless network, and a remote-control center. The control center leverages the TagoIO cloud IoT platform (version 2.18.0) for data transfer, storage, and visualization. TagoIO was chosen due to its comprehensive suite of IoT data management features, including real-time alerts, data visualization, storage capabilities, and seamless integration with other platforms.

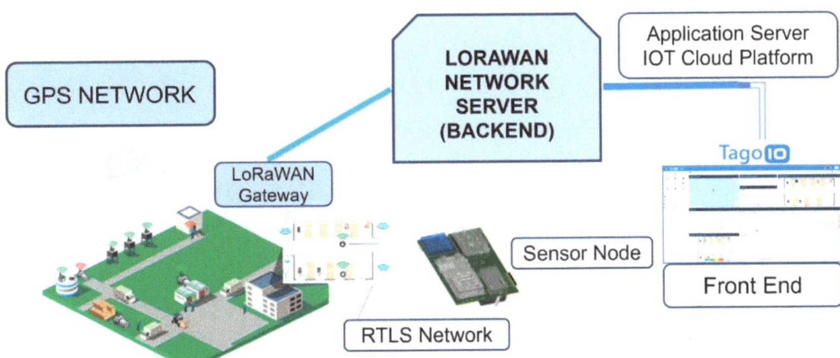

Figure 1. General architecture of the system from smart PPE to control room.

This architecture allows for remote worker status monitoring via a custom-designed visual interface within the TagoIO platform.

2.1. Signal Processing and Visualization

The node continuously monitors various sensor parameters, ensuring their compliance with predefined limits. Upon anomaly detection, the node transmits alarm or telemetry messages along with the worker's location via an LoRa network.

The communication protocol leverages the standardized Cayenne Low-Power Payload (CayenneLLP) coding scheme. This approach guarantees compliance with payload size restrictions and enables the node to transmit diverse sensor data within a single message. Each data point is preceded by two bytes: a data channel identifier and a data type identifier. Table 2 details the payload structure.

Table 2. Payload coding.

1 Byte	1 Byte	N Byte	1 Byte	1 Byte	M Byte	...
Data 1 Ch.	Data 1 Type	Data 1	Data 2 Ch.	Data 2 Type	Data 2	...

During normal operation, the packet size is minimized to optimize transmission efficiency. Alarm status is represented by a 16-bit register, where each bit flag corresponds to a specific alarm type (refer to Table 3). This register is then embedded within a CayenneLLP packet that also includes worker location information and, in telemetry mode, the actual sensor value that triggers the alarm. This minimized payload size facilitates faster transmission, reduces on-network packet collisions, and allows for future expansion of the message format to accommodate additional data.

The received raw buffer, representing the encoded message, requires decoding to reach a human-readable format. This is achieved by employing a dedicated "Payload Formatter" software module (version 1.1), designed to decode newly received messages. The system dashboard (Figure 2) displays both alarm notifications generated by the node and a clear map depicting the worker's location within the monitored environment. In the telemetry mode, the dashboard additionally displays data from various sensors alongside the standard information (alert, worker position, and timestamp).

Table 3. List of alarm coding.

Alarm Type	Register Value
Temperature alarm	xxxxxxxxxxxxxxx1
Humidity alarm	xxxxxxxxxxxxxx1x
Pressure alarm	xxxxxxxxxxxxx1xx
VOC alarm	xxxxxxxxxxxx1xxx
CO_2	xxxxxxxxxxx1xxxx
PMS alarm	xxxxxxxxxx1xxxxx
Audio alarm	xxxxxxxxx1xxxxxx
BLE broadcast alarm	xxxxxxxx1xxxxxxx
Fall alarm	xxxxxxx1xxxxxxxx
User alarm	xxxxxx1xxxxxxxxx
-Unused-	xxxxx1xxxxxxxxxx
-Unused-	xxxx1xxxxxxxxxxx
-Unused-	xxx1xxxxxxxxxxxx
POZYX timeout	xx1xxxxxxxxxxxxx
POZYX sensor failures	x1xxxxxxxxxxxxxx
PMS sensor failures	1xxxxxxxxxxxxxxx
No alarms- No sensor failures	0000000000000000

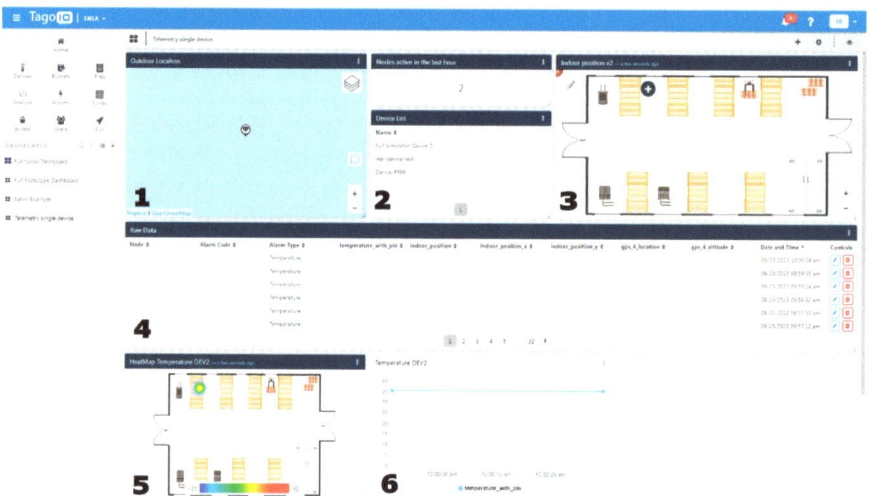

Figure 2. This picture shows the telemetry dashboard with the following info: 1. outdoor position, 2. active nodes, 3. indoor position, 4. raw sensor data, alarm type, and timestamp, 5. heat map, 6. data plot.

2.2. Sensor Node Description

The IoT node's core architecture comprises a multi-sensor node (MSN) and a communication board (CB). The MSN integrates sensors for environmental and motion monitoring, including temperature–humidity (HTS221), pressure (LPS22DF), and 6-axis inertial (LSM6DSOX) measurement. The CB incorporates the Teseo-LIV3F GNSS module for outdoor positioning and the STeval-STRKT01 LoRa connectivity module. To enhance air quality monitoring, PM (Plantower PMS 7003), VOC, CO_2, CO, and O_2 sensors are integrated. Indoor positioning is achieved through an external board in conjunction with fixed anchors. A dedicated audio board (STLCS01V1) processes noise levels, triggering alarms when they exceed safety thresholds. Instantaneous alerts are provided via buzzer, LED, and vibration motor.

Prioritizing a compact, wearable form factor, a custom two-layer PCB minimizes external cables. The bare sensor node measures approximately 12 cm × 13 cm × 5 cm.

Figure 3 illustrates the hardware architecture and the final packaged personal protective equipment (PPE), measuring 20 cm × 16 cm × 10 cm and weighing approximately 800 g with the 8000 mAh battery.

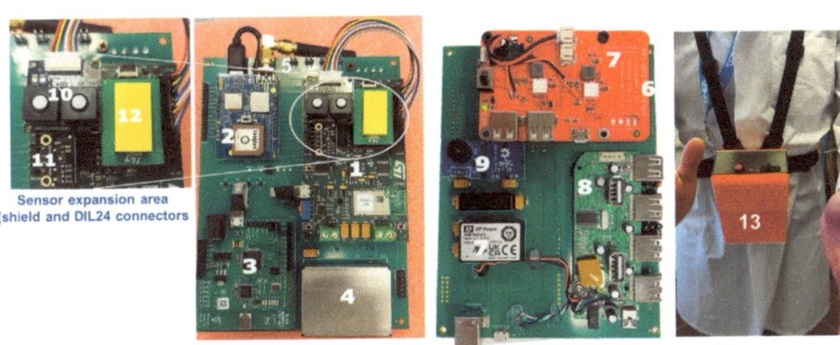

Figure 3. This picture shows hardware characteristics (block scheme and node photo); (1) MSN—multi sensor sub-node (STM customized board with sensor expansion area, shield, and DIL24 connectors); (2) CB—communication board, gps, and LoraWan tracker (STM STEVAL-STRKT01); (3) RTLS tag (POZYX tag); (4) PM—particulate matter sensor (PlantowerPMS7003); (5) PL, SL, MPS: power/status LEDs and main power switch connectors; (6) AB—smart audio board (STM STEVAL-STLCS01V1); (7) PMS—power management system (management and distribution); (8) HU—hub unit (power and data distribution unit); (9) BZ—buzzer; VM—vibration motor; (10) O_2 and CO electrochemical solid polymer sensors (EC sense ES1-O2-25%, EC sense ES1-CO-1000 ppm); (11) VOC MOX sensor (Sensirion SGP40, 0–1000 ppm eth. Eq.); (12) experimental NDIR CO_2 sensors; (13) final wearable sensor node. The PCB serves the primary function of interconnecting all the aforementioned boards and integrating the power management system (PMS). Individual power switches facilitate power-saving algorithms by enabling control over the power supply to each board. This allows for selective activation, such as engaging the real-time locating system (RTLS) solely during alarm situations. A lithium-ion battery with a capacity of 8000 mAh furnishes power to the system.

3. Results

The sensor node adopts a modular design, integrating dedicated sections for gas sensing, power management, localization, particulate matter detection, audio capture, and inertial measurement onto a single wiring motherboard. The node acquires data at predetermined sampling rates for various parameters and transmits it to a central control unit for visualization and data analysis.

To facilitate testing, the node's functionalities are divided between the communication board (CB) and the measurement sensor node (MSN). This architecture enables independent testing of communication and sensor components but necessitates a synchronized communication protocol. An asynchronous leader–follower protocol is implemented, with the CB as the leader, initiating data exchange with the MSN.

Power consumption optimization is crucial due to varying sensor demands. The RTLS and communication modules consume the most power. To extend battery life, a hardware switch deactivates the RTLS when inactive, and the communication module enters a low-power state.

Initial testing revealed noise interference from the audio sensor, which was addressed by strategically positioning sensors to minimize disturbances. The audio sensor is placed away from noise sources, gas sensors are positioned for optimal airflow, and the temperature/humidity sensor is located in a cooler area.

3.1. Gas Detection Capability

The node incorporates commercial sensors. For CO and O_2 sensing, EC sensors' solid-state electrochemical sensors were selected based on TLV-STEL limits of 100 ppm for CO and 16% v/v for O_2 [7]. The ES1-CO-1000 ppm and ES1-O2-25% models offer compact dimensions and low power consumption. For VOC sensing, the Sensirion MOX SGP40 model, with a range of 0–1000 ppm [8], was chosen. To meet the CO_2 TLV-STEL of 30,000 ppm [9], a suitable sensor was selected. The gas sensor array is shown in Figure 3.

To evaluate sensor performance, extensive laboratory tests were conducted using a gas sensing characterization system (GSCS). This system simulates hazardous environments by controlling gas concentrations within a test chamber. The chamber, equipped with temperature, humidity, and pressure sensors, allows for precise atmosphere control. An artificial atmosphere is generated by mixing dry and wet air with the target gas, controlled by software (version 1.1) and C++ libraries. The sensor node was installed in the test chamber (Figure 4) and exposed to a constant airflow of controlled atmosphere at $22 \pm 2\ °C$ and 500 sccm.

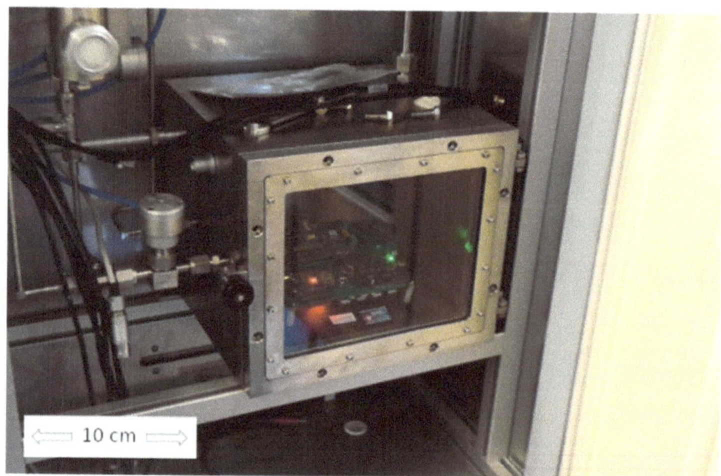

Figure 4. Sensor node installed in the gas sensing test chamber inside the GSCS equipment.

To assess gas sensor performance and verify alarm threshold accuracy, three tests were conducted, as illustrated in Figure 5. The tests involved sequentially exposing the sensor node to CO concentrations up to 250 ppm, reducing oxygen levels to 10%, and exposing it to ethanol concentrations exceeding 2000 ppm as a VOC representative. The red lines in the graphs indicate the sensor node's alarm thresholds. Custom firmware was developed to extract data from each gas sensor for analysis.

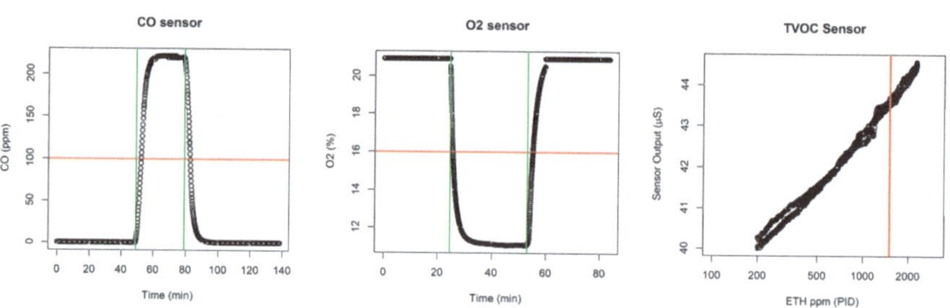

Figure 5. Sensor node gas sensors test in controlled environment: on the (**left**), see the time evolution of the CO-calibrated sensor response under an exposure of a concentration of 250 ppm of CO in synthetic air humified at 50% RH. The vertical green lines indicate the start injection and the stop injection while the horizontal red line indicates the alarm threshold; similarly, in the (**center**), see the O_2 calibrated sensor response to an oxygen decrease in the artificial atmosphere; on the (**right**), see the ethanol sensor sensitivity curve from 200 ppm to 2000 ppm correlated to a calibrated PID. The red vertical line indicates the alarm threshold.

The sensor node uses a prototype of an experimental NDIR CO_2 sensor. The NDIR gas sensors have traditionally been larger in size compared to some other types of gas sensors, due to the need of longer optical path length and to place the source and the detector face to face, increasing the sensitivity. The gas sensor described in this work followed a different approach. It has been designed to fit into a DIL24 (dual in line) adapter board in order to be integrated into the IoT node.

The developed solution includes a flat lower internal wall, made of PCB (printed circuit board), to which the detector and the emitter are fixed, and a biconical upper internal wall which is configured to reflect the IR radiation issued by the emitter. Both the emitter and the detector face the upper internal wall so that the IR radiation is reflected by the upper internal wall before reaching the detector. Thanks to this reflection, the optical path is no longer formed by a single straight stretch, as mentioned before, and its total length is greater than the mutual distance between the emitter and the detector. This allows the size and bulk of the container body to be reduced while maintaining high absorption of IR radiation.

Thanks to the 3D simulation, the shape and size of the two reflectors have been optimized to achieve maximum sensitivity without increasing the overall dimensions. The shape of the lid is a biconical surface and coupling this surface occurs with a flat reflector on top of a PCB, which permits multiple reflections. Regardless of the direction of the emissions, every ray launched from the source hits at the detector and then is absorbed.

Since the IR sensor itself is not capable of providing a radiation measurement that is frequency-selective, the first IR sensor also includes the optical filter, which allows the frequency of the IR radiation entering the sensor to be selected. In the case of CO_2 detection, where the absorption wavelength of the main absorption peak is approximately 4.3 μm, the optical filter used is a bandpass filter from 4.2 μm to 4.4 μm.

The second (reference) IR sensor differs from the first regarding the absorption range of the optical filter; in particular, at the reference wavelength, the absorption peaks of one or more gases to be detected are absent. For example, the reference wavelength can be equal to 3.89 μm; therefore, the absorption range of the optical filter of the second IR sensor can be between 3.79 μm and 3.99 μm approximately.

Reflecting the surface of the lid is achieved by standard gold plating, while the reflector on the PCB is achieved by rectangular metalization with NiAu plating.

Figure 6 shows the DIL24 with a reflective surface and the internal shape of the lid.

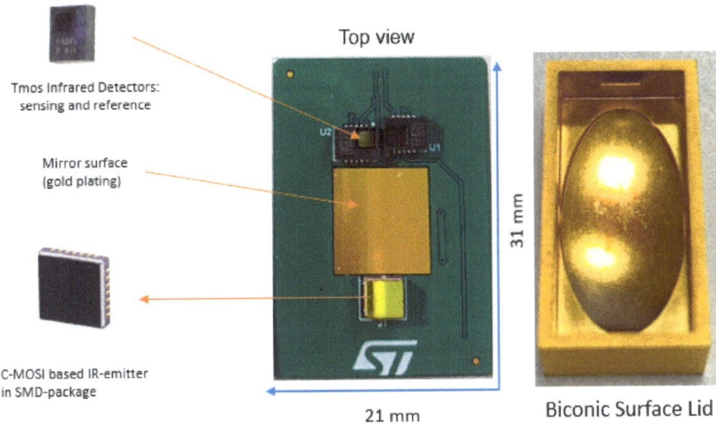

Figure 6. (**Left**): DIL24 PCB with reflecting surface; (**right**): internal shape of the lidded 3D rendering of the final wearable sensor node.

To evaluate the performance of this sensor and thus understand whether it can meet the requirements of the scenario in which the multi-sensor node is to operate, we used the same instrumentation as for the characterization of the commercial gas sensors previously seen.

The node was exposed to a concentration of 10,000 ppm CO_2 and its response in terms of resolution and noise was then evaluated.

Figure 7a shows two different calibration curves to which we subjected the sensor. As can be seen, the sensor response time was determined to be significantly faster than the chamber's estimated 1200 s response time, indicating rapid response to the target gas. This result allows us to state that the sensor responds to the analyte gas within a few seconds, and this is a first result that meets the specifications of the scenario.

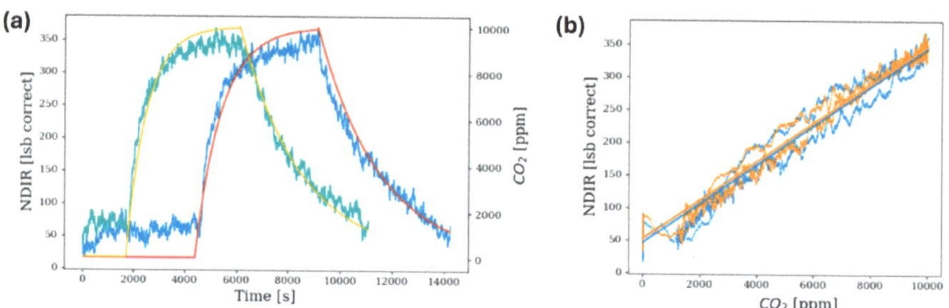

Figure 7. Calibration curves over time: (**a**) yellow and red show the concentration trends in the controlled chamber; green and blue show the relative sensor responses; (**b**) scatter plot between chamber concentration and sensor response in the two analyte injections considered, respectively, in orange and blue.

On the other hand, as can be seen from the scatter plot in Figure 7b, the sensor exhibits a strong linear relationship with CO_2 concentration within the tested range. The correlation coefficient between the values recorded in the chamber and the sensor response is close to 1. The sensitivity recorded is 30 ppm per appreciable lsb of the sensor. However, in terms of noise, we estimated the noise from the baseline; therefore, when the sensor is in a 'resting' state (not exposed to the analyte), it is around 30 lsb.

The sensor's response to common disturbances such as temperature and humidity was also evaluated. Considering the temperature variation, a drift was recorded, which was easily corrected with a firmware recalibration, while no drift was recorded when the sensor was exposed to a relative humidity variation between 20 and 70%.

Considering the data obtained, the sensor under examination is certainly capable of meeting the specifications dictated by the scenarios in which it will operate, even if noise rejection can be improved working on the hardware components.

3.2. Power Management

To improve battery life, this study investigates and implements tailored power management strategies that address the node's diverse operational scenarios.

Beyond employing low-power sensor solutions (e.g., electrochemical sensors, MEMS with onboard complex gesture recognition), component activation policies were introduced to ensure on-demand operation, minimizing unnecessary power consumption. Testing revealed that localization devices, specifically TAG RTLS, significantly impacted overall power consumption. Consequently, a key strategy involved minimizing their active mode time to strictly necessary periods for localization tasks. Similarly, communication operations were optimized by limiting data transmission to sampled values and enabling programmable transmission frequency based on application requirements, further reducing energy consumption.

To determine the optimal balance between operational endurance and battery capacity, a series of empirical trials were conducted. These trials measured the operating time from full battery charge to complete discharge following some power management policies (e.g., RTLS module continuously active vs. active on-demand, as shown in Table 4) while capturing a single alarm event. The trials were performed under the same operational mode to isolate the impact of specific policies. The final battery selection (8000 mAh) aimed to ensure a monitoring session of approximately one workday, even in the worst-case scenario without any power-saving features enabled. Implementing these features demonstrably extends operational endurance, allowing for extended operation periods in mobility. Further energy savings can be achieved by leveraging the sleep and deep sleep modes of specific node modules. Due to the independent operating states of these modules (e.g., GPS tracking and LoRaWAN™ communication devices), utilizing their native low-power and ultra-low-power states can significantly reduce current draw. For instance, switching from active operation to a low-power state can decrease current consumption from 60 mA to 200 uA, representing a substantial 36% energy saving compared to operation without power-saving policies. This highlights the significant potential for further optimization through strategic use of these built-in functionalities.

Table 4. Battery comparison (RTLS module continuously active vs. active on-demand).

RTLS	Battery Capacity (mAh)	Endurance (h)	Battery Weight (g)
Continuously ON	8000	>7	121
Continuously ON	3800	<4	58
Continuously ON	6600	<6	97
ON-DEMAND	8000	>9	121
ON-DEMAND	3800	>5	58
ON-DEMAND	6600	>7	97

3.3. Localization

Localization is achieved through a dual technology system combining RTLS UWB for indoor and GPS for outdoor positioning. This approach enhances localization precision, enabling sub-meter accuracy even in complex outdoor environments.

The RTLS UWB localization system is based on a commercial tag (POZYX developer tag [10]) incorporated in the node. The whole system, supporting the tag in the node, is a commercial product (Pozyx developer kit), which works through the dialogue of a master tag connected to a PC (both constituting the localization engine) and a constellation of anchors which are used as references for location measurements. Figure 8 illustrates the POZYX developer kit block diagram (source: POZYX product sheet).

Figure 8. RTLS block scheme.

In order to assess the precision and reliability of the localization system, a series of experiments were conducted within a controlled indoor environment. In the outdoor case the locating capabilities are entrusted to the performance of GPS system in a two-dimensional plane. The indoor case contemplates three-dimensional positioning. The test area (Figure 9) was a rectangular room measuring 6.2 m by 3.4 m, characterized by the presence of typical indoor obstacles such as chairs, cabinets, and various tools. The room is quite noisy in terms of electromagnetic radiation. To establish a reference frame for the localization data, a Cartesian coordinate system was defined within the room. The origin of this coordinate system was situated at the lower left corner of the room. The pavement corresponds to the x–y plane and the height corresponds to the z-axis. The x-axis coincides with the south wall and the y-axis coincides with the west wall. Five anchors have been placed in the locations shown in Figure 9. Two distinct test positions, labeled T1 and T2, were established within the room to evaluate the system's performance under different conditions, while the anchors have been indicated with red cross and the capital letter A. The anchors have different heights from the pavement, A1 @ 1.5 m, A2 @ 1.6 m, A3 @ 1.5 m, A4 @ 1.9 m, and A5 @ 1.9 m. T1 (the node position for the test 1) is placed 133 cm from the north wall and 133 cm from the west wall and the sensor node has been elevated to 60 cm from the pavement. T2 is placed 66 cm from the south wall and 80 cm from the east wall and has been elevated to 180 cm.

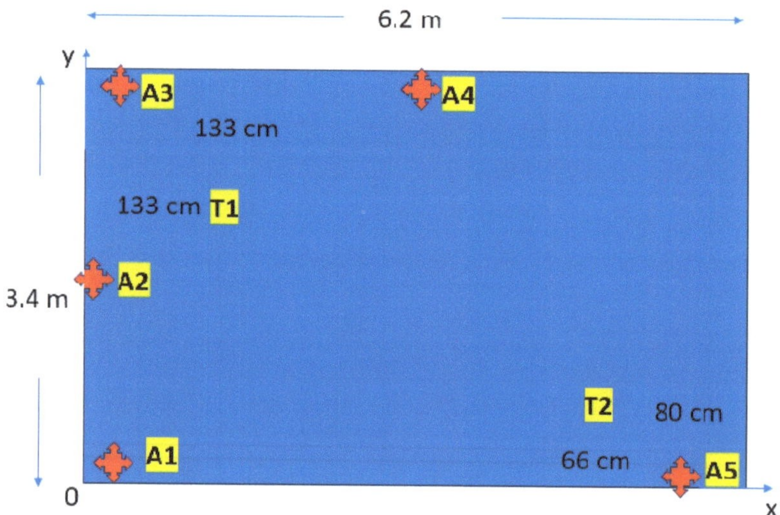

Figure 9. Two-dimensional map of the test room. In red, the anchors, T1 and T2 are the test points in the room.

Data analysis was performed using Python-based packages (v.3.6) including Seaborn, Matplotlib, NumPy, and Pandas. Initially, the zero-position offset was determined to remove systematic errors from subsequent position measurements. Analysis revealed accurate zero reference in the x–y plane, with errors within a 10 cm interval, but larger errors exceeding 10 cm in the z-axis (Figure 10a). Two node positions were then tested. In both cases, high precision was observed across all axes, as indicated by low variability and interquartile ranges in the box plots (Figure 10b–d). Considering the system's application in worker localization and activity recognition, an error tolerance of 60 cm per axis was deemed acceptable for distinguishing between normal and abnormal working postures. Average measured positions were compared to true values (Table 5). While precision was satisfactory, accuracy was found to be limited, primarily due to the number and placement of anchors. The POZYX system manufacturer claims sub-10 cm accuracy under optimal conditions. To enhance accuracy, optimizing anchor positions is recommended.

Table 5. Average of the measured samples versus true values.

	X	Y	Z
T1	84.66 vs. 133	198.82 vs. 207	50.22 vs. 60
T2	485.03 vs. 540	52.5 vs. 66	169.25 vs. 180

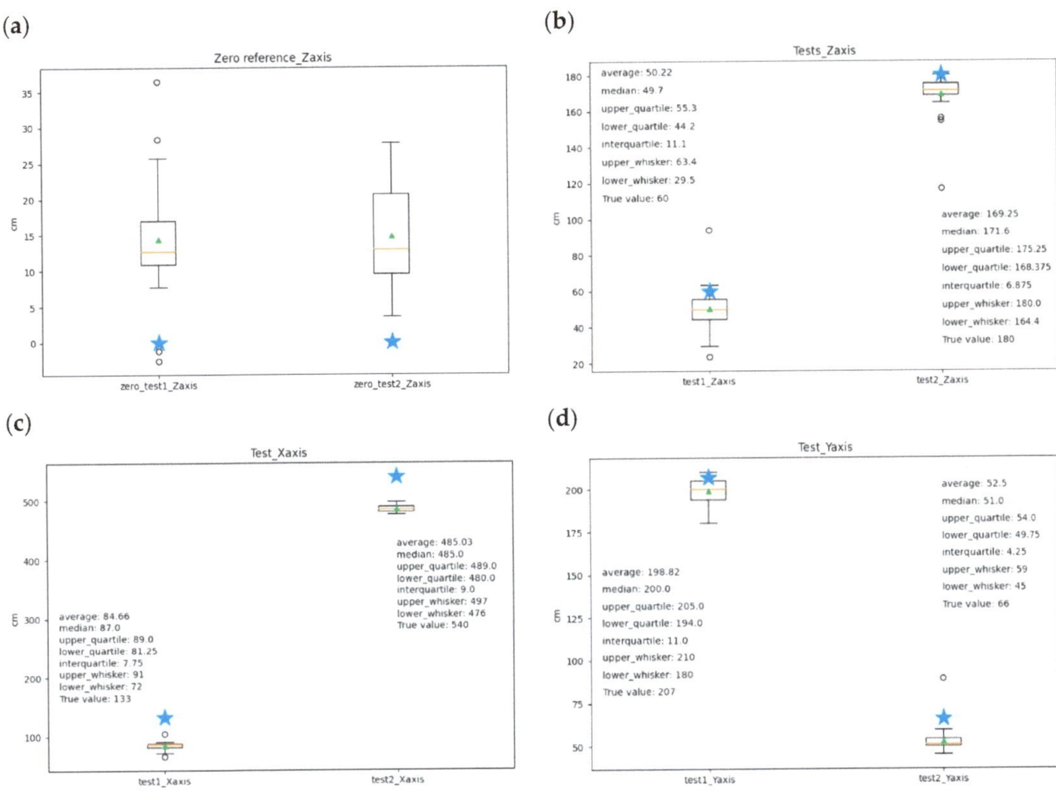

Figure 10. Boxplot representations of the test: the star is the true value, the triangle is the average value of the measures, and the line in the box is the median (2 quartile). Zero reference for the *z*-axis (**a**). Results for two tests on *z*, *x*, and *y*-axes, respectively (**b**–**d**).

3.4. Particulate Matter

To evaluate performance in high-particulate-matter conditions, the sensor node, equipped with a commercial Plantower PMS7003 dust sensor, was placed within a 500 L sealed chamber. A variable amount of certified dry Arizona Test Dust (ISO 12103-1) was injected into the chamber using a TOPAS SAG411 generator. A certified DustTrack 8533 reference instrument was concurrently installed to provide a benchmark measurement of particulate concentration.

The sensor node's accuracy and precision were assessed across two PM2.5 concentration intervals: 0–500 $\mu g/m^3$ and 500–2000 $\mu g/m^3$. Figure 11 (top left and right) illustrates the linear correlation between measured and reference PM2.5 values. The sensor provides two calibrated outputs: SP and AE, reflecting different calibration standards. While the sensor demonstrates precision in high dust concentrations, accuracy diminishes, indicated by the increasing discrepancy between measured and reference values. This loss of sensitivity is likely attributed to the high concentration of submicron particles attenuating the laser beam used for measurement. Following recalibration, measurement error was reduced to an average of less than 30%, as shown in Figure 11 (bottom left and right).

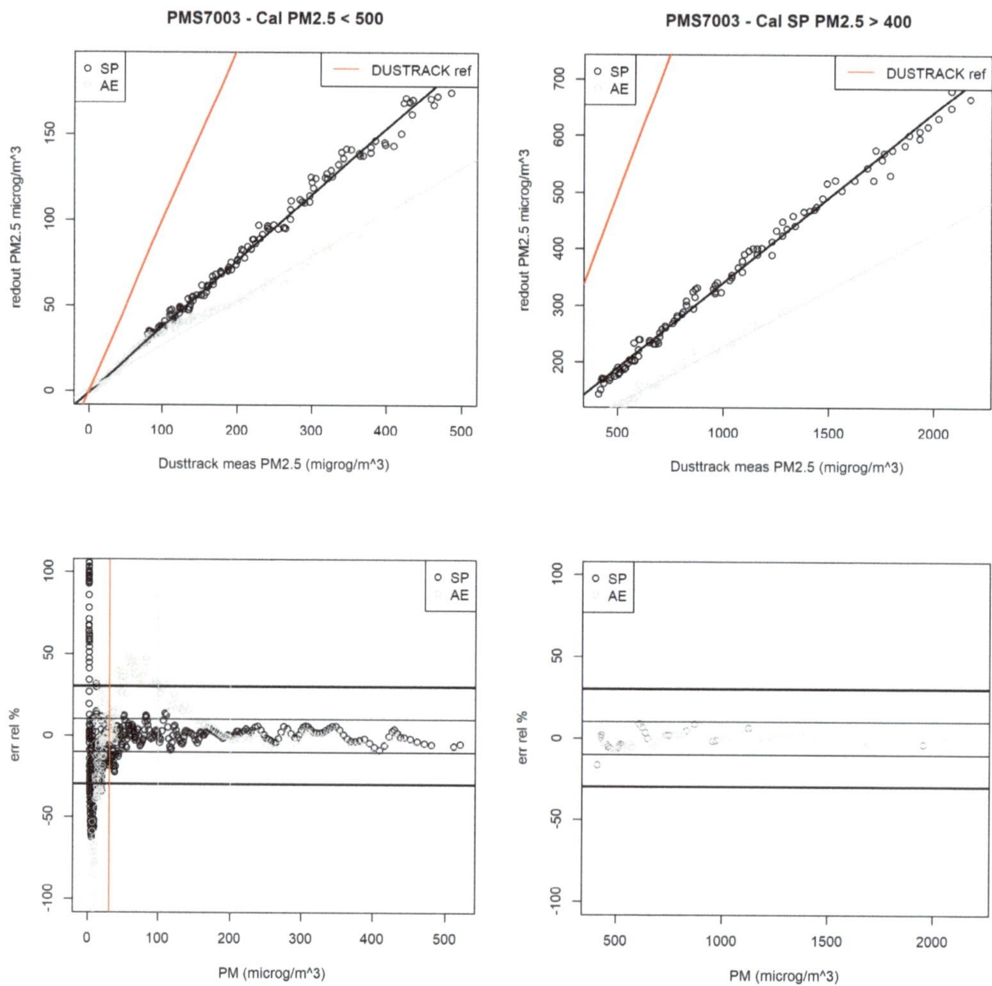

Figure 11. Sensor node PM measurement accuracy and precision: compared to reference instrument. The (**a,b**) images show how the Plantower PMS7003, installed on the sensor node, estimates the real value of PM2.5, measured by the reference instrument in two ranges: on the (**a,c**) from 0 $\mu g/m^3$ to 500 $\mu g/m^3$ and on the (**b,d**) from 500 $\mu g/m^3$ to 2000 $\mu g/m^3$. The two graphs at the (**c,d**) refer to the relative error that the PMS7003 made in a PM2.5 measurement after the recalibration. SP and AE refer to two different types of calibrated output.

3.5. Audio

The sensor node incorporates an audio section for real-time environmental noise monitoring. A commercially available sensor tile board (STLCS01V1) equipped with a low-power, omnidirectional, digital MEMS microphone (MP34DT05-A) is employed for this purpose. The microphone's capacitive sensing element, integrated circuit interface, low distortion, and high signal-to-noise ratio (64 dB) contribute to precise sound pressure detection and accurate signal representation. Notably, the guaranteed sensitivity of -26 dB $\pm$ 3 dB ensures a wide dynamic range for capturing diverse sound levels. The microphone's reliable operation across a temperature range of $-40\ °C$ to $+85\ °C$ guarantees consistent performance in various environments. The audio section autonomously measures ambient sound pressure, triggering an alert when exceeding a predefined threshold. This threshold corresponds to noise levels mandating personal protective equipment (PPE) use. The sensor tile board enables not only real-time monitoring but also potential expansion to advanced audio analysis, adapting the sensor node to evolving requirements and enabling more sophisticated noise characterization. The audio section's performance was evaluated in a dedicated test chamber (Figure 12).

A test setup was established comprising the sensor node, a reference-grade Deltha Ohm HD2010UC phonometer, and a Bluetooth speaker for generating controlled sound stimuli. The speaker emitted sound frequencies ranging from 500 Hz to 4000 Hz in 500 Hz increments. Sound pressure levels were simultaneously monitored by both the phonometer and sensor node. The sensor node demonstrated exceptional performance, accurately detecting all sound events exceeding the 85 dB(A) threshold across the tested frequency range. These results confirm the sensor node's reliable sound pressure detection capabilities.

Figure 12. Audio test setup.

3.6. Accelerometer and Gyroscope

The LSM6DSOX, a 3D digital accelerometer and gyroscope, was selected for its configurable full-scale acceleration ($\pm$2 g, $\pm$4 g, $\pm$8 g, $\pm$16 g) and angular rate ($\pm$125 dps, $\pm$250 dps, $\pm$500 dps, $\pm$1000 dps, $\pm$2000 dps) ranges, high mechanical shock tolerance, and embedded complex gesture recognition capabilities. The sensor's integrated pre-programmed free-fall detection feature, which leverages a dedicated register configuration to identify zero-gravity conditions, was of particular interest. To evaluate this feature, the sensor node was dropped from a height of 30 cm thirty times. Free-fall was accurately detected in all trials, triggering an appropriate alarm on the dashboard.

4. Discussion

This work introduces a novel personal protective device designed to enhance workplace safety and risk awareness. The device collects real-time data to enable data-driven security, monitoring worker actions and utilizing historical data with intelligent algorithms to prevent accidents. Given the safety-critical nature of the application, the hardware incorporates redundancies to ensure accurate environmental data.

Previous sections detailed the sensor node's performance in individual components. Gas sensor tests, conducted in a controlled chamber, successfully detected CO levels exceeding the 100 ppm threshold. Similar tests using ethanol as a TVOC marker demonstrated strong correlation between sensor and reference instrument outputs, enabling TVOC threshold detection.

Localization, which is essential for worker tracking, was achieved through a combination of RTLS and GPS technologies. Indoor accuracy was influenced by anchor placement, while outdoor performance relied on GPS. A target localization accuracy of 5 m was established to support worker rescue and context-aware safety measures.

The sensor node's particulate matter sensor accurately measured PM2.5 concentrations compared to a reference instrument, enabling the identification of excessive exposure levels.

Early detection of elevated particulate matter allows for timely interventions to mitigate air pollution and protect worker health.

The integrated audio section monitored noise levels, triggering alerts when noise exceeded the predefined thresholds. This enables timely interventions to protect workers from excessive noise exposure. Additionally, the accelerometer and gyroscope detected falls and abnormal movements, initiating appropriate emergency responses.

Comprehensive system tests simulated real-world scenarios, including exposure to high noise levels and particulate matter concentrations. The system accurately detected these hazards, triggered alarms, and transmitted relevant data to the control center. Fall detection tests were also successful, demonstrating the system's ability to identify and respond to emergency situations.

While the system demonstrated promising results, further optimization of anchor placement is necessary to improve indoor localization accuracy. Overall, the sensor node effectively collects and processes environmental data, enabling real-time monitoring and timely interventions to enhance worker safety.

Security and Privacy Issues

The implementation of a system that monitors workers and their environment presents challenges beyond technical considerations. Paramount among these are the security of collected information and the protection of workers' personal data. These factors are crucial in the widespread adoption of this technology in workplaces.

Smart PPE, integrating computing, communication, and artificial intelligence, offers dynamic hazard adaptation. To effectively manage emergencies, the system necessitates confidential, integral, and accessible data. Data confidentiality is essential to prevent unauthorized access and tampering, ensuring accurate decision making and intervention. Data integrity guarantees reliable and representative information for algorithm training and operational effectiveness. Data availability is crucial for timely responses to critical situations.

Privacy concerns arise from the device's monitoring of worker position, movement, and audio. The collection of data ranging from device access logs to personal profiles raises questions about remote worker control. While legal exceptions exist for workplace safety and asset protection, navigating the evolving legal landscape and maintaining compliance can be complex and costly, particularly when relying on third-party software.

Comprehensive solutions addressing data security, privacy, and legal compliance are imperative for the widespread adoption of this technology. This paper focuses on technical aspects, recognizing the need for further research and development in these critical areas to fully realize the potential of smart PPE in enhancing workplace safety.

5. Conclusions

This work introduces a novel IoT sensor node for enhancing workplace safety. The system effectively collects and records environmental data, facilitating rapid intervention and worker alerts in hazardous situations. In emergency scenarios, it provides crucial information to rescue teams. Promising test results indicate its potential to significantly improve workplace safety when implemented and validated.

The sensor node incorporates a comprehensive suite of sensors, prioritizing wearability and extended battery life. Its dual-technology localization system enables precise indoor and outdoor positioning, while the multi-sensor array provides a holistic environmental assessment. These capabilities, combined with potential augmented reality integration, position the device as a pioneering solution within the emerging field of active PPE.

Future research will focus on device-to-device communication protocols, advanced AI algorithms for complex pattern recognition, and the development of a robust security and privacy framework to address legal and ethical considerations. By overcoming these challenges, this technology can be fully realized as a transformative tool for enhancing workplace safety.

Author Contributions: Hardware and software: conceptualization, methodology, design, software, validation; writing—original draft preparation, writing—review and editing, F.F., A.D.G. and M.D.; Sensor array design: conceptualization, methodology, design, validation; writing—original draft preparation, writing—review and editing, E.M., L.B., M.D. and F.F.; supervision, F.F.; project administration, G.D.F. All authors have read and agreed to the published version of the manuscript.

Funding: This research was funded by the following project: multiSensore per il monitoraggio degli Ambienti di LaVOro (S. A. L. V. O.), MSE PON 2014–2020, CUP B48I20000050005.

Institutional Review Board Statement: Not applicable.

Informed Consent Statement: Not applicable.

Data Availability Statement: No data available for disclosure.

Acknowledgments: The authors thank FCM Technology Srl electronics manufacturing, Italy, and its general manager Fabio C. Burgarello for the technical support during the executive design and implementation phases.

Conflicts of Interest: Authors Michele Dellutri and Luigi Barretta were employed by the company STMicroelectronics. The remaining authors declare that the research was conducted in the absence of any commercial or financial relationships that could be construed as a potential conflict of interest.

References

1. INAIL Annual Report. 2022. Available online: https://www.inail.it/cs/internet/comunicazione/pubblicazioni/rapporti-e-relazioni-inail/relazione-annuale-anno-2022.html (accessed on 10 June 2024).
2. Sánchez, M.S.; Rodriguez, C.; Manuel, J. Smart Protective Protection Equipment for an accessible work environment and occupational hazard prevention. In Proceedings of the 10th International Conference on Cloud Computing, Data Science & Engineering, Online, 28 February 2021. [CrossRef]
3. Kanan, R.; Elhassan, O.; Bensalem, R. An IoT-based autonomous system for workers' safety in construction sites with real-time alarming, monitoring and positioning strategies. *Autom. Constr.* **2018**, *88*, 73–86. [CrossRef]
4. Wu, F.; Wu, T.; Yuce, M.R. An internet-of-things (IoT) network system for connected safety and health monitoring applications. *Sensors* **2019**, *19*, 21. [CrossRef] [PubMed]
5. Del Giudice, A.; Dellutri, M.; Di Francia, G.; Formisano, F.; Loffredo, G. "SALVO: Towards a Smart Personal Protective Equipment" Sensors and Microsystems. In Proceedings of the AISEM 2021, Rome, Italy, 10–11 February 2021; Lecture Notes in Electrical Engineering; Springer: Cham, Switzerland, 2023; Volume 918. [CrossRef]
6. Formisano, F.; Del Giudice, A.; Dellutri, M.; Di Francia, G.; Loffredo, G.; Picardi, A.; Salvatori, S. *A Novel Sensor Node for Smart Personal Protective Equipment*; Part of the Lecture Notes in Electrical Engineering Book Series; Lecture Notes in Electrical Engineering; Springer: Cham, Switzerland, 2022; Volume 999. [CrossRef]
7. Allegato XXXVIII Testo Unico Sicurezza, DM 18/5/2021. Available online: https://www.ambientesicurezzaweb.it/wp-content/uploads/sites/5/2021/06/Allegato.pdf (accessed on 10 June 2024).
8. INAIL. Liliana Frusteri, Seminario: Gli Ambienti Confinati e i Rischi per la Salute e Sicurezza, Bologna 4 Maggio 2011. 2011. Available online: http://www.sistemaambiente.net/Materiali/IT/Spazi_confinati/Ambienti_confinati_e_rischi_derivanti_da_sostanze_e_atmosfere_pericolose.pdf (accessed on 10 June 2024).
9. CO_2 Safety Data Sheet Online. Available online: https://chemicalsafety.ilo.org/dyn/icsc/showcard.display?p_card_id=0021&p_version=1&p_lang=it (accessed on 10 June 2024).
10. POZYX Developer Tag. Available online: https://www.pozyx.io/products/hardware/tags/developer-tag (accessed on 10 June 2024).

MDPI AG
Grosspeteranlage 5
4052 Basel
Switzerland
Tel.: +41 61 683 77 34

Sensors Editorial Office
E-mail: sensors@mdpi.com
www.mdpi.com/journal/sensors

9 783725 823796